Springer Proceedings in Physics

Volume 155

For further volumes:
http://www.springer.com/series/361

Ahmet Yavuz Oral · Zehra Banu Bahsi
Mehmet Ozer
Editors

International Congress on Energy Efficiency and Energy Related Materials (ENEFM2013)

Proceedings, Antalya, Turkey,
9–12 October 2013

Springer

Editors
Ahmet Yavuz Oral
Department of Materials Science and Engineering
Gebze Institute of Technology
Gebze, Kocaeli
Turkey

Mehmet Ozer
Department of Physics
Istanbul Kultur University
Istanbul
Turkey

Zehra Banu Bahsi
Department of Environmental Engineering
Gebze Institute of Technology
Gebze, Kocaeli
Turkey

ISSN 0930-8989 ISSN 1867-4941 (electronic)
ISBN 978-3-319-34643-4 ISBN 978-3-319-05521-3 (eBook)
DOI 10.1007/978-3-319-05521-3
Springer Cham Heidelberg New York Dordrecht London

Printed on acid-free paper

Springer is part of Springer Science+Business Media (www.springer.com)

Preface

The first International Congress on Energy Efficiency and Energy Related Materials (ENEFM2013) provided all scientists the opportunity to meet, present their work, discuss, and mutually interact in order to enhance and promote their research work.

This volume, published by Springer, includes all the papers presented at this Congress, held in Antalya, Turkey, October 9–12, 2013.

On behalf of the organizing committee we would like to thank all the plenary and invited speakers for their valuable contributions.

We would also like to thank TURA Tourism for their support in the organization of the Congress, as well as our sponsor Doga Nanobiotech Inc. Finally, we would like to thank the publishers for the quality of this edition.

Gebze, Turkey — Ahmet Yavuz Oral
Gebze, Turkey — Zehra Banu Bahsi
Istanbul, Turkey — Mehmet Ozer

Preface

Organization

International Scientific Committee

Zoran Bojkovic	University of Belgrade, Serbia
Siew-Hwa Chan	Nanyang Technological University, Singapore
Metin Colak	Ege University, Turkey
Mehmet Oguz Guler	Sakarya University, Turkey
Lenny Koh	The University of Sheffield, UK
Satoshi Konishi	Kyoto University, Japan
Tayfur Ozturk	Middle East Technical University (METU), Turkey
Sanjib Kumar Panda	National University of Singapore, Singapore
Saban Patat	Erciyes University, Turkey
Jing Shi	North Dakota State University, USA
Eugenijus Ušpuras	Lithuanian Energy Institute, Lithuania

Organizing Committee

Zehra Banu Bahsi	Co-Chair, Gebze Institute of Technology, Turkey
Nermin Demirkol	Kocaeli University, Turkey
Ersin Kayahan	Kocaeli University, Turkey
Ahmet Yavuz Oral	Co-Chair, Gebze Institute of Technology, Turkey
Mehmet Ozer	Istanbul Kultur University, Turkey
Jan Treur	VU University Amsterdam, The Netherlands

Sponsors

- Doga Nanobiotech Inc.

Doga Nanobiotech

Conference Organizing Company

TURA TOURISM LTD.

TURA
TURİZM

Cumhuriyet Cad. No: 109/A Elmadag—Sisli/Istanbul
E-mail: enefm2013@turaturizm.com.tr
Phone: +90-212-241 27 00
Fax: +90-212-241 29 89
Web: http://www.turaturizm.com.tr

Contents

Contributors

Seddik-El-Hak Abaidia Faculté des Sciences, University of M'hamed Bougara, Boumerdès, Algeria

Hamed Abbasi Faculty of Physics, Department of Photonics, University of Kashan, Kashan, Iran

Bahy Abdel-Mesih Centre for Renewable Energy (CRE), Mechanical Engineering, The British University in Egypt (BUE), Sherouk, Cairo, Egypt

Tarek I. Abdelsalam Centre for Renewable Energy (CRE), Mechanical Engineering, The British University in Egypt (BUE), Sherouk, Cairo, Egypt

Yassine Agguine Unité Matériaux et Transformations—UMET (UMR CNRS N°8207), Bâtiment C6, Université Lille 1—Sciences et Technologies, Villeneuve d'Ascq Cedex, France

Rashidah Zainal Alam Institute for Environment and Development (LESTARI), Universiti Kebangsaan Malaysia (UKM), Bangi, Selangor, Malaysia

Mohammed R. AlShareef Department of Electrical and Computer Engineering, University of Waterloo, Waterloo, ON, Canada

Ramazan Atilgan Faculty of Aeronautics and Astronautics, Anadolu University, Eskisehir, Turkey

Hakan Aydin TUSAS Engine Industries, Eskisehir, Turkey

Bashar A. Bdaiwi Department of Mechanical Engineering, College of Engineering, Al-Mustansirya University, Baghdad, Iraq

Ahmed Bekka Laboratoire de synthèse et caractérisation des oxydes, Université des Sciences et de la Technologie d'Oran (USTO), Oran, Algeria

Djamel Bellala Control Engineering and Manufacturing Laboratory, University of Batna, Batna, Algeria

A. Bellifa Laboratory of Materials, Application and Environment, University of Mascara, Mascara, Algeria

Khalil A. Benabbou Laboratoire physico-chimie des matériaux-catalyse et environnement, Université des Sciences et de la Technologie d'Oran (USTO), Oran, Algeria

Chahrazed Benseghir Faculty of Technology, Department of Mechanical, University of Batna, Batna, Algeria

Kahina Bentaleb Laboratoire physico-chimie des matériaux-catalyse et environnement, Université des Sciences et de la Technologie d'Oran (USTO), Oran, Algeria

Yacine Berbar Centre de Recherche en Technologie des Semi-conducteurs pour l'Energétique (CRTSE), Algiers, Algeria

Olivier Bethoux LGEP CNRS UMR 8507/SPEE Labs, Gif-sur-Yvette Cedex, France

Pablo Bonelli Programa de Investigación y Desarrollo de Fuentes Alternativas de Materia Prima y Energía (PINMATE), Facultad de Ciencias Exactas y Naturales, Departamento de Industrias, Universidad de Buenos Aires Ciudad Universitaria, Buenos Aires, Argentina

Zohra Bouberka Laboratoire physico-chimie des matériaux-catalyse et environnement, Université des Sciences et de la Technologie d'Oran (USTO), Oran, Algeria

Abd El-Ghani Boucheham Research Center in Semi-conductors Technology for the Energetic, Algiers, Algeria

Djoudi Bouhafs Research Center in Semi-conductors Technology for the Energetic, Algiers, Algeria

Milan Brož Institute of Rock Structure and Mechanics, Academy of Science Czech Republic, v.v.i., Prague, Czech Republic

Marija Burinskienė Department of Urban Engineering, Vilnius Gediminas Technical University, Vilnius, Lithuania

Licia Capodiferro Department of Information Processing and Management General Requirements, Fondazione Ugo Bordoni, Rome, Italy

Matěj Černý PROGEO s.r.o., Roztoky, Czech Republic; Faculty of Environmental Sciences, Czech University of Life Sciences Prague, Suchdol, Czech Republic

Sanjib Chakraborty Department of Management and Engineering, Linköping University, Linköping, Sweden

Guorong Chen Department of Materials Science and Engineering, East China University of Science and Technology, Shanghai, China

Tetiana Cherniavska Chuiko Institute of Surface Chemistry, National Academy of Sciences of Ukraine, Kyiv, Ukraine

A. Choukchou-Braham Laboratory of Catalysis and Synthesis in Organic Chemistry, University of Tlemcen, Tlemcen, Algeria

Javier Cubas Instituto Universitario de Microgravedad "Ignacio da Riva", ETSI Aeronáuticos, Universidad Politécnica de Madrid, Madrid, Spain

Ziad Darwish Enactus BUE, Sherouk, Cairo, Egypt

Carlos de Manuel Instituto Universitario de Microgravedad "Ignacio da Riva", ETSI Aeronáuticos, Universidad Politécnica de Madrid, Madrid, Spain

Paolo Del Vecchio Department of Engineering, University of Roma TRE, Rome, Italy

M. Dolatyari School of Engineering-Emerging Technologies, University of Tabriz, Tabriz, Iran

Nadjib Drouiche Centre de Recherche en Technologie des Semi-conducteurs pour l'Energétique (CRTSE), Algiers, Algeria

Olesia Dudik Chuiko Institute of Surface Chemistry, National Academy of Sciences of Ukraine, Kyiv, Ukraine

Said Eddarir Laboratoire de Chimie Bioorganique et Macromoléculaire (LCBM), Faculté des Sciences et Techniques, Université Cadi Ayyad, Guéliz, Marrakech, Morocco

Ceyhan Erdem Department of Aerospace Engineering, Middle East Technical University, Ankara, Turkey

Ergin Erdem Department of Industrial and Manufacturing Engineering, North Dakota State University, Fargo, ND, USA

Vasile Farcas Faculty of Civil Engineering, Department of Structures, Technical University of Cluj Napoca, Cluj Napoca, Romania

Héctor José Fasoli Pontificia Universidad Católica Argentina de los Buenos Aires, Buenos Aires, Argentina

Radia Imane Fertout Laboratoire de Matériaux et Catalyse, Faculté des Sciences, Université Djillali Liabes, Sidi Bel-Abbes, Algeria

Juan Isidro Franco Departamento de Investigación y Desarrollo de Energías Renovables (CITEDEF-EST), Buenos Aires, Argentina

Jan Franěk Czech Geological Survey, Prague, Czech Republic

Georg Frey Chair of Automation, Saarland University, Saarbrücken, Germany

Bing Gao Department of Materials Science and Engineering, East China University of Science and Technology, Shanghai, China

Lei Gao State Key Laboratory of Power Systems, Department of Electrical Engineering, Tsinghua University, Beijing, China

Madani Ghelamallah Laboratoire de Matériaux, Applications et Environnement, Faculté des Sciences et de la Technologie, Université de Mascara, Mascara, Algeria

Giovanni Giannoccaro Dipartimento di Ingegneria Elettrica e dell'Informazione, Politecnico di Bari, Bari, Italy

Emmanuel Godoy Automatic Control Department, Supélec Systems Sciences (E3S), Gif-sur-Yvette Cedex, France

Fatma Pınar Gökdemir Department of Physics, Yıldız Technical University, Esenler, Istanbul, Turkey

Krzysztof Górecki Department of Marine Electronics, Gdynia Maritime University, Gdynia, Poland

Adeleh Granmayeh Rad Faculty of Sciences, Roudehen Branch, Department of Physics, Islamic Azad University, Roudehen, Iran

Djamel Haddad Faculty of Technology, Department of Mechanical, University of Batna, Batna, Algeria

N. M. A. Hadia Faculty of Science, Department of Physics, Sohag University, Sohag, Egypt

Boualem Hamdi LCVR, ENSMAL, Algiers, Algeria; USTHB, LEPCMAE, Algiers, Algeria

Safia Hamdi LCVR, ENSMAL, Algiers, Algeria; USTHB, LEPCMAE, Algiers, Algeria

Ali H. Hasan Department of Mechanical Engineering, College of Engineering, Al-Mustansirya University, Baghdad, Iraq

Tarek M. Hatem Mechanical Engineering, The British University in Egypt (BUE), Sherouk, Cairo, Egypt

Juma Haydary Faculty of Chemical and Food Technology, Institute of Chemical and Environmental Engineering, Slovak University of Technology, Bratislava, Slovakia

Hakan Hayta Department of Mechanical Engineering, University of Gaziantep, Gaziantep, Turkey

Jinliang He State Key Laboratory of Power Systems, Department of Electrical Engineering, Tsinghua University, Beijing, China

H. Heidarzadeh Photonics and Nanocrystal Research Lab (PNRL), Department of Electrical and Computer Engineering, University of Tabriz, Tabriz, Iran

Mikko Helle Thermal and Flow Engineering Laboratory, Åbo Akademi University, Turku, Finland

Radim Hladký Division Geotechnika, ARCADIS CZ a.s., Prague, Czech Republic

Rubayet Hosain Department of Management and Engineering, Linköping University, Linköping, Sweden

Jun Hu State Key Laboratory of Power Systems, Department of Electrical Engineering, Tsinghua University, Beijing, China

Nicoleta Ilies Faculty of Civil Engineering, Department of Structures, Technical University of Cluj Napoca, Cluj Napoca, Romania

Gerardo Martín Imbrioscia Departamento de Investigación y Desarrollo de Energías Renovables (CITEDEF-EST), Buenos Aires, Argentina

Neda Jalali Farahani Plasma Physics Research Center, Science and Research Branch, Islamic Azad University, Tehran, Iran

Ľudovít Jelemenský Faculty of Chemical and Food Technology, Institute of Chemical and Environmental Engineering, Slovak University of Technology, Bratislava, Slovakia

Soufi Kacimi Laboratoire de Chimie Appliquée, Institut des Sciences et de la Technologie, Centre Universitaire d'Ain Temouchent, Ain Temouchent, Algeria

Adomas Kanopka Mykolas Romeris University, Vilnius, Lithuania

C. Kappenstein Laboratory of Catalysis in Organic Chemistry, University of Poitiers, Poitiers, France

Aissa Kefaifi Centre de Recherche en Technologie des Semi-conducteurs pour l'Energétique (CRTSE), Algiers, Algeria

Bahadır Keskin Department of Physics, Yıldız Technical University, Esenler, Istanbul, Turkey

Basima S. Khalaf Department of Mechanical Engineering, College of Engineering, Al-Mustansiriyah University, Baghdad, Iraq

Nabil Khelifati Research Center in Semi-conductors Technology for the Energetic, Algiers, Algeria; Faculté des Sciences, University of M'hamed Bougara, Boumerdès, Algeria

Abdelkrim Kheloufi Centre de Recherche en Technologie des Semi-conducteurs pour l'Energétique (CRTSE), Algiers, Algeria

Yusuf Kışlak Informatics Institute, Istanbul Technical University, Maslak, Istanbul, Turkey

Valentin Kosov Joint Institute for High Temperatures of the Russian Academy of Sciences (JIHT RAS), Moscow, Russia

Vladimir Kosov Joint Institute for High Temperatures of the Russian Academy of Sciences (JIHT RAS), Moscow, Russia

Tetiana Kulik Chuiko Institute of Surface Chemistry, National Academy of Sciences of Ukraine, Kyiv, Ukraine

Kubilay Kutlu Department of Physics, Yıldız Technical University, Esenler, Istanbul, Turkey

Vladimír Lachman ISATech s.r.o., Prague, Czech Republic

Nadjia Laouedj Laboratoire de synthèse et caractérisation des oxydes, Université des Sciences et de la Technologie d'Oran (USTO), Oran, Algeria

María José Lavorante Departamento de Investigación y Desarrollo de Energías Renovables (CITEDEF-EST), Buenos Aires, Argentina

Filiberto Lembo School of Engineering, University of Basilicata, Potenza, Italy

Yang Li State Key Laboratory of Power Systems, Department of Electrical Engineering, Tsinghua University, Beijing, China

Norasikin Ahmad Ludin Solar Energy Research Institute (SERI), Universiti Kebangsaan Malaysia (UKM), Bangi, Selangor, Malaysia

Rickard Magnéli Machine Design Division, Department of Management and Engineering, Linköping University, Linköping, Sweden

Qasim S. Mahdi Department of Mechanical Engineering, College of Engineering, Al-Mustansiriyah University, Baghdad, Iraq

Mohammed H. Mahmoud Alhamdo Department of Mechanical Engineering, College of Engineering, Al-Mustansirya University, Baghdad, Iraq

Federica Mangiatordi Department of Engineering, University of Roma TRE, Rome, Italy; Department of Information Processing and Management General Requirements, Fondazione Ugo Bordoni, Rome, Italy

Francesco Paolo R. Marino School of Engineering, University of Basilicata, Potenza, Italy

Vytautas Martinaitis Department of Building Energetics, Vilnius Gediminas Technical University, Vilnius, Lithuania

Ulrich Maschke Unité Matériaux et Transformations—UMET (UMR CNRS N°8207), Bâtiment C6, Université Lille 1—Sciences et Technologies, Villeneuve d'Ascq Cedex, France

Aicha Medjghou Department of Industrial Engineering, University of Batna, Batna, Algeria

H. A. Mohamed Faculty of Science, Department of Physics, Sohag University, Sohag, Egypt; Department of Physics, Teachers College, King Saud University, Riyadh, Saudi Arabia

Abdelouahab Nadim Unité Matériaux et Transformations—UMET (UMR CNRS N°8207), Bâtiment C6, Université Lille 1—Sciences et Technologies, Villeneuve d'Ascq Cedex, France

Zbigniew Nahorski Systems Research Institute, Polish Academy of Sciences, Warsaw, Poland

Mohsen Namakian Machine Design Division, Department of Management and Engineering, Linköping University, Linköping, Sweden

Younane Nassif Department of Automatic Control, Supélec Systems Sciences (E3S), Gif-sur-Yvette Cedex, France

Marco Nesarajah Chair of Automation, Saarland University, Saarbrücken, Germany

Jaroslav Nosek Technical university of Liberec, Liberec, Czech Republic

Johan Ölvander Machine Design Division, Department of Management and Engineering, Linköping University, Linköping, Sweden

Muammer Ozbek Faculty of Mechanical, Maritime and Materials Engineering, Engineering Dynamics Section, Delft University of Technology, Delft, The Netherlands

Orhan Özdemir Department of Physics, Yıldız Technical University, Esenler, Istanbul, Turkey

Baya Palahouane Research Center in Semi-conductors Technology for the Energetic, Algiers, Algeria

Borys Palianytsia Chuiko Institute of Surface Chemistry, National Academy of Sciences of Ukraine, Kyiv, Ukraine

Gražvydas Mykolas Paliulis Department of Urban Engineering, Vilnius Gediminas Technical University, Vilnius, Lithuania

Emiliano Pallotti Department of Engineering, University of Roma TRE, Rome, Italy; Department of Information Processing and Management General Requirements, Fondazione Ugo Bordoni, Rome, Italy

Vittorio M. N. Passaro Dipartimento di Ingegneria Elettrica e dell'Informazione, Politecnico di Bari, Bari, Italy

Santiago Pindado Instituto Universitario de Microgravedad “Ignacio da Riva”, ETSI Aeronáuticos, Universidad Politécnica de Madrid, Madrid, Spain

L. Pirault-Roy Laboratory of Catalysis in Organic Chemistry, University of Poitiers, Poitiers, France

Ahmead Fazle Rabbi Department of Management and Engineering, Linköping University, Linköping, Sweden

Weronika Radziszewska Systems Research Institute, Polish Academy of Sciences, Warsaw, Poland

Toufiqur Rahman Department of Management and Engineering, Linköping University, Linköping, Sweden

Omar Ramahi Department of Electrical and Computer Engineering, University of Waterloo, Waterloo, ON, Canada

Daniel J. Rixen Institute of Applied Mechanics, Technische Universität München, Garching, Germany

Ivan Roche PSA Peugeot Citroën, Vélizy-Villacoublay Cedex, France

A. Rostami Photonics and Nanocrystal Research Labaratory (PNRL), Department of Electrical and Computer Engineering, University of Tabriz, Tabriz, Iran; School of Engineering-Emerging Technologies, University of Tabriz, Tabriz, Iran

G. Rostami School of Engineering-Emerging Technologies, University of Tabriz, Tabriz, Iran

Paulius Rudzkis Vilnius Gediminas Technical University, Vilnius, Lithuania

Edris Safavi Machine Design Division, Department of Management and Engineering, Linköping University, Linköping, Sweden

Tahar Sahraoui Centre de Recherche en Technologie des Semi-conducteurs pour l’Energétique (CRTSE), Algiers, Algeria

Saad M. Saleh Mechanical Engineering Department, College of Engineering, Baghdad University, Baghdad, Iraq

Henrik Saxén Thermal and Flow Engineering Laboratory, Åbo Akademi University, Turku, Finland

Ying She School of Economics and Management, Nanchang Hangkong University, Nanchang, Jiangxi Province, China

Chao Shen Department of Materials Science and Engineering, East China University of Science and Technology, Shanghai, China

Jing Shi Department of Industrial and Manufacturing Engineering, North Dakota State University, Fargo, ND, USA

Marco Sinagra Dipartimento di Ingegneria Civile, Ambientale, Aerospaziale, dei Materiali, Università degli Studi di Palermo, Palermo, Italy

Vladimir Sinelschikov Joint Institute for High Temperatures of the Russian Academy of Sciences (JIHT RAS), Moscow, Russia

Tim Sirén Machine Design Division, Department of Management and Engineering, Linköping University, Linköping, Sweden

Giedrius Šiupšinskas Department of Building Energetics, Vilnius Gediminas Technical University, Vilnius, Lithuania

Chamhuri Siwar Institute for Environment and Development (LESTARI), Universiti Kebangsaan Malaysia (UKM), Bangi, Selangor, Malaysia

Hacene Smadi Control Engineering and Manufacturing Laboratory, University of Batna, Batna, Algeria

Mehmet Sait Söylemez Department of Mechanical Engineering, University of Gaziantep, Gaziantep, Turkey

Wesam Taha Department of Electrical Engineering, The Petroleum Institute, Abu Dhabi, UAE

Liangliang Tang Department of Thermal Science and Energy Engineering, University of Science and Technology of China, Hefei, People's Republic of China

Adem Tekin Informatics Institute, Istanbul Technical University, Maslak, Istanbul, Turkey

Oksana Tonkha National University of Life and Environmental Sciences of Ukraine, Kyiv, Ukraine

Jan Treur Agent Systems Research Group, VU University Amsterdam, Amsterdam, The Netherlands

Tullio Tucciarelli Dipartimento di Ingegneria Civile, Ambientale, Aerospaziale, dei Materiali, Università degli Studi di Palermo, Palermo, Italy; Sezione di Idraulica e efficienza delle risorse, Istituto Euro-Mediterraneo di Scienza e Tecnologia, Palermo, Italy

Onder Turan Faculty of Aeronautics and Astronautics, Anadolu University, Eskisehir, Turkey

Mustafa Tutar Department of Mechanical and Manufacture, Mondragon Goi Eskola Politeknikoa, Arrasate, Spain; IKERBASQUE, Basque Foundation for Science, Bilbao, Spain

Jan Uhlík PROGEO s.r.o., Roztoky u Prahy, Czech Republic

Jūratė Venckauskaitė Department of Urban Engineering, Vilnius Gediminas Technical University, Vilnius, Lithuania

Carl-Mikael Wiklund Thermal and Flow Engineering Laboratory, Åbo Akademi University, Åbo, Finland

Yunxia Yang Department of Materials Science and Engineering, East China University of Science and Technology, Shanghai, China

Hong Ye Department of Thermal Science and Energy Engineering, University of Science and Technology of China, Hefei, People's Republic of China

İbrahim Halil Yılmaz Department of Mechanical Engineering, University of Gaziantep, Gaziantep, Turkey

Shuanglong Yuan Department of Materials Science and Engineering, East China University of Science and Technology, Shanghai, China

Recep Yumrutaş Department of Mechanical Engineering, University of Gaziantep, Gaziantep, Turkey

Jurgis Zagorskas Department of Urban Engineering, Vilnius Gediminas Technical University, Vilnius, Lithuania

Victor Zaichenko Joint Institute for High Temperatures of the Russian Academy of Sciences (JIHT RAS), Moscow, Russia

Janusz Zarębski Department of Marine Electronics, Gdynia Maritime University, Gdynia, Poland

Talie Zarei Science and Research Branch, Plasma Physics Research Center, Islamic Azad University, Tehran, Iran

Vygantas Žėkas Department of Building Energetics, Vilnius Gediminas Technical University, Vilnius, Lithuania

Mostefa Zeroual Laboratory LPEA, Faculty of Science, University of Batna, Batna, Algeria

Bariza Zitouni Faculty of Technology, Department of Mechanical, University of Batna, Batna, Algeria

Chapter 1
Study of the Effect of Transition Metals on Titanium Dioxide Phase Transformation

A. Bellifa, A. Choukchou-Braham, C. Kappenstein and L. Pirault-Roy

Abstract MTiX samples with different atomic metal percentage were synthesised by sol-gel method and calcined at 400 °C under air. The anatase-rutile transformation in TiO_2 in the presence of transition metals (Cr, V and Mn) was investigated. The kinetics of anatase-rutile transformations were determined by XRD over the temperature range 500–800 °C. It was found that the presence of V and Mn accelerate the transformation anatase-to-rutile. However, the anatase phase stability increases as the chromium content increases.

1.1 Introduction

There are many studies on the anatase-to-rutile transformation [1–6]. This transition depends on several parameters such as the preparation techniques used, preparation conditions, thermal treatments, alkoxide nature, particle size, aging, nature and content of possible doping agents [7–11]. The doping metal oxides can be added by different ways such as further impregnations or mixed powders [12–14], but the best effect is obtained by a close contact between the stabilizing element and oxide. To optimize this interaction, many studies focused on doped TiO_2 prepared by various methods. Li et al. [15], have prepared titania-silica samples by sol-gel hydrothermal (SGH) method and sol-gel (SG) method. The results showed that the

A. Bellifa (✉)
Laboratory of Materials, Application and Environment, University of Mascara, Mascara, Algeria
e-mail: abellifa@yahoo.fr

A. Choukchou-Braham
Laboratory of Catalysis and Synthesis in Organic Chemistry, University of Tlemcen, Tlemcen, Algeria

C. Kappenstein · L. Pirault-Roy
Laboratory of Catalysis in Organic Chemistry, University of Poitiers, Poitiers, France

A. Y. Oral et al. (eds.), *International Congress on Energy Efficiency and Energy Related Materials (ENEFM2013)*, Springer Proceedings in Physics 155,
DOI: 10.1007/978-3-319-05521-3_1, © Springer International Publishing Switzerland 2014

samples prepared by SGH had better thermal stability compared with the SG samples which retard the anatase-rutile transformation. On the other hand, Zhang [7] has used the titanium oxide doped with silica for photodegradation of toluene. It has found that the embedding of small amount of silica into anatase titania matrix enhanced the thermal stability of nanophase titania particle resulting in the suppression of the phase transformation from anatase to rutile phase. A clear understanding of the key factors that control the phase stability, the growth, and the phase transformation kinetics in materials is critical to quantification of materials behaviour. Consequently, new understanding of the factors that dictate the sequence of the phase transformations may provide insights into how phase composition, microstructures, and properties of titania-based materials can be manipulated [3]. In this, paper we report the sol-gel synthesis of MTiX mixed oxides with $X = 0, 5, 10$ and 20 of Cr, V and Mn atom percentage. We study the effect of the weight metals content on the temperature of the anatase-rutile transition and the kinetics of the transformation by in situ XRD.

1.2 Experimental

1.2.1 Synthesis Procedure

The preparation of MTiX was carried out in a 100 mL beaker. Titanium butoxide was dissolved in absolute ethanol. The mixture was stirred for 30 min with addition of an aqueous solution of metal and then a solution of ethanol and nitric acid was added drop ways in the mixture. A white gel was obtained which is dried and calcined in a muffle furnace at 400 °C for 4 h under air atmosphere. The samples were called MTi5, MTi10 and MTi20 for 5, 10 and 20 of metal atom percentage respectively.

1.2.2 Samples Characterization

Micrometrics Tristar 3,000 instrument was using to measure adsorption-desorption isotherms at liquid nitrogen to determine specific area and calculation of mean pore size from BJH methods.

The crystalline phases of the samples were identified by X-ray powder diffraction (XRD) using Bruker D8-ADidentified VANTED diffractometers. The diffractograms were obtained under the following conditions: dwell time: 2 s; step: 0.04° 2θ; divergence slit: 0.3°. The crystallite sizes were determined from the Scherrer equation [16] using the integrated width corrected from apparatus using LaB_6 as a standard.

Table 1.1 Surface area, porous volume and pore size of the samples

Sample	SBET m^2g^{-1}	Pore volume cm^3g^{-1}	Average pore size/nm
TiO_2	68	0.06	3.60
CrTi5	180	0.22	5.00
CrTi10	230	0.19	3.30
CrTi20	152	0.10	2.60
VTi5	76	0.07	3.50
VTi10	129	0.11	3.50
VTi20	98	0.10	4.00
MnTi5	102	0.10	3.60
MnTi10	204	0.33	5.75
MnTi20	246	0.36	5.00

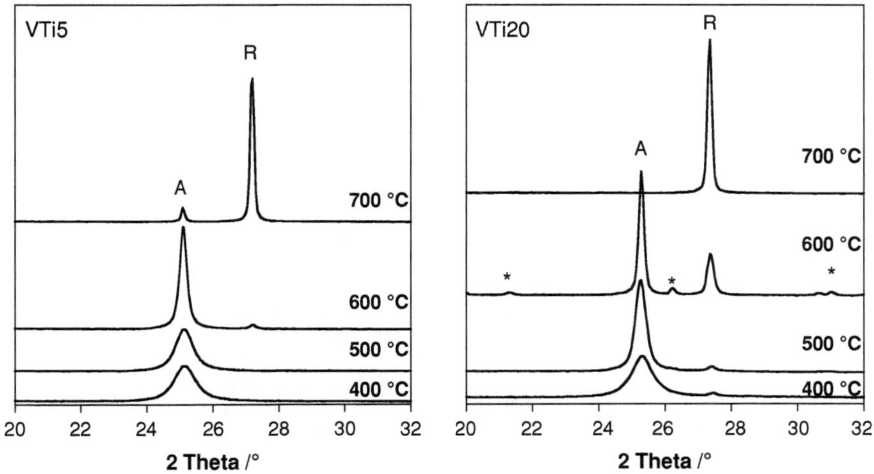

Fig. 1.1 X-ray diffraction (XRD) patterns of VTiX; *A* anatase; *R* rutile; *asterisk* V_2O_5

1.3 Results and Discussion

1.3.1 Textural Data

The experimental results on BET surface area, mean pore diameter, and total pore volume of all synthesized materials are shown in Table 1.1. As obviously seen, the surface area increased with increasing of Mn content from 5 to 20 %. In the other hand the surface area increased with increasing both V and Cr content from 5 to 10 %. With increased Cr and V content (20 %), the observed loss in surface area could be attributed to probable segregation of chromium and vanadium [17–19]. The average pore size of anatase phase decrease with increasing of Cr content. In the case of VTiX samples, the average pore size remains constant with increasing of V content. Wearies an increasing of average pore size was observed with

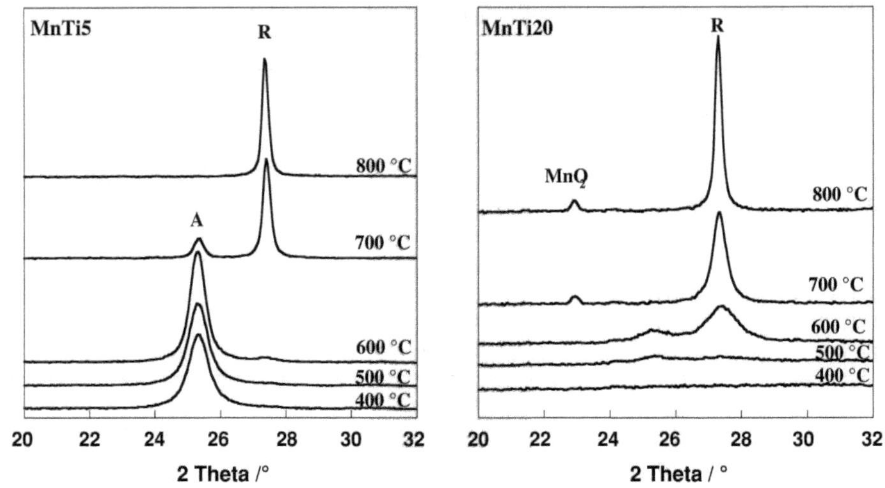

Fig. 1.2 X-ray diffraction (XRD) patterns of MnTiX; *A* anatase; *R* rutile

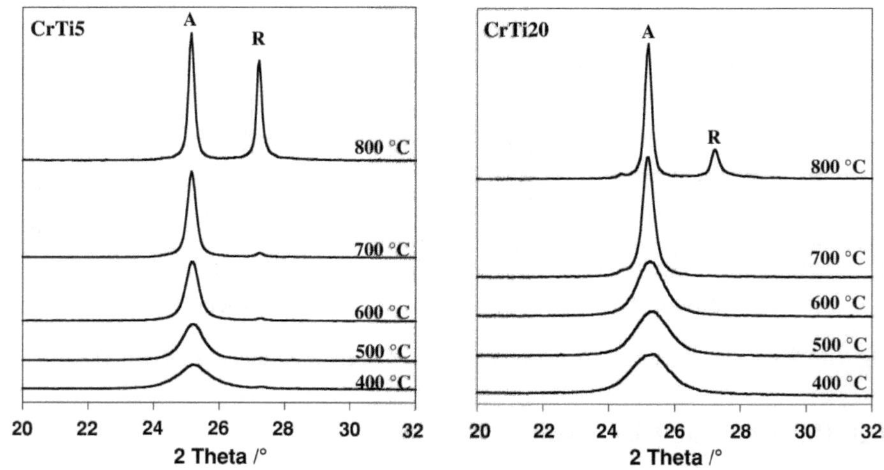

Fig. 1.3 X-ray diffraction (XRD) patterns of CrTiX; *A* anatase; *R* rutile

increasing of Mn content. The pore volume varied with varying metal and metal content. The biggest pore volume was signalled for MnTi20. So the presence of Mn content increased both pore size and pore volume.

1.3.2 X-ray Diffraction

The results of XRD analysis are shown in Figs. (1.1, 1.2 and 1.3). The diffractograms were recording in situ for MTiX samples. In this work we present only the

Table 1.2 Crystallite size versus temperature

Crystallite size (nm)										
T/°C	400		500		600		700		800	
Sample	A	R	A	R	A	R	A	R	A	R
TiO_2	10	/	16	/	27	41	37	36	/	40
CrTi5	07	/	10	/	16	/	22	32	32	37
CrTi10	06	/	13	/	13	/	22	/	32	39
CrTi20	06	/	07	/	09	/	20	/	30	35
VTi5	09	/	11	/	24	34	41	42	/	47
VTi10	09	/	12	/	37	40	/	57	/	60
VTi20	10	/	19	28	36	31	/	40	/	41
MnTi5	12	/	12	/	15	/	24	31	/	37
MnTi10	11	/	11	/	14	13	/	25	/	36
MnTi20	/	/	/	/	18	09	/	18	/	35

diffractograms of MTi5 and MTi20. Anatase phase was observed for all samples at 400 and 500 °C excepted for MnTi20 with amorphous phase. The partial anatase-rutile transformation is at 600, 800, 600 and 600 °C for TiO_2 (not presented), CrTi20, VTi20 and MnTi20 respectively. The total transformation anatase-rutile is at 700, >800, 700 and 600 °C for TiO_2, CrTi20, VTi20 and MnTi20 respectively. However, the rutile fraction is higher for VTiX and MnTiX than TiO_2. The rutile phase is absent after treatment at 400–500 °C and decreases with increase of chromium content. Wearies the rutile fraction increases with increasing vanadium and manganese content. The Cr_2O_3 phase, MnO_2 phase and V_2O_5 phase appeared at 700, 600 and 700 °C for CrTi20, VTi20 and MnTi20 respectively. This last results leads that the segregation remains modest.

In Table 1.2 are illustrated the variation of the size of crystallite as a function of the temperature. For all samples a sintering of the anatase phase takes place with increasing of temperature from 400 to 800 °C. On the other hand, Cr content retards the sintering of the anatase phase whereas Mn and V content accelerate the sintering of the anatase phase. The anatase phase remains stable when the crystallite size is smaller than 14, 19 and 22 nm for MnTiX, VTiX and CrTiX respectively. This results are similar with results founded by Zhu et al [4], were the anatase phase is stable with small particle size wearies rutile phase is stable for particle size larger than 30 nm.

1.4 Conclusions

The MTiX mixed oxides with X = 0, 5, 10 and 20 of atomic metal content was synthesised by sol-gel method. All the samples were characterized by means of some techniques (XRD and BET). The metal content, temperature and particle size are crucial to the anatase phase stability and to the transformation anatase-to-rutile. Without addition of metal the anatase phase is stable at low temperature (400–500 °C). In the presence of chromium with different content, the anatase phase remains stable at 800 °C. Chromium ions incorporated in TiO_2 anatase phase delay the transformation to rutile and preserve smaller crystallite sizes thus inducing better catalytic activities such as hydrocarbon oxidation. The transformation anatase-to-rutile is accelerated with increasing both V and Mn content. The stability of anatase phase is favored with excess of Cr content and the crystallite size is less than 30 nm. Wearies the rutile phase is favored with V and Mn content.

References

1. G. Cruciani, M.C. Dalconi, M. Dondi, C. Meneghini, F. Matteucci, A. Barzanti, G. Lorenzi, G. Baldi, J. Nanopart. Res. **13**, 711–719 (2011)
2. Y. Hu, H.L. Tsai, C.L. Huang, Mater. Sci. Eng. A **344**, 209–214 (2003)

3. H. Zhang, J.F. Banfield, J. Phys. Chem B **104**, 3481–3487 (2000)
4. K.R. Zhu, M.S. Zhang, J.M. Hong, Z. Yin, Mater. Sci. Eng. A **403**, 87–93 (2005)
5. Y. Djaoued, S. Badilescu, P.V. Ashrit, D. Bersani, P.P. Lottici, J. Robichaud, J. Sol-Gel, Sci. Technol **24**, 255–264 (2002)
6. H.E. Chao, Y.U. Yun, H.U. Xingfang, A. Larbot, J. Eur. Ceram. Soc. **23**, 1457–1464 (2003)
7. R.B. Zhang, J. Non-Cryst. Solids **351**(24–26), 2129–2132 (2005)
8. S. Mahanty, S. Roy, S. Sen, J. Crys, Growth **261**, 77–81 (2004)
9. H. Zhang, J.F. Banfield, Chem. Mater. **17**, 3421–3425 (2005)
10. E. Setiawati, K. Kawano, J. Alloy, Compd **451**, 293–296 (2008)
11. Y. Hu, H.L. Tsai, C.L. Huang, Mater. Sci. Eng. A **344**, 209–214 (2003)
12. S. Rajesh Kumar, S.C. Pillai, U.S. Hareesh, P. Mukundan, K.G.K. Warrier, Mater. Lett. **43**, 286–290 (2000)
13. R. Arroyo, G. Cordoba, J. Padilla, V.H. Lara, Mater. Lett. **54**, 397–402 (2002)
14. D.J. Reidy, J.D. Holmes, C. Nagle, M.A. Morris, J. Mater. Chem. **15**, 3494–3500 (2005)
15. Z. Li, B. Hou, Y. Xu, D. Wu, Y. Sun, W. Hu, F. Deng, J. Solid, State. Chem **178**, 1395–1405 (2005)
16. A.F. Popa, L. Courtheoux, E. Gautron, S. Rossignol, C. Kappenstein, Eur. J. Inorg. Chem. **2005**, 543–554 (2005)
17. S. Pavasupree, Y. Suzuki, S. Pivsa-Art, S. Yoshikawa, J. Solid, State. Chem **178**, 128–134 (2005)
18. H.E. Chao, Y.U. Yun, H.U. Xingfang, A. Larbot, J. Eur. Ceram. Soc. **23**, 1457–1464 (2003)
19. R. Arroyo, G. Cordoba, J. Padilla, V.H. Lara, Mater. Lett. **54**, 397–402 (2002)

Chapter 2
Green Synthesized CdSe Quantum Dots Capped by 3-Mercaptopropionic Acid Sensitized Solar Cells

Bing Gao, Chao Shen, Yunxia Yang, Shuanglong Yuan and Guorong Chen

Abstract In this paper, the green synthesized CdSe quantum dots (QDs) capped by 3-mercaptopropionic acid (MPA) sensitized solar cells are fabricated. Glycerol is chosen as the solvent to prepare the CdSe QDs, which makes the whole reaction is environmental-friendly. After MPA ligand exchanging, the as-synthesized CdSe QDs maintain the original absorbance at ca. 500 nm, coherence to the maximum absorbance of the solar spectrum. The strong photoluminescence quenching of the MPA capped QDs increases the nonradiative decay processes and is beneficial to the electron transfer. Then the MPA ligand exchanged QDs were assembled onto mesoscopic TiO_2 film to integrate into QD sensitized solar cells. The power conversion efficiency of the QD-sensitized solar cells reaches 0.12 % under sun illumination (AM 1.5, 100 mW cm^{-2}) and relatively mechanism is discussed.

B. Gao · C. Shen · Y. Yang (✉) · S. Yuan · G. Chen (✉)
Department of Materials Science and Engineering, East China University of Science and Technology, Shanghai 200237, China
e-mail: yangyunxia@ecust.edu.cn

G. Chen
e-mail: grchen@ecust.edu.cn

B. Gao
e-mail: gaobing8581@163.com

C. Shen
e-mail: paulin15@163.com

S. Yuan
e-mail: shuanglong@ecust.edu.cn

A. Y. Oral et al. (eds.), *International Congress on Energy Efficiency and Energy Related Materials (ENEFM2013)*, Springer Proceedings in Physics 155,
DOI: 10.1007/978-3-319-05521-3_2,

2.1 Introduction

Dye-sensitized solar cells (DSSC) are considered to be a low-cost alternative to conventional solid-state solar cells. DSSC are based on photosensitization of mesoporous TiO_2 films by adsorbed sensitizers, such as Ruthenium complexes and organic dyes with the power conversion efficiency up to 11 % [1]. Besides, semiconductor nanocrystals-quantum dots (QDs) such as CdS, CdSe, PbS, PbSe and InP [2–10], which absorb light in the visible region, can also serve as sensitizer of DSSCs. The specific advantages of QDs over organic materials in light harvest come from the quantum confinement effect due to its size-dependent bandgap. Moreover, by using QDs, it is possible to utilize hot electrons to generate multiple electron-hole pairs per photon through the impact ionization effect [11]. QD-sensitizer also has a higher extinction coefficient than conventional dyes, which can reduce the dark current and increase the overall efficiency of a solar cell. Since QDs' proper energy level is suitable for electrons effectively transfer to the conduction band of TiO_2 or ZnO upon visible light irradiation, the maximum efficiency theoretically predicted for quantum dots-sensitized solar cells (QD-SSC) (44 %) is higher than that for DSSC using organic sensitizers (33.5 %) and for traditional Si solar cells (35 %).

Among various semiconductor QDs, CdSe QDs are suitable to integrate to solar cells because of its absorption in the visible region. Over the past two decades, various approaches have been explored to synthesize CdSe QDs. Although Zhang et al. [12] synthesized high quality scalable CdSe QDs, the trioctylphosphine (TOP) used as the solvent is highly toxic. More recently, focus of investigations turns to the water-phase synthesis of CdSe QDs [13, 14] in order to realize green synthesis of this QDs.

It is known that QD-SSC's performance largely benefit from the solubility and surface modification of the QD sensitizers which significantly influence charge transfer behavior [15]. In most cases, the surfactants used for preventing aggregation during the growth of the QDs contain long alky chains, like olic acid (OA) [16]. In order to improve the charge transfer between nanocrystals and polymer, as well as electron transfer between QDs, extensive investigations on the surface modification of nanocrystals have been reported based on short insulating ligands, which leads to a relatively improved efficiency of solar cells. For example, pyridine treatment of the QDs is the most commonly used and effective procedure for improving device efficiency [17]. Olson et al. reported on a CdSe/P3HT blended device using butylamine as a shorter capping ligand for CdSe nanocrystals (NCs) [18]. Considering that 3-mercaptopropionic acid (MPA) is a kind of short-chain ligand and according to the well-known hard and soft acids and bases theory [19], the hard-hard or soft-soft molecules or ions can form a strong binding. The Cd^{2+} and Se^{2-} ions are classified as soft ions and have stronger interaction with soft "–SH (from MPA)" but not hard "–COOH (from OA)" So the MPA can form rather strong bonding with CdSe.

In this paper, we choose water phase glycerol as solvent to prepare the CdSe QDs. The reaction is environmental-friendly and energy saving, with which we have realized the in situ observation and control of the whole procedure. Then the long chain ligands olic acid (OA) of CdSe QDs was exchanged by the short chain ligands MPA. The MPA-capped CdSe QDs with absorbance of ca. 500 nm, which is coherence to the maximum absorbance of sun spectrum, were assembled onto mesoscopic TiO_2 film, choosing platinum as the counter electrode and fabricated into QD sensitized solar cells. With the increase of MPA, the energy conversion efficiency (ECE) increases from 0.08 to 0.12 % under the sun illumination (AM 1.5, 100 mW cm^{-2}).

2.2 Experimental

The preparation of CdSe quantum dots: The Se precursor for the present work was prepared by mixing the selenium powder (0.005 mol) (A. R, Shanghai Meixing Chemical Co., Ltd.), sodium sulfite (0.01 mol) (A.R, Sinopharm Chemical Reagent Co., Ltd.), sodium hydroxide (0.01 mol) (A.R, Shanghai Aijian ready-made Reagent Co. Ltd.) and deionized water (50 ml) with N_2 protection and magnetic stirring. The mixture was heated to boiling while the solution turned from red brown to the black precipitated. Then sodium hydroxide (0.01 mol) was added, and after returning for 4 h, the reactants were cooled down to the room temperature followed by nitrogen protection for 30 min.

The Cd precursor was prepared by mixing water, ethanol and oleic acid (A.R, Jiangsu Yonghua Fine Chemicals Co., Ltd.) together in volume ratio of 10:30:8. Cadmium acetate 0.001 mol (A.R, Sinopharm Chemical Reagent Co. Ltd.) and sodium hydroxide (0.7 g) were added as precipitation agents to the mixed solvent (48 ml). The mixture was under the N_2 protection until it became clear and transparent. The reaction temperature was 120 °C.

The CdSe QDs were synthesized by adding 10 ml Se precursor solution to the 48 ml Cd precursor solution, under N_2 protection and magnetic stirring. The reactions were carried out at 120 °C for 5, 10, 15, 20, 25 and 30 min, respectively. The samples were centrifuged with ethanol, and dispersed in hexane.

The ligand exchange of olic acid (OA) acid-capped CdSe QDs to 3-mercaptopropionic acid (MPA)-capped CdSe QDs: Mix 3-Mercaptopropionic acid (MPA) (with molar ratio of 5,10, 20 and 30 comparing to CdSe QDs), 0.3 mL H_2O and 1 mL methanol, adjusting the pH value of the solution around 12 with 40 wt% of NaOH solution. Then, add 5 ml CdSe solution (in hexane, at 120 °C in 30 min). After stirring the above mixed solution for 2 h, 7 mL H_2O was added. The whole solution are kept standstill for 5 min, which is layered with the bottom layer of yellow mixed CdSe QDs and MPA and the up layer of transparent hexane. The bottom layer is used in the solar cell fabrication. And the four samples (molar ratio MPA/QDs = 5, 10, 20 and 30) are named I, II, III and IV, respectively.

The preparation of CdSe QDs sensitized solar cells-The preparation of anode: The fluorine-doped SnO_2 (FTO) glass (7 Ω/□) was first cleaned in a detergent solution using an ultrasonic bath for 10 min and then washed with water and ethanol with the same method. A layer of TiO_2 past (P25) was coated on the FTO glass plates by screen-printing. Then, the FTO glass was kept in a clean box for 3 min and then dried for 6–8 min at 125 °C. This procedure is repeated for 6 times and the final thickness of TiO_2 film is 12 μm and area is 25 mm^2. The electrodes coated with the TiO_2 pastes were gradually heated under an airflow at 125 °C for 20 min, 325 °C for 5 min, 375 °C for 5 min, 450 °C for 15 min and finally at 500 °C for 15 min. The FTO glass with the TiO_2 film is treated at 70 °C in 30 min with 40 mM $TiCl_4$ solution, then washed with water and ethanol and sintered at 500 °C for 30 min (2 °C/min), and then kept at room temperature for 20–24 h. After that, the TiO_2 anode was immersed in the MPA-capped CdSe QDs for 24 h.

The preparation of the counter electrode: To prepare the counter electrode, a hole (d = 1 mm) was drilled in the FTO glass by sandblasting. The perforated FTO glass was washed with H_2O as well as with 0.1 M HCl solution in ethanol by ultrasonic bath for 10 min. After wiping off residual organic matter by heating in air for 15 min at 400 °C, the Pt past was deposited on the FTO glass by screen-printing technique, and sintered at 400 °C for 30 min.

The encapsulation of the CdSe QDs sensitized solar cells: Seal the anode and Pt-counter-electrode, and the polysulfide electrolyte (consisting of 2 M Na_2S and 3 M S) was infused into the solar cells.

The high resolution transmission electron microscopy (HRTEM) (JEM–201, American FEI Company) was used to characterize the morphology of the CdSe QDs. The optical absorption spectra (a Lambda 950 UV-Visible Spectrophotometer, Perkin Elmer, Waltham, MA) and the photoluminescent (PL) spectra (model Fluorolog-3-P, France JobinYvoncompany, model Cary 500, American Varian company) were used to test the optical properties of CdSe QDs. Current density-voltage measurements under AM 1.5 G 100 mW/cm^2 illumination are used to characterize the efficiency of the CdSe QDs sensitized solar cells.

2.3 Results and Discussion

Characterization of CdSe QDs: Figure 2.1(a–d) presents HRTEM, absorption spectra and luminescence spectra of the synthesized CdSe QDs. From Fig. 2.1a, b it is seen that oval shaped CdSe QDs show an excellent monodispersity with the size distribution mainly close to 3–4 nm as shown in the inset of Fig. 2.1a, while the existence of lattice fringe indicates that the products have excellent crystallinity. With the prolonged reaction time from 5 to 30 min, the color of real samples changed from yellow, orange to red (the inset of Fig. 2.1b). From the absorption spectra of the CdSe QDs it is seen that absorption peaks appear at 475–500 nm, coherence to the maximum remittance region of the solar spectrum. An obvious

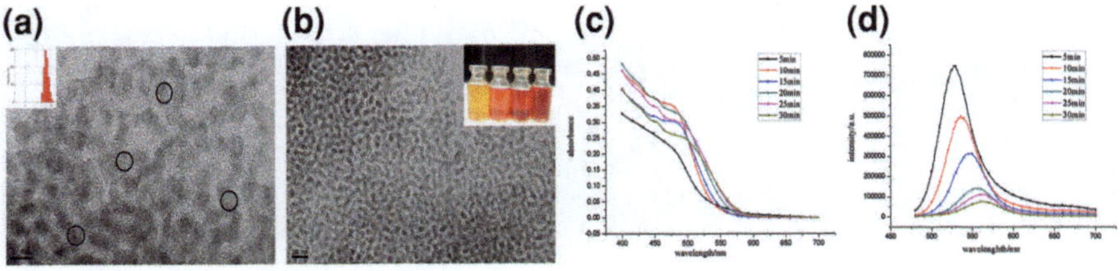

Fig. 2.1 **a** HRTEM of CdSe QDs (scale bar 5 nm) and its size distribution, **b** HRTEM of CdSe QDs (scale bar 20 nm) and the real sample of CdSe QDs, **c** absorption spectra of the QDs at 120 °C in different reaction time, and **d** emission spectra of the QDs at 120 °C in different reaction time excited at 460 nm

blue shift of the absorption peak comparing with that of the bulk CdSe (717 nm) proves occurrence of the band-gap broadening phenomenon as a consequence of the quantum size effect. Furthermore, with the increased reaction temperature, the size of products grew larger; the quantum size effect thus reduces, leading to the red shift of the absorption peak due to narrowing of the band-gap. This tunable band-gap of CdSe QDs is suitable to be applied in solar cells as sensitizers. Figure 2.1d shows the normalized emission spectra of CdSe QDs. Under the excitation of blue light (460 nm), a broad emission band is observed peaking in the green to yellow region and showing obvious red shift with the increasing reaction time also due to the quantum size effect.

Characterization of MPA capped-CdSe QDs: Because MPA contains short alky chains, while olic acid has long alky chains, after MPA capping, samples I, II, III and IV show the monodispersity (Fig. 2.2a–d) inferior to that of olic acid capped CdSe QDs (Fig. 2.1a, b). With the increased MPA/QDs molar ratio (5, 10, 20 and 30), the aggregation of the CdSe QDs becomes more severe. MPA can be a large driving force in nanocrystal aggregation which improves the charge transport [20–22]. It is expected to increase relatively power conversion efficiency of solar cells [23, 24].

Figure 2.3 illustrates the absorption spectra of MPA capped samples I, II, III and IV where samples I and II shows the similar spectra to that of CdSe QDs capped by olic acid (Fig. 2.1c), i.e., ligand exchanging doesn't affect the QDs absorption. However, with the further increased quality of MPA (samples III and IV), absorption peaks become undetectable; proving the aggregation among MAP capped QDs again, coinciding with the result of HRTEM (Fig. 2.2).

In contrast, strong photoluminescence quenching was observed in the case of MPA replaced QDs, as shown in Fig. 2.4, comparing to the MPA-free CdSe QDs' (Fig. 2.1d). With the increased molar ratio MPA/QDs = 5, 10, 20 and 30, emission intensity decreased obviously. This phenomenon could be attributed to the removal of excess oleic acid ligands accumulated around the QDs, resulting in the increase of nonradiative decay processes [25–27], which is beneficial to the electron transferring from the conduction band of CdSe to TiO_2's [28–31].

Current and voltage measurements: The current density (J) and voltage (V) were measured under AM 1.5 G 100 mW/cm^2 illumination on samples I, II, III

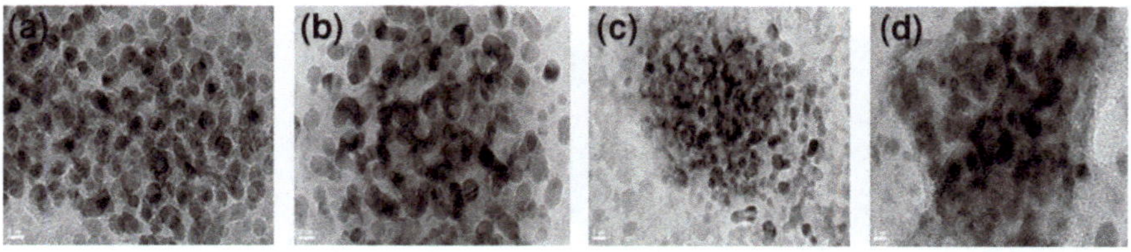

Fig. 2.2 HRTEM photos of samples I, II, III and IV (**a**, **b**, **c** and **d**)

Fig. 2.3 Absorption spectra of samples I, II, III and IV

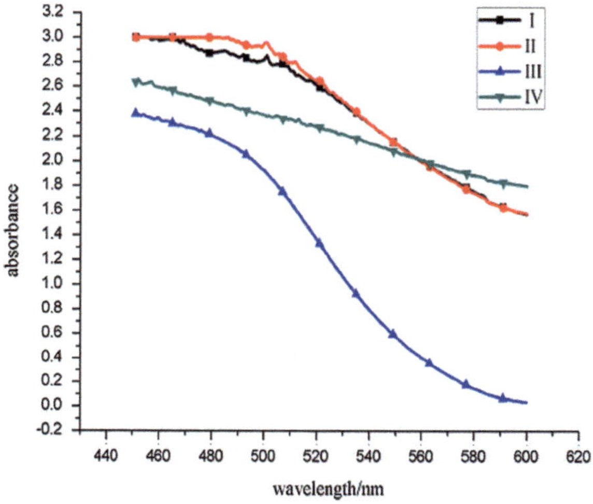

Fig. 2.4 Emission spectra of samples I, II, III and IV excited at 460 nm

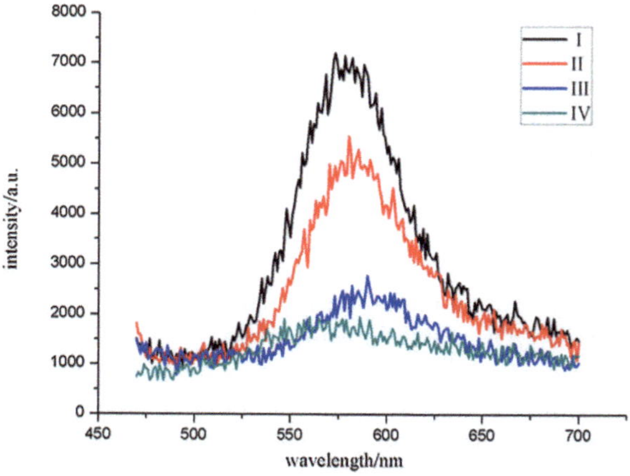

and IV being assembled onto the mesoscopic TiO_2 film. As shown in Fig. 2.5 and Table 2.1, the optimal results have been obtained from the sample IV, that is, an open-circuit voltage (V_{oc}) of 300.11 mV, a short-circuit current density (J_{sc}) of 1.12 mA/cm^2, a fill factor (FF) of 0.36, and a PCE of 0.12 %.

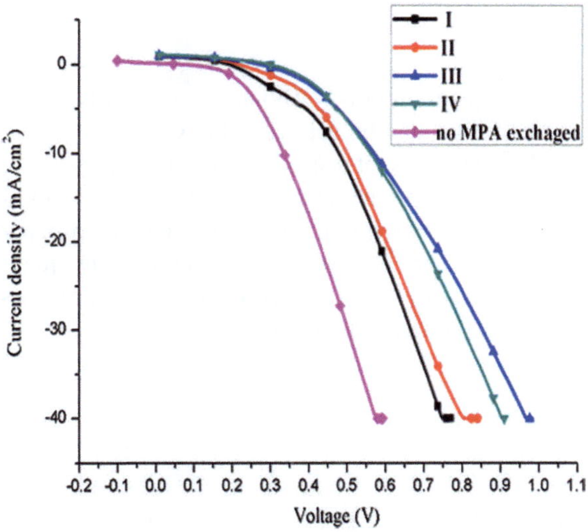

Fig. 2.5 the sealed solar cells' I–V *curves* of sample I, II, III and IV and no MPA exchanged CdSe QDs

Table 2.1 parameters obtained from the I–V curves of QD-SSCs with different mol ratio of MPA/CdSe QDs

Sample number	Different mol ratio of MPA/CdSe QDs	η (%)	FF	V_{oc} (mV)	J_{sc} (mA cm^{-2})
No MPA exchanged CdSe QDs	0:1	0.00	0.26	69.81	0.12
I	5:1	0.08	0.45	191.11	0.90
II	10:1	0.10	0.46	215.12	0.99
III	20:1	0.11	0.41	275.82	1.00
IV	30:1	0.12	0.36	300.11	1.12

From Fig. 2.5 and Table 2.1 it is also seen that the PCE of devices is sensitive to the molar ratio of MPA/CdSe QDs. With the increase of the MPA loading, the increase of PCE can be observed while the MPA free CdSe QDs has no such an effect. This is because when oleic acid ligands accumulated around the CdSe QDs, their monodispersity is excellent, and this is not beneficial for the electron transfer between QDs and leads to the increase of radiative of the electrons as proved by the strong emissions, resulting in no power conversion efficiency of solar cells. However, when olic acid is replaced by MPA, especially, with the increase of MPA (samples II, III and IV), the aggregation among QDs occurs. Therefore, electron transfer and the nonradiative decay processes are enhanced, i.e. after the solar illumination, the electrons preferentially transfer to the conduction band of TiO_2 from the conduction band of CdSe QDs. The efficiency of the solar cells increases thus to 0.08, 0.10, 0.11 and 0.12 %, respectively.

2.4 Conclusion

In this paper, the CdSe quantum dots (QDs) are green synthesized by the glycerol solvent. The long chain ligands olic acid of CdSe QDs was replaced by the short chain ligands of MPA. The as-synthesized CdSe QDs with the absorbance at ca. 500 nm, coherence to the maximum absorbance of sun spectrum, were assembled onto mesoscopic TiO_2 film, choosing platinum as the counter electrode to prepare the CdSe QDs sensitized solar cells. With the increase of the quality of MPA, the power conversion efficiency of the QD-sensitized solar cell reaches 0.12 % under the sun illumination (AM 1.5, 100 mW cm^{-2}).

References

1. S. Hwang, J.H. Lee, C. Park, H. Lee, C. Kim, C. Park, M.-H. Lee, W. Lee, J. Park, K. Kim, N.-G. Park, C. Kim, A highly efficient organic sensitizer for dye-sensitized solar cells. Chem. Commun. 4887–4889 (2007)
2. L.M. Peter, D.J. Riley, E.J. Tull, K.G. Upul Wijayantha, Photosensitization of nanocrystalline TiO2 by self-assembled layers of CdS quantum dots. Chem. Commun. 1030–1031 (2002)
3. Y.-L. Lee, Y.-S. Lo, Highly efficient quantum-dot-sensitized solar cell based on co-sensitization of CdS/CdSe. Adv. Funct. Mater. **19**, 604–609 (2009)
4. S.-C. Lin, Y.-L. Lee, C.-H. Chang, Y.-J. Shen, Y.-M. Yang, Quantum-dot-sensitized solar cells: Assembly of CdS-quantum-dots coupling techniques of self-assembled monolayer and chemical bath deposition. Appl. Phys. Lett. **90**, 143517 (2007)
5. I. Robel, V. Subramanian, M. Kuno, P.V. Kamat, Quantum dot solar cells. Harvesting light energy with CdSe nanocrystals molecularly linked to mesoscopic TiO2 films. J. Am. Chem. Soc. **128**, 2385–2393 (2006)
6. I. Shin, H. Seo, M.-K. Son, J.-K. Kim, H. Kim, Characteristic of (Pb1-xZnx)S tandem structured quantum dot-sensitized solar cell having wide light absorbance. Phys. Status Solidi C. **8**, 631–633 (2011)
7. R.D. Schaller, V.M. Agranovich, V.I. Klimov, High-efficiency carrier multiplication through direct photogeneration of multi-excitons via virtual single-exciton states. Nat. Phys. **1**, 189–194 (2005)
8. R.D. Schaller, V.I. Klimov, High efficiency carrier multiplication in PbSe nanocrystals: implications for solar energy conversion. Phys. Rev. Lett. **92**, 186601 (1–4) (2004)
9. X. Ma, Y. Shen, Q. Wu, T. Shen, M. Cao, F. Gu, L. Wang, Free-standing TiO2 nanotube arrays for front-side illuminated CdS quantum dots sensitized solar cells. J. Inorg. Organomet. Polym. **23**, 798–802 (2013)
10. J.L. Blackburn, D.C. Selmarten, R.J. Ellingson, M. Jones, O. Micic, A.J. Nozik, Electron and hole transfer from indium phosphide quantum dots. J. Phys. Chem. B, **109**(7), 2625–2631 (2005)
11. A.J. Nozik, Exciton multiplication and relaxation dynamics in quantum dots: applications to ultrahigh-efficiency solar photon conversion. Inorg. Chem. **44**, 6893 (2005)
12. W. Zhang, H. Zhang, Y. Feng, X. Zhong, Scalable single-step noninjection synthesis of high-quality core/shell quantum dots with emission tunable from violet to near infrared. Nano Lett. **6**(12), 11066–11073 (2012)
13. L. Liu, Q. Peng, Y. Li, Preparation of CdSe quantum dots with full color emission based on a room temperature injection technique. Inorg. Chem. **47**, 5022–5028 (2008)

14. X. Chen, J.L. Hutchison, P.J. Dobson, G. Wakefield, Highly luminescent monodisperse CdSe nanoparticles synthesized in aqueous solution. J. Mater. Sci. **44**, 285–292 (2009)
15. D.S. Ginger, N.C. Greenham, Efficient CdSe/CdS quantum dot light-emitting diodes using a thermally polymerized hole transport layer. J. Appl. Phys. **87**, 1361 (2000)
16. A.J Nozik, Quantum dot solar cells. Phys. E **14**,115–120 (2002)
17. J. Seo, W.J. Kim, S.J. Kim, K.S. Lee, A.N. Cartwright, P.N. Prasad, Polymer nanocomposite photovoltaics utilizing CdSe nanocrystals capped with a thermally cleavable solubilizing ligand. Appl. Phys. Lett. **94**, 133302 (2009)
18. N.C. Greenham, X. Peng, A.P. Alivisatos, CdSe nanocrystal rods/poly (3-hexylthiophene) composite photovoltaic devices. Phys. Rev. B, **54**, 17628 (1996)
19. R.G. Pearson, Hard and soft acids and bases. J. Am. Chem. Soc. **85**, 3533 (1963)
20. J.D. Olson, G.P. Gray, S.A. Carter, Optimizing hybrid photovoltaics through annealing and ligand choice. Solar Energy Mater. Solar Cells. **93**, 519–523 (2009)
21. B. Sun, N.C. Greenham, Improved efficiency of photovoltaics based on CdSe nanorods and poly(3-hexylthiophene) nanofibers. Phys. Chem. Chem. Phys. **8**, 3557 (2006)
22. B. Sun, H.J. Snaith, A.S. Dhoot, S. Westenhoff, N.C. Greenham, Vertically segregated hybrid blends for photovoltaic devices with improved efficiency. J. Appl. Phys. **97**, 014914 (2005)
23. I. Gur, N.A. Fromer, C. Chen, A.G. Kanaras, A.P. Alivisatos, Hybrid solar cells with prescribed nanoscale morphologies based on hyperbranched semiconductor nanocrystals. Nano Lett. **7**, 409 (2007)
24. W.U. Huynh, J.J. Dittmer, W.C. Libby, G.L. Whiting, A.P. Alivisatos, Controlling the morphology of nanocrystal-polymer composites for solar cells. Adv. Funct. Mater. **13**, 73 (2003)
25. C.H. Chang, Y.L. Lee, Chemical bath deposition of CdS quantum dots onto mesoscopic TiO_2 films for application in quantum-dot-sensitized solar cells. Appl. Phys. Lett. **91**, 053503 (2007)
26. P. Yu, K. Zhu, A.G. Norman, S. Ferrere, A.J. Frank, A.J. Nozik, J. Phys. Chem. B **110**, 25451 (2006)
27. Y.-L. Lee, C.-H. Chang, Efficient polysulfide electrolyte for CdS quantum dot-sensitized solar cells. J. Power Sources **185**, 584–588 (2008)
28. S.-Q. Fan, D. Kim, J.-J. Kim, D.W. Jung, S.O. Kang, J. Ko, Highly efficient CdSe quantum-dot-sensitized TiO2 photoelectrodes for solar cell applications. Electrochem. Commun. **11**, 1337–1339 (2009)
29. O. Niitsoo, S.K. Sarkar, C. Pejoux, S. Ruhle, D. Cahen, G. Hodes, High efficiency of CdSe quantum-dot-sensitized TiO2 inverse opal solar cells. J. Photochem. Photobiol. A **181**, 306 (2006)
30. S. Nishimura, N. Abrams, B.A. Lewis, L.I. Halaoui, T.E. Mallouk, K.D. Benkstein, J. van de Lagemaat, A.K. Frank, Standing wave enhancement of red absorbance and photocurrent in dye-sensitized titanium dioxide photoelectrodes coupled to photonic crystals. J. Am. Chem. Soc. **125**, 6306 (2003)
31. I. Robel, V. Subramanian, M. Kuno, P.V. Kamat, Size-dependent electron injection from excited CdSe quantum dots into TiO_2 nanoparticles. J. Am. Chem. Soc. **128**, 2385 (2006)

Chapter 3
Study of the Thermal Decomposition of Some Components of Biomass by Desorption Mass Spectrometry

Borys Palianytsia, Tetiana Kulik, Olesia Dudik, Tetiana Cherniavska and Oksana Tonkha

Abstract The investigation of thermal transformations of lignin samples have been carried out using temperature programmed desorption mass spectrometry method (TPD-MS). Main stages and products of lignin pyrolysis have been identified. The first stages (T_{max} = 230 °C and T_{max} = 300 °C) are attributed to thermal transformations of lignin peripheral polysaccharide fragments such as hemicellulose and cellulose respectively. The second stage (T_{max} = 335 °C) is associated with desorption of lignin structural elements in the molecular forms as a result of depolymerization processes of polymeric blocks of lignin. The third stage (T_{max} = 370 °C) correspond to a deeper decomposition of lignin and characterized by desorption of smaller structural fragments in molecular forms (*m/z* = 110, pyrocatechol). Pressure–temperature curves of pyrolysis of lignin samples have been analyzed.

3.1 Introduction

Increased demand in transport fuels, environmental concerns, and depletion of fossil fuels requires the development of efficient conversion technologies for second-generation biofuels, which being received by means of different pyrolytic methods [1–5]. Pyrolysis utilizes biomass to produce a product, used both as second-generation energy source and as a feedstock for chemical production [1–5]. Virtually any form of biomass can be considered in fast pyrolysis. Cellulose,

B. Palianytsia · T. Kulik (✉) · O. Dudik · T. Cherniavska
Chuiko Institute of Surface Chemistry, National Academy of Sciences of Ukraine, 17 General Naumov Str. Kyiv 03169, Ukraine
e-mail: tanyakulyk@i.ua

O. Tonkha
National University of Life and Environmental Sciences of Ukraine, Kyiv Ukraine

A. Y. Oral et al. (eds.), *International Congress on Energy Efficiency and Energy Related Materials (ENEFM2013)*, Springer Proceedings in Physics 155,
DOI: 10.1007/978-3-319-05521-3_3,

hemicellulose and lignocellulose are the main raw materials for production of second generation biofuel by pyrolysis [6–9].

Lignin is the second most common renewable natural raw material and the most widespread aromatic phenolic polymer. Lignin content in wood tissues varies from 18 to 30 % of dry weight of wood [6, 8, 10]. Hydroxycinnamic acids (coumaric, ferulic and sinapic) and their corresponding alcohols (p-coumaryl alcohol or 4-(3-hydroxy-1-propenyl)phenol; coniferyl alcohol or 4-(3-hydroxy-1-propenyl)-2-methoxyphenol; and sinapyl alcohol or (4-(3-hydroxy-1-propenyl)-2,6-dimethoxyphenol) are intermediates of both synthesis of lignin and its biodegradation resulted in the formation of humic acids in soils [11–14].

The chemical composition and structure of lignin is heterogeneous. It is very complex compound. But this complexity is not defined by the large numbers of different monomer units. All monomer units in lignin molecule are phenylpropan derivatives (mainly coniferyl alcohol). Lignin of coniferous woods consists chiefly of coniferyl alcohol. Lignin of deciduous woods contains coniferyl and sinapyl alcohols, and lignin of gramineous plants contains p-coumaryl alcohol. Phenylpropanoid units in lignin molecule are connected between each other by ether and carbon–carbon bonds.

The specified complexity of lignin structure is caused by the fact, that the most complex processes under thermochemical conversion of biomass are observed under lignin pyrolisis. Therefore the understanding of lignin pyrolysis processes is very important for development of new technologies of thermochemical processing of lignocellulosic biomass. The present work describes the investigations of lignin pyrolysis by TPD MS method to identify main products of pyrolysis and basic paths of thermal transformations of lignin separated from pine needles. Such type of lignin has one of the simplest structures as its macromolecule consist mainly of monomer units of coniferyl alcohol.

3.2 Experimental

Temperature Programmed Desorption Mass Spectrometry. The experiments were performed in an instrument with an electron impact ionization and an MKh-7304A monopole mass spectrometer (Sumy, Ukraine), adapted for thermo desorption measurements. Samples weighing 0.1–20 mg were first placed in a quartz-molybdenum ampoule (5.4 mm in diameter, a length of 20 cm and the volume of 12 ml) and degassed to $\sim 5 \times 10^{-5}$ Pa at ~ 20 °C. Then they were heated up to ~ 750 °C at 0.17 °C s^{-1} using a programmed linear heating schedule. T and t are related by dT/dt = b, in which b is the heating rate. Volatile pyrolysis products passed through a high-vacuum valve and into the ionization chamber of the mass-spectrometer where they were ionized and fragmented by electron impact. After mass separation in the mass analyzer, the ion current due to desorption and thermolysis was amplified with a VEU-6 secondary-electron multiplier. The mass spectra and the *P–T* pressure-temperature curves (*P*—the pressure of volatile

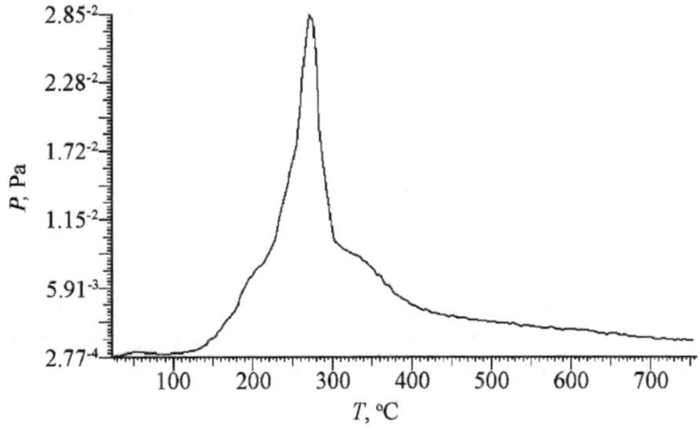

Fig. 3.1 Temperature-pressure *(P–T) curve. P* pressure of the volatile products; *T* temperature of the *lignin* sample

pyrolysis products, *T*—temperature of samples) were recorded and analyzed using a computer-based data acquisition and processing setup. The pressure of volatile pyrolysis products was measured with a pressure vacuum gauge VMB-14. The mass spectra were recorded for a mass range of 1–210 amu. During each TPD-MS experiment, ~240 spectra were recorded and averaged.

Lignin (LC) separated from pine needles contained 93–95 % of lignin and 5–7 % of cellulose was used in the work.

3.3 Results and Discussion

It can be seen from analyzing pressure of volatile products—temperature of sample pyrolysis curve (P–T), that the thermal decomposition of main part of lignin under conditions of TPD MS experiment occurs over broad temperature range 160–450 °C because of absence of oxygen and processes of oxidative destruction respectively (Fig. 3.1).

Owing to the absence of oxygen the pyrolysis is a result of consistent breaking weak chemical bonds with release of stable compounds in molecular form. The analysis of TPD-curves allowed to determine basic stages of pyrolysis (Figs. 3.2, 3.3 and 3.4) and identify products of thermal transformations, which are formed on these stages. Therefore pyrolysis stages of LC: Ia – $T_{max} \approx 230$ °C and Ib – $T_{max} \approx 300$ °C are caused by thermal transformations of polysaccharide residues, which are included as a functional part of macromolecular complex, hemicellulose and cellulose respectively (Fig. 3.2). It has been known, that basic products of hemicellulose pyrolysis are furan derivatives and other low-molecular products such as water, carbon oxides, methanol, acetic acid [15]. Stage Ib – $T_{max} \approx 300$ °C is connected with destruction of pyranose cycles of glucose, from which cellulose macromolecule is composed (Fig. 3.2). And this stage is characterized by the presence of peaks on TPD-curves for ions with m/z = 60, 97, 98, 84. Mass spectra of pyrolysis products of glucose, mono-, oligosaccharides, β-cyclodextrin [16, 17] contain exactly such characteristic set of ions.

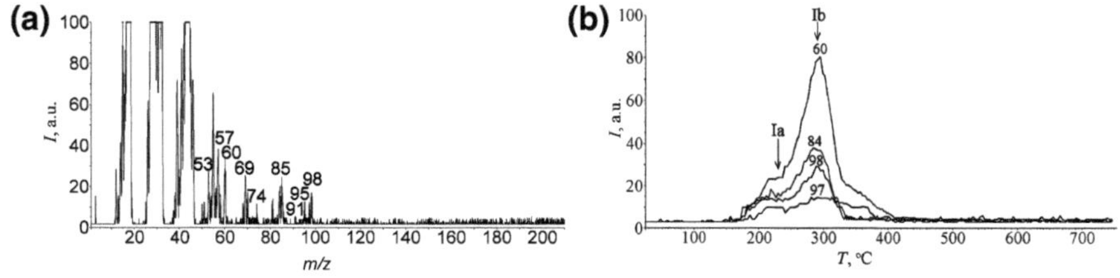

Fig. 3.2 Pyrolysis of a lignin sample. **a** Mass-spectrum of pyrolysis products at 230 °C, obtained after electron impact ionization. **b** TPD traces of the ions with m/z = 98, 97, 84, 60 attributed to the destruction of the polysaccharide components of hemicellulose and cellulose (Ia, b stages)

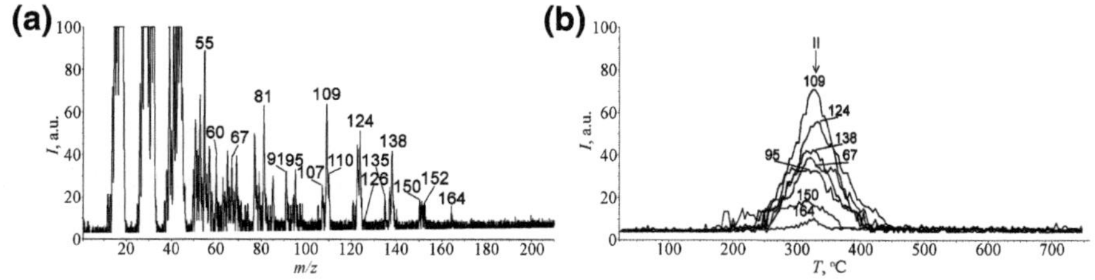

Fig. 3.3 Pyrolysis of a lignin sample. **a** Mass-spectrum of pyrolysis products at 333 °C, obtained after electron impact ionization. **b** TPD traces of the ions with m/z = 164, 150, 138, 124, 109, 95, 67 attributed to the thermal decomposition of lignin macromolecule (stage II)

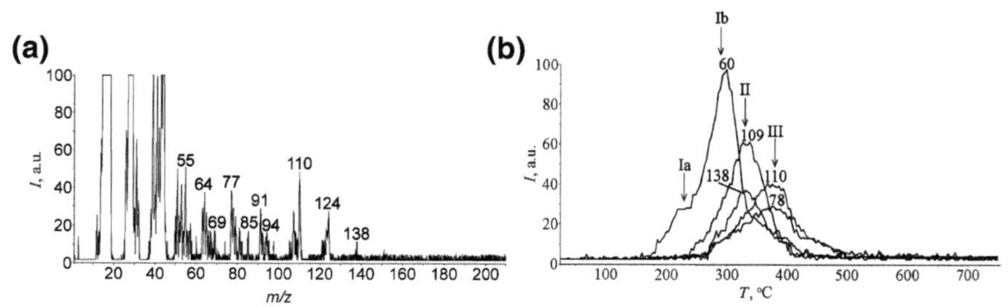

Fig. 3.4 Pyrolysis of a lignin sample. **a** Mass-spectrum of pyrolysis products at 370 °C, obtained after electron impact ionization. **b** TPD trace of the fragment ion with m/z = 60 attributed to the thermal transformations of polysaccharide fragments (stages Ia, b); TPD traces of the ions with m/z = 138, 109 attributed to the thermal transformations of peripheral phenylpropanoid structures of lignin (stage II); TPD traces of the ions with m/z = 110, 78 attributed to the thermal transformations of inner phenylpropanoid structures of lignin (stage III)

Stage II ($T_{max} \approx 335$ °C) is characterized by presence of peaks on TPD-curves for ions with m/z = 164, 150, 138, 124, 109, 95, 67 (Fig. 3.3). Analysis of mass-spectrometric information and its comparison with data obtained from catalog of mass-spectra [18] and literature sources [19, 20] indicates that the stage II is caused by thermal transformations of phenylpropanoid blocks of lignin. Ion with m/z = 164 is identified as fragment of coniferyl alcohol. This product is desorbed

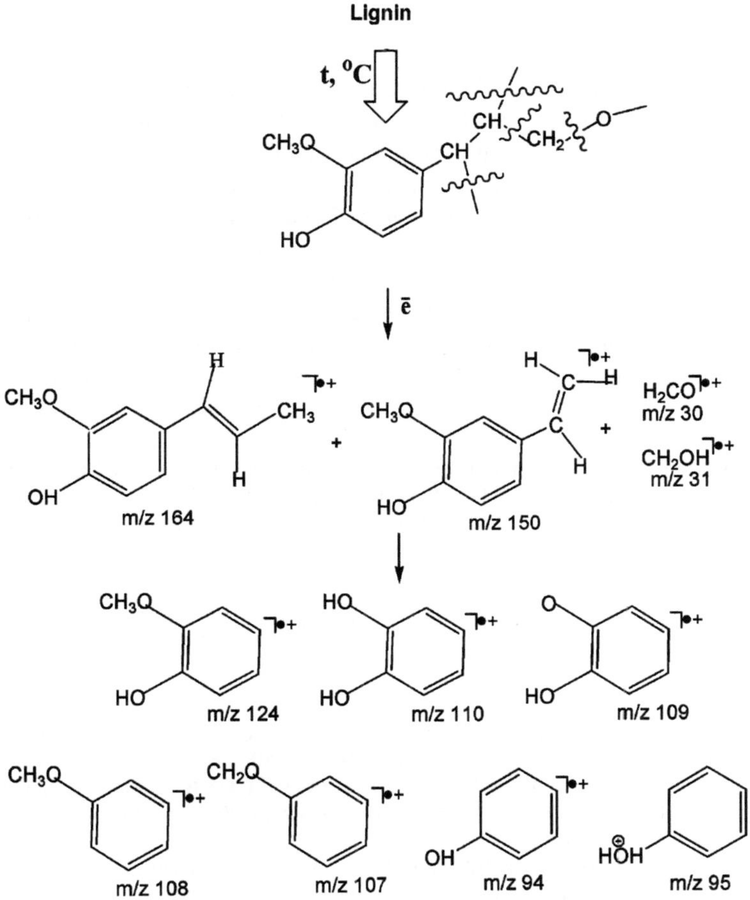

Scheme 3.1 Structural composition of ions, observed in mass-spectra at temperatures, corresponding to stage II of pyrolysis of lignin (T_{Makc} = 330 °C)

in molecular form as a result of its splitting-off from lignin molecule (Scheme 3.1). Ion with m/z = 150 is based on desorption of 4-hydroxy-3-methoxy-vinylbenzene, which is registered during pyrolysis of ferulic acid at temperature above 350 °C [21, 22].

Stage III (T_{max} ≈ 370 °C) is described by peaks for ions with m/z = 110, 108, 78 (Fig. 3.4). Its presence is a result of deeper decomposition of lignin. And this stage is characterized by elimination of smaller structural units of lignin in molecular form, such as pyrocatechol (Mr = 110 Da, m/z = 110), cresol (Mr = 108 Da, m/z = 108, 107), benzene (Mr = 78 Da, m/z = 78). Obviously, the presence of two basic stages of thermal transformations of lignin can be explained by the presence of two basic types of phenylpropanoid blocks (inner and peripheral) in lignin molecule.

3.4 Conclusions

The basic stages and products of lignin thermolysis are identified. The first two stages at $T_{max} \approx 230$ °C and $T_{max} \approx 300$ °C are connected with thermal transformations of peripheral polysaccharide fragments of hemicelluloses and cellulose respectively. Stage at $T_{max} \approx 335$ °C is caused by elimination of structural elements of lignin in molecular form, such as 4-(1-propenyl)-2-methoxyphenol (m/z = 164) and 4-hydroxy-3-methoxy-vinylbenzene (m/z = 150), as a result of processes of destruction of phenylpropanoid polymer blocks of lignin. Stage at $T_{max} \approx 370$ °C is due to deeper decomposition of lignin and characterized by elimination of smaller structural units of lignin in molecular form, such as pyrocatechol (Mr = 110 Da, m/z = 110), cresol (Mr = 108 Da, m/z = 108, 107), benzene (Mr = 78 Da, m/z = 78).

Acknowledgments This publication is based on work supported by a grant UKC2-7072-KV-12 from the U.S. Civilian Research and Development Foundation (CRDF Global) with funding from the United States Department of State and by a grant M/299-2013 from the State Agency of Ukraine for Science, Innovation and Information.

References

1. A. Oasmaa, Y. Solantausta, V. Arpiainen, E. Kuoppala, K. Sipila, Fast pyrolysis bio-oils from wood and agricultural residues. Energy Fuels **24**(2), 1380–1388 (2010)
2. M. Calonaci, R. Grana, E. Barker Hemings, G. Bozzano, M. Dente, E. Ranzi, Comprehensive kinetic modeling study of bio-oil formation from fast pyrolysis of biomass. Energy Fuels **24**(10), 5727–5734 (2010)
3. S. Thangalazhy-Gopakumar, S. Adhikari, R.B. Gupta, S.D. Fernando, Influence of pyrolysis operating conditions on bio-oil components: a microscale study in a pyroprobe. Energy Fuels **25**(3), 1191–1199 (2011)
4. C.A. Mullen, A.A. Boateng, K.B. Hicks, N.M. Goldberg, R.A. Moreau, Analysis and comparison of bio-oil produced by fast pyrolysis from three barley biomass/byproduct streams. Energy Fuels **24**(1), 699–706 (2010)
5. D. Mohan, C.U. Pittman, P.H. Steele, Pyrolysis of wood/biomass for bio-oil: a critical review. Energy Fuels **20**(3), 848–889 (2006)
6. H. Yang, R. Yan, H. Chen, D.H. Lee, C. Zheng, Characteristics of hemicellulose, cellulose and lignin pyrolysis. Fuel **86**(12–13), 1781–1788 (2007)
7. H. Ben, A.J. Ragauskas, Pyrolysis of kraft lignin with additives. Energy Fuels **25**(10), 4662–4668 (2011)
8. A. Gandini, The irruption of polymers from renewable resources on the scene of macromolecular science and technology. Green Chem. **13**, 1061–1083 (2011)
9. A. Corma, S. Iborra, A. Velty, Chemical routes for the transformation of biomass into chemicals. Chem. Rev. **107**(6), 2411–2502 (2007)
10. K.V. Sarkanen, *Lignins: Occurrence, Formation, Structure and Reactions* (Wiley, New York, 1971)
11. R. Gross, A. Kumar, B. Kalra, Polymer synthesis by in vitro enzyme catalysis. Chem. Rev. **101**(7), 2097–2124 (2001)
12. S. Burton, Oxidizing enzymes as biocatalyst. Trends Biotechnol. **21**, 543–549 (2003)

13. Y.N. Kupriyanovich, S.A. Medvedeva, A.V. Rokhin, L.V. Kanitskaya, Regioselectivity of ferulic acid polymerization catalyzed by oxidases. Russian J. Bioorg. Chem. **33**(5), 516–522 (2007)
14. H. Shindo, T. Marumoto, T. Higashi, Behavior of phenolic substances in the decaying process of plants. Soil Sci. Plant Nutr. **25**(4), 591–600 (1979)
15. A.A. Rosatella, S.P. Simeonov, R.F.M. Frade, C.A.M. Afonso, 5-Hydroxymethylfurfural (HMF) as a building block platform: Biological properties, synthesis and synthetic applications. Crit. Rev. Green Chem. **13**, 754–793 (2011)
16. T.V. Kulik, Mass spectrometry of carbohydrate fragments—terminal groups of receptors molecules in condensed state and in adsorbed state on the surface of ultrafine silica. Ph.D. Thesis, Institute of Surface Chemistry of the NAS of Ukraine, 2000
17. L.A. Belyakova, A.M. Varvarin, O.V. Hora, B.B. Palyanytsya, T.V. Kulik, Thermodestruction of inclusion compounds of "β-cyclodextrin–benzene carboxylic acid". Russ. J. Mass Spectrom. **6**(1), 47–52 (2009)
18. SDBSWeb: http://riodb01.ibase.aist.go.jp/sdbs/. National Institute of Advanced Industrial Science and Technology (date of access 27.05.2012)
19. J. Cho, S. Chu, P.J. Dauenhauer, G.W. Huber, Kinetics and reaction chemistry for slow pyrolysis of enzymatic hydrolysis lignin and organosolv extracted lignin derived from maplewood. Green Chem. **14**, 428–439 (2012)
20. B. Hwang, J.R. Obst, Basic studies on the pyrolysis of lignin compounds, in *Proceedings of the IAWPS International Conference on Forest Products: Better Utilization of Wood for Human, Earth and Future*. The Korean Society of Wood Science and Technology International Association of Wood Products Societies, Daejeon, Korea, vol. 2, pp. 1165–1170, 21–24 April 2003
21. T.V. Kulik, V.N. Barvinchenko, B.B. Palyanytsya, N.A. Lipkovska, O.O. Dudik, Thermal transformations of biologically active derivatives of cinnamic acid by TPD MS investigation. J. Anal. Appl. Pyrol. **90**(2), 219–223 (2011)
22. T.V. Kulik, N.A. Lipkovska, V.N. Barvinchenko, B.B. Palyanytsya, O.A. Kazakova, O.A. Dovbiy, V.K. Pogorelyi, Interactions between bioactive ferulic acid and fumed silica by UV-VIS spectroscopy, FT-IR, TPD MS investigation and quantum chemical methods. J. Colloid Interface Sci. **339**(1), 60–68 (2009)

Chapter 4
Thermal Properties of Algerian Diatomite, Study of the Possibility to Its Use in the Thermal Insulation

Boualem Hamdi and Safia Hamdi

Abstract The chemical and physical properties of a Algerian diatomite were given before and after heat treatment and chemical with an aim of a use in the heat insulation of constructions. The preliminary results obtained showed that this material is extremely porous (porosity >70 %), characterized of a low density and a very low thermal conductivity. These promising properties support the use of this local material in the thermal insulation.

4.1 Introduction

Under contemporary economic conditions in order to provide competitiveness domestic metallurgical enterprises need to increase the production efficiency by reducing energy consumption, increasing the service life of heating units, and reducing specific consumption for output of finished product. In this case, the use of porous materials is a recommended solution. Among the most used porous materials, silica with its two particulate and amorphous forms were identified like very good insulators. With the nano structuring of this type of material, it was developed of super insulators. In spite of synthetic silica have a very low thermal conductivity related to its high porosity and its form nanoparticulate, its use current is problematic because of its high manufacturing cost [1]. This situation led, these last years, many laboratories to be been interested the search of new

B. Hamdi (✉) · S. Hamdi
LCVR, ENSMAL, BP 19 Dely Ibrahim, 16320 Algiers, Algeria
e-mail: bhamdi_99@yahoo.fr

S. Hamdi
e-mail: shamdi_01@yahoo.fr

B. Hamdi · S. Hamdi
USTHB, LEPCMAE, BP 32El Alia Bab Ezzouar Algiers, Algeria

A. Y. Oral et al. (eds.), *International Congress on Energy Efficiency and Energy Related Materials (ENEFM2013)*, Springer Proceedings in Physics 155,
DOI: 10.1007/978-3-319-05521-3_4,

porous materials of origin naturalness. This work falls under this prospect, it relates to the valorization of diatomaceous earth (diatomite) very abundant in Algeria in the field of the thermal insulation. This material has a great interest and it has several uses such as: water filtration and purification, porous absorbent in environments, manufacture of antibiotics, some pharmaceutical syrups and it is qualified like ecological material. With an aim of predicting the potential use of the natural diatomite in the thermal insulation, we undertook a systematic study of characterization of its chemical and physics properties.

4.2 Experimental

4.2.1 Materials

The basic raw material used in this study was obtained from Sig deposit (west of Algeria). Algeria possesses important diatomaceous beds in the northwest region of Sig. The Sig deposit is made of three distinct layers called massive, bedded and inert or third category layer. The first two layers correspond to rich rocks partially associated with carbonates. The sample was taken from the intermediate (bedded) layer in various accessible parts of the Gana 1 gallery. This sample is light grey in color, compact and very light and shows a bedded texture which is due probably to the presence of marl sub layers. After dry state crushing and grinding of the material in a toothed roll crusher, a representative sample, designated ND of size 40–100 μm was selected. This raw material has already been described in previous papers [2].

4.2.2 Geology of the Layer

The diatomite layer of a reserve estimated at several million tons is located at Sig at the west of Algeria (Fig. 4.1), currently exploited by Diatale, one of the subsidiary companies of the ENOF (Company National of the nonferrous Materials).The series with diatomite's rests on beige argillaceous limestone's several million years of age. This series is upwards made up of three members. The first member is made up approximately 20 m of gray marls with gray-blue, sometimes rolled, in which at the top the first benches of diatomite appear.

Fig. 4.1 Sig (Algeria) deposit site

4.3 Results and Discussion

4.3.1 Chemical Composition Analysis

Table 4.1 show that the natural product is essentially composed of silica with CaO as the main impurity (19 %) which contaminates the surface of that raw material.

4.3.2 Physico-Chemical Analysis

The physic-chemical properties of the Diatomite are presented in Table 4.2. From these results, we can see that the diatomite material has a high porosity, approximately 72 %.

4.3.3 Thermal Analysis

Thermal analysis of the sample was depicted in Fig. 4.2. Showed an endothermic pattern due to the release of diatomite absorbed water at the temperature range 100–200 °C, with a loss in sample weight reaching 5.315 %. At temperature range 600–700 °C, another endothermic reaction due to the transformation calcite according to the following reaction [3]:

Table 4.1 Chemical composition of natural diatomite

Diatomite	SiO_2	Al_2O_3	Fe_2O_3	CaO	MgO	K_2O	Na_2O
W (%)	73.86	1.81	1.48	18.89	1.85	0.77	1.16

Table 4.2 Physico-chemical analysis of natural diatomite

Characterization	Color	pH	S_{N2} (m^2/g)	Vp (cm^3/g)	Pt (%)
ND	White	7.9	21.0	1.18	72.17

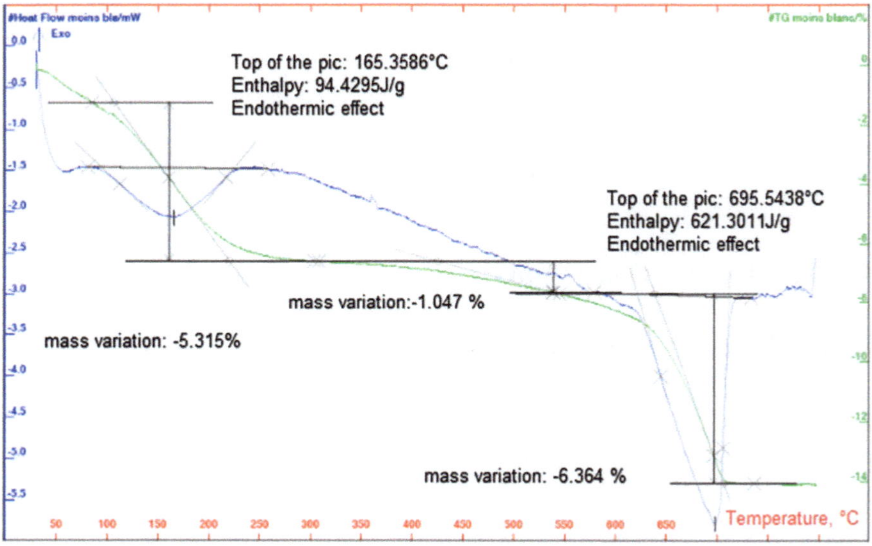

Fig. 4.2 Thermal analysis of natural diatomite

$$CaCO_3 \rightarrow CaO + CO_2$$

This endothermic reaction caused a loss in weight reaching 5.55 %. An intense endothermic peak at the temperature range 600–750 °C due to calcite mineral break down, with a loss in weight reaching 6.36 %, was depicted (Fig. 4.2).

4.3.4 SEM Analysis

SEM has been used to study the different forms of porosity and the structure of the diatomite (Fig. 4.3). We can note that the pores are predominantly in circular form. The fragments of the diatomite are in slice form with two faces opened and we can also see the cylindrical nature of the pores.

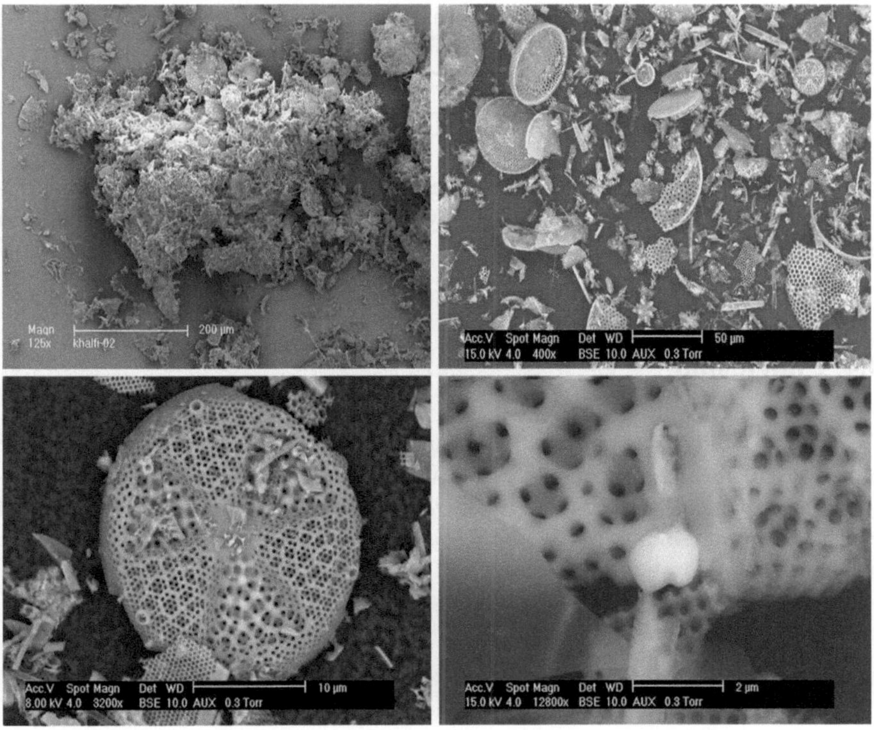

Fig. 4.3 SEM images of natural diatomite

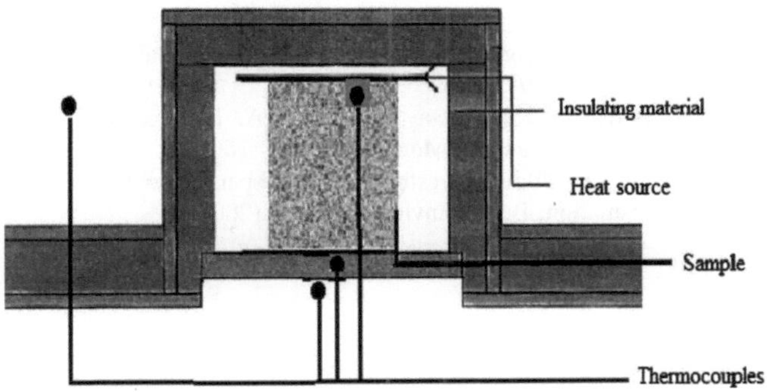

Fig. 4.4 Experimental device used for the measurement of thermal conductivity

4.3.5 Thermal Conductivity

The technique of measurement of thermo physical characteristics used in this work is the box-counting method. The dimensions of the rectangular bricks are samples whose thickness e is from 2 to 6 cm, they are determined according to the measurement tool used in Fig. 4.4 (Table 4.3).

Table 4.3 Apparent thermal conductivity of diatomite of various origins

Diatomite	Algeria	Morocco [4]	Turkish [5]
λ (W/m K)	0.097	0.18	0.23

In the light of the results obtained, we note that the Algerian diatomite decalcified has a very low thermal conductivity. This result is related to the increase in porosity.

4.4 Conclusion

During this study, we undertook a work of thorough characterization of the properties of a diatomaceous earth of Algerian origin before and after physic-chemical treatment. On the whole, the results of measurements of porosity and measurements relating to conductivity show that this material can be used as heat insulator in the construction industry.

References

1. B. Yuan, S. Ding, D. Wang, G. Wang, H. Li, Heat insulation properties of silica aerogel/glass fiber composites fabricated by press forming. Mater. Lett. **75**, 204–206 (2012)
2. B. Hamdi, Z. Kessaissia, J.B. Donnet, T.K. Wang, Analytical study of physico-chemical, textural and structural of kieselguhr. Analusis J. **26**, 8–12 (1998)
3. C. Rodriguez-Navarro, E. Ruiz-Agudo, A. Luque, A. Rodriguez-Navarro, M. Ortega-Huertas, Am. Mineral. **94**, 578–593 (2009)
4. N. Aakka, H. Ezbaakhe, A. Benmoussa, T. Ajzoul, A. El Bakkourri, 12èmes Journées Internationales de Thermique, Tangier, Morocco, 15–17 (2005)
5. O. Unal, T. Uygunoglu, A. Yildiz, Investigation of properties of low-strength lightweight concrete for thermal insulation. Build. Environ. **42**, 584 (2007)

Chapter 5
Optimal Resource Allocation in Steel Making Using Torrefied Biomass as Auxiliary Reductant

Carl-Mikael Wiklund, Henrik Saxén and Mikko Helle

Abstract Steelmaking is an energy intensive industrial sector and being largely coal-based it gives rise to 5–6 % of the global CO_2 emissions. Energy use for producing 1 ton of crude steel has been reduced by 50 % since 1975, but the annual production rate of crude steel has been increasing more strongly. Since 2002, the production rate has increased by almost 80 % amounting to 1,510 Mt in 2012, and this trend seems to continue in the future. Therefore, making the iron production itself more efficient is not enough to reduce carbon dioxide emissions. A possible remedy is to replace part of the fossil reductants by renewables and to optimize the entire production chain from ores to steel, allowing more beneficial resource allocation in the processes involved. The present study focuses on the use of biomass as auxiliary reductant in the blast furnace, also paying attention to the effect of the introduction on the material and energy flows of the whole steel plant using a simulation model. Substituting part of the fossil coke or injected hydrocarbon by biomass may result in reduced fossil carbon dioxide emissions, as long as the biomass is harvested, transported and pre-processed in a sustainable way. As the biomass may need upgrading before it is used, a torrefaction model is included in the steel plant model. Results are presented from studies where the entire system is optimized with respect to costs, considering a penalty for CO_2 emissions.

5.1 Introduction

Whether or not global warming is occurring, the concern for it has further increased public awareness of anthropogenic activities with negative impacts. Previously it was e.g. acid rains, pesticides and ozone depletion, now it is, among

C.-M. Wiklund (✉) · H. Saxén · M. Helle
Thermal and Flow Engineering Laboratory, Åbo Akademi University,
20500 Åbo, Finland
e-mail: cwiklund@abo.fi

A. Y. Oral et al. (eds.), *International Congress on Energy Efficiency and Energy Related Materials (ENEFM2013)*, Springer Proceedings in Physics 155,
DOI: 10.1007/978-3-319-05521-3_5,

others, carbon dioxide (CO_2) emissions. Substituting some of the fossil fuels used with "sustainable biomass" can lower the fossil CO_2 emissions. The amount by which the emissions are lowered however depends strongly on harvesting methods, transport related issues as well as on processing of the biomass. Therefore, long-term impact on the environment should be considered before a certain biomass is used as a substitute fuel. In this study, the use of a variety of biomasses as auxiliary reductants in steel production is considered. For this purpose, a mathematical model is used to optimize raw material streams and process parameters in the steel plant in order to minimize production cost of raw steel, also considering costs of emissions. Before the biomass is injected in the blast furnace it is taken to be torrefied and grinded to make it suitable as an auxiliary reductant in the process.

Torrefaction of biomass includes heating in a furnace under a non-oxidizing atmosphere and at modest temperatures, i.e. $\leq$300 °C. This form of mild pyrolysis [1] can be divided into three temperature zones: light (220 °C), mild (250 °C) and severe torrefaction (280 °C) [2]. Depending on the degree of torrefaction, biomass used and residence time different ratios of condensable, non-condensable and higher quality solids are acquired. In general, a higher torrefaction temperature improves fuel quality due to increase in heating value. At the same time, however, the solids suffer from yield loss [3]. The loss of mass at light torrefaction is mainly caused by degradation of hemicellulose, whilst both hemicellulose and cellulose accounts for it at mild to severe torrefaction [2]. Lignin does not yet degrade significantly at these temperatures. Biomass and grinding data used in this study are taken from the literature, where available. Here the main focus will be on analysing the suitability of the solid fraction deriving from the different biomasses. Future work will concentrate on the energy yields of the condensable and non-condensable fractions.

5.2 Steel Plant Model

The mathematical description of the plant consists of models of the blast furnace (BF) as well as of the basic oxygen furnace (BOF), coke oven (CP), grinding unit (GU), hot stoves (HS), power plant (PP), sinter plant (SP) and a torrefaction unit (TU). These units are described below. For an indicative schematic of the system studied, please refer to [4].

A semi-linearized surrogate BF model was created using eight input values, expressing 13 outputs (Table 5.1). The input values were varied within admissible regions for a large number of experimental runs with the original, more sophisticated model [5], where only the feasible solutions were retained [3, 5]. In earlier work, this approach was made to study resource allocation in steelmaking using one type of biomass as an auxiliary reductant [4, 6]. In the present study, the model was adapted to account for different types of biomass used by redeveloping the surrogate model on a large number of (feasible) points generated by the original model.

Table 5.1 Input and output variables for BF model (where units m^3n and t_{hm} refer to normal cubic meter and ton of hot metal, respectively)

Input variable	Range	Output variable	Range
X_1: Blast volume	0–140 km^3n/h	Y_1: Production rate	120–160 t_{hm}/h
X_2: Oxygen volume	0–40 km^3n/h	Y_2: Coke rate	≥0 kg/t_{hm}
X_3: Oil rate	0–120 kg/t_{hm}	Y_3: Flame temperature	2,000–2,300 °C
X_4: Blast temperature	850–1,100 °C	Y_4: Top gas temperature	100–250 °C
X_5: Pellet rate	0–1,200 kg/t_{hm}	Y_5: Bosh gas volume	170–200 km^3n/h
X_6: Limestone rate	0–100 kg/t_{hm}	Y_6: Residence time of solids	6.0–9.0 h
X_7: Oxygen flow with biomass	≥0 kg/t_{hm}	Y_7: Slag basicity, X_{CaO}/X_{SiO2}	1.00–1.20
X_8: Energy flow with biomass	≥0 MJ/t_{hm}	Y_8: Slag rate	≥0 kg/t_{hm}
		Y_9: Top gas volume	≥0 km^3n/h
		Y_{10}: Top gas CO content	≥0 %
		Y_{11}: Top gas CO_2 content	≥0 %
		Y_{12}: Top gas H_2 content	≥0 %
		Y_{13}: Heating value of top gas	≥0 MJ/m^3n

Blast furnace top gas requirement for reaching a desired blast temperature at a given blast volume is determined by the hot stoves model, which considers the stove set as a single continuous counter-current heat exchanger. The total oxygen demand is dictated by the oxygen enrichment of the blast and by the oxygen needed for steel conversion in the BOFs. Remaining BF top gases, coke oven gases and half of the BOF gas are used for steam production in the power plant. Low pressure steam from turbines is used in a condenser, providing heat for district heating. Electricity produced, but not used by the grinding unit and for compression of the cold blast, is taken to be sold to the grid. The models of the coke plant, sinter plant and BOF are simple linear models based on the overall behaviour of these units in a Finnish steel plant used as reference for this work. The power demand for grinding the pre-treated biomass is based on data from literature. Here it was available for only two of the biomasses studied, namely pine-wood chips (Bio1) and logging residue (Bio2). An approximately ten-fold reduction in power demand could be seen when the biomass was torrefied at 300 °C in comparison to oven dried mass. The grinded biomass is injected into the BF. It should be noted that injecting untreated biomass into the BF would lower the production rate and flame temperature, so some pre-treatment is needed. However, a high torrefaction temperature results in unnecessary loss of yield [4]. For the sake of simplicity, the torrefaction unit is here considered to be self-sufficient; i.e. heat deriving from combusting product gases and liquids is assumed adequate for drying and torrefying the incoming biomass. A more accurate modelling of the heat demand of the torrefaction unit will be developed in future work.

5.3 Objective Function and Simulation Setup

The objective function comprises of raw material costs, penalty fee for fossil CO_2 emissions as well as credits for electricity and district heating sold.

$$F = \left(\sum \dot{m}_i c_i + \psi \dot{m}_{CO_2} c_{CO_2} - P_{c_{el}} - \dot{Q}_{dh} c_{dh}\right) / \dot{m}_{ls} \quad (5.1)$$

where $\dot{m}$ is the mass flow rate, c is a cost factor, ψ expresses the mass ratio of fossil versus total CO_2, P is electrical power produced (reduced by the power demand of blast compression and grinding), $\dot{Q}_{dh}$ is the district heat and subscript ls stands for liquid steel. Raw materials indicated by i are biomass, coal, coke (external), limestone, oil, ore, oxygen, pellets and scrap iron. Due to nonlinear terms in the surrogate BF model [4], the resulting optimization problem was solved by mixed integer linear programming problem (MILP) using the CPLEX software [7].

A set of five different possible biomasses (Bio1–5) were used in this study: pine wood chips (PC) and logging residue (LR) [8], loblolly pine (LP) [9], bagasse (B) [10] and reference wood (RW) [11]. Data for these are reported in Table 5.2. The grinding energy demand of Bio1 and Bio2 is given in Table 5.3.

Torrefaction time normally ranges from a few minutes to 1 or 2 h [1, 2]. Biomass 1 and 2 had a reported torrefaction time of 30 min, whilst the following two had a torrefaction time of 2.5 and 60 min, respectively. An increase in the residence time reduces the hydrogen and oxygen content, while the carbon content is increased. It can also cause secondary reactions between volatiles and solids [1], which make comparison of the biomasses more difficult. However, many authors concluded that the effect of residence time is of less importance than that of temperature [1, 2, 12]. Another problem, pointed out by Ohlinger et al. [1], is the drying method used on the biomass samples, where, e.g., a remoistened sample could lead to modified binding forces within the sample. Therefore, the reported performance of the biomasses should be treated with certain caution. The system simulated is a steel plant with one BF and with an overall production of 160 t_{ls}/h. The coke plant can produce 55 t/h of coke and the BF is operated with a blast temperature of 1,100 °C. Four cases were simulated, as indicated in Table 5.4. In the four cases investigated, the biomass feed rate, oxygen availability and oil rate to the blast furnace (expressed in specific quantities per ton of hot metal) were varied. Raw material prices and energy prices, as well as the carbon dioxide emission cost are given in Table 5.5.

5.4 Results

Optimization results for the cases using the biomasses are presented below, where the optimal values at lower or upper bounds are reported in bold Italic and bold Roman fonts, respectively. Table 5.6 presents results for Case 1, with and without

Table 5.2 Data for biomass used in simulations (bold font: linearized values)

	T (°C)	Bio1-PC	Bio2-LR	Bio3-LP	Bio4-B	Bio5-RW
C (%-wt)	–	47.21	47.29	50.5	46.38	50.1
	225	49.47	50.15	**53.73**	**51.85**	60.15
	250	51.46	54.91	**54.44**	53.04	70.6
	275	54.91	53.24	**55.38**	**59.32**	72.25
	300	63.67	66.07	57.3	65.59	73.2
H (%-wt)	–	6.64	6.2	6.26	4.68	6
	225	6.07	6.1	**6.03**	**4.92**	5.78
	250	5.86	5.87	**5.98**	4.97	5.2
	275	6.2	5.39	**5.92**	**4.64**	5.02
	300	5.58	4.92	5.79	4.31	4.9
O (%-wt)	–	45.76	45.19	42.6	44.11	43.9
	225	44.03	42.74	**39.51**	**36.08**	32.39
	250	42.02	40.96	**38.84**	34.34	24.3
	275	38.17	40.12	**37.92**	**26.23**	22.8
	300	29.99	27.34	36	18.11	21.9
HHV (MJ/kg)	–	18.46	18.79	20	18.32	20.02
	225	19.48	19.79	**21.94**	**20.46**	24.35
	250	20.08	21.21	**22.36**	20.94	28.12
	275	21.82	22.03	**22.92**	**22.14**	28.63
	300	25.38	26.41	24	23.37	28.91
Yield	–	100	100	100	100	100
	225	89	88	**88.14**	**67.96**	78.79
	250	82	81	**85.56**	60.99	65.2
	275	73	70	**82.23**	**48.99**	56.93
	300	52	52	75.9	36.98	51.4

Table 5.3 Grinding energy for Bio1 and 2

	T °C	Bio1 PC	Bio2 LR
Grinding energy (kWh/t)	–	237.7	236.7
	225	102.6	113.8
	250	71.4	110.4
	275	52	78
	300	23.9	37.6

Table 5.4 Experimental plan

Cases	1	2	3	4
Biomass (t/h)	30	15	15	15
Oxygen (km^3 n/h)	20	20	10	20
Oil (kg/t hm)	120	120	120	60

grinding for Bio1 and Bio2. This allows for a comparison of the effect of the grinding power on the optimization. For indicative top gas composition, please refer to [4].

Table 5.5 Raw material prices

Biomass	60 €/t
Ore	80 €/t
Pellets	100 €/t
Coal	145 €/t
Coke	300 €/t
Oil	150 €/t
Limestone	30 €/t
Oxygen	50 €/km^3 n
Electricity sold	50 €/MWh
District heating sold	10 €/MWh
CO_2 tax	20 €/t
Scrap	100 €/t
Electricity bought	100 €/MWh

As seen from Table 5.6, the maximum amount of biomass (120 kg of torrefied biomass/t_{hm}) is injected for all materials, except for Bio4 (bagasse), which is not used at all. The reason for this is its high material loss during heating. Bio5 also suffers also from low yield. However, this is compensated by its higher heating value and thus, higher energy yield. Additionally, BF operation with Bio5 uses less blast than for the other biomasses, while the oxygen enrichment is at maximum, i.e., the blast oxygen is 32 %. All samples use maximum oil input (16.8 t/h) and are also forced to use pellets (56.5 t/h), since the sinter production limit (153.6 t/h) is reached. In addition, no coke is bought for cases 1–3 nor is any limestone used in the BF. For the biomass with reported grinding energy, consideration of this only affects the production price and not the optimal operation state, since the cost is marginal. As for the production price between the biomasses, the most beneficial one is Bio5, while the most expensive alternative is to use no biomass (which was the case where Bio4 was an option).

In cases 2–3, where biomass availability is further restricted, no Bio4 is used, while the maximum amount of other biomasses (15 t/h) is utilized. In Case 2, Bio5 again requires maximum blast oxygen and the torrefaction temperature remains the same as in Case 1. Limiting the oxygen flow to 10 km^3n/h (Case 3) reduces the blast oxygen content for Bio5. For Cases 2–3, the optimal torrefaction temperatures for the other biomasses (Bio1–Bio3) vary greatly compared to Case 1. Here the temperatures are much lower; Bio3 is used un-torrefied in the furnace, as is also the case for Bio1 and Bio2 when not grinded. A reason for this behaviour isthe lowered amount biomass available, so a maximum or near-maximum yield is more beneficial for maximizing the objective function. As for production cost, Bio5 is no longer the most beneficial one, but instead un-grinded Bio2. In comparison, the grinded Bio2 is 1.1 (€/t_{ls}) more expensive than Bio2 without grinding. Bio1 follows a similar trend. Production prices at the optimized states are shown in Fig. 5.1 (left) and biomass torrefaction temperatures for all cases in (right).

Halving the oil injection rate (Case 4) increases the production cost with all biomass alternatives, as the oil is the cheapest reductant. Surprisingly, the use of

Table 5.6 Simulation results from Case 1—Bio1 and 2 with and without Grinding (Gr.)

Case 1	Bio1 (Gr./No gr.)		Bio2 (Gr./no gr.)		Bio3	Bio4	Bio5 (ref)
Biomass (t/h)	25	25	21	21	22	***0***	**30**
Coking coal (t/h)	60.9	60.9	59.7	59.7	65.8	77.6	61.1
Oxygen (km^3n/h)	7.3	7.3	4.9	4.9	7.6	7.5	15.7
Electricity sold (MW)	21.8	21.8	20.6	20.6	23.9	23.8	32.0
District heat sold (MW)	58.1	58.1	57.2	57.2	61.4	69.5	165.5
Fossil CO_2 emissions (t/t_{ls})	1.37	1.37	1.35	1.35	1.45	1.65	1.37
Blast volume (km^3n/h)	138.9	138.9	**140.0**	**140.0**	135.3	**140.0**	112.9
Pyrolysis temperature (°C)	283	283	250	250	292	272	276
Compressor power (MW)	7.4	7.4	7.6	7.6	7.2	7.4	5.9
Flame temperature (°C)	2,021	2,021	2,070	2,070	2,023	2,116	2,010
Bosh gas volume (km^3n/h)	**200.0**	**200.0**	196.7	196.7	**200.0**	197.7	185.0
Residence time (h)	8.8	8.8	8.9	8.9	8.5	7.6	8.9
Slag basicity (-)	1.08	1.08	1.08	1.08	1.06	1.04	1.07
Top gas temperature (°C)	246	246	243	243	231	195	209
Coke rate (kg/t hm)	250.6	250.6	244.2	244.2	274.9	333.4	251.2
Slag rate (kg/t hm)	209.2	209.2	208.7	208.7	211.0	215.2	204.8
CO_2 emissions (€/t steel)	27.4	27.4	27.0	27.0	29.0	32.9	27.5
Liquid steel (€/t steel)	237.8	237.6	234.7	234.1	241.6	247.6	232.3
Injected biomass (kg/t hm)	**120**	**120**	**120**	**120**	**120**	**0**	**120**
Grinding energy (MWh)	0.7	N/A	1.9	N/A	N/A	N/A	N/A

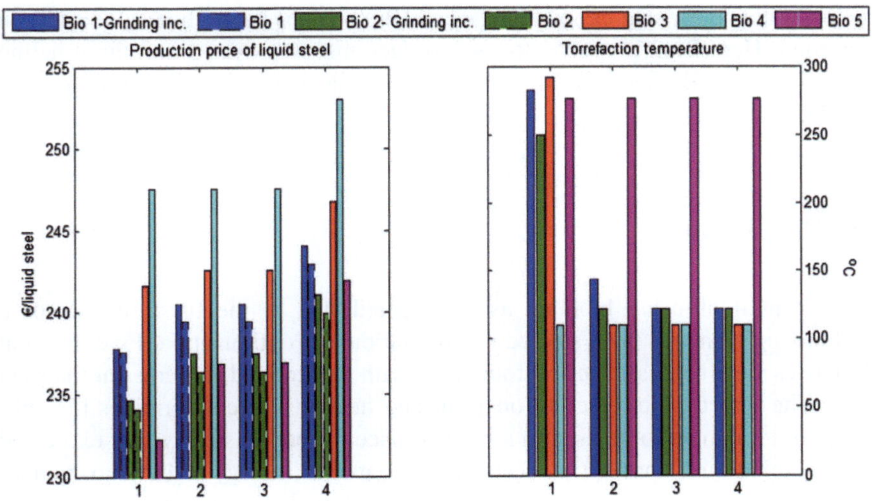

Fig. 5.1 Liquid steel production costs depending on biomass used (*left*). Biomass torrefaction temperature; Bio1 and 2 with grinding (*right*)

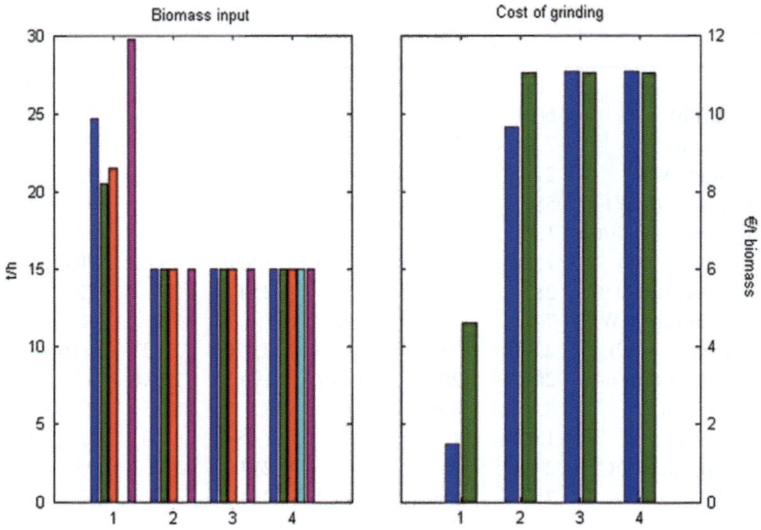

Fig. 5.2 Biomass input (*left*). Grinding costs (*right*)

bagasse (Bio4) becomes beneficial for this case. It is, like for Bio3, used un-torrefied. Along with the reduction of the oil rate, external coke is needed with Bio3–Bio5, because of the restriction on own coke availability (55 t/h). This, in turn, makes it more beneficial to replace a small part of the sinter with pellets and the lower bound for slag basicity is reached (Bio3–Bio5) but no limestone addition is needed. Here Bio2 gives the lowest production costs, even when the grinding cost is included. Biomass use and grinding effect as well as costs are presented in Fig. 5.2.

5.5 Conclusion

The potential of using biomass as an injected fuel in the blast furnace of a steelmaking plant has been studied by simulation and optimization. Five different biomasses were evaluated, using four cases with constrained resources at the steel plant. The effect of torrefaction on yield and heating value determines the suitability of a biomass to be used in a blast furnace, where feasibility was expressed in economic terms, considering costs of raw materials, energy and the taxation on CO_2 emissions. The study showed that it is difficult to draw definite conclusions on the biomass which would be the most beneficial one, since the optimal choice changes with the process conditions. For instance, the use of bagasse was not economical until the oil content was severely restricted. The optimization also showed that in some cases it was better to conduct a strong torrefaction, while in other cases the use of un-torrefied biomass was optimal. Using information about the grinding energy, available for pine chips and logging residue (Bio1 and Bio2), increased the production price in the Cases 2–4 with about 1.1 €/ton of liquid steel. Therefore, the effect of grinding should be considered, in particular since the injection of non-grinded biomass would be very challenging in practice. A crucial factor is the condensable and non-condensable fractions in the torrefaction process and their energy yields. Consideration of these is likely to affect the optimal raw material streams and production price. Additionally, they affect the dimensioning of the torrefaction unit, which, in turn, has its own limitations. These problems are to be addressed in future work.

Acknowledgments I would like to express my gratitude to Frank Pettersson and Anders Skjäl for their assistance with the mathematical model. Additionally, I would like to thank Oskar Karlström for good discussions concerning the torrefaction of biomass. Support from the Academy of Finland in the Symbiosis research project is gratefully acknowledged.

References

1. A. Ohliger, M. Förster, R. Kneer, Fuel **104**, 607–613 (2012)
2. W.-H. Chen, H.-C. Hsu, K.-M. Lu, W.-J. Lee, T.-C. Lin, Energy **36**, 3012–3021 (2011)
3. H. Helle, M. Helle, F. Pettersson, H. Saxén, ISIJ Int. **49**, 1316–1324 (2009)
4. C.-M. Wiklund, F. Pettersson, H. Saxén, Metall. Mater. Trans. B **44**(2), 447–458 (2013)
5. H. Helle, M. Helle, H. Saxén, Chem. Eng. Sci. **66**, 6470–6481 (2011)
6. C.-M. Wiklund, F. Pettersson, H. Saxén, ISIJ Int. **52**, 35–44 (2012)
7. http://www-03.ibm.com/ibm/university/academic/pub/page/ban_ilog_programming 18.09.2013
8. M. Phanphanich, S. Mani, Bioresour. Technol. **102**, 1246–1253 (2011)
9. J. Meng, J. Park, D. Tilotta, S. Park, Bioresour. Technol. **111**, 439–446 (2012)
10. W.-H. Chen, S.-W. Du, C.-H. Tsai, Z.-Y. Wang, Bioresour. Technol. **111**, 433–438 (2012)
11. J. Ranta, M. Sc. Thesis, Helsinki University of Technology (1994)
12. M.J.C. van der Stelt, H. Gerhauser, J.H.A. Kiel, K.J. Ptasinski, Biomass Bioenergy **35**, 3748–3762 (2011)

Chapter 6
Numerical Simulation of Cavity with an Upper Free Surface

Chahrazed Benseghir, Bariza Zitouni, Djamel Haddad and Mostefa Zeroual

Abstract In this paper, 3D numerical study of Bénard-Marangoni instabilities in a horizontal liquid layer for a Rayleigh number Ra = 0 and aspect ratio A = 20. The layer is heated from below and cooled from above (vertical temperature gradient). The upper surface is assumed to be free and non-deformable. The surface tension at the free surface is linearly dependent on the temperature. Silicon oil with Prandtl number (Pr = 880) has been used as a working liquid. A parametric study has been carried out by considering the following parameters: the thermal Marangoni number (Ma), the Biot number (Bi). The governing equations were discretized by the finite volume method. The resolution of the coupling (pressure-velocity) was done with the projection method. A code has been elaborated with *FORTRAN 6.6*. Velocity vectors and temperature fields on the upper free surface are obtained.

6.1 Introduction

The Bénard-Marangoni convective instability (BMI) appears when the liquid layer is heated from below and the upper free surface is cooled from above. Beyond the onset of convection, corresponding to a critical value of the temperature gradient, polygonal cells are generally formed. The flows induced by the surface tensions

C. Benseghir (✉) · B. Zitouni · D. Haddad
Department of Mechanical, Faculty of Technology, University of Batna, 05000 Batna, Algeria
e-mail: chahrazed_information@yahoo.fr

B. Zitouni
e-mail: zitounibariza@yahoo.fr

M. Zeroual
Laboratory LPEA, Faculty of Science, University of Batna, 05000 Batna, Algeria

A. Y. Oral et al. (eds.), *International Congress on Energy Efficiency and Energy Related Materials (ENEFM2013)*, Springer Proceedings in Physics 155,
DOI: 10.1007/978-3-319-05521-3_6, © Springer International Publishing Switzerland 2014

have a several applications as well in nature as in the industrial field, such as crystalline growth processes, film coating processes and convection in the stars (solar granulation).

This problem was studied the first time analytically by Pearson [1] which neglected the effects of buoyancy and by Nield [2] which included the forces of gravity in its analysis. The two authors took into account the approximation of Boussinesq. Scriven et al. [3] proposed the first analysis of the effects free surface deformation on the thermocapillary convection. Davis et al. [4] found that the deflection of the free surface stabilizes the system if buoyancy is dominant and destabilizes the system if the effects of surface tension are dominant. The effect of the variation of viscosity with the temperature has been studied by Selak et al. [5]. Thess and Orszag [6] studied numerically under microgravity the effect of surface tension on the infinite Prandtl number. Schwabe [7] studied experimentally under the microgravity a liquid layer of the Silicone oil. Its objective was to observe the beginning of the hydrothermal patterns. Cerisier et al. [8] studied experimentally in a circular container with different aspect ratio. They found that more dynamics are induced by increasing the Biot number. Rahal and Cerisier [9] studied (BMI) experimentally for various Prandtl and Marangoni numbers in small aspect ratio geometries. They found that for given values of the Prandtl number and aspect ratio, mono-periodic, bi-periodic and chaotic states were successively observed as the Marangoni number was increased. Nezar and Rahal [10] studied numerically the effect of inclined temperature gradient (both horizontal and vertical temperature were applied). They found that the morphology of patterns is changed completely from hexagonal cells to stationary longitudinal rolls for a specific value of the liquid layer thickness.

In the present work a 3D numerical simulation is used to study a liquid layer at Rayleigh number ($\mathrm{Ra} = 0$) and for a large Marangoni number.

6.2 Problem Definition

The studied configuration (Fig. 6.1) has an upper free surface. The cavity contains silicon oil ($\mathrm{Pr} = 880$). The liquid-air interface is assumed to be non-deformable. The layer has a thickness (d) and it is heated from below and cooled from above (vertical temperature gradient).

6.3 Mathematical Model

Thermophysical properties of the fluid are assumed constant, except the density variation in the buoyancy term, i.e., Boussinesq approximation is valid. The surface tension at the free surface is linearly dependent on the temperature: $\sigma = \sigma_0 - \gamma(\theta - \theta_0)$. The heat transfer in the liquid-air interface is defined across

Fig. 6.1 Studied configuration

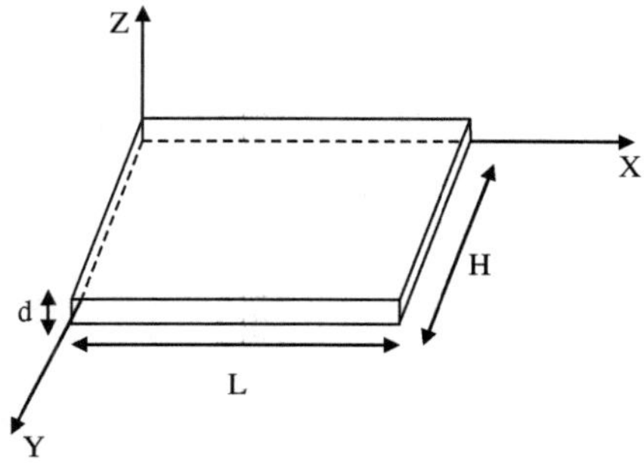

the Biot number. The dimensionless equations for the conservation of mass, momentum, and energy equations are:

$$\frac{\partial U}{\partial X}+\frac{\partial V}{\partial Y}+\frac{\partial W}{\partial Z}=0 \tag{6.1}$$

$$\frac{\partial U}{\partial \tau}+U\frac{\partial U}{\partial X}+V\frac{\partial U}{\partial Y}+W\frac{\partial U}{\partial Z}=-\frac{\partial P}{\partial X}+\Pr\left(\frac{\partial^2 U}{\partial X^2}+\frac{\partial^2 U}{\partial Y^2}+\frac{\partial^2 U}{\partial Z^2}\right) \tag{6.2}$$

$$\frac{\partial V}{\partial \tau}+U\frac{\partial V}{\partial X}+V\frac{\partial V}{\partial Y}+W\frac{\partial V}{\partial Z}=-\frac{\partial P}{\partial Y}+\Pr\left(\frac{\partial^2 V}{\partial X^2}+\frac{\partial^2 V}{\partial Y^2}+\frac{\partial^2 V}{\partial Z^2}\right) \tag{6.3}$$

$$\frac{\partial W}{\partial \tau}+U\frac{\partial W}{\partial X}+V\frac{\partial W}{\partial Y}+W\frac{\partial W}{\partial Z}=-\frac{\partial P}{\partial Z}+\Pr\left(\frac{\partial^2 W}{\partial X^2}+\frac{\partial^2 W}{\partial Y^2}+\frac{\partial^2 W}{\partial Z^2}\right)+\mathrm{Ra}\cdot\Pr\cdot\mathrm{T} \tag{6.4}$$

$$\frac{\partial T}{\partial \tau}+U\frac{\partial T}{\partial X}+V\frac{\partial T}{\partial Y}+W\frac{\partial T}{\partial Z}=\frac{\partial^2 T}{\partial X^2}+\frac{\partial^2 T}{\partial Y^2}+\frac{\partial^2 T}{\partial Z^2}+W \tag{6.5}$$

The dimensionless variables U, V, W, P and T are the velocity components in the X, Y and Z direction, pressure and perturbation temperature, respectively.

Parameters A, Pr, Ra, Ma and Bi are aspect ratio, Prandtl, Rayleigh, Marangoni and Biot numbers, respectively. These are defined as:

$$\mathrm{A}=\mathrm{L}/\mathrm{d},\ \Pr=\upsilon/\alpha,\ \mathrm{Ra}=\mathrm{g}\beta\Delta\theta\mathrm{d}^3/\upsilon\alpha,\ \mathrm{Ma}=\gamma\Delta\theta\mathrm{d}/\rho_0\upsilon\alpha,\ \mathrm{Bi}=\mathrm{hd}_{\mathrm{liquid}}/\lambda_{\mathrm{liquid}} \tag{6.6}$$

where quantities ρ_0, υ, α, β, $\Delta\theta$, λ, h and d are the density, kinematic viscosity, thermal diffusivity, thermal expansion coefficient of the liquid, temperature variation, thermal conductivity of the liquid, thermal exchange coefficient liquid-air and thickness of the liquid layer respectively.

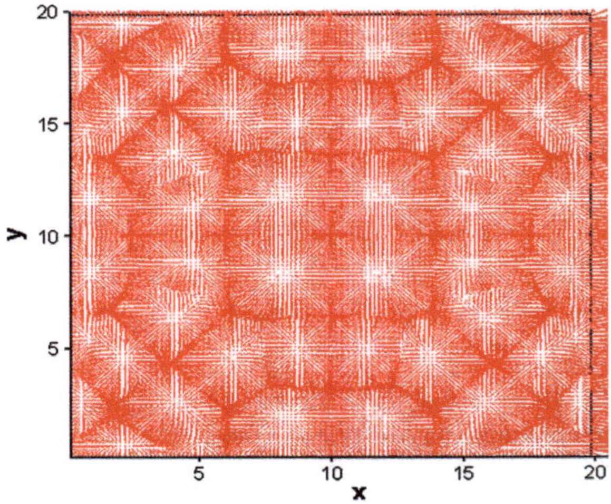

Fig. 6.2 Velocity vectors for Ma = 2,000

The dimensionless boundary conditions are:

- For the hot wall of the cavity:

$$U = V = W = 0,\ T = 1 \tag{6.7}$$

- For the free surface:

$$\frac{\partial V}{\partial Z} = -Ma\frac{\partial T}{\partial Y} \tag{6.8}$$

$$\frac{\partial V}{\partial Z} = -Ma\frac{\partial T}{\partial Y} \tag{6.9}$$

$$W = 0 \tag{6.10}$$

$$\partial T/\partial Z + Bi \cdot T = 0 \tag{6.11}$$

- For the insulated walls:

$$U = V = W = \partial T/\partial \xi = 0 \tag{6.12}$$

6.4 Results and Discussions

The obtained results present the velocity vectors (Figs. 6.2, 6.3 and 6.4) and the temperature fields (Figs. 6.5, 6.6 and 6.7) in the case (Bi = 0 and Ma = 2,000, 3,000 and 4,000). The plots are arranged with ascending of the (Ma) value.

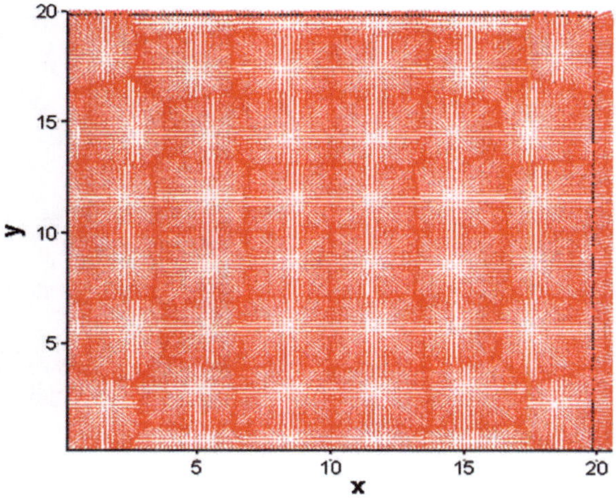

Fig. 6.3 Velocity vectors for Ma = 3,000

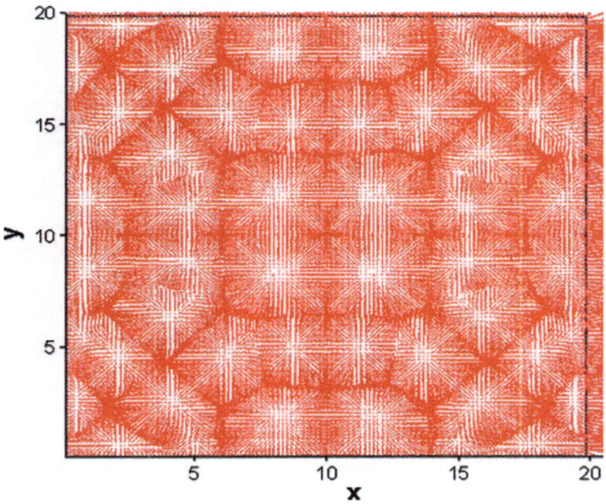

Fig. 6.4 Velocity vectors for Ma = 4,000

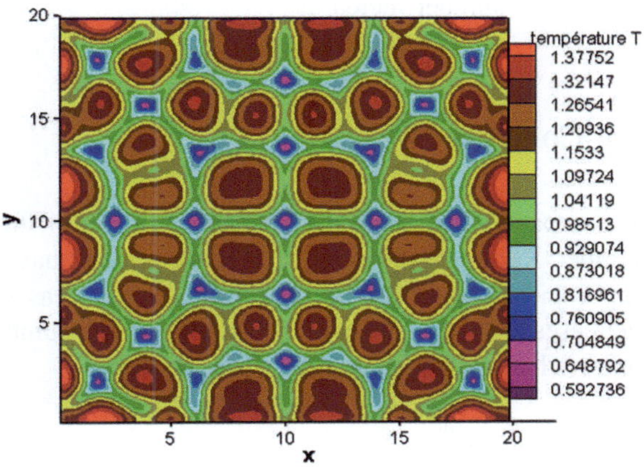

Fig. 6.5 Isotherms for Ma = 2,000

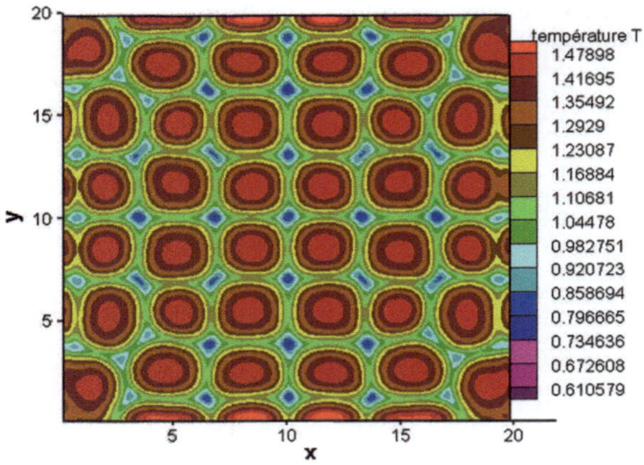

Fig. 6.6 Isotherms for Ma = 3,000

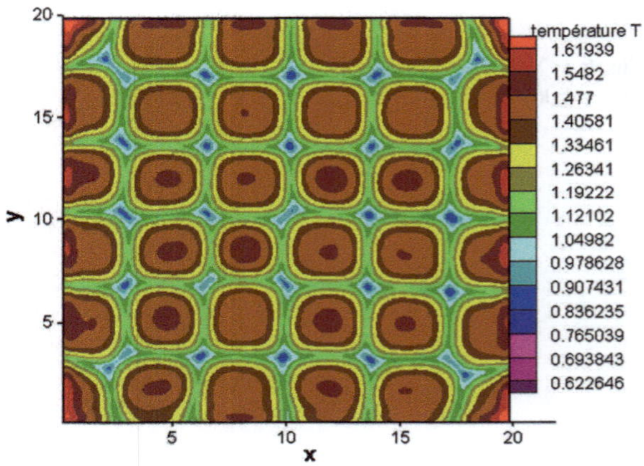

Fig. 6.7 Isotherms for Ma = 4,000

By observing all the cases, it is found that the number of cells decreases with the increase of the Marangoni number and the sizes of the cells are larger when the Marangoni number increases.

6.5 Conclusions

In this paper, 3D numerical simulation of a horizontal liquid layer for Rayleigh number Ra = 0, only surface tension effects, has been carried out. The most important obtained results can be summarized as follows: the number of cells decreases with the increase of the Marangoni number; the sizes of the cells are larger when the Marangoni number increases.

References

1. J.R.A. Pearson, On convection cells induced by surface tension. J. Fluid Mech. **4**, 489–500 (1958)
2. D.A. Nield, Surface tension and buoyancy effects in cellular convection. J. Fluid Mech. **19**, 41–352 (1964)
3. L.E. Scriven, C.V. Sterling, On cellular convection induced by surface tension gradients: effects of mean surface tension and viscosity. J. Fluid Mech. **19**, 321–340 (1964)
4. S.H. Davis, G.M. Homsy, Energy stability theory for free Surface problems buoyancy-thermocapillarity layers. J. Fluid Mech. **98**, 527–553 (1980)
5. R. Selak, G. Lebon, Instabilité thermoconvective de Bénard-Marangoni en présence d'une viscosité dépendant de la température. J. Phys. II France. **3**, 1185–1199 (1993)
6. A. Thess, S.A. Orszag, Surface-tension-driven Bénard convection at infinite Prandtl number. J. Fluid Mech. **283**, 201–230 (1995)
7. D. Schwabe, Hydrodynamic instabilities under microgravity in a differentially heated long liquid bridge with aspect ratio near the Rayleigh limit: experimental results. Adv. Space Res. **36**, 36–42 (2005)
8. P. Cerisier, S. Rahal, H. Azuma, Pattern dynamics of the Bénard-Marangoni instability in a medium aspect ratio container. J. Phys. Conf. Ser. **64**, 012004 (2007)
9. S. Rahal, P. Cerisier, Bifurcation to chaos in the Bénard-Marangoni instability in a confined geometry. J. Phys. Conf. Ser. **64**, 012005 (2007)
10. D. Nezar, S. Rahal, Numerical simulation of convective instabilities in a liquid layer submitted to an inclined gradient of temperature. Energy Procedia. **36**, 380–385 (2013)

Chapter 7
Walk Optimization of a Drilling Operation Carried Out by a Two Link Robot Arm Using Heuristic Approach

Djamel Bellala, Hacene Smadi and Aicha Medjghou

Abstract Exact solutions for the TSP problem are typically difficult from computational point of view, because of their size and time complexities. That is why, heuristics are substituted to exact algorithms in order to provide a good solution to the problem. In this paper two heuristics, the nearest-neighbor and the subtour-reversal algorithms, are used to solve an industrial problem. The first algorithm gives birth to an optimal tour by which the industrial process can be carried out while the second algorithm generally provides an improvement to the previous optimal tour.

7.1 Introduction

Because of the lucrative character of the manufacturing companies, these always seek to improve their manufacturing process by improving the quality of their products with the lowest possible cost. This cost minimization can be obtained; among other things; by reducing the manufacturing time of each product, what makes the delivery deadline more competitive. A start-up, small and medium-sized enterprise (SME) intends to manufacture a series of a new electronic toy for children. The printed circuit board (PCB) of every toy is fitted with holes for mounting different electronic components. These holes have to be drilled with a

D. Bellala (✉) · H. Smadi
Control Engineering and Manufacturing Laboratory, University of Batna, Batna, Algeria
e-mail: bellala_djamel@yahoo.co.uk

H. Smadi
e-mail: h.smadi@hotmail.fr

A. Medjghou
Department of Industrial Engineering, University of Batna, Batna, Algeria
e-mail: amedjghou@yahoo.fr

A. Y. Oral et al. (eds.), *International Congress on Energy Efficiency and Energy Related Materials (ENEFM2013)*, Springer Proceedings in Physics 155,
DOI: 10.1007/978-3-319-05521-3_7,

robot arm according to the operation mode which will be developed later in this paper. The aim of the enterprise is to carry out the drilling process in a shortest possible time. In order to respond favorably to the enterprise requirements the following steps are successfully carried out.

1. Design of the circuit board using simple PCB software.
2. Determination of the robot arm trajectory in between every holes pair.
3. Setting up the distance matrix by evaluating the displacement time of the robot arm of every pair of holes.
4. Application of the suggested metaheuristics in order to optimize the tour carried out by the robot arm during the drilling process.

7.2 PCB Preparation

The software used for the circuit board is free CAD software. It includes *ExpressSCH* for drawing schematics and *ExpressPCB* for circuit board layout. The process starts by using the *ExpressSCH* schematic design program according to the following steps:

Step 1. component selection
Step 2. component position
Step 3. wiring
Step 4. edition of the schematic
Step 5. linking the schematic and the PCB.

The *ExpressPCB* circuit board layout program will now continue the design process according to the following steps:

Step 1. component selection
Step 2. component position
Step 3. adding the traces
Step 4. editing the layout
Step 5. ordering the PC boards.

The obtained printed circuit board is shown in (Fig. 7.1).

Using Matlab system the coordinates of each drilling point have been accurately determined, and so the distance matrix is obtained. The number of the drilling points is 45. All the drilling points are numbered from 1 to 45 as shown in (Fig. 7.2). The lower left corner point of Fig. (7.2) is taken as the origin of a Cartesian reference frame.

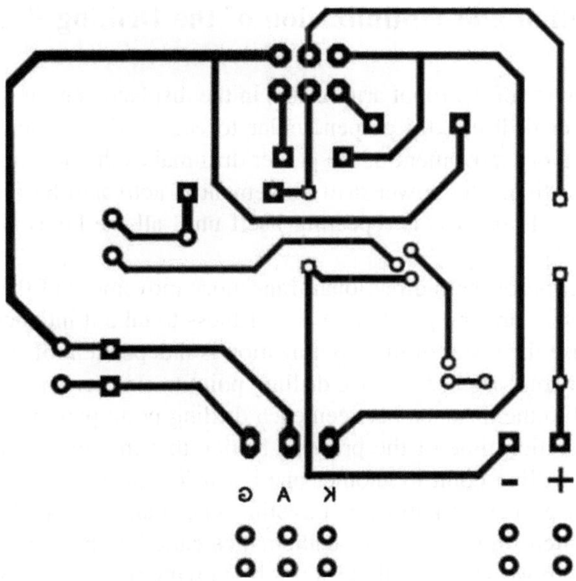

Fig. 7.1 Copper side of the obtained PCB

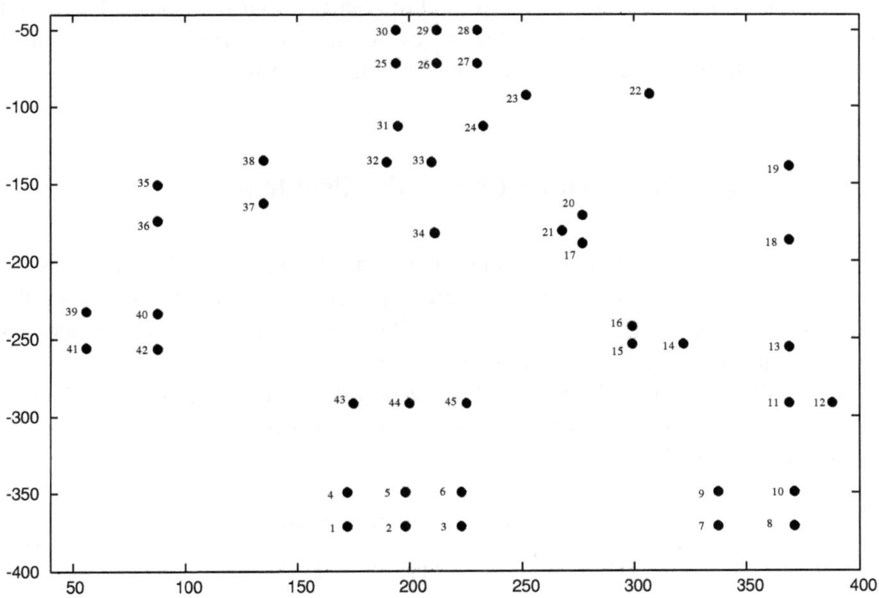

Fig. 7.2 Drilling points

7.3 Description and Optimization of the Drilling Process

The motion control of the robot arm results in the displacement of the end-effector holding a power drill exactly perpendicular to each drilling point. Then a bidirectional translatory movement of the power drill makes the hole and returns back to its initial position. The power drill movement is activated by a programmable counter. The whole process is repeating itself until all the holes are successfully carried out [1, 2].

The time period of the bidirectional translatory movement of the power drill is the same for every drilling point, so it is worthless to take it into account it in this problem because the result of this optimization is independent of that time. In fact only the time displacement from one drilling point to another one which in its turn is proportional to the distance between each drilling point pairs is the decisive for the overall operation time of the process. Notice that the movement of the robot arm from one drilling point to another one is carried out without obstacles [1, 2].

This type of problems; known as traveling salesman problem TSP; falls down under a very interesting branch of mathematics called graph theory. In this optimization problem we seek to drill 45 holes by moving through every drilling point once and only once in a shortest period of time. The data of the problem is a time (or distance) matrix representing the time necessary to move from each pair of points. Graphically the system is represented as a network of nodes [3].

The aim of this research paper is to find the optimal tour that should be carried out by the robot arm, i.e. the tour that gives the minimum time (minimization of the cost function) of the robot arm when visiting all the points.

7.4 Mathematical Formulation of the Problem

TSP is a combinatorial optimization problem and it belongs to an NP-hard class of problems but it has the power of representing a large class of problems. Among the combinatorial optimization problems the TSP is the most important one, since it is very easy to describe, but very difficult to solve.

The TSP problem deals with finding the shortest (closed) tour in an n-node situation where each node is visited exactly once. Specifically, in an n-node situation, we define [4, 5]:

$$x_{ij} = \begin{cases} 1, & \textit{if node } j \textit{ is reached from node } i \\ 0, & \textit{otherwise} \end{cases} \tag{7.1}$$

Given that d_{ij} is the distance from node i to j, the TSP model is given as

$$\text{Minimize } z = \sum_{i=1}^{n}\sum_{j=1}^{n} d_{ij}x_{ij}, d_{ij} = \infty \quad \text{for all } i = j \tag{7.2}$$

Subject to

$$\sum_{j=1}^{n} x_{ij} = 1, \quad i = 1, 2, \ldots, n \tag{7.3}$$

$$\sum_{i=1}^{n} x_{ij} = 1, \quad j = 1, 2, \ldots, n \tag{7.4}$$

$$x_{ij} = (0, 1) \tag{7.5}$$

7.5 The Nearest-Neighbor Algorithm

As the name of the heuristic suggests, a "good" solution of the TSP problem can be found by starting with any city (node) and then connecting it with the closet one. The just-added city is then linked to its nearest unlinked city (with ties broken arbitrarily). The process continues until a tour is formed [4, 6].

7.6 Sub-tour Reversal Heuristic

In an n-city situation, the sub-tour reversal heuristic starts with a feasible tour and then tries to improve on it by reversing 2-city sub-tours, followed by 3-city sub-tours, and continuing until reaching sub-tours of size n–1 [4, 6].

7.7 Results and Conclusion

After running the appropriate Matlab program, the optimal tour is obtained as follows:

The first heuristic gives the following shortest tour:

18-19-17-16-15-14-13-12-11-10-9-8-7-6-5-4-1-2-3-45-44-43-42-40-41-37-39-38-36-35-33-34-32-30-31-29-22-21-20-23-26-27-28-25-24-18.

This tour corresponds to a travelled distance of **361,875** units.

The second heuristic which is an improvement of the first one gives the following shortest tour:

18-17-16-15-14-13-12-11-9-10-8-7-6-5-4-1-2-3-45-44-43-42-40-41-37-39-38-36-35-33-34-32-30-31-29-22-21-23-26-27-28-25-24-20-19-18.

This tour corresponds to a total distance of **356,59**.

The two heuristics are executed within one Matlab program in exactly **1053** iterations.

The improvement gained using the second heuristic is rated 1.5 %. Even though the rate of the improvement is small, it still of great importance especially when the number of PCB that will be fabricated is very large.

References

1. D.B. Marghitu, *Mechanisms and Robots Analysis with MATLAB*. British Library Cataloguing in Publication Data. (Springer, London, 2009)
2. B.I. Kazem, A.I. Mahdi, A.T. Oudah, Motion planning for a robot arm by using genetic algorithm. Jordan J Mech Ind Eng **2**(3), 131–136 (2008). ISSN 1995-6665
3. F. Greco, *Travelling Salesman Problem*. (In-Teh is Croatian branch of I-Tech Education and Publishing KG, Vienna, Austria, 2008)
4. H.A. Taha, *Operations Research an Introduction*, 8th edn. (Pearson International Edition, 2007)
5. R. Bronson, G. Naadimuthu, *Operations Research*, 2nd edn. Schaum's Outline Series. (McGraw-Hill, 1997)
6. F.S. Hillier, G.J. Lieberman, *Introduction to Operations Research*, 9th edn. (McGraw-Hill International Edition, 2010)

Chapter 8
Design and Evaluation of Airborne Wind Turbine Utilizing Physical Prototype

Edris Safavi, Mohsen Namakian, Tim Sirén, Rickard Magnéli and Johan Ölvander

Abstract Moving towards renewable sources of energy has become one of the most important energy-related strategies in recent decades. High-altitude wind power (HAWP) has been discovered in 1833 as a source of useful energy. Wind power density (Watts/m^2) can significantly increase ($\sim$6 times) by going from 80 to 500 m altitude. The global capacity of 380 TW (terawatt) as well as abundance, strength, and relative persistency of wind in higher altitude are eye-catching points to consider HAWP as a reliable energy source in the future. A research project called "THOR" has been initiated at Linköping University by a group of master students (soon to graduate) as proof of concept of airborne wind energy (AWES). THOR is about feasibility analysis of different concepts of HAWP and proof of concept of balloon based AWES as one of the appropriate existing concepts. THOR is intended to be a research platform at Linköping University for further development of AWES concepts in future.

8.1 Introduction

The term "renewable energy" covers all types of energy that come from natural resources which are continually replenished e.g. water, winds, tides, sun, and biomass etc. Although using renewable energies is not as straightforward as for example fossil fuels, technological advancements on one side and natural resource depletion on the other side have caused human to think about renewable energy more and more. Therefore moving towards renewable sources of energy has become one of the most important energy-related strategies in recent decades.

E. Safavi (✉) · M. Namakian · T. Sirén · R. Magnéli · J. Ölvander
Department of Management and Engineering, Machine Design division,
Linköping University, 58181 Linköping, Sweden
e-mail: edris.safavi@liu.se

A. Y. Oral et al. (eds.), *International Congress on Energy Efficiency and Energy Related Materials (ENEFM2013)*, Springer Proceedings in Physics 155,
DOI: 10.1007/978-3-319-05521-3_8, © Springer International Publishing Switzerland 2014

On first hand the world has faced rigorous environmental degradation which draws attention to clean solutions. On the other hand, the limited availability of the traditional energy resources like petroleum and the huge increasing demands make the cost of these resources higher. Furthermore greenhouse effect is another driving force which makes it necessary to find other available resources for energy that are renewable and clean.

In recent years wind power has gotten increasingly attention as a source of renewable and green energy. World total installed capacity has grown nearly ten times (from 24,836 to 239,485 MW) during 10 years (2001–2011). This growth rate is significantly more than the average annual growth in global energy consumption (2.5 %) [1]. However, achieving this growth rate is very respectable but comparison of total harvesting of wind energy to the total capacity of the easy accessible wind power (approximately 160 Giga Watt (GW) is expected to generate more than 331 TWh in 2010) shows massive potential of this field to grow even further [2]. On the other hand, the mentioned wind energy capacity can cover just 1.6 % of global electricity consumption which needs to be almost 12 % in 2050 according to the International Energy Agency. It is important to take into consideration that by reaching to this level the annual CO_2 emissions can be reduce by 2.8 giga tone (Gt) which is massive reduction compared to e.g. annual Sweden CO_2 emissions (0.049 Gt) in 2008 [3].

Winds are created over the earth due to pressure gradients caused by temperature difference between the equator and the poles. Humans, on the earth, feel a small part of the winds since they are in the lower part of a boundary layer called "No slip condition" where the air touches the earth with almost zero velocity. Therefore it can be concluded that in general there is no wind on the surface compared to higher altitudes. Exceptional cases are where the wind is guided to the surface. These are the places that are known as "wind prone" areas where the conventional wind turbines are normally installed.

As illustrated in Fig. 8.1, wind speed is getting stronger at higher altitudes. Although, increasing the efficiency of conventional wind turbine is always on demand among companies and research institutes. A preliminary result from a European project namely KitVes shows that 800–1200 m is a sweet spot with less challenging altitude but strong winds where the power efficiency is 4 times higher than current 5–6 MW wind turbine at 80 m [4]. This massive increasing in efficiency seems almost non accessible for conventional wind turbines. Therefore, the concept of high altitude wind power is considered as a convincing concept to obtain the wind energy at higher altitude.

Airborne Wind Energy Systems (AWEs) have presented recently as revolutionary alternatives for conventional wind turbines to capture high altitude wind energy (HAWE). The general idea of AWEs is to lift up a wind turbine to the desired altitude using a lifting device such as lifting balloon, kite, propeller etc. Energy generated by the AWEs may be used aloft or sent to the ground using specific devices such as conducting cables, mechanical force through a tether etc. There are several concepts in the market to harvest wind power at higher altitudes. However, most of them are at the first stage of development. So, they might be not

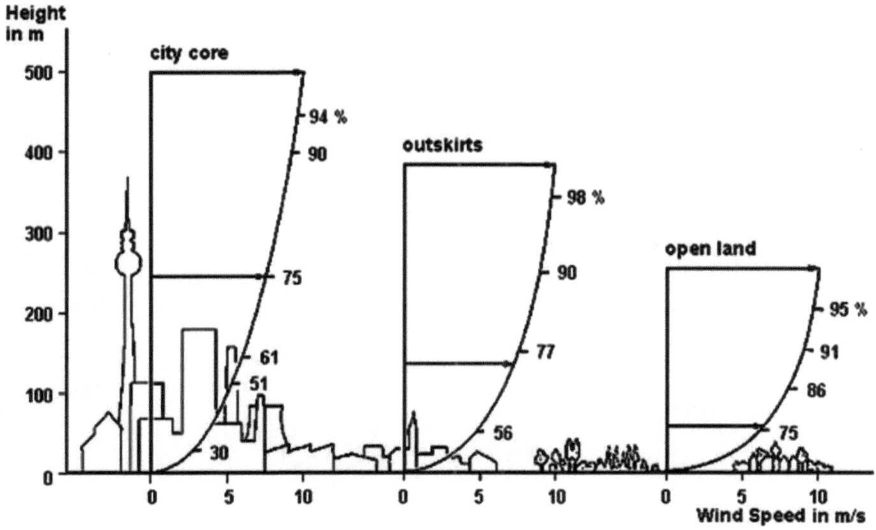

Fig. 8.1 Approximation of the wind speed profile based on the roughness of the area (adapted, from BaumBach [13])

well optimized and cost efficient yet. In general, existing concepts can be categorized as:

- **Propeller based designs**: Using "tethered wing" or "flying electric generator" which passes a circular path aloft [5, 6].
- **Kite based designs**: Kites are employed to convert the wind energy to electric energy [7].
- **Balloon based designs**: Balloons are used to keep the system aloft. These designs have been initiated in some project e.g. MARS [8] or airborne wind turbine developed by Altaeros Energies [9].

8.2 Design Study

This study is initiated by assessment of existing concepts. However, due to the lack of both theoretical and practical information about HAWP and especially about AWEs, this study is initiated mostly from scratch. In this case, rapid prototyping has been proven to be a promising method used early in the design process, not just to get a better feeling of the final product but also to obtain more validated information about the concept under evaluation and models. THOR is created to be used as a proof of concept of AWEs which can be rapidly fabricated and tested. The achieved information from THOR can be used later to more investigate the balloon based concept e.g. scale up the concept to the commercial size.

Design Specifications	
Max Altitude	*150m*
Max Weight	24 Kg
Max Power	100 Watt
Working Temperature	*(-30 C)-(+40 C)*
Max wind speed	*25 m/s*

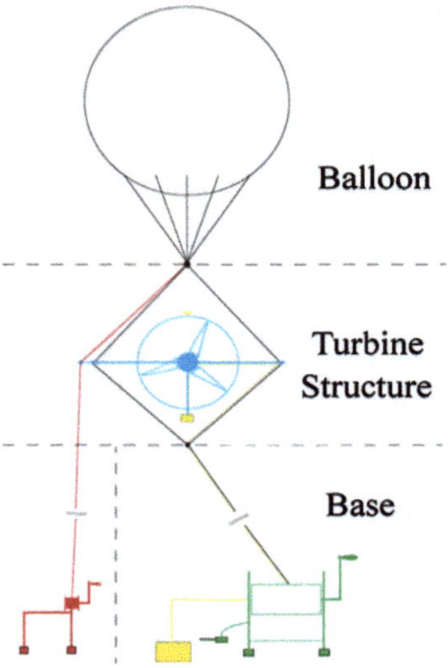

Fig. 8.2 Schematic view and main specifications of THOR Project

8.2.1 Design Requirements

Apparently all the proposed concepts available in the market have many pros and cons which make them hard to compare. In the first phase of THOR project an assessment is performed among the different mentioned concepts to select the best well-fitted concept to the project requirements. The main requirements are listed as:

- High reliability
- High safety
- Easy to manufacture (specially for students)
- Less design complexity
- Less environmental impact.

Taking the above mentioned factors into consideration, a range of designs was investigated. The balloon based design is selected as a best fitted concept to the requirements. However, Douglas J. Amick (2008) suggests a well aerodynamically designed balloon having a small conventional turbine installed on it [10]. Frederick D. Ferguson (2008) proposes a design which is now fully developed and sold in the market by Magenn Power Inc. [11]. There are more complicated designs, like Meller (2009) or simpler ones like Kling (1978), [12–17]. After assessment of different concepts, the initial designed concept is consisted of a small wind turbine, attached (via a turbine structure), to spherical helium filled balloon, to keep the system up in the air. The electric power is transferred down using an electric system. The entire system is held by a base structure which could winch the turbine up and down. Another safety system is designed to assure the safety of the systems in the air (Fig. 8.2).

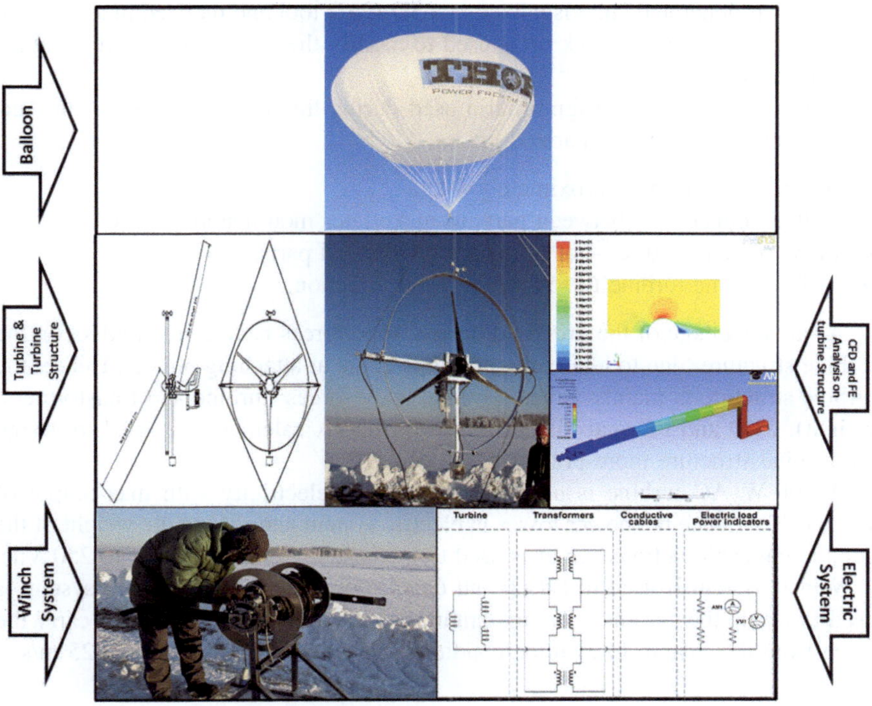

Fig. 8.3 Different components of the THOR concept including examples of CFD and FE analysis

8.2.2 Detail Design

As illustrated in Fig. 8.2, THOR concept consists of 4 main parts; *Helium balloon, Turbine and Turbine structure, Electric system, Winch system.*

The function of the helium balloon was to lift up the turbine to the desired altitude. A spherical balloon with diameter of 4.5 m was designed, due to higher lifting force per weight of the balloon. Balloon was chosen to build from PU Nylon material among three different options (PVC, Latex, PU Nylon) offered by the suppliers due to light weight, good stiffness and higher temperature resistance.

The loads applied on the system come mainly from drag force. To increase safety, an extreme condition (max. wind velocity equal to 25 m/s) was assumed. Using CFD simulation for spherical object, 4.5 m diameter (Fig. 8.3) an upper limit for the drag force was calculated to be 1,500 N. This force together with the maximum weight and drag force are used to select the proper rope to keep the system up in the air. In order to use the turbine in an efficient way the most important requirement is the ability of structures to follow the wind direction and to keep the turbine horizontal and in-line with the wind direction. The initial concept was generated and visualized in a 3D-CAD tool and realized in a 1:1 scale mock-up. The result of mock up is used to modify the models and more evaluate the concepts.

The most important design criteria used during the detailed design of the turbine structure can be summarized in:

- A target weight of approximately 4 kg.
- Bolted connections between parts to make later modifications easier.
- Easy to manufacture—not to many complicated parts.
- Ability of the turbine to follow the wind direction.

As an advantage of the proposed design, rope forces have a small effect on the turbine structure (due to relatively high rope angle at attaching points). Hence, the turbine structure is mainly affected by the body forces (turbine and transformers' weight). CFD and FE analysis as well as handbook calculation is used to design the turbine structure more precisely, see Fig. 8.3.

A 100 W AC turbine is used to generate the electricity with max-output of 15 Vac. To reduce the power losses in electric system and reduce the weight of the conducting cable, a transformer is used to increase the output voltage to 250 Vac.

Winch system is used to lift up and down the whole system. Which system is used to lift up the system to the maximum altitude of 150 m? It can also carry the force from the system in extreme condition e.g. wind velocity equal to 25 m/s.

8.3 Results and Discussion

To more evaluate the concept, two separate tests have performed in December 2012 and May 2013, at Östergötland-Sweden. The result of the first test shows that while there was a little breeze at ground level (0.3–0.7 m/s), at 100 m altitude winds, strong enough for running the turbine, was observed (3.4–4.1 m/s). The turbine started functioning under no electrical load at 3.4 m/s and with an electrical load (an 18 W bulb lamp) at 3.7 m/s. The power produced was calculated to be in the range 4.5–7.8 W in the first test, depending to the wind velocity (3.7–4.1 m/s). However during the second test the power increased to higher value e.g. 50 watt, when there was stronger wind on the ground (3 m/s), see Fig. 8.4.

The maximum power gained practically was equal to 48 W, which create a satisfactory power factor for the entire system. However the difference between the theoretical calculated power and measured power in the field can be resulted from three main factors explained as following: Low efficiency of turbine, Instabilities in the turbine structure, Losses in electric system.

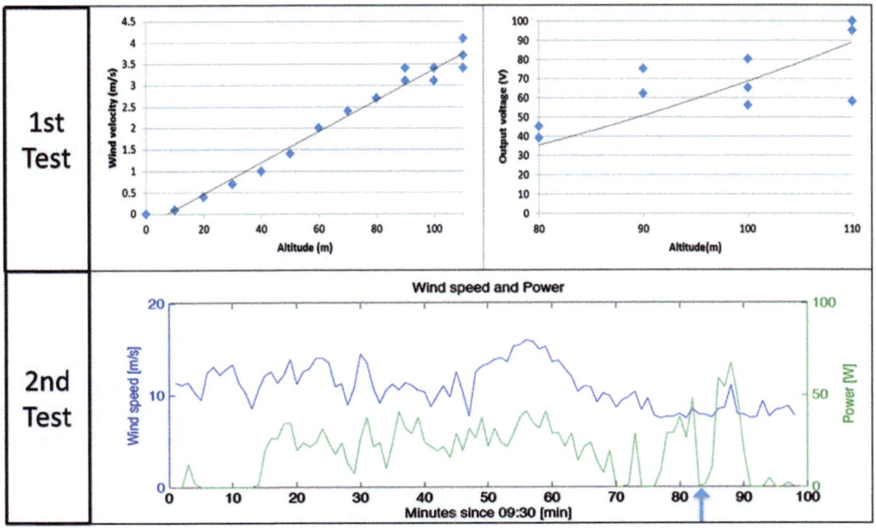

Fig. 8.4 Wind speed and output power versus altitude (*up*), wind speed and power versus time (*bottom*)

8.4 Conclusions

An Airborne Wind Energy System was successfully designed and tested. The project provided a good foundation with theoretical and practical knowledge, and different perspectives for further developments. The THOR project holds a great potential to be developed. As a necessary step for all following suggested future works, long time site tests at higher altitudes and under different wind situations could be conducted in order to gain more functioning-related data. However the result was satisfactory for the selected turbine, but the result can be improved by using a more efficient turbine, more stable turbine structure and better electric system to avoid losses. Weight reduction approaches can be also applied on all aloft parts, taking part's function into consideration. For instance a more efficient turbine can be employed, while in this case efficiency means higher power factor per unit weight. Also turbines with high voltage single-phase outputs are more suitable as transformers would not be needed and also lighter conductive cable (e.g. two core wires instead of three). Furthermore, a control system needs to be developed. The primary function of this control system is to sense the weather condition and locate the turbine at the best altitude and responding to all major risks. The possibility of using hydrogen, instead of helium, which has a higher lift to cost ratio, needs to be investigated. However, hydrogen is explosive, and is more suitable for remote areas e.g. offshore applications.

References

1. BP, *Statistical Review of World Energy* (online). (2012) Available from: http://www.bp.com/statisticalreview
2. M.Z Jacobson, C.L Archer, in *Saturation Wind Power Potential and its Implications for Wind Energy*. Proceedings of the National Academy of Sciences. pp. 15679–15684 (10 Sept 2012). doi:10.1073/pnas.1208993109
3. Wind Energy Roadmap, International Energy Agency, 2010. www.iea.org
4. Kvitas, Seventh Framework Programme (FP7) www.kvitas.com. 2010
5. D.J Amick Thethered wind turbine, U.S. patent US2008/0048453A1 (2008)
6. SkySails GmbH (n.d) SkySails GmbH—Power www.skysails.info/index.php?id=11&L2012
7. Makani Power Inc. (n.d) Makini power airborne wind energy. www.makanipower.com
8. Sky WindPower Corporation, Sky WindPower Corporation www.skywindpower.com/ww
9. Kite Gen Research S.r.l (n.d) KiteGen Research http://www.kitegen.com/en/. 2013
10. Magenn Power Inc. (n.d) Magenn Power Inc. Available from: http://www.magenn.com/
11. M Meller, Airborne wind turbine electricity generating system, U.S. Patent US7,582,981B1 (2009)
12. A Kling, Wind driven power plant, U.S. patent 4,073,516 (1978)
13. G. Baumbach, *Luftreinhaltung* (Springer, Berlin, 1991)
14. Altaeros energies, high altitude wind turbine http://www.altaerosenergies.com. 2012
15. Magenn Power Inc. (n.d) Magenn Power Inc. http://www.magenn.com/. 2012
16. O.J Kushto, Thethered lighter than air turbine, U.S. Patent 4,207,026 (1980)
17 M Meller, Airborne wind turbine electricity generating system, U.S. Patent US7,582,981B1 (2009)

Chapter 9
Comparison of Two ARMA-GARCH Approaches for Forecasting the Mean and Volatility of Wind Speed

Ergin Erdem, Jing Shi and Ying She

Abstract In this study, we develop two ARMA-GARCH models for predicting the mean and volatility of wind speed. The first model employs the standalone ARMA-GARCH model for modeling the mean wind speed and the variance simultaneously. For the second model, in the first step, the current wind vector is decomposed into lateral and longitudinal components by using the prevailing wind direction. The mean and variance of the two components are then modeled using two separate ARMA-GARCH processes. Thereafter, the two components are combined back to form the resultant single wind vector. A large wind dataset is employed for model building and prediction so that the two approaches can be compared. It shows that the standalone ARMA-GARCH model is more accurate for predicting the wind speed, whereas the component ARMA-GARCH model performs better for predicting the wind variance.

9.1 Introduction

One of the fastest growing renewable energy sources is the wind energy. To date, most wind energy is obtained through wind turbines, which convert the kinetic energy of the wind into the mechanical energy. The recent interest in the

E. Erdem (✉) · J. Shi
Department of Industrial and Manufacturing Engineering, North Dakota State University, Fargo, ND 58108, USA
e-mail: ergin.erdem@my.ndsu.edu

J. Shi
e-mail: jing.shi@ndsu.edu

Y. She
School of Economics and Management, Nanchang Hangkong University, Nanchang 330063 Jiangxi Province, China
e-mail: sheying1980@hotmail.com

A. Y. Oral et al. (eds.), *International Congress on Energy Efficiency and Energy Related Materials (ENEFM2013)*, Springer Proceedings in Physics 155,
DOI: 10.1007/978-3-319-05521-3_9,

renewable energy sources sparks enormous development in wind energy. Starting from 1980s, the current installed wind capacity has been experiencing a sharp growth. The growth rate for the installed wind energy capacity is at around 25 % each year worldwide. The total installed energy capacity as of December 2012, surpassed 280,000 MW worldwide. In 2012, the installed capacity exceeded 100,000 MW in Europe, while it passed the mark of 60,000 MW in the U.S. (American Wind Energy Association [1]). In 2010, the current energy obtained from wind related sources constitutes 2.5 % of the total electrical energy usage worldwide. In terms of cost, the energy obtained from the wind is becoming cheaper and moving closer than ever to that obtained from natural gas or coal.

A U.S. Department of Energy report [2] estimates that wind energy have the potential to contribute to the 20 % of the energy supply by 2030 in the scenarios assuming modest growth trajectory. One of the important aspects of wind energy generation is that the wind energy is an intermittent energy source, in which the energy output to the large extent depends on the wind speed. This presents a challenge for generating electricity from the wind. The problem is especially aggravated if the electricity obtained from wind-based sources constitutes a significant portion (e.g., 15 %) of the total electricity generation. The intermittent wind energy generation can be handled well if the wind speed can be predicted accurately in advance. For instance, given the accurate forecast of wind speed, it will be easier to arrange the operation schedules of wind farms.

Wind speed prediction carries two aspects of meaning. One is the prediction of the mean of wind speed. The other is the prediction of the volatility of wind speed. Unfortunately, the existing research in literature has been dominated by the first aspect, while the second aspect has not been paid enough attention. For the short-term mean wind speed forecasting, a large number of methods have been developed. Besides the physical models, there exist a long list of statistical models, including autoregressive (AR) models, autoregressive moving average models (ARMA), autoregressive integrated moving average (ARIMA) models, Bayesian methods, vector autoregression (VAR) methods, general optimization tools (i.e., neural networks), techniques borrowed from control theory, generalized impulse response analysis method and other non-conventional time-series models [3–6]. In addition, hybrid approaches have been developed in the literature for forecasting wind speed [7, 8].

However, those approaches cannot consider the volatility of wind speed. In other words, it is assumed that the turbulence of wind speed is constant (homoscedastic). However, the variation in wind speed might not be constant and might be subject to change with respect to time. Although the point estimates of mean wind speed is usually not biased, the variation in wind speed might cause the error estimates to be underestimated, and thus affect the operation of wind turbines [9]. The prevalent tools that might be used for modeling the wind volatility is the autoregressive conditional heteroskedasticity (ARCH) and generalized autoregressive conditional heteroskedasticity (GARCH). Tol [10] proposed a GARCH model to capture the heteroscedastic features associated with the wind speed. It is indicated that incorporating the wind speed variance improves the wind speed forecast. In a similar token, Ewing et al. [11] introduced the Garch-in-mean (GARCH-M) models for forecasting the wind speed. The analysis shows that the variability of wind speed exists for all heights, however, at higher altitudes; the wind volatility is more persistent as compared to the lower altitudes. In a follow-up study, Ewing et al. [12] employed the GARCH-M methodology to investigate the 26 regional wind energy markets. It was concluded that the wind speed exhibits time-based volatility in all the locations. Payne and Carroll [13] extended the works of Ewing et al. to investigate the ARMA-GARCH-M framework for the energy markets in China, and demonstrated that the wind volatility and its relation with the wind speed is based on the geographical region. Payne [14] used the component ARMA-GARCH-M model to differentiate the transitory and permanent components of the wind volatility. The results show that the transitory components play a significant role in determining the wind volatility.

Another consideration is the analysis of wind direction and how it affects the prediction of wind speed. Incorporating the wind direction information in the forecasting process might potentially improve the forecasts of wind speed. The important distributions for modeling wind direction include the uniform distribution, the cardioid distribution, the wrapped Cauchy, and wrapped normal distributions, along with the von Mises distribution [15]. In literature, various approaches are undertaken for analyzing wind direction. For instance, Carta et al. [16] used a mixture of von Mises distributions to characterize the wind direction for Canary Islands, Spain. Only a few studies have been conducted to forecast wind direction. Garcia-Rojo [17], based on Measure-Predict-Correct algorithm, identified the discrete bivariate statistical distribution to assess the long-term wind climate. Weber [18] generalized the isotropic Gaussian model, and lifted equal variance assumption. In this way, an anisotropic Gaussian model was produced. Johnson and Wehrly [19] developed a method based on the angular-linear distributions to represent wind speed and direction. Erdem and Shi [20] compared three different approaches for forecasting the wind speed and wind direction. The results indicate that for the wind speed forecasting, the traditional linked ARMA models is a better approach compared to the component approach.

This paper focuses on the ARMA-GARCH methodology for modeling and predicting wind speed and its variance, as well as the incorporation of component model to study the effects of wind direction on the wind speed forecasts.

9.2 Methodology

Bollerslev [21] developed the GARCH based approaches to generalize the ARCH processes develop by Engle [22]. In the ARCH approach, the time based changing error terms associated with the stochastic process can be represented by,

$$\varepsilon_t = \sqrt{v_t} z_t \tag{9.1}$$

where ε_t is the corresponding error term at time t, and z_t is the white noise sequence denoted by the variate which is based on the standardized normal distribution with mean 0, and variance 1. In that regard, v_t represents the time changing variance of the underlying process that can be denoted by the linear combination of the squared errors of order l. This leads to the ARCH (l) model, denoted by,

$$v_t = \varsigma_0 + \eta_1 \varepsilon_{t-1}^2 + \eta_2 \varepsilon_{t-2}^2 + \eta_3 \varepsilon_{t-3}^2 + \cdots + \eta_l \varepsilon_{t-l}^2 \tag{9.2}$$

where ς_0 and η_i are constant terms. If the model is ARCH (l), then only error terms of the first order (i.e., one period prior) is involved. It is possible to generalize this process and express the variance term in terms of the past errors and variances. This leads to the GARCH (k, l) model, represented by,

$$v_t = \varsigma_0 + \sum_{i=1}^{k} \varsigma_i v_{t-i} + \sum_{i=1}^{l} \eta_j v_{t-j} \tag{9.3}$$

If the orders associated with the variance and error terms (i.e., k and l) are specified to be 1, then the model is called standard GARCH, namely, GARCH (1, 1).

The component model approach is developed from the concept of anisotropic Gaussian model. The first step for the wind direction is finding the prevailing wind direction. The schematic representation of the prevailing wind direction and the axes is shown in Fig. 9.1. θ indicates the angle of a given wind vector with the north axis. θ_D represents the prevailing wind direction with the north axis. θ' denotes the angle of a wind vector angle with the axis of prevailing wind direction. y' and x' denote the longitudinal and lateral axes with respect to prevailing wind direction, respectively.

For finding the prevailing wind direction, the following formula can be utilized [16],

$$\theta_D = \tan^{-1}\left(\frac{\bar{S}}{\bar{C}}\right), \quad \bar{S} > 0, \bar{C} > 0 \tag{9.4}$$

$$\theta_D = \tan^{-1}\left(\frac{\bar{S}}{\bar{C}}\right) + \pi, \quad \bar{C} < 0 \tag{9.5}$$

$$\theta_D = \tan^{-1}\left(\frac{\bar{S}}{\bar{C}}\right) + 2\pi, \quad \bar{S} < 0, \bar{C} > 0 \tag{9.6}$$

where $\bar{S}$ is the summation of sine values of the angles with respect to the north direction, and $\bar{C}$ is the average of the cosine values of the angles with respect to the north axis. After finding the prevailing wind direction, the wind speed vector can be decomposed into lateral and longitudinal components using the following,

Fig. 9.1 Schematic of prevailing wind direction [20]

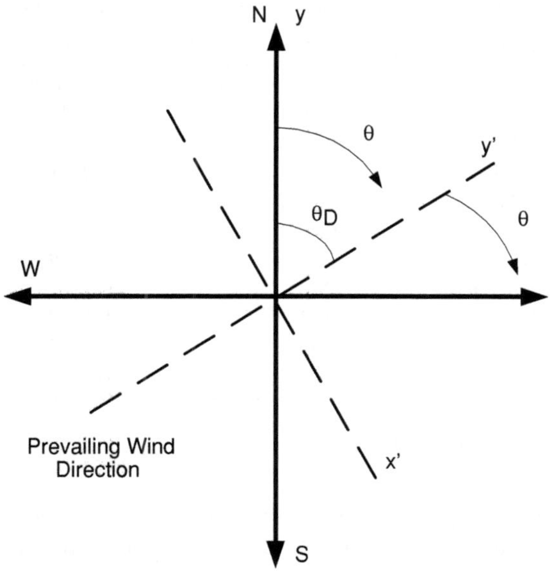

$$v_{y'} = v \cdot \cos(\theta - \theta_D) \tag{9.7}$$

$$v_{x'} = v \cdot \sin(\theta - \theta_D) \tag{9.8}$$

Using a single prevailing wind direction helps us to obtain a more parsimonious model. Based on the decomposition, the correlation between these two variables is assumed to be 0. After the lateral and longitudinal components are obtained, two separate ARMA-GARCH processes are employed to forecast the mean wind speed and volatility of the two speed components, respectively. After that, the lateral and longitudinal components are combined back to form the resultant vector and volatility.

The fundamental difference between the traditional ARMA-GARCH model for forecasting the wind speed alone and this approach is that: we incorporate the wind direction indirectly in the model for forming lateral and longitudinal components. Incorporating the wind direction might bring benefits in terms of forecasting the wind volatility.

9.3 Data Collection and Analysis

For wind data collection, we resorted to one wind observation site in Colorado, U. S. and obtained the minute data for wind direction and speed for the entire year of 2012. The hourly averages of the wind speed and direction are obtained based on the minutely data. Additionally, using the minutely data, the standard deviation for

Table 9.1 Parameter values associated with the prediction models

GARCH specification	ARMA(2, 2)-GARCH	ARMA(3,0)-GARCH	ARMA(1, 3)-GARCH
Mean term	Standalone wind speed	Longitudinal component	Lateral component
AR [1]	1.4673**	0.9243**	1.6679**
AR [21]	−0.5041**	−0.1366**	−0.7774**
AR [16]		0.0151*	0.0863**
MA [1]	0.5543**		0.8802**
MA [21]	0.1945**		
Constant	3.7288**	1.0459**	−0.1188
Variance term			
ARCH	0.2128**	0.1667**	0.2323**
GARCH	0.7346**	0.8225**	0.7323**
Constant	0.2351**	0.1833**	0.6245**

* Significant at 0.05 confidence level
** Significant at 0.01 confidence level

the wind speed and wind direction are obtained and converted to the variance for providing input for the GARCH component of the ARMA-GARCH model. The ARMA component of the ARMA-GARCH model is used for predicting wind speed, whereas the GARCH component is used for predicting wind volatility.

Two different ARMA-GARCH models are compared. The first model involves using the ARMA-GARCH model for the wind speed. The second model involves employing two separate ARMA-GARCH models to model the lateral and longitudinal components. After the lateral and longitudinal components and associated volatility are predicted, as previously mentioned, the two components are combined together to form a single value for the wind speed associated with the corresponding volatility. For identification of model structure, the minimum information criterion is used. Based on the moving average and autoregressive values, the ARMA-GARCH models are formed, and the parameter values for each GARCH model are given in Table 9.1.

9.4 Results and Discussion

The first 420,000 minutely data points are used for model building (i.e., the period between 01/01/2012–18/10/2012). The hourly values are computed based on the minutely data, which resulted in 7,000 hourly data values. After the models are built, the remaining data points (i.e., 93,600 minutely data points, and 1,560 hourly data points) are used for the evaluation purposes. For this purpose, the mean absolute error (i.e., MAE) values are used. Table 9.2 represents the performance measures with respect to both ARMA-GARCH processes.

By examining the results, it can be seen that the standalone ARMA-GARCH model performs better as compared with the component model for forecasting the mean wind speed. The MAE value for the standalone ARMA-GARCH model is

Table 9.2 Performance comparison of the two ARMA-GARCH methodologies

	MAE (for wind speed)	MAE (for variance)
Component ARMA-GARCH	1.493	1.6438
Standalone ARMA-GARCH	1.3525	1.7018

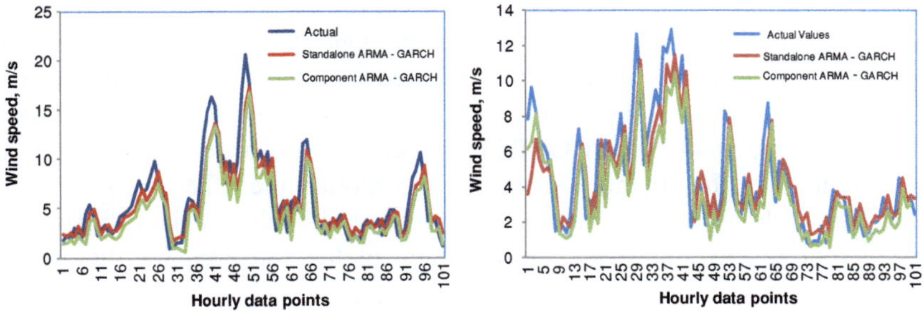

Fig. 9.2 Comparison of actual and predicted for wind speed values for two samples of 100 hourly data points

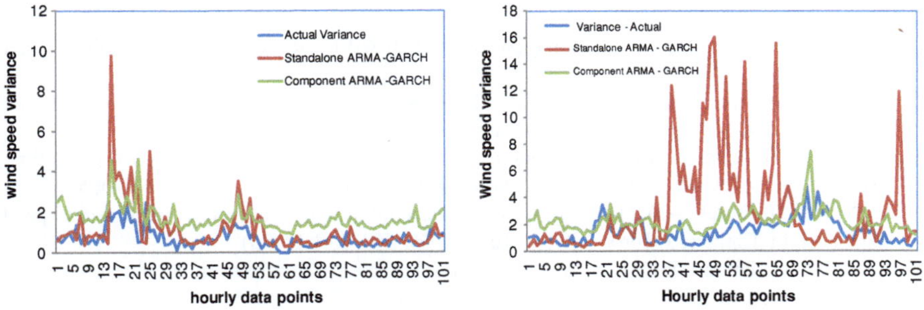

Fig. 9.3 Comparison of actual and predicted for wind sped variance for two samples of 100 hourly data points

1.3525, while the component ARMA-GARCH model has a value of 1.493. This corresponds to an improvement of approximately 7.39 % on forecasting accuracy. However, if the variance terms are considered, the component ARMA-GARCH model has the upper hand. The MAE value for the variance turns out to be 1.6438 for the component ARMA-GARCH model, while the value is 1.7018 for the standalone ARMA-GARCH model. As such, the component model is more accurate by about 3.4 % for predicting the volatility of wind speed.

In order to visually analyze the forecasted variance versus the predicted values for the variance model, the graphs pertaining to actual and predicted values for the component and standalone ARMA-GARCH models are drawn. The mean wind speed and wind volatility are shown in Figs. 9.2 and 9.3, respectively. Two representative samples that have a length of 100 data points are shown to provide the comparison of the proposed two models. Basically, the observations made from Table 9.2 can be verified by the two figures.

9.5 Conclusion

In this paper, we investigate the feasibility of predicting the mean and volatility of wind speed based on component modeling approach. The wind vector is decomposed into lateral and longitudinal components based on wind direction. The two components are predicted and combined back to form a single wind forecast. The results are compared with the traditional standalone ARMA-GARCH model for modeling the mean wind speed and the variance simultaneously. It shows that the standalone ARMA-GARCH model is more accurate for predicting the wind speed, whereas the component ARMA-GARCH model performs better for predicting the wind variance. In the future, we plan to incorporate the other model building approaches such as multivariate auto regressive models for investigating the possibility of further improving the accuracy of the forecasts for the ARMA-GARCH models.

References

1. American Wind Energy Association (AWEA) (2013) AWEA 4th Quarter 2012 Public market report
2. U.S. Department of Energy (DOE) (2008) 20 % Wind Energy by 2030. http://www.nrel.gov/docs/fy08osti/41869.pdf. Accessed 20 Aug 2013
3. M. Fuentes, L. Chen, J. Davis, G. Lackmann, Modeling and predicting complex space-time structures and patterns of coastal wind fields. Environmetrics **16**(5), 449–464 (2005)
4. H. Nfaoui, J. Buret, A.A.M. Sayigh, Stochastic simulation of hourly average wind speed sequences in Tangiers (Morocco). Sol. Energy **56**, 301–314 (1995)
5. J. Torres, A. Garcia, M. Deblas, A. Defrancisco, Forecast of hourly average wind speed with ARMA models in Navarre (Spain). Sol. Energy **79**(1), 65–77 (2005)
6. J. Zhou, J. Shi, G. Li, Fine tuning support vector machines for short-term wind speed forecasting. Energy Convers. Manag. **52**(4), 1990–1998 (2011)
7. G. Li, J. Shi, Bayesian adaptive combination of short-term wind speed forecasts from neural network models. Renew. Energy **36**(1), 352–359 (2011)
8. J. Shi, J. Guo, S. Zeng, Evaluation of hybrid forecasting approaches for wind speed and power generation time series. Renew. Sustain. Energy Rev. **16**(5), 3471–3480 (2012)
9. R.F. Engle, GARCH101: The use of ARCH/GARCHmodels in applied econometrics. J. Econ. Perspect. **15**(4), 157–168 (2001)
10. R.S.J. Tol, Autoregressive conditional heteroscedasticity in daily wind speed measurements. Theoret. Appl. Climatol. **56**(1–2), 113–122 (1997)
11. B.T. Ewing, J.B. Kruse, J.L. Schroeder, D.A. Smith, Time series analysis of wind speed using VAR and the generalized impulse response technique. J. Wind Eng. Ind. Aerodyn. **95**(3), 209–219 (2007)
12. B.T. Ewing, J.B. Kruse, M.A. Thompson, Analysis of time-varying turbulence in geographically-dispersed wind energy markets. Energy Sources Part B **3**, 340–347 (2008)
13. J.E. Payne, B. Carroll, Modeling wind speed and time-varying turbulence in geographically dispersed wind energy markets in China. Energy Sources Part A **31**(19), 1759–1769 (2009)
14. J.E. Payne, Further evidence on modeling wind speed and time-varying turbulence. Energy Sources Part A **31**(13), 1194–1203 (2009)

15. K.V. Mardia, P.E. Jupp, Directional Statistics (Wiley Series in Probability and Statistics) (Wiley, New York, 2008)
16. J.A. Carta, P. Ramirez, C. Bueno, A joint probability function of wind speed and direction for wind energy analysis. Energy Convers. Manag. **49**, 1309–1320 (2008)
17. R. García-Rojo, Algorithm for the estimation of the long-term wind climate at a meteorological mast using a joint probabilistic approach. Wind Eng. **28**(2), 213–223 (2004)
18. R. Weber, Estimator for the standard deviation of wind direction based on moments of the Cartesian components. J. Appl. Meteorol. **30**(9), 1341–1353 (1991)
19. R.A. Johnson, T. Wehrly, Measures and models for angular correlation and angular–linear correlation. J. Roy. Stat. Soc. B **39**(2), 222–229 (1977)
20. E. Erdem, J. Shi, ARMA based approaches for forecasting the tuple of wind speed and direction. Appl. Energy **88**(4), 1405–1414 (2011)
21. T. Bollerslev, Generalized autoregressive conditional heteroskedasticity. J. Econometrics **31**, 307–327 (1986)
22. R.F. Engle, Autoregressive conditional heteroscedasticity with estimates of variance of United Kingdom inflation. Econometrica **50**(4), 987–1000 (1982)

Chapter 10
Residential Consumption Scheduling Based on Dynamic User Profiling

Federica Mangiatordi, Emiliano Pallotti, Paolo Del Vecchio and Licia Capodiferro

Abstract Deployment of household appliances and of electric vehicles raises the electricity demand in the residential areas and the impact of the building's electrical power. The variations of electricity consumption across the day, may affect both the design of the electrical generation facilities and the electricity bill, mainly when a dynamic pricing is applied. This paper focuses on an energy management system able to control the day-ahead electricity demand in a residential area, taking into account both the variability of the energy production costs and the profiling of the users. The user's behavior is dynamically profiled on the basis of the tasks performed during the previous days and of the tasks foreseen for the current day. Depending on the size and on the flexibility in time of the user tasks, home inhabitants are grouped in, one over *N*, energy profiles, using a k-means algorithm. For a fixed energy generation cost, each energy profile is associated to a different hourly energy cost. The goal is to identify any bad user profile and to make it pay a highest bill. A bad profile example is when a user applies a lot of consumption tasks and low flexibility in task reallocation time. The proposed energy management system automatically schedules the tasks, solving a multi-objective optimization problem based on an MPSO strategy. The goals, when identifying bad users profiles, are to reduce the peak to average ratio in energy demand, and to minimize the energy costs, promoting virtuous behaviors.

F. Mangiatordi (✉) · E. Pallotti · P. Del Vecchio
Department of Engineering, University of Roma TRE, Via della Vasca Navale 84 00146 Roma, Italy
e-mail: fmangiatordi@ieee.org

E. Pallotti
e-mail: e.pallotti@ieee.org

P. Del Vecchio
e-mail: paolo.delvecchio@uniroma3.it

F. Mangiatordi · E. Pallotti · L. Capodiferro
Department of Information Processing and Management General Requirements, Fondazione Ugo Bordoni, Viale del Policlinico 147 00161 Roma, Italy
e-mail: lcapodiferro@fub.it

A. Y. Oral et al. (eds.), *International Congress on Energy Efficiency and Energy Related Materials (ENEFM2013)*, Springer Proceedings in Physics 155,
DOI: 10.1007/978-3-319-05521-3_10, © Springer International Publishing Switzerland 2014

10.1 Introduction

Recent studies on energy consumptions reveal the high power demand of electrical appliances in buildings sector, which includes single- and multi-family residences and commercial buildings [1]. Accordingly, it becomes important the design and the implementation of integrated intelligent systems and networks of sensor to monitor and to optimize the consumption of energy made by households electric devices. The general goal is to limit the environmental impact and to improve energy efficiency.

To achieve the energy saving is necessary the direct involvement of the end-users in the process of load control by means of appropriate pricing strategies and planning of their electrical tasks. Dynamic pricing, implemented on a daily basis, reflects the variation in the cost of energy production and the variations of the demand for electricity hour by hour. On the other hand, the stimulus of increased economic burden, in terms of high generation and operational costs, encourages the consumers to reduce the domestic appliances use and to shift them to more convenient time during the day. This results in an overall improvement of efficiency and peak to average ratio of electricity demand inside buildings.

The variety of patterns of energy households consumptions leads to: analyze the data of power requirements of electrical appliances, extract typical features and cluster the users respect to their pattern of energy consumption, to develop new pricing strategies and to take advantages from the competitive electricity market [2]. Morover user profiling can be helpful setting automatically system parameters in distribution planning and demand management. If accurate estimates can be made for the maximum or minimum load of each hour, day, month of the year, the utility companies can make significant economies in setting the operating reserve, in maintenance scheduling, and in fuel inventory.

This paper deals with an intelligent energy management system able to control the day-ahead electricity demand in a neighborhood by automatically scheduling the households tasks through a multi-particle swarm optimization strategy. The considered objective functions are the reduction of peak load of the power system and the minimization of the bill paid by the electricity consumers when a dynamic pricing is applied. In this study the profiling of the users is employed to apply different pricing strategies. These strategies may include price incentives or penalties according to the class of profile to which a user belongs.

10.2 Dynamic User Profiling

Various studies have shown that the dynamic tariffs tied closely to the peak load and variations in the marginal cost of generating electricity, are necessary conditions to incentivize the users to proactively contribute to the improvement of energy efficiency adjusting their consuming habit and shifting their demands from peak to off-peak periods [3–5]. This motivate the adoption of pricing structures that increase or decrease the economic burden for each time period on the basis of the cost of generating electricity and the features of the electricity load profile of each groups of users.

For a specific geographical region the classification of the energy load profiles can be adaptively created by using the data about the power needs of daily tasks performed by the users. Moreover this work considers the temporal availability for the execution of each task.

Let $\{TS_{j,i}, j = 1 \ldots k_i\}$ be the set of j daily tasks of the user i of a given buildings sector. Assuming the day divided in T equal time period, each task can be associated with two vectors that define its power requirements and temporal attributes as follows:

$$\overline{X_{TS_{ji}}} = \begin{bmatrix} x^1_{TS_{ji}} & \ldots & x^t_{TS_{ji}} & \ldots & x^T_{TS_{ji}} \end{bmatrix} \tag{10.1}$$

$$\overline{Z_{ji}} = \begin{bmatrix} E_{ji} & d_{ji} & tp_{ji} & ts_{ji} & tf_{ji} \end{bmatrix} \tag{10.2}$$

In (10.1) $\overline{X_{TS_{ji}}}$ is the pattern of electrical consumption to perform the task TS_{ji}; $E_{ji} = \sum_{t=1}^{T} X_{TS_{ji}}$ is the electric power required to perform the task; $d_{j,i}$ is the duration of the task TS_{ji}; tp_{ji} is the start time preferred by user i to run the task TS_{ji}; $[ts_{ji}, tf_{ji}]$ is the maximum time window admitted to run the task TS_{ji}.

The daily load profile of each user i for each time period can be represented by the vector of total energy consumption over the time slots as:

$$\overline{l_i} = \begin{bmatrix} l^1_i & \ldots & l^t_i & \ldots & l^T_i \end{bmatrix} \tag{10.3}$$

where $l^t_i = \sum_{\{TS_{ji,j=1\ldots k_i}\}} x^t_{TS_{ji}}$ is the total load of user i for each time slot. If n consecutive day are considered, with $n = -1, -2, \ldots, -m$, the classification of energy profile of a user i can be done considering two kind of attributes: the *temporal availability* A^n_i defined in (10.4) and the *daily energy demand* ED^n_i defined in (10.5).

$$A^n_i = \frac{1}{k^n_i} \sum_{j=1}^{k^n_i} \left(tf^n_{ji} - ts^n_{ji} - d^n_{ji} \right) \tag{10.4}$$

$$ED^n_i = \sum_{j=1}^{k^n_i} E^n_{ji} \tag{10.5}$$

The clustering method is based on the k-means algorithm [5] choosing a Euclidean distance in the space of attributes. This unsupervised technique partitions the objects (represented by the tuples of the attributes) into N groups so that objects within a group (cluster) are similar to one another and dissimilar to objects

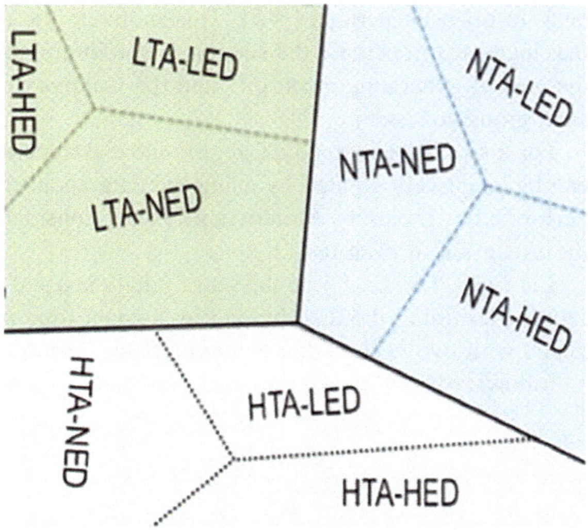

Fig. 10.1 Hierarchical clustering of the user load profiles

in other clusters. Similarity is defined in terms of Euclidean distance in the *n*-dimensional space of attributes.

This work, it is adopted a hierarchical clustering based on the following two steps:

- Partitioning of the energy consumers into three exclusive clusters depending on the temporal availability A_i^n; These clusters are classified and ordered on the basis of the decreasing values of their centroids to assign them the labels: *High Temporal Availability (HTA), Normal Temporal Availability* (*NTA*), *Low Temporal Availability* (*LTA*);
- Partitioning of the users of each cluster according to the daily energy demand attributes ED_i^n in order to identify three clusters, labeled *High Energy Demand* (*HED*), *Normal Energy Demand* (*NED*), *Low Energy Demand* (*LED*), on the basis of the increasing values of their centroids.

The two set of labels {LTA, NTA, HTA} and {HED, NED, LED} categorizes nine class of user's load profile as shown in Fig. 10.1. Each class of user is subjected to a economic incentive or penalty when the daily bill is computed, as explained in the subsequent section.

10.3 MPSO Planning of User Consumption

The Energy Management System (EMS) of residential area schedules the daily tasks $\{TS_{j,i}\}$ with the aim of reducing the pick loads in energy demand and the bill paid by each users. The resulting multi-objective optimization problem is solved

Table 10.1 Economic incentives/penalties for each cluster of user load profiles

HTA–LED	HTA–NED	HTA–HED	NTA–LED	NTA–NED	NTA–HED	LTA–LED	LTA–NED	LTA–HED
−0.3	−0.2	0.1	0.2	0	−0.2	0.1	0.2	0.3

with a heuristic approach running a MPSO algorithm. The MPSO method mimics the social behavior of a bird flock finding the food. Each individual moves in a multidimensional solution space and is characterized by a *position* (X_i) and velocity (V_i). The position represents a possible solution of the optimization problem while the velocity is the rate of change of the current position. The MPSO technique implements the concept of *Pareto Dominance*. So a solution X_1 is said to dominate another solution X_2 if the following conditions are true:

- the solution X_1 is no worse than X_2 in all the objectives functions;
- the solution X_1 is strictly better than X_2 in at least one objective function value.

In this paper the multi-objective scheduling problem is formulated considering two objective functions. The first function represents the daily electricity cost and can be written as :

$$\begin{aligned} f_1 &= \sum_i \sum_t CoE_{t,p} \cdot l_i^t \\ CoE_{t,p} &= \alpha_t B_{pi} \\ B_{pi} &= 1 \pm \delta_{pi} \end{aligned} \tag{10.6}$$

In (10.6) α_t is the retail price of electricity and is defined by energy provider; B_p denotes the economic incentive/penalty according with the class of dynamic load profile of user; In fact, δ assumes one of the values in Table 10.1.

The second objective function is a measure of the smoothness of the energy demand and is expressed as:

$$\begin{aligned} f_2 &= \max\left\{\sum_t (AvgP - P^t)^2\right\} \\ P^t &= \sum_t l_i^t \\ AvgP &= \frac{1}{T}\sum_t P^t \end{aligned} \tag{10.7}$$

The MPSO algorithm uses an external repository in which the Pareto flight experiences (non dominated solutions) are stored. The external archive is employed to identify a leader to guide the search.

The main steps of MPSO are described in the rest of this section.

MO PSO Algorithm

Initialize population and the external repository by:

- *Setting randomly the initial positions of each individual of the swarm;*
- *Setting to zero the initial velocity of the swarm;*
- *Computing the position (i.e. the fitness values) of each individual in objective space;*
- *Archiving the non dominated-solutions;*
- *Generating the hyper-cubes of the explored search- space by the division of the objective axes in K equal intervals;*

While the maximum iteration or ideal fitness is not attained do:

- *Evaluate the velocity of each individual, Vi, and new position with the expressions*

$$\begin{cases} V_i = \alpha V_i + c_1 \cdot rand_1(\cdot)(P_i - X_i) + c_2 \cdot rand_2(\cdot)(REP_h - X_i) \\ X_i = X_i + V \end{cases} \quad (10.8)$$

- *If the new position falls out the admissible regions for the solution space, locate it on the boundary of the admissible regions and reverse the direction of the velocity;*
- *Apply the mutation operator to each individual;*
- *Evaluate the new positions in objective space and update the external repository with non-dominated solutions;*

10.4 Case Study and Experimental Results

The performance of the presented energy management system (EMS) is tested considering a case study of a residential area with 272 household users. Each building is equipped with a terminal unit controller (TCU), which monitors the consumptions of the electrical devices and transmits them to EMS. In each building the user programs from 10 to 15 electrical tasks selected randomly from Table 10.2. The EMS schedules the task obtaining an average reduction of peak to average ratio of 18 % and of the electricity bill of 20 %. In Fig. 10.2 it is shown the energy demand of the area when the 2,600 different electrical tasks are planned. The red curve represents the energy demand of the area when the tasks are efficiently scheduled by EMS.

Table 10.2 Electrical appliances

Task	Power (kW/h)	ts	tf	d (h)
Dish washer	1.0	7	19	3
Washing machine	1	6	24	2
Spin dryer	2.9	13	18	1
Cooker hob	3.0	7	9	1
Cooker oven	1.7	18	21	1
Cooker microwave	1.7	6	9	1
Interior lighting	0.2	18	24	6
Laptop	0.1	18	24	2
Desktop	0.3	18	24	3
Fridge	0.03	0	24	24
Electrical car	3.5	18	8	4
Boiler	0.8	15	22	2
Iron	1.2	10	19	
Air conditioning system	1.5	10	16	4
Vacuum cleaner	1.2	9	17	1

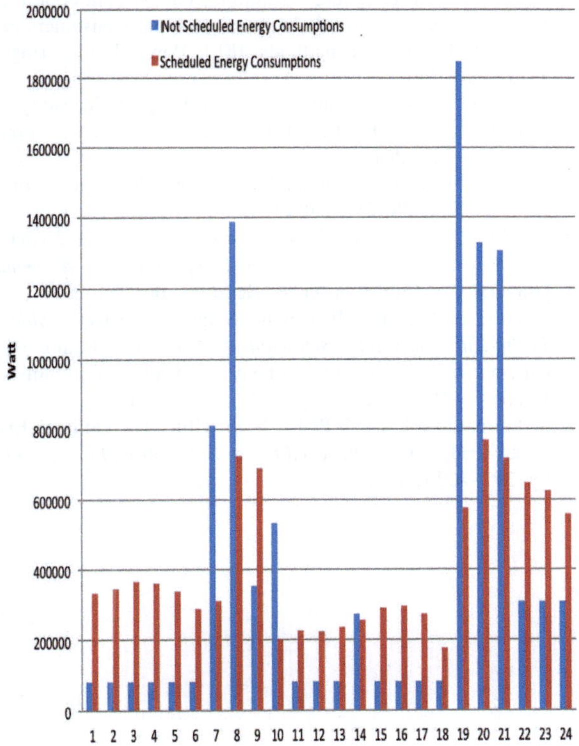

Fig. 10.2 Energy demand of residential area with scheduled tasks and clustering

10.5 Conclusions

This paper presents an intelligent energy management system for the efficient control of the electricity demand in residential buildings. The proposed strategy is based on MPSO algorithm and adopts dynamic pricing. Consumptions data of electrical appliances are used to classify the load profiles of users with k-mean technique. Based on the class of user profiles, the Energy Management System modulates daily the cost of energy provided by the utility company, to encourage efficient behaviors in electricity consumptions. Simulations on realistic situation employing practical cost functions show a sensible reduction of energy payments to utility company as well as a reduction of the peak to average ratio in demand curve. This derives clear advantages for the environment, by preventing the carbon emissions due to the installation of new generation power plants.

References

1. US Department Of Energy: Buildings energy data book (2011)
2. D. Gerbec, S. Gasperic, I. Smon, F. Gubina, Consumers load profile determination based on different classification methods. IEEE Power Engineering Society General Meeting, vol. 2, 13–17 July 2003
3. P. Joskow, C.D. Wolfram, Dynamic pricing of electricity. Am. Econ. Rev. (2011)
4. A. Conejo, J. Morales, L. Baringo, Real-time demand response model. IEEE Smart Grid Trans. **1**(3), 236–242 (2010)
5. M.A. Crew, C.S. Fernando, P.R. Kleindorfer, The theory of peak-load pricing: a survey. J. Regul. Econ. **8**, 215–248 (1995)
6. J.B. MacQueen, Some Methods for Classification and Analysis of Multivariate Observations, in *Proceedings of 5th Berkeley Symposium on Mathematical Statistics and Probability*. University of California Press, Berkeley, pp. 281–297
7. J. Torriti, M. Leach, P. Devine-Wright, in *Demand Side Participation: Price Constraints, Technical Limits and Behavioural Risks*. The Future of Electricity Demand: Customers, Citizens and Loads. Department of Applied Economics Occasional Papers (Cambridge University Press, Cambridge, 2011), pp. 88–105
8. S. Salinas, L.M. Li, L.P. Li, Multi-Objective Optimal Energy Consumption Scheduling in Smart Grid, in *IEEE Proceedings of International Conference on Smart Grid Communications*, pp. 397–402 (2010)

Chapter 11
Transforming a Brutalist Monument into an Energy Efficient Building Without Destroying the Formal Appealing: The Example of the Mediterranean Bank in Potenza (Italy)

Filiberto Lembo

Abstract In years 1980 the "*brutalist*" and "*monumentalist*" architecture went even in Italy, and in Potenza, and it finds some interesting examples, such as the building designed as the home of the Mediterranean Bank. A monumental building, all in reinforced facing concrete and curtain walls, with refined proportional relationships, and so devoid of insulation (annual heat demand of 69 kWh/m^3 year), that its management has become unbearable from the economic point of view, so it was abandoned a few years ago. Aim of this work was to define a design methodology that preserves all the qualities of architecture, but at a cost that is economically bearable and made the building energetically efficient. This was done by allying a very thick and efficient isolation, protected by a ventilated continues rainscreen, finished with several layers of a thin plaster with a *bèton*-like effect, which determines a morphology that, while different, recalls the original. The curtain walls were doubled with a double skin façade, whose performance has been optimized with a purpose created software. The huge skylight roof of the interior atrium has been doubled with a new, of different trend and thermally more effective. The roof was covered with photovoltaic panels. The result is an annual heat and refrigeration demand of 17 kWh/m^3 year, at a cost of 20,000 €/m^2, quickly depreciating, due to savings in operating costs for air conditioning (from €104,800.00 to €14,700.00 per year).

F. Lembo (✉)
School of Engineering, University of Basilicata, Viale dell'Ateneo Lucano n.10, 85100 Potenza, Italy
e-mail: filiberto.lembo@unibas.it

A. Y. Oral et al. (eds.), *International Congress on Energy Efficiency and Energy Related Materials (ENEFM2013)*, Springer Proceedings in Physics 155,
DOI: 10.1007/978-3-319-05521-3_11,

11.1 Introduction

In the 80s the Brutalist movement became famous in Italy, it was born in England in 1954 with Alison and Peter Smithson, theorized by Reyner Banham [1], and practiced by Le Corbusier, Atelier 5 and Denys Lasdun (in Italy by Vittorio Viganò and Carlo Scarpa) and spread with the accent given to it by Paul Rudolph, who contaminated the surface with monumentalist suggestions derived from the work of Louis Kahn, and turned it into an easy to learn and use language, especially with its appealing and Art and Architecture Building in New Haven (1963).

The headquarter of Banca Mediterranea, designed in 1976 by Arch. Dante B. Maggio, and built in the years between 1982 and 1985, incorporates these suggestions, in the clear contrast between solid walls in "*béton brut*" and curtain wall façades in black aluminum and glass mirrors, and it joins the explicit reference to the nearby castle of Lagopesole built by the Normans and later enlarged by Frederick II of Swabia, of which it has the orientation, the module design of the "cubit" (55 cm) and the four corner towers (Figs. 11.1, 11.2).

The building (with a surface bounding the heated volume of 6,410 m^2, a heated gross volume of 20,500 m^3—S/V = 0.31—and a useful floor area of 11,400 m^2), spreads around the public room, which is triangular and three floors high, covered by a skylight of degrading cable cubic elements made of stainless steel, covered with nearly horizontal tempered glass (Fig. 11.3) [2].

Only 16 years after its completion, circa during the 2001, the building was abandoned, due to the excessive cost of its exercise. In fact, in a site like Potenza (latitude 40°38′, longitude 15°48′, 819 m osl, 2,472 degree-days, Climatic Zone E, heating days 183, external temperature reference −3 °C), the building is practically devoid of insulation: its reinforced concrete walls, the lower and the upper border have a U of 3.07 W/(m^2 K), its windows a U_w value of 2.64 W/(m^2 K), when 5.5 W/(m^2 K) as in the case of the stained-glass window on the lobby, and its annual heat demand is 69 kWh/m^3 year. In addition, the covers have proved to be poorly designed and have become a quagmire, with losses in several places, as well as the stained glass windows of the lobby coverage (Fig. 11.4a, b), while the low concrete reinforcement cover resulted in spalling of the concrete "*béton brut*" at several points (Fig. 11.4c, d).

11.2 Aim of the Research

In the research has been identified a method of intervention to recover and upgrade, especially on the thermal point of view, all cases of "*brutalist*" architecture, which in itself Potenza and its province count several schools, a psychiatric hospital, and the Court (buildings which, in greater or lesser extent, have the same problems of the building under study), with the constraint to obtain a formal result that would preserve the values of the textural image of the building.

Fig. 11.1 View of the North-East tower

Fig. 11.2 Southern view of the building

Fig. 11.3 Views of the central atrium glass roof

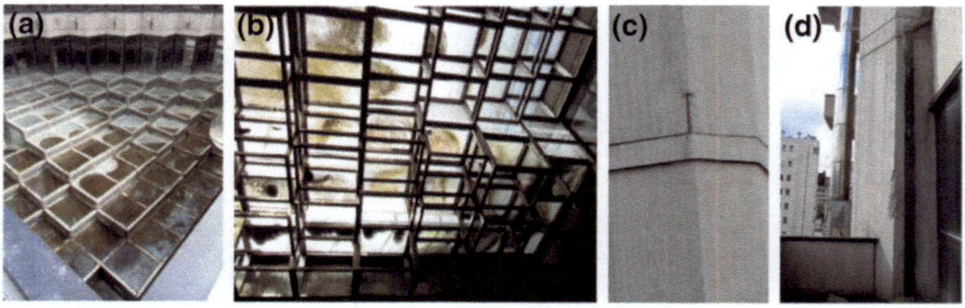

Fig. 11.4 Coverage degradation (view from *outside* and *inside*, first and second image from the *left*); corrosion of reinforcing steel (third image from the *left*); exposure of reinforcing *bars*

11.3 Content of the Research

The first requirement was to provide a heavily insulated and ventilated casing, in order to allow the heating and cooling of the building (air conditioned) as possible in a "passive" and "natural" manner, enhancing the great thermal inertia of its heavy concrete walls (750 kg/m^2). Therefore we started looking on the building systems market for a more advanced rainscreen: a continuous ventilated (to remove the heat irradiation summer), and durable one, which could materially be treated in a similar way to a concrete-face view. After searching among all products certified by the Official Laboratories, which are available on the international market, we have identified as most suitable to solve the problems of the particular case the StoVentec Render system, object of the Approval of Construction General by the Deutsches Institut fur Bautechnik Berlin (Zulassung) Z-33.2-394. It is based on the use of sustaining plaster recycled glass plates, 1.2 cm thick, screwed with stainless steel screws to a substructure in extruded aluminum profiles, supported by stainless steel shelves, such as hinges or operated trolleys and connection to the support wall (Fig. 11.5).

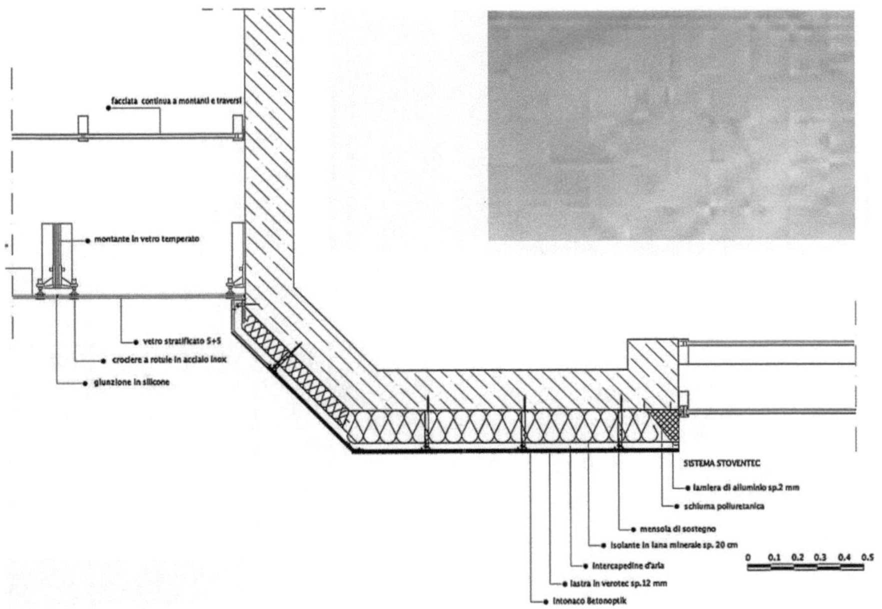

Fig. 11.5 Recovery project: double skin façade; Betonoptik finish

On the plates is arranged an organic plaster, armed with a glass fibers network, which provides, among its different finishes, that "Betonoptik" (Fig. 11.5), which is obtained by arranging on the StoArmat reinforcement render a fine grain finish render Stolit MP 1.5 K in color of reinforced concrete (RAL 7023) and, on this, a further finishing plaster Stolit Milan, treated with a special technique of smoothing with steel spatula and subsequently sanded, obtaining a finish that, both in the views from afar that in those closely, is very similar in color, grain and texture, to "*béton brut*" (Fig. 11.6).

Naturally, the ventilated rainscreen is mounted after the concrete surface is cured flaking with the usual methods for the rehabilitation of reinforced concrete surfaces, and the application with plugs of 20 cm of mineral wool 035, with which is obtained a U of 0,166 W (m^2 K). Regarding the curtain wall surfaces, leaving in place the existing ones, it is expected to double them, forming a ventilated practicable interspace, of the double skin façade type, useful both to prevent overheating in the summer, which to reduce winter heat losses.

With regard to the DSF, the author has developed a calculation software [3], which allows to evaluate their optimum configuration, in relation to the schemes of operation, and that has been implemented for this specific restoration project. In this case, it is optimized in the width 60 cm, and the DSF is sectioned horizontally, in correspondence of each floor, for partitioning it in function of fire protection (Fig. 11.7). The result is a dramatic improvement in performance, both summer and winter, of the transparent surface as measured by ECOTECT.

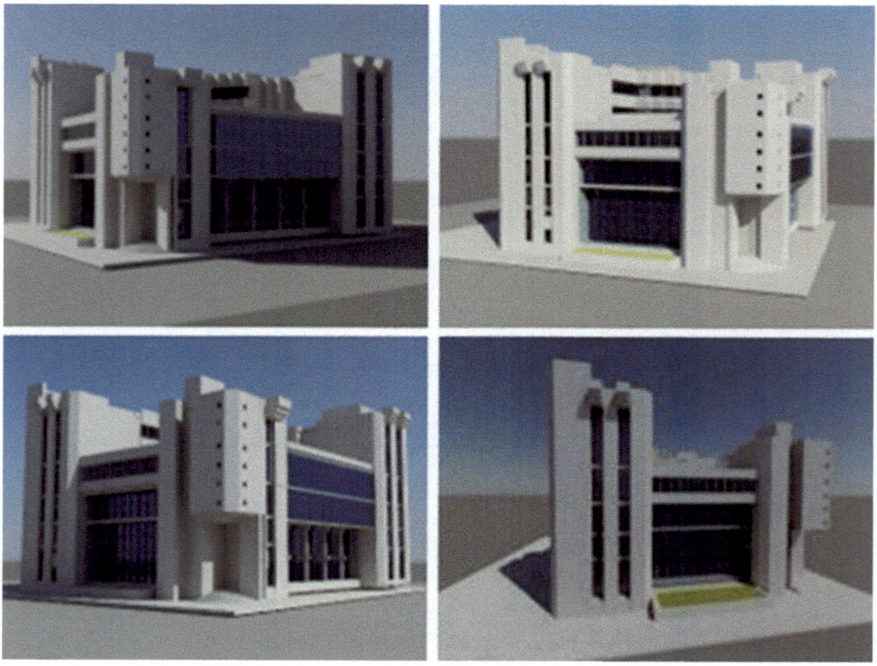

Fig. 11.6 Recovery project: axonometric views

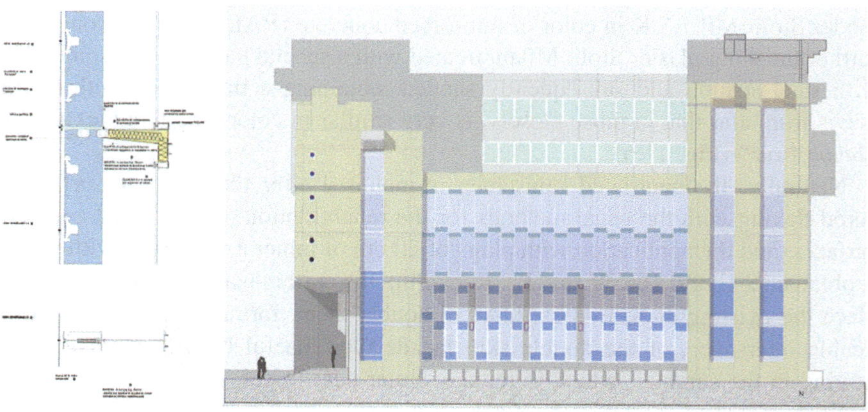

Fig. 11.7 Recovery project: *vertical* and *horizontal* section of double skin façade (on the *left*); overall view of the West elevation (on the *right*)

It was also possible to add the production of energy, using photovoltaic panels inserted anywhere architecture permits. To solve the problems of the glass roof of the central atrium, it is expected to leave in place the existing one, after it has been

Fig. 11.8 Recovery project: glass roof of the central atrium

cleaned, and to realize a new cover, different and softer in terms of configuration, completely outside and above the atrium, which can download the rainwater directly on the cover, with additional downpipes (Fig. 11.8), and that it connotes as current intervention. Although this intervention made with high-performance glazing, allows realizing fundamental reductions in consumption and improvements in the performances in summer and winter. On covers, rehabilitated and isolated until U = 0.16 W/(m^2 K), the placement of solar p.v. panels has been scheduled until the free-sun surface. The replacement of the heat generator has been provided with a more modern and efficient, and the revision of the refrigeration units.

11.4 Conclusion

The building, like healed, is in Class A, with an Epi 17 kWh/m^3 year. Expected the exemplary value of the intervention, was carried out a careful assessment of their economic costs and depreciation. Its consumption of natural gas decreased from 141,033 to 14,698 kWh/year. The savings of methane is €90,089.17 per year, a savings compounded over 15 years of €1,385,987.30.

The duration of life of the building passes from 90 to 120 years. The building has a current market value of €7,914,000.00. The processing cost is €2,250,000.00

The transformation value is €4,679,000.00. The project is therefore economically viable and beneficial. And above all, perfectly compatible with the conservation of the formal characteristics of the building, an important element of the architectural heritage of the 900 in the city of Potenza (Italy).

References

1. R. Banham, The new brutalism. Archit. Rev. **708**, 357 (1955)
2. D.B. Maggio, *Il palazzo della Banca Popolare di Pescopagano 1976–1985, testi di Gillo Dorfles e Benedetto Gravagnuolo* (Mazzotta Editore, Milano, 1985)
3. F. Lembo, F.P.R. Marino, G. Lacava, in *Double Skin Façades: Definition of an Ideal Depth for Cavities, 4th International Research Symposium (SCRI)*, ed. by C. Abbott, G. Aquad, M. Kagioglu, L. Ruddock (Salford (Greater Manchester), University of Salford, UK, 2007), pp. 147–160, ISBN 9781905732210
4. F.P.R. Marino, M. Grieco, *La certificazione energetica degli edifici. D.Lgs. 192/2005 e 311/2006—IV edizione aggiornata alle UNI TS 11300—Algoritmi di calcolo ed esperienze internazionali. Edifici ad alta efficienza* (EPC Libri, Roma, 2009), p. 720, ISBN 978-88-6310-113-3

Chapter 12
The Re-invention of the Tower House for the Construction of Green Buildings NZEB, Integrated With the Vertical Axis Small Wind System

Francesco Paolo R. Marino

Abstract Nowadays the cultural and economic context aims to create a sustainable "carbon zero" society through energy-efficient green buildings NZEB, but it has so far overlooked a construction type widely spread throughout Europe, especially in the Middle Ages, and that in Italy still characterizes the most beautiful landscapes of Tuscany and other cities: the tower-house. The aim of the research was to verify the possibility of reinventing the type of the familiar tower-house, which is intrinsically directed to conquer the height and therefore higher wind conditions, assuming the installation on the top of a small wind system to use wind energy, to make the building energetically self-sufficient. This building is designed from a wooden structure of a deciduous tree widespread in the Italian region of Basilicata, the Turkish Oak, which, subject to processes of hygrothermal conditioning, can be transformed into the base material to compose laminated timber beams and pillars, able to guarantee a load of exercise, to bending stress, equal to 40.9 N/mm^2, as followed by tests in the Laboratory of Engineering of the University of Basilicata, Potenza. With normal wind conditions in the city of Potenza (average of 6.5 m/s), a 5 kW wind turbine mounted at 25 m tall on a 13 m high building is able to provide all the energy the building needs, with its attractive tapered oval top that minimizes turbulence. Entirely made with structures, finishes and natural insulation, the building is a sign in the landscape, history and future together.

F. P. R. Marino (✉)
School of Engineering, University of Basilicata, Viale dell'Ateneo Lucano n.10, 85100 Potenza, Italy
e-mail: francesco.marino@unibas.it

A. Y. Oral et al. (eds.), *International Congress on Energy Efficiency and Energy Related Materials (ENEFM2013)*, Springer Proceedings in Physics 155,
DOI: 10.1007/978-3-319-05521-3_12,

12.1 Introduction

In the landscape of the renewable sources, the exploitation of wind energy presents itself as the most dynamic factor in the market.

In Italy, according to the GSE (Electricity Manager), the target of 26.4 % of electricity from renewable sources by 2020, has already been reached. "*In 2012, the production of electricity from renewable sources was 92,222 GWh, with an installed capacity of 47,345 MW. At the first place hydropower, which provided 41,875 GWh [45.4 %], followed by solar (photovoltaic) with 18,862 GWh [20.5 %], and with 13,407 GWh from wind sources [14.5 %]. Bioenergy have provided 12,487 GWh [13.5 %] and geothermal 5,592 GWh [6.1 %]*".[1]

The Global Wind Energy Council estimated that by 2020 the wind power could easily cover up to 12 % of electricity world production, avoiding the emission of about 10 billion tons of carbon dioxide.

All analysts are convinced that, in the short run, no other renewable sources can offer a contribution on a global scale than wind power in reducing "*climate-altering*" emissions. The spread of such wind power plants is favored, as well as to the possibility of installation of offshore fields (the seas are areas of wind with the greatest speed, even in Italy), by reducing the size of the installations, their typological evolution, with the spread of vertical axis turbines, more easily integrated into a building structure, and their study as an object not only functional, but of design, after which they make up appendages consciously formalized in the profile of the building against the sky (see Fig. 12.1).

There are some research that place in relation the efficiency and performances of different types of wind turbines, depending on the speed of the wind expected in the implantation site, and it is therefore possible to optimize the choice of the system in function of the specific characteristics of the project (Fig. 12.2 displays Johnson's performance comparison of power coefficient versus tip speed ratio for some of the most common wind turbine designs).

12.2 Aim of the Research

The European Community, with the Directive 2010/31/EU, transposed into Italian with DL n. 63 of 4 June 2013, decided that starting from the 1st January 2019 all public buildings, and from the 1st January 2021 all private, must be *Net Zero Energy Building*, which, despite being connected to the network, must have reduced consumption of power for heating, cooling and domestic hot water production, and will have to produce renewable electricity they need for their operation.

[1] www.infoprogetto.it/202020/ date 17/09/2013.

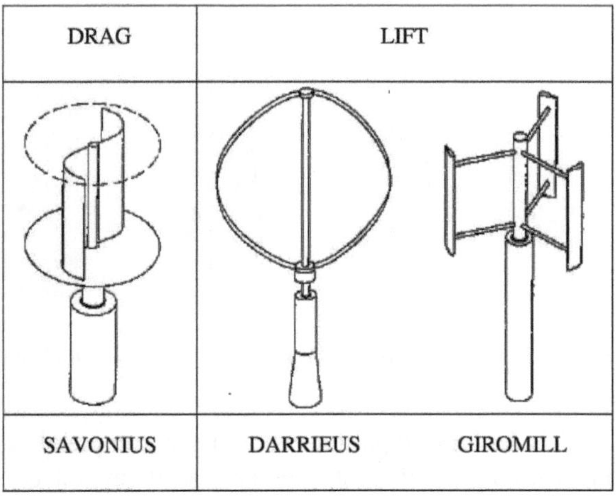

Fig. 12.1 Wind turbine classification according to the mechanism of operation

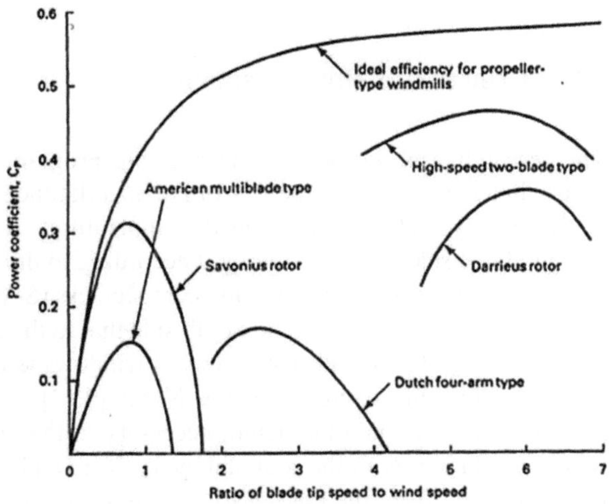

Fig. 12.2 Typical performances of wind energy systems (wind turbine $C_p - \lambda$ comparison) [1]

For this reason it's probable that in the future mini-wind turbines integrated into buildings will develop, and it seems necessary to rethink the building types in this direction, as it has done for the integration of photovoltaic panels in the building volumes.

That purpose became obvious, although until now anyone attempted it, designers have the objective of re-thinking and re-inventing the building type of tower-houses, which grew steadily higher in the Middle Ages both for defense needs and social status, to have a profound effect on the creation of urban landscapes of cities, which are now famous because of their tower-houses, especially in Tuscany and Emilia-Romagna, such as S. Gimignano, Monteriggioni, Lucca, but also Pavia, Bologna, and many other Italian cities.

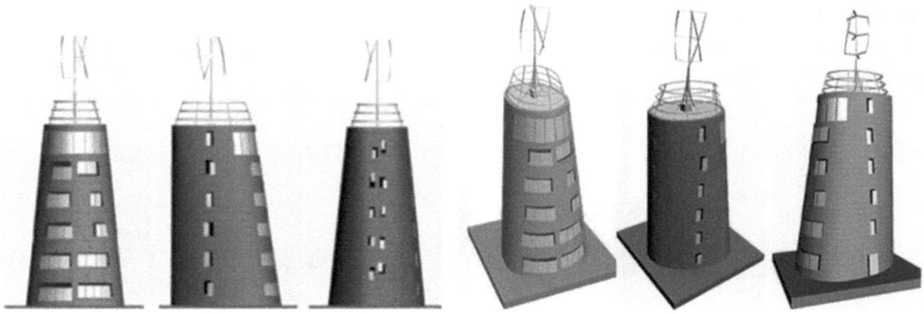

Fig. 12.3 Elevations (from the *left*: Sud, Est-West, Nord) and axonometric views

This building type has been almost completely abandoned after the spread of firearms, in the fifteenth century, as it proved too vulnerable for them, to the benefit of other types, such as that of the palace.

12.3 Content of the Research

The research was focused on the building program of a single-family detached house, in a real context in the city of Potenza (latitude 40°38′43″0.08, climate zone E, degree days 2,472, design outside temperature −3 °C, heating days 183, height 819 m above sea level), designing it according to the principles of *bioclimatic* and *passive solar architecture*: extreme compactness (S/V ratio of 0.5) (see Fig. 12.3), trapezoidal profile, wider in the South side that to the North, to maximize free solar gains (see Fig. 12.4), with large convex windows (supported by external shading) to the South, almost closed in the North [2, 3]; overlap in height of different internal spaces, to gain the height necessary to the optimal operation of the mini-wind installation, with the roof slab at 19.50 m and turbine height of 25 m; ovoid profile tapered from the bottom to the top, to minimize turbulences, and to create advantage to the wind turbine blade operation.

From structural point of view, all design has been performed in accordance with EN 1995-1-1 2004 Code, with antiseismic beam-column wooden structure in Turkish Oak glued laminated timber [4], modified trough a purpose-made thermo-hygrometric treatment, which gives it incredible structural qualities, with a load of exercise, to bending stress, equal to 40.9 N/mm^2, as followed by tests in the Laboratory of Engineering of the University of Basilicata, Potenza, Italy (see Fig. 12.5).

The housing is designed to be that of a “passive house”, with an optimal behavior in both winter and summer (see Fig. 12.6): an envelope of 20 cm oak as blockhouse, 20 cm of super-compressed high density cork, and a ventilated rainscreen for the protection in winter from rain and in summer from the sun, with a U transmittance of 0.167 K/(m^{2}K).

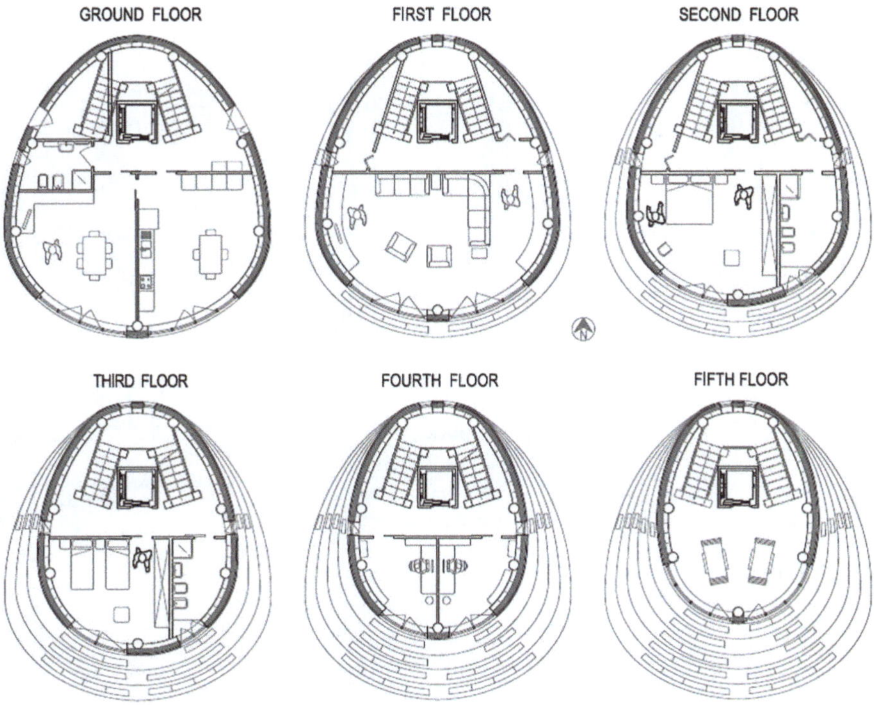

Fig. 12.4 Floors

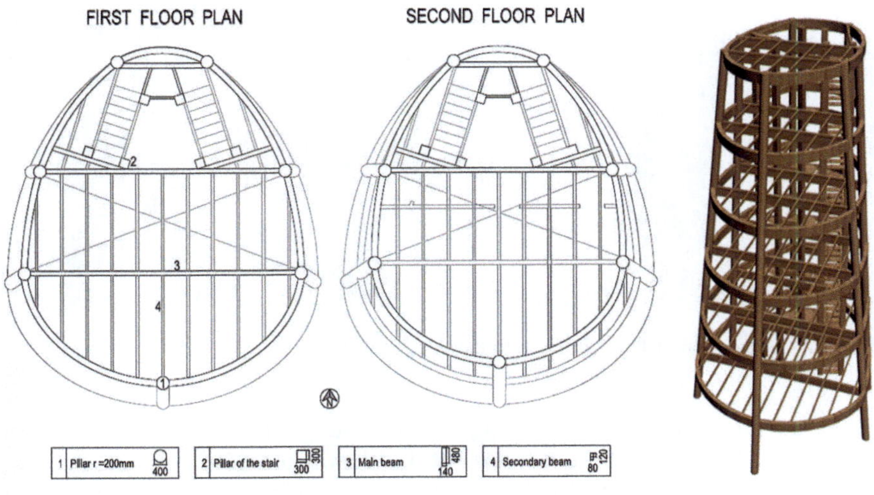

Fig. 12.5 Structure

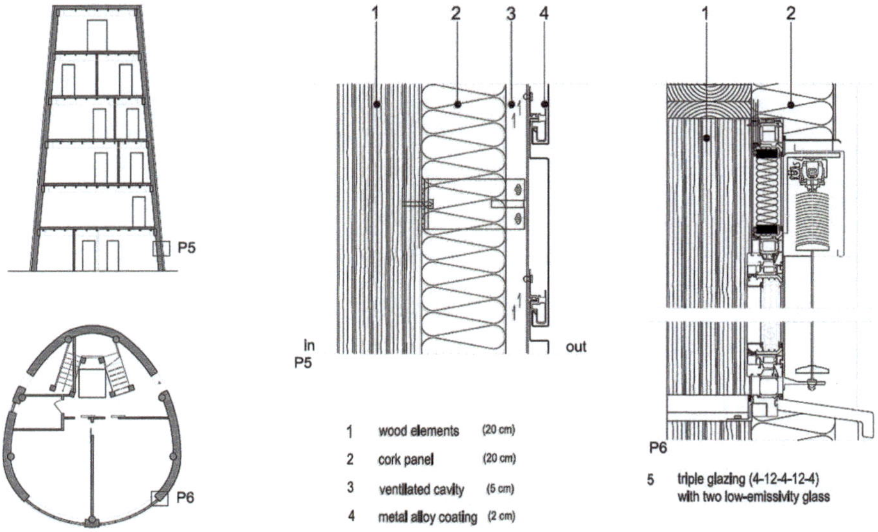

Fig. 12.6 Façade

Fig. 12.7 Integration of the building into the urban context

For the production of heat (heating and hot water), a heat air to air pump, with the yield of 16 kW and the absorption of 4 kW, was provided.

Primary energy demand EPI = 10 kWh/m^2 per year. Annual consumption of energy according to UNI TS 11300, about 3,000 kWh for heating, 3,050 kWh for cooling and 2,370 kWh for the production of domestic hot water [5]. All the requirement may be provided by a wind generator with a rated power of 5 kW, and the surface of 7.50 m^2, which provides an efficiency of 40 %, with the mean wind speed 6.5 m/s, Weibull factor 2.5, and can produce annually, considering this site, 8,550 kWh.

12.4 Conclusion

The research has shown that it is possible to rediscover and enhance the tower-house type to architecturally integrate mini-wind turbines on vertical axis, that are able to produce all the energy the building needs throughout the year and that with the mode of "*on-site metering*", balance the energy produced and fed into the grid with the one supplied from the power-distribution network.

A design careful to the sustainability aspects of the project, both during construction and in the management, and in that of disposal/recycling, allows to minimize the environmental impact of the building (see Fig. 12.7).

References

1. G.L. Johnson, Wind Energy Systems, Electronical Edition. Manhattan, KS, (2001)
2. B. Cody, Low energy apartment building in Berlin. Arup J, ed. by D.J. Brown, 33(3), 14–17 (1998)
3. F. Lembo, F.P.R. Marino, C. Calcagno, Semi underground house models as new concepts for urban sustainable environment. Procedia Eng 21, 570–579. ISSN 1877-7058, doi:10.1016/j.proeng.2011.11.2052 (2011)
4. F. Lembo, F.P.R. Marino, Floors realized in Quercus Cerris (Turkish Oak) laminated timber in buildings with structure in masonry, performing anti-seismic functions. In *8th International Seminar on Structural Masonry (ISSM08)*, ed. by B.P. Sinha, L. Tanaçan, Istanbul Technical University, pp.441-448, ISBN 978-975-561-342-0 (2008)
5. F.P.R. Marino, M. Grieco, La certificazione energetica degli edifici. D.Lgs. 192/2005 e 311/2006—IV edizione aggiornata alle UNI TS 11300—Algoritmi di calcolo ed esperienze internazionali. Edifici ad alta efficienza, *EPC Libri*, Roma, pp. 720. ISBN 978-88-6310-113-3 (2009)

Chapter 13
Analysis and Simulation of Superlattice GaN/InGaN p-i-n Solar Cells

Giovanni Giannoccaro and Vittorio M. N. Passaro

Abstract Indium gallium nitride (InGaN) is becoming a promising semiconductor material for fabrication of solar cells due to its high absorption coefficient (about 10^5 cm^{-1}) and tunable (by its In content) direct band-gap, from 0.71 eV (E_{InN}) to 3.43 eV (E_{GaN}). Solar cells based on structures with variable In content should show a reduction of thermalization losses, absorbing almost the whole of solar spectrum. Unfortunately, these advantages are partially lost and InGaN solar cells performance reduced due to technological issues (low quality of InGaN layer growth on GaN substrate) and wurtzite nature of InGaN semiconductor (spontaneous and piezoelectric polarizations), in particular in case of In-rich ternary alloy. This paper is focused on the parametric analysis and simulation of an intermediate band (IB) superlattice GaN/InGaN p-i-n solar cell on GaN quasi-bulk substrate. Photovoltaic device performances are investigated and simulated as a function of both In concentration in GaN/InGaN quantum wells (QWs) and QW number. For a superlattice with 28 QWs, an efficiency of 1.05 % with a short-circuit current density of 1.193 mA/cm^2 has been achieved. Furthermore, some considerations on the possibility to obtain a more efficient InGaN solar cell with a wider absorption spectrum using InN quantum dots (QDs) in a photovoltaic device structure are also derived.

13.1 Introduction

Today, a reduction in solar energy conversion costs is needed for having a widespread availability of photovoltaic systems for exploitation of solar sustainable energy in the world. It will be possible if solar cell efficiency should increase.

G. Giannoccaro · V. M. N. Passaro (✉)
Dipartimento di Ingegneria Elettrica e dell'Informazione,
Politecnico di Bari, 70125 Bari, Italy
e-mail: vittorio.passaro@poliba.it

A. Y. Oral et al. (eds.), *International Congress on Energy Efficiency and Energy Related Materials (ENEFM2013)*, Springer Proceedings in Physics 155,
DOI: 10.1007/978-3-319-05521-3_13,

Thermalization loss (higher in low band gap semiconductors) and missed absorption of low energy photons (in high band gap materials) represent mean reasons for relatively low efficiency of conventional single junction (SJ) solar cells.

Other possible working principles for semiconductor solar cells, able to reduce these problems, are based on multi-junction (MJ) and intermediate band devices. The former can achieve high efficiency with complex heterostructures as InGaP/GaAs/InGaAs inverted metamorphic triple junction (TJ) solar cells, having world record of 37.7 and 43.5 % without and with use of sunlight concentrators for terrestrial applications [1], respectively. Nowadays, the latter could theorically exhibit high efficiency with one junction, but some issues should be solved yet. Recently InGaN, widely employed in high performance LED and laser fabrication, is becoming a promising material for photovoltaic application for MJ and intermediate band solar cells (IBSCs). In fact, InGaN band gap can be tuned from 0.71 eV (E_{InN}) to 3.43 eV (E_{GaN}) by varying its In content. Then, its absorption spectrum can cover almost the whole solar one.

In this work, firstly InGaN optical and physics properties and some technological issues about its use in solar cells are briefly described. Secondly, a p-i-n GaN/InGaN on a GaN quasi-bulk substrate [2] with a finite GaN/InGaN superlattice intrinsic layer (i-layer) is investigated, analysing its performance as a function of In content in InGaN QWs and QW number. Finally, some considerations on possibility to achieve a more efficient InGaN solar cell with a wider absorption spectrum using InN QDs are derived.

13.2 InGaN Properties and Its Technological Issues

$In_xGa_{1-x}N$ is a direct band gap ($E_{InxGa1-xN}$) semiconductor with a high light absorption coefficient (about 10^5 cm^{-1} [3]). Its band gap changes as a function of In content (x) according to (13.1):

$$E_{In_xGa_{1-x}N} = x \cdot E_{InN} + (1 - x) \cdot E_{GaN} - b \cdot x(1 - x) \tag{13.1}$$

where b = 1.43 eV (bowing parameter). For high In content, InGaN has a small band gap (13.1), useful for absorption of low energy photons in near infrared region of solar spectrum.

These advantages are partially suppressed by difficult heteroepitaxial growth on GaN bulk buffer on Al_2O_3 (sapphire) or also Si (silicon) or SiC (silicon carbide) substrate without defects and by its wurtzite crystal structure. Effectively, a wurtzite crystal structure shows spontaneous pyroelectric polarization along c crystallographic axis as well as piezoelectric one (due to crystal stress) along the whole device, inducing large built-in internal electric field of several MV/cm [4]. This high electric field shifts electrons and holes energy states to lower and higher values, respectively (QCSE—quantum confined Stark effect), reducing the energy gap for optical transitions (red shift in light emission and absorption) as well as light absorption (solar cell efficiency loss). Furthermore, heteroepitaxial growth leads to a high lattice mismatch (maximum value of 11 % when InN is grown on GaN [4]) and, consequently, a biaxial induced compressive stress (in pseudomorphic growth) that increases with InGaN thickness and/or In content. When InGaN thickness becomes larger than critical value for In-rich three dimensional growth (spontaneous QDs formation) and/or for threading dislocations (TDs) formation (density with an order of magnitude of 10^8 cm^{-2}), InGaN grows on GaN as relaxed layer (without any strain) [5]. In solar cells, high defects density reduces the shunt resistance (leakage currents increase). Detrimentally, both material defects density and pyroelectric (higher in technological well-established c-plane (0001) crystal growth orientation) and piezoelectric polarizations reduce the open-circuit voltage (V_{oc}) and, consequently, the solar cell efficiency. The photocurrent is reduced, too. In fact, defects and In-rich QDs, in which carriers localization occurs, become carriers non-radiative recombination centers and reduce their lifetime. In our simulations the carrier lifetime is assumed equal to 1 ns [6].

Recently, new native GaN substrates with different growth orientation with respect to the c-plane, and GaN quasi-bulk substrate [2] or InGaN free-substrate (thin film structures) [7], have been demonstrated in order to reduce lattice mismatch and to have a higher quality material in terms of polarization reduction and defect density. GaN quasi-bulk substrate is obtained by hydride vapor phase epitaxy (HVPE) or low-pressure metal-organic vapor phase epitaxy (MOVPE) on sapphire substrate, later delaminated by laser lift-off (LLO) technique.

13.3 Superlattice GaN/InGaN p-i-n Thin Film Solar Cells

Each QW of a MQW (multi quantum well) shows localized wave functions in conduction band (CB) and valence band (VB). If barriers thickness is thin enough, these wave functions could have a significant overlap in the barriers and quantum confined carriers could have a finite probability to pass through the barrier from one to the neighbor QW (quantum tunneling current). Then, localized wave functions become delocalized along the whole structure. In a finite MQW with a large QW number or, ideally, in an infinite MQW, delocalized wave functions could cover a continuous range of permitted carrier energy levels (IB) placed into the highest band gap of semiconductor hetero-nano-structure (superlattice). When a superlattice is used in a solar cell (namely, IBSC), not only photons with energy values higher than the maximum band gap could be absorbed (by electron transitions from VB to CB), but also photons with lower energy by electron transitions from VB to IB or from IB to CB, ideally without increasing the non-radiative carriers recombination (SHR—Shockley Hall Read recombination) through IB permitted states, due to wave functions delocalization.

GaN/InGaN p-i-n solar cell growth on c-plane sapphire substrate with a GaN/InGaN finite superlattice (28 QWs), used as i-layer, has already been demonstrated

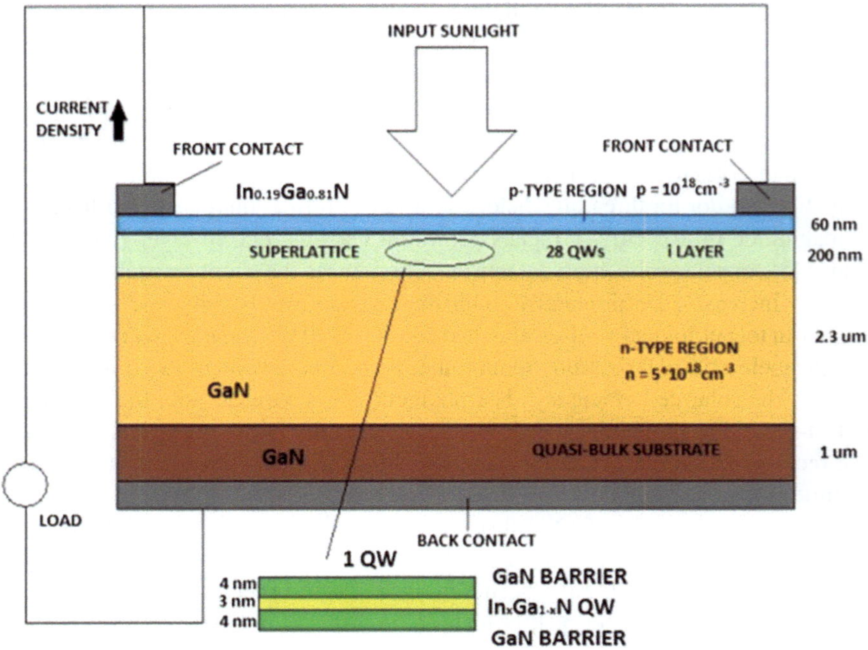

Fig. 13.1 Superlattice GaN/InGaN p-i-n thin film solar cell

[8]. GaN ($E_{GaN} = 3.43$ eV) barriers and $In_{0.25}Ga_{0.75}N$ ($E_{InGaN} = 2.48$ eV) wells thicknesses of 4 and 3 nm, respectively, have been assumed. These small thicknesses reduce the defect density of InGaN epitaxial layers grown on GaN (they are smaller than the critical value h_c for TD generation [5]). On the other hand, total superlattice thickness is equal to 200 nm (device photocurrent and efficiency increase with QW number). Cell is formed by undoped GaN buffer on c-plane sapphire substrate, GaN n-type layer (doping concentration of 5×10^{18} cm^{-3}), superlattice and $In_{0.19}Ga_{0.81}N$ p-type layer (doping concentration of 10^{18} cm^{-3}) with thicknesses of 1 μm, 2.3 μm, 200 nm and 60 nm, respectively. A part of top total cell area (15.8 %) is covered by front electrodes.

In this work, two-dimensional simulation studies on a structure without sapphire substrate (superlattice p-i-n GaN/InGaN on a c-plane GaN quasi-bulk substrate, see Fig. 13.1) under one-sun AM1.5 (air mass 1.5) terrestrial global solar illumination (total power densities of 963.56 W/m^2) have been carried out. A symmetric flat band profile is used in QW band calculation, as well as parabolic dispersion model and non-parabolic approximation are assumed for CB and VB, respectively. Both bands are characterized by own different carrier effective masses for perpendicular and parallel directions to the c-growth one (wurtzite crystal material). Both 6×6 Hamiltonian, used to calculate heavy hole (HH), light hole (LH) and crystal field split hole (CH) subbands wave functions of strained VB, and simplified dipole moment calculation, used for photon absorption evaluation, are derived by k·p method.

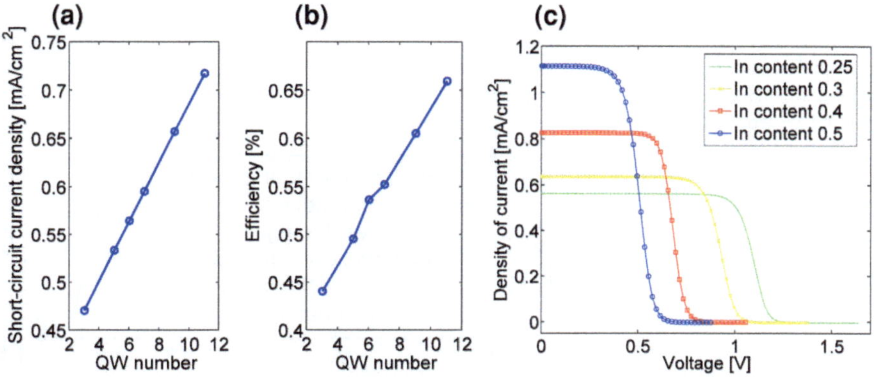

Fig. 13.2 MQW GaN/InGaN p-i-n solar cells with different QW number. **a** Short-circuit current density (J_{sc}). **b** Efficiency. **c** I-V characteristics of MQW GaN/InGaN p-i-n solar cells with 6 QWs as a function of QW In content (0.25, 0.3, 0.4 and 0.5)

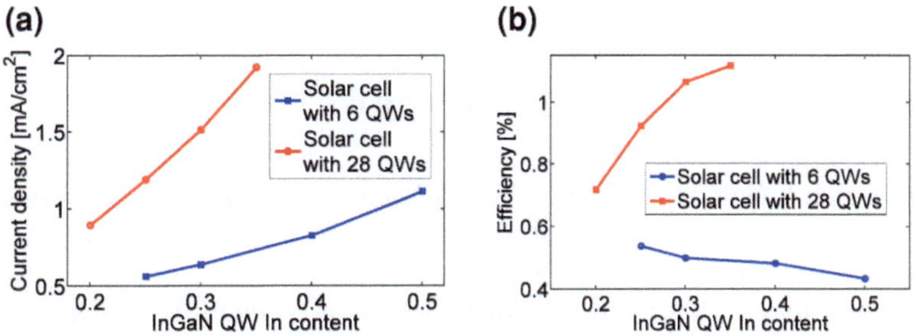

Fig. 13.3 MQW GaN/InGaN p-i-n solar cells with 6 and 28 QWs with different QW In content. **a** Photocurrent density. **b** Efficiency

In simulations leakage currents due to TD have not been considered because In content x is low (under 0.5). For example, if x is equal to 0.25, h_c is about 10 nm while in this work InGaN QW thickness is only 3 nm (Fig. 13.2a, b). Simulation of device with 28 QWs presents an efficiency of 1.05 % with a short-circuit current density (J_{sc}) of 1.193 mA/cm^2. Calculated efficiency and J_{sc} are higher than those experimentally exhibited by structures on sapphire (J_{sc} = 0.8 mA/cm^2 and efficiency 0.58 %) [8]. This is due to different structure (GaN quasi-bulk substrate) and to presence of leakage currents in the real sample. Photon absorption occurs in superlattice and p-type emitter (first layer crossed by sunlight). Simulation without absorption in p-type layer shows an efficiency of 0.62 % and J_{sc} = 0.597 mA/cm^2.

MWQs GaN/InGaN p-i-n solar cell has been also investigated when wells In concentration changes in structure with 6 (Fig. 13.2c) and 28 QWs. In both situations a greater In content increases photocurrent (Fig. 13.3a) because smaller $In_xGa_{1-x}N$ energy gap permits low energy photons absorption. On the other hand,

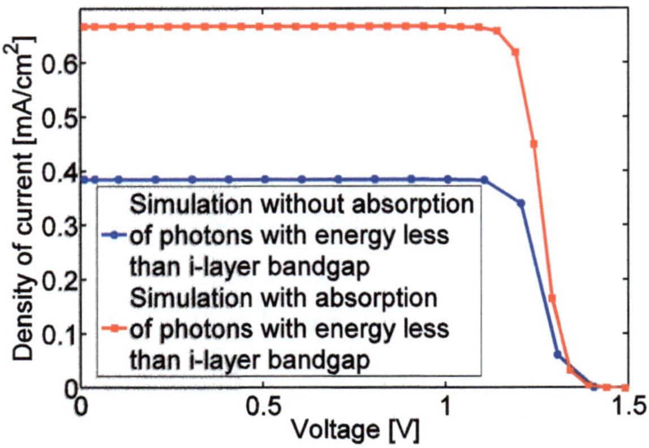

Fig. 13.4 I-V characteristics of GaN/InGaN p-i-n solar cell (undoped $In_{0.45}Ga_{0.55}N$ i-layer with InN QDs embedded) when absorption of photons with energies smaller than $In_{0.45}Ga_{0.55}N$ band gap occurs or not

only in the second structure efficiency raises (Fig. 13.3b). For 6 QWs heterointerface polarization charge reduces V_{oc} more quickly than J_{sc} increase with QWs In content.

13.4 InGaN Solar Cells With InN QDs

During In-rich InGaN layer growth, InN QDs could be spontaneously formed due to In phase segregation. Theoretical analysis of In nucleation and critical thickness of three dimensional growth (QDs formation) as a function of layer growth In content are shown in literature [5]. An InGaN solar cell with InN QDs with their band gap of 0.71 eV could absorb photons of solar spectrum with low energy increasing photocurrent. In simulations, a GaN/InGaN p-i-n solar cell grown on GaN quasi-bulk substrate with a single undoped $In_{0.45}Ga_{0.55}N$ i-layer (thickness of 200 nm) is assumed. In this layer, cylindrical InN QDs with a size of 4×4 nm, are embedded with a surface density of 5×10^{12} cm^{-2}. Only in i-layer photon absorption is imposed in order to study effect of embedded QDs. Thus, photocurrent density and efficiency shift from 0.39 to 0.67 mA/cm^2 and from 0.44 to 0.78 %, respectively, when absorption of photons with energy less than 1.85 eV ($In_{0.45}Ga_{0.55}N$ band gap) occurs (Fig. 13.4). In these simulations, localized QD-confined wave functions are assumed isolated to each other. However, if QDs have similar dimensions and regular distribution, each localized QD-confined wave function could couple to each other and form an IB in the structure (recently named supercrystal), strongly broadening its absorption spectrum. QDs formation could be partially controlled in strain induced Stranski–Krastanov (SK) growth mode [4].

13.5 Conclusion

Our simulations on GaN/InGaN superlattice p-i-n solar cell show an improvement with respect to already demonstrated structures due to use of GaN quasi-bulk substrate, but further optimization could be based on increase of coupling between QWs wave functions. A well-controlled In-rich InGaN QDs spontaneous formation could be exploited to make a supercrystal increasing InGaN IBSCs efficiency.

References

1. M.A. Green, K. Emery, Y. Hishikawa, W. Warta, E.D. Dunlop, Solar cell efficiency tables (version 41). Prog. Photovolt. Res. Appl. **21**, 1–11 (2013)
2. T. Paskova, D.A. Hanser, K.R. Evans, GaN substrates for III-Nitride devices. Proc. IEEE **98**(7), 1324–1338 (2010)
3. V.Y. Davydov, A.A. Klochikhin, V.V. Emtsev, D.A. Kurdyukov, S.V. Ivanov, V.A. Vekshin, F. Bechstedt, J. Furthmüller, J. Aderhold, J. Graul, A.V. Mudryi, H. Harima, A. Hashimoto, A. Yamamoto, E.E. Haller, Band gap of hexagonal InN and InGaN alloys. Phys. Stat. Sol. B **234**(3), 787–795 (2002)
4. N. Grandjean, M. Ilegems, Visible InGaN/GaN quantum-dot materials and devices. Proc. IEEE **95**(9), 1853–1865 (2007)
5. W. Zhao, L. Wang, J. Wang, Z. Hao, Y. Luo, Theoretical study on critical thicknesses of InGaN grown on (0 0 0 1) GaN. J. Cryst. Growth **327**, 202–204 (2011)
6. G.F. Brown, J.W. Ager III, W. Walukiewicz, J. Wu, Finite element simulations of compositionally graded InGaN solar cells. Sol. Energy Mater. Sol. Cells **94**, 478–483 (2010)
7. C.R. Miskys, M.K. Kelly, O. Ambacher, M. Stutzmann, Freestanding GaN-substrates and devices. Phys. Stat. Sol. C **0**(6), 1627–1650 (2003)
8. J.-K. Sheu, C.-C. Yang, S.-J. Tu, K.-H. Chang, M.-L. Lee, W.-C. Lai, L.-C. Peng, Demonstration of GaN-based solar cells with GaN/InGaN superlattice absorption layers. IEEE Electron Device Lett. **30**(3), 225–227 (2009)

Chapter 14
Improvement in the Efficiency of Thin Film CdS/CdTe Solar Cells Using Different TCO Materials

H. A. Mohamed and N. M. A. Hadia

Abstract CdS/CdTe heterojunction based solar cells have been considered one of the main candidates for terrestrial energy production. This work represents the theoretical results of using ZnO and its alloys as a front contact in CdS/CdTe solar cell as alternative material to expensive and not abundant ITO. The calculation of optical losses is carried out based on the multi-reflections effect and absorption in TCO and CdS layers. Both the front and back surfaces recombination of the CdTe layer are taken into account to describe the recombination losses. It has been found that using the multi-reflections effect leads to increase the ratio of transmitted light reaching the absorber layer. Both the internal and external quantum efficiency are strongly depending on the width of space-charge region. ZnO and its alloys are considered suitable alternative materials to ITO when used as front electrode in CdS/CdTe cells. ZnO:Al has the maximum short-circuit current density of 22.64 mA/cm^2 at space-charge width of 0.11 μm and the corresponding optical (reflection and absorption) and recombination (front and back) losses are about 27 %. The efficiency of CdS/CdTe solar cell using ZnO:Al is about 17.9 % at certain parameters of absorber layer.

H. A. Mohamed (✉) · N. M. A. Hadia
Physics Department, Faculty of Science, Sohag University, Sohag 82524, Egypt
e-mail: hussein_abdelhafez2000@yahoo.com

N. M. A. Hadia
e-mail: nomery_abass@yahoo.com

H. A. Mohamed
Physics Department, Teachers College, King Saud University, Riyadh 11148, Saudi Arabia

A. Y. Oral et al. (eds.), *International Congress on Energy Efficiency and Energy Related Materials (ENEFM2013)*, Springer Proceedings in Physics 155,
DOI: 10.1007/978-3-319-05521-3_14,

14.1 Introduction

For the last decades CdS/CdTe heterojunction based solar cells have been considered as one of the main candidates for large-scale production and application for terrestrial energy production [1, 2]. Recently, the highest efficiency for CdTe thin-film solar cells is about 18.3 % which also achieved large area module efficiency of 15.3 % [3], although this value is still far from the calculated efficiency 28–30 % [4, 5]. After the optical losses result from reflection at different interference layers in cell, absorption in ITO and CdS layers and recombination losses (front and pack) were taken into account, a good agreement between practical and theoretical results were achieved [6–8].

Development the efficiency of CdS/CdTe solar cells is a great goal of scientists in all decades. Despite decades of these development, solar cells are still relatively expensive. Using indium tin oxide as a front contact in CdS/CdTe solar cell is considered one of the reasons that make solar cells technology is expensive. It is known that, the front contact in CdS/CdTe solar cell is made of transparent conducting oxides (TCO) materials that must be have transmission more than 85 % in visible region and sheet resistance less than 10 Ω/square at room temperature as well as good adhesion to glass substrate [8, 9]. Although indium tin oxide (ITO) is considered one of these promising materials, it is expensive and not abundant. Therefore, ZnO has been actively investigated as an alternative material to ITO because ZnO is non-toxic, inexpensive and abundant material. It is also chemically stable so it is used for the production of solar cells [10–12].

The main objective of this paper is to assess the effects of using ZnO and its alloys such as ZnO:Sn, ZnO:Al and ZnO:In as front electrode on the calculation of CdS/CdTe solar cell efficiency. Both optical and recombination losses will be taken into consideration. The calculation of optical losses is carried out based on the multiple reflections effect and absorption in TCO and CdS layers. Both the front and back surfaces recombination of the CdTe layer are used to describe the recombination losses.

14.2 Calculation of Transmittance Using Multi Reflections Effects

As the photons are incident upon a boundary between materials of different refractive indices, a reflection occurs, known as the Fresnel reflection. For near-normal incidence angles at the boundary, Fresnel Power Reflection Coefficient can be expressed as:

$$R_f = \frac{(n_1 - n_2)^2}{(n_1 + n_2)^2} \tag{14.1}$$

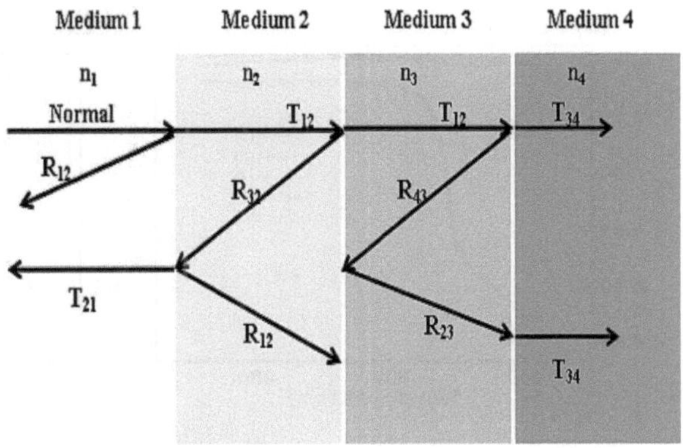

Fig. 14.1 Diagram of the total transmission of four layers structure due to multi reflection effect

Or in terms of transmittance it may be written as the Fresnel Power Transmission coefficient:

$$T_f = 1 - R_f = \frac{4n_1n_2}{(n_1 + n_2)^2} \tag{14.2}$$

where n_1 and n_2 are the refractive indices of the two media. The total transmittance for L number of layers can be calculated from:

$$T_{tot} = T_{12} \prod_{j=2}^{L-1} \frac{T_{j,j+1}}{(1 - R_{j,j-1}R_{j,j+1})} \tag{14.3}$$

which expands to

$$T_{tot} = 4\frac{n_1n_2}{(n_1 + n_2)^2} \prod_{j=2}^{L-1} \frac{4\frac{n_jn_{j+1}}{(n_j+n_{j+1})^2}}{\left(1 - \frac{(n_j-n_{j-1})^2(n_j-n_{j+1})^2}{(n_j+n_{j-1})^2(n_j+n_{j+1})^2}\right)} \tag{14.4}$$

These calculations can be found elsewhere [13, 14]. Figure 14.1 shows a four layers structure and a more layers can be assumed by the same way.

When the absorption effect which takes place into TCO and CdS layers is taken into account, the transmission coefficient T(λ) can be written in the form:

$$T(\lambda) = T_{tot}\,(e^{-\alpha_1 d_1})(e^{-\alpha_2 d_2}) \tag{14.5}$$

where $\alpha_1, \alpha_2, d_1, d_2$ is the absorption coefficient and thickness of ITO and CdS layers, respectively. The absorption coefficient is calculated from the following equation:

$$\alpha(\lambda) = \frac{4\pi}{\lambda}\,k(\lambda) \tag{14.6}$$

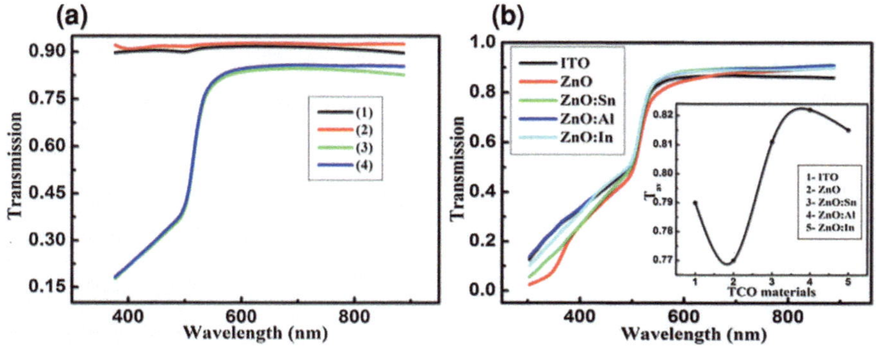

Fig. 14.2 Calculated transmission of ITO. **a** Considering only reflection (curves *1*, *2*) as well as absorption (curves *3*, *4*). Curves (*1*, *3*) and (*2*, *4*) represent the calculated transmission according to Ref. 8 and the current work, respectively. Calculated transmission of different TCOs. **b** Using the current calculations. The inset figure shows the average transmission (T_{av}) in the wavelength range 500–700 nm for different TCOs

where k is the extinction coefficient of the used materials (TCO, CdS and CdTe). Using the values of optical constant (n, k), the transmission coefficient can be calculated for different TCO materials. The data of n and k of ITO, ZnO, ZnO:Sn, ZnO:Al, ZnO:In, CdS and CdTe are taken from Refs. [15, 16, 17, 18, 19, 20, 21], respectively. The extinction coefficient (k) value of glass substrate was taken as $k = 0$, while the Sellmeier dispersion equation has been applied for calculating the refractive index of glass substrate [22].

Figure 14.2a shows the comparison between the calculated transmission of ITO using the present equations (14.1–14.4) which take into account the multi reflections effect (curves 1, 3) and those calculated from Ref. [8] (curves 2, 4). In this figure, curves 1, 2 represent the calculated transmission resulting from the reflection from all interfaces (air-glass, glass-ITO, ITO-CdS and CdS-CdTe). While, curves 3, 4 represent the calculated transmission due to reflection from all interfaces and absorption into ITO and CdS layers. It is clear that the variation of transmission coefficient with wavelength has the same behavior of both causes. The multi reflections effect leads to slight increase in transmission by a ratio of 1 %. Where the average transmission of the current cause about 81 % in the wave length range 500–700 nm, while the transmission coefficient which was calculated in Ref. [8] is about 80 %. These values are based on the losses due to reflection and absorption (optical losses).

Figure 14.2b shows the calculated transmission spectra (considering 14.1–14.5) that reach to the absorber layer after passing through glass substrate, TCOs and CdS layers. The transmission of various transparent conducting oxides such as ITO, ZnO, ZnO:Sn, ZnO:Al and ZnO:In is plotted in this figure. It is clear that, apart from the transmission of ZnO all other TSOs have values of transmission greater than the transmission of ITO. This result mainly depends on the optical constants of the used material and hence the conditions of preparation. Therefore any material of them can be used as alternative of ITO material, particularly ZnO:Al that has the maximum average transmission of 82 % in wavelength range 500–700 nm as shown in the inset figure.

14.3 Spectral Distribution of Quantum Efficiency

The total internal quantum efficiency η_{int} of solar cell is the sum of the drift (η_{drift}) and diffusion (η_{dif}) components of quantum efficiency. This quantity is used in calculating the short-circuit current density. The drift component (η_{drf}), which takes into account recombination at the CdS-CdTe interface (front recombination), is governed by the following expression [5]:

$$\eta_{drift} = \frac{1 + \frac{S}{D_P}\left(\alpha + \frac{2}{W}\frac{\varphi_0 - qv}{kT}\right)^{-1}}{1 + \frac{S}{D_P}\left(\frac{2}{W}\frac{\varphi_o - qv}{kT}\right)^{-1}} - \exp(-\alpha W) \tag{14.7}$$

where S is the recombination velocity at the heterojunction interface, D_{p} is the diffusion coefficient of holes, α is the absorption coefficient of CdTe at a given wavelength, φ_0 is the barrier height at the semiconductor side, v is the applied voltage, q is the electron charge and k is the Boltzmann constant.

The dependence of width W of space-charge region (depletion layer) on the concentration of uncompensated acceptors (N_a–N_d) is given by:

$$W = \sqrt{\frac{2\varepsilon\varepsilon_0(\varphi_0 - qv)}{q^2(N_a - N_d)}} \tag{14.8}$$

where ε is the relative permittivity of the semiconductor and ε_0 is the permittivity of free space. In current calculations ($\varphi_0 - qv$) is taken as 1 eV, $D_{\text{p}} = 2.5\ \text{cm}^2/\text{s}$, ε is taken as 10.6 and $S = 10^7$ cm/s, these values are taken from Ref. [23]. The diffusion component η_{dif} of the internal quantum efficiency, which takes into account recombination at the back surface of the solar cell, is given by the following expression [24, 25]

$$\eta_{dif} = \frac{\alpha L_n}{\alpha^2 L_n^2 - 1}\exp(-\alpha W) \times \left\{ \alpha L_n - \frac{\frac{S_b L_n}{D_n}\left[\cos\left(\frac{d-W}{L_n}\right) - \exp(-\alpha(d-W))\right] + \sinh\left(\frac{d-W}{L_n}\right) + \alpha L_n \exp(-\alpha(d-W))}{\frac{S_b L_n}{D_n}\sinh\left(\frac{d-W}{L_n}\right) + \cosh\left(\frac{d-W}{Ln}\right)} \right\} \tag{14.9}$$

where $L_{\text{n}} = (\tau_{\text{n}} D_{\text{n}})^{1/2}$ is the electron diffusion length, τ_{n} is electron lifetime, D_{n} is the electron diffusion coefficient, S_{b} is the recombination velocity at the back surface of CdTe layer, d is the thickness of CdTe. In the present calculations,

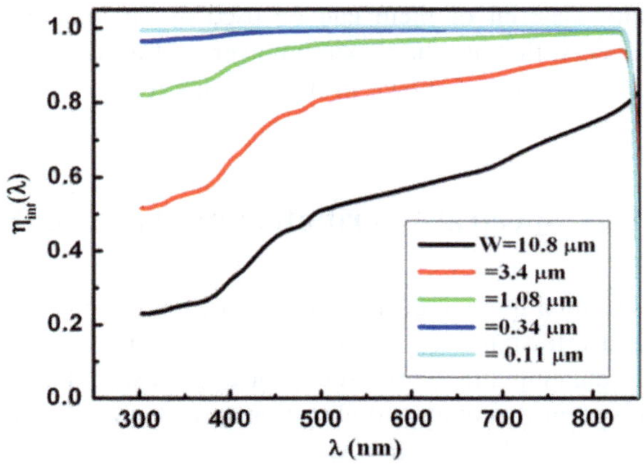

Fig. 14.3 Total internal quantum efficiency spectra (η_{int}) for different values of space-charge width (W)

$\tau_n = 10^{-9}$ s, $D_n = 25$ cm^2/s, $S_b = 10^7$ cm/s and $d = 5$ μm. Figure 14.3 shows the total internal quantum efficiency spectra (η_{int}) for different values of space-charge width (W). It is clear that η_{int} is strongly depending on the width of space-charge region and hence on the uncompensated acceptors (N_a–N_d). When the space-charge region width is wide, the internal quantum efficiency records low values due to the electric field in the space charge region is weak and then the front recombination is strongly takes place. At narrowing width of space charge region, the internal quantum efficiency represents higher values than the above case. Where the strong electric field prevents recombination of carriers generated near the CdTe surface [26]. Therefore, the values of short-circuit current will be high at small values of the width of space charge region as will be seen in the following section.

The external quantum efficiency η_{ext} of a solar cell is defined as the ratio of the number of charge carriers that form the photocurrent I_{ph} to the number of photons of a given energy impinging on the solar cell [23]:

$$\eta_{ext} = \frac{I_{ph}/q}{P_{opt}/hv} \tag{14.10}$$

where P_{opt} is the optical power at a given wavelength. Besides, η_{ext} is related to η_{int} according to the following formula [23]:

$$\eta_{ext} = F\ T\ \eta_{int} \tag{14.11}$$

Where F is the shade factor of the frontal contact and it is taken by unity in these calculations.

Using (14.7–14.9, 14.11), the external quantum efficiency η_{ext} of CdS/CdTe cell is calculated and plotted in Fig. 14.4 for different TCO materials. This figure shows that the response observed below 500 nm is due to the small thickness of CdS (70 nm). This thin layer allows a fraction of photons with energy above its band-gap (2.42 eV) to be transmitted to the CdTe absorber layer, contributing to an increased

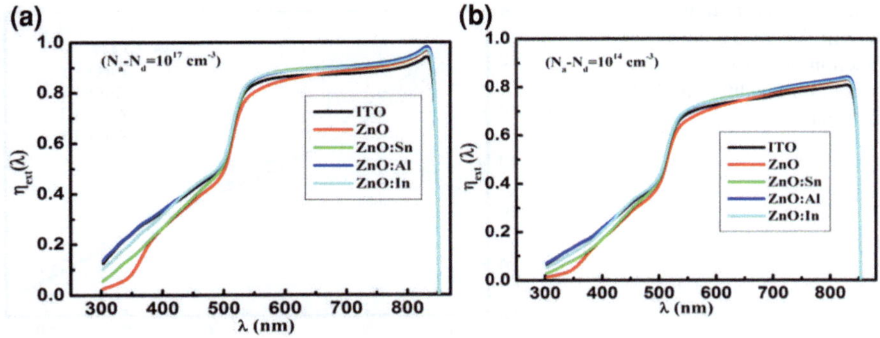

Fig. 14.4 External quantum efficiency spectra (η_{ext}) for different TCO materials at N_a–$N_d = 10^{17}$ cm^{-3} (**a**) and N_a–$N_d = 10^{14}$ cm^{-3} (**b**)

photocurrent [27]. With further increase in wavelength an increase in η_{ext} is observed. The cutoff wavelength observed at 855 nm corresponds to the cadmium telluride band-gap. One observes that the external quantum efficiency curves for all TCO material are similar in shape and ZnO:Al has the highest value comparing with other TCO materials in most wavelength range. When the N_a–$N_d = 10^{17}$ cm^{-3} which corresponds to $W = 0.11$ μm (Fig. 14.4a), η_{ext} represent higher values than for $N_a - N_d = 10^{14}$ cm^{-3} which corresponds to $W = 10.83$ μm (Fig. 14.4b).

14.4 Short-Circuit Current and Cell Efficiency

Using (14.5, 14.7–14.9) the short-circuit current density can be calculated from the following formula:

$$J_{SC} = q \sum_i T(\lambda) \frac{\varphi_i(\lambda_i)}{h v_i} \eta_{int}(\lambda_i) \Delta\lambda_i \tag{14.12}$$

where Φ_i is the spectral power density (mWcm^{-2}μm^{-1}) and $\Delta\lambda_i$ is the interval between the two neighboring values λ_i. The calculations will be done for AM1.5 solar radiation using Tables ISO 9845-1:1992 (Standard ISO, 1992) [28]. In order to study the effect of reflection losses that calculated from this work (considering the multi-reflection effect) and those calculated from Ref. [8], the values of J_{SC} is calculated under the same conditions. According to Ref. [8], J_{SC} records a value of 28.61 indicating that the losses due to this type of reflection are about 8 %. When the multi-reflections method is taken into account, the contribution of reflection loss is about 7 % where $J_{SC} = 28.86$ mA/cm^2. This indicates there is no big difference between the two methods, however the multi-reflections method is more accurate and it will effect on the efficiency of CdS/CdTe solar cell. Figure 14.5 represents the short-circuit current density J_{SC} as a function of space-charge width

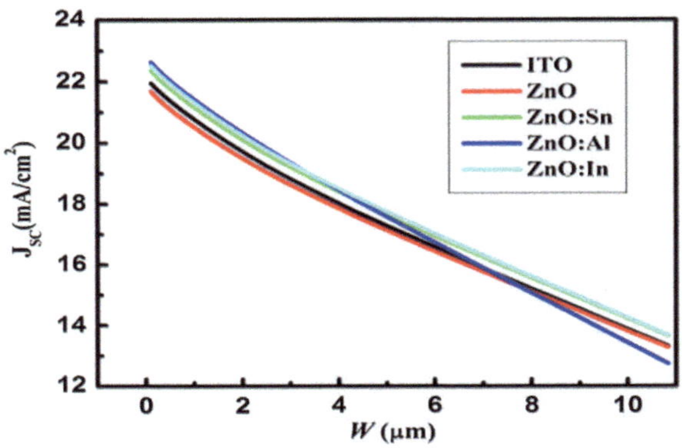

Fig. 14.5 Short-circuit current density J_{SC} as a function of space-charge width (*W*) for different TCO materials

(*W*) for different TCO materials. It can be seen that J_{SC} strongly depends on the width of space-charge region. At low width, J_{SC} attains it maximum value for all TCOs. In case of ZnO:Al, the value of J_{SC} is greater than the others TCOs and this value is about 22.64 mA/cm^2 for $W = 0.11$ μm. Moreover, these results show that the optical and recombination losses are about 27 % at the lowest value of the width space-charge region and this ratio is increased to more than 59 % at $W = 10.83$ μm since the calculated current density is 12.75 mA/cm^2.

14.5 Current-Voltage Curve and the Cell Efficiency

The *I-V* characteristic under illumination of CdS/CdTe solar cells can be presented as:

$$J(V) = J_d - J_{ph} \tag{14.13}$$

where $J_d(V)$ is the dark current density and J_{ph} is the photocurrent density. The dark current density is the sum of the generation-recombination current density J_{gr} and the over-barrier current density J_n and can be written in the form.

$$J_d(V) = J_{gr}(V) + J_n(V) \tag{14.14}$$

More details of calculation J_{gr} and J_n can be found elsewhere [4, 5, 29, 30].

The typical I-V characteristic curve for CdS/CdTe solar cell at different TCO materials is shown in Fig. 14.6. Some important parameters such as the maximum voltage V_m, maximum current density J_m and open circuit voltage V_O can be determined from this figure as shown in Fig. 14.6b for ZnO:Al. These parameters and other related parameters such as cell power density $P_{max}(P_{max} = V_m J_m)$ are estimated and listed in Table 14.1. It is clear that most TCO materials have very close values of open circuit voltage. Besides, the maximum voltage and current

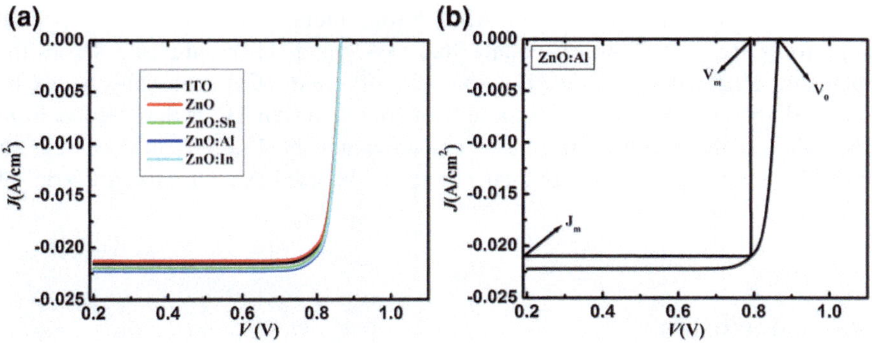

Fig. 14.6 I-V characteristic for different TCOs (**a**) and for ZnO:Al (**b**)

Table 14.1 The maximum voltage V_m, maximum current density J_m, open circuit voltage V_0 and CdS/CdTe solar output power P_{max} for different TCO materials

P_{max} (mW/cm^2)	V_m (mv)	J_m (mA/cm^2)	V_0 (mv)	TCO material
16.11	783	20.58	863	ITO
15.74	774	20.34	860	ZnO
16.52	787	20.99	864	ZnO:Sn
17.2	790	21.77	866	ZnO:Al
16.58	787	21.07	864	ZnO:In

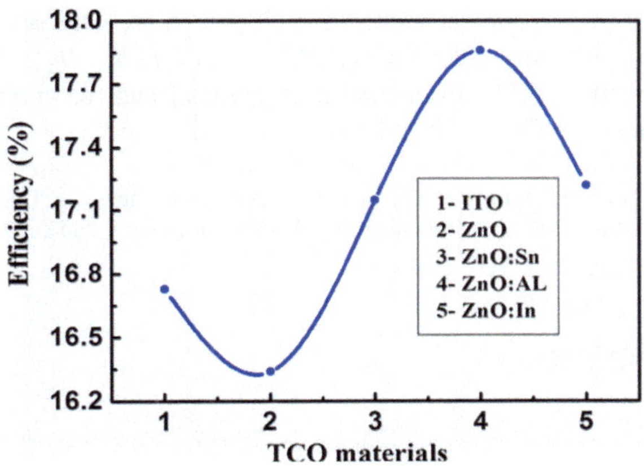

Fig. 14.7 The efficiency of CdS/CdTe solar cells for different TCO materials as a front contact

density of 790 mV and 21.77 mA/cm^2 is observed for ZnO:Al: and consequently the corresponding maximum power density is 17.2 mW/cm^2.

The CdS/CdTe solar cell efficiency can be expressed by:

$$\eta = {}^{P_{\max}}/_{P_{inc}} \qquad (14.15)$$

where P_{inc} is the density of the total AM 1.5 solar radiation power over the spectral range hv $\geq E_{gCdTe} = 1.46$ eV equals 96.3 mW/cm^2 [31]. Figure 14.7 shows the calculated efficiency of CdS/CdTe solar for different TCO materials. It can be observed, the efficiency of cell is more than 16.3 % for all TCOs that used as front electrodes in these cells. The maximum efficiency of 17.86 % is observed for ZnO:AL indicating ZnO or any their alloys can be used as alternative material of ITO.

14.6 Conclusions

ITO has unique properties to be used as front electrode in CdS/CdTe solar cells, however ITO is expensive not abundant and consequently this type of solar cells is still relatively expensive. This work studies the using of ZnO and some alloys such as ZnO:Sn, ZnO:Al and ZnO:In as alternative materials of ITO and then develop the efficiency of CdS/CdTe solar cells. The calculation of optical losses is carried out based on the multi-reflections effect and absorption in TCO and CdS layers. The results show that the multi-reflections method is considered more accurate method to calculate the total transmission of incident light that will reach to the absorber layer. ZnO and its alloys are considered good alternative materials of ITO that used as a front contact in CdS/CdTe cells. The internal and external quantum efficiency strongly depend on the width W of space-charge region and record highest values at width of 0.11 μm. In the case of ZnO:Al, the calculated short-circuit current density is 22.64 mA/cm^2 at N_A–N_d = 10^{17} cm^{-3} which corresponds to $W = 0.11$ μm indicating the optical and recombination losses are about 27 % and the CdS/CdTe efficiency is about 17.9 %.

Acknowledgments The author would like to thank the Deanship of scientific research, King Saud University, Riyadh, Saudi Arabia, for funding and supporting this research.

References

1. A. Goetzberger, C. Hebling, H. Schock, Photovoltaic materials, history, status and outlook. Mater. Sci. Eng. **40**, 1–46 (2003)
2. T. Surek, Crystal growth and materials research in photovoltaics. J. Cryst. Growth **275**, 292–304 (2005)
3. M.A. Green, K. Emery, Y. Hishikawa, W. Warta, E.D. Dunlop, Solar cell efficiency tables (version 41). Prog. Photovoltaics Res. Appl. **21**, 1–11 (2013)
4. S. Sze, *Physics of Semiconductor Devices*, 2nd edn. (Wiley, New York, 1981)
5. L.A. Kosyachenko, A.I. Savchuk, E.V. Grushko, Dependence of efficiency of thin-film CdS/CdTe solar cell on parameters of absorber layer and barrier structure. Thin Solid Films **517**, 2386–2391 (2009)
6. H.A. Mohamed, Influence of the optical and recombination losses on the efficiency of CdS/CdTe solar cell at ultrathin absorber layer. Can. J. Phys. (2014)

7. H.A. Mohamed, Study towards high efficiency CdS/CdTe solar cells. J. Optoelectron. Adv. Mater. (2014)
8. H.A. Mohamed, Dependence of efficiency of thin-film CdS/CdTe solar cell on optical and recombination losses. J. Appl. Phys. 113(1–6), 093105 (2013)
9. A.M. Acevedo, Thin film CdS/CdTe solar cells: research perspectives. Sol. Energy **80**, 675–681 (2006)
10. S.H. Jeong, S.B. Lee, J.-H. Boo, The insert of zinc oxide thin film in indium tin oxide anode for organic electroluminescence devices. Curr. Appl. Phys. **4**, 655–658 (2004)
11. A. Goyal, S. Kachhwaha, ZnO thin films preparation by spray pyrolysis and electrical characterization. Mater. Lett. **68**, 354–356 (2012)
12. K. Kim, S. Kim, S.Y. Lee, Effect of excimer laser annealing on the properties of ZnO thin film prepared by sol-gel method. Curr. Appl. Phys. **12**, 585–588 (2012)
13. F.W. Mont, J.K. Kim, M.F. Schubert, E.F. Schubert, R.W. Siegel, High-refractive-index TiO2-nanoparticle-loaded encapsulants for light-emitting diodes. J. Appl. Phys. 103(1-6), 083120 (2008)
14. F.W. Mont, J.K. Kim, M.F. Schubert, H. Luo, E.F. Schubert, Encapsulants with an optimized scattering coefficient. In Proceedings of SPIE, ed. by R.W. Siegel, vol. 6486, 64861C (2007)
15. http://homepages.rpi.edu/schubert/Educational-resources/Materials-Refractive-index-and-extinction-coefficient.pdf (E.F. Fred Schubert, Educational Resources. Refractive index and Extinction Coefficient of Materials, Rensselaer Polytechnic Institute, NY, USA), (2004)
16. S.W. Xue, X.T. Zu, W.L. Zhou, H.X. Deng, X. Xiang, L. Zhang, H. Deng, Effects of post-thermal annealing on the optical constants of ZnO thin film. J. Alloy. Compd. **448**, 21–26 (2008)
17. E. Çetinörgü, Characteristics of filtered vacuum arc deposited ZnO–SnO2 thin films on room temperature substrates. Opt. Commun. **280**, 114–119 (2007)
18. Q. Xu, R.D. Hong, H.L. Huang, Z.F. Zhang, M.K. Zhang, X.P. Chen, Z.Y. Wu, Laser annealing effect on optical and electrical properties of Al doped ZnO films. Opt. Laser Technol. **45**, 513–517 (2013)
19. G.C. Xie, L. Fang, L.P. Peng, G.B. Liu, H.B. Ruan, F. Wu, C.Y. Kong, Effect of In-doping on the optical constants of ZnO thin films. Phys. Procedia **32**, 651–657 (2012)
20. S. Ninomiya, S. Adachi, Optical properties of wurtzite CdS. J. Appl. Phys. 78(1–11), 1183 (1995)
21. P.D. Paulson, X. Mathew, Spectroscopic ellipsometry investigation of optical and interface properties of CdTe films deposited on metal foils. Sol. Energ. Mat. Sol. C. **82**, 279–290 (2004)
22. S.O. Kasap, *Optoelectronics and Photonics: Principles and Practice* (Prentice Hall, New Jersey, 2000), p. 45
23. V.V. Brus, On quantum efficiency of nonideal solar cells. Sol. Energy **86**, 786–791 (2012)
24. S.M. Sze, K.N.G. Kwok, *Phys. Semicond. Devices* (Wiley, New Jersy, 2007)
25. W.J. Yang, Z.Q. Ma, X. Tang, C.B. Feng, W.G. Zhao, P.P. Shi, Internal quantum efficiency for solar cells. Sol. Energy **82**, 106–110 (2008)
26. L.A. Kosyachenko, X. Mathew, V.Y. Rshko, E.V. Grushko, Optical absorptivity and recombination losses: the limitations imposed by the thickness of absorber layer in CdS/CdTe solar cells. Sol. Energ. Mat. Sol. C. **114**, 179–185 (2013)
27. L.R. Cruz, W.A. Pinheiro, R.A. Medeiro, C.L. Ferreira, R.G. Dhere, J.N. Duenow, Influence of heat treatment and back contact processing on the performance of CdS/CdTe thin film solar cells produced in a CSS in-line system. Vacuum **87**, 45–49 (2013)
28. Reference Solar Spectral Irradiance at the Ground at Different Receiving Conditions, Standard of International Organization for Standardization ISO 9845-1 (1992)
29. C.T. Sah, R. Noyce, W. Shokley, Carrier generation and recombination in P-N junctions and P-N junction characteristics. Proc. IRE **45**, 1228–1243 (1957)

30. L.A. Kosyachenko, V.M. Sklyarchuk, O.F. Sklyarchuk, V.A. Gnatyuk, Features of generation-recombination processes in CdTe-based Schottky diodes. Semicond. Sci. Technol. **22**, 911–918 (2007)
31. T. Toshifumi, S. Adachi, H. Nakanishi, K. Ohtsuka, Optical constants of Zn1-xCdxTe Ternary alloys: experiment and modeling. Jpn. Appl. Phys. **32**, 3496–3501 (1993)

Chapter 15
Effect of Dopant Concentrations on Conversion Efficiency of SiC-Based Intermediate Band Solar Cells

H. Heidarzadeh, A. Rostami, M. Dolatyari and G. Rostami

Abstract It was recognized that the introducing of a narrow metallic band states in the crystal structure of semiconductors make materials that they can be used as intermediate band materials for improving the power conversion efficiency of high band gap single junction solar cells. In these structures intermediate bands would serve as a "stepping stone" for photons with different energies to excite electrons from the valence to the conduction bands. Low-energy photons can be captured by this method that would pass through a conventional solar cell. An optimal IBSC (intermediate band solar cells) has a total band gap of about 1.95 eV and 3C-SiC has the closest band gap to this value (band gap of 2.2 eV). Excellent electronic properties of 3C-SiC such as high electron mobility and saturated electron drift velocity and its suitable band gap makes it an important alternative material for light harvesting technologies instead of conventional semiconductors like silicon. In this condition detailed balance analysis predicts a limiting efficiency of more than 55 % for an optimized, single junction intermediate band solar cell that it is higher than efficiency of an optimized two junction tandem solar cell. In this study we have analyzed Fe doped 3C-SiC by ab initio calculations for Fe concentration of 1.05, 1.85, 3.22, and 5.55 %. The results show conversion efficiency for designed solar cell change with altering in Fe contents. The maximum efficiency has been obtained for crystals with 3 % Fe^{3+} as dopant in 3C-SiC structure.

H. Heidarzadeh · A. Rostami (✉)
Photonics and Nanocrystal Research Labaratory (PNRL), Department of Electrical and Computer Engineering, University of Tabriz, Tabriz, Iran
e-mail: Rostami@tabrizu.ac.ir

A. Rostami · M. Dolatyari · G. Rostami
School of Engineering—Emerging Technologies, University of Tabriz, Tabriz, Iran

A. Y. Oral et al. (eds.), *International Congress on Energy Efficiency and Energy Related Materials (ENEFM2013)*, Springer Proceedings in Physics 155,
DOI: 10.1007/978-3-319-05521-3_15,

15.1 Introduction

Single junction silicon solar cells have an efficiency of about 20 %. In order to increase efficiency of solar cells, multi-junction thin film solar cells with different bandgaps is one of the most promising approaches. The best efficiency of such solar cells demonstrated on the research scale is 43 %. However, the challenges in fabrication of multi-junction solar cells lie in the growth of multi-stacked material and balance of junction currents. Cubic silicon carbide(3C-SiC) due to excellent electronic prosperities such as high electron mobility [1, 2] in a single material which is doped during growth, pave the way for more efficient solar cell concepts.

Research on intermediate band solar cell is motivated by high theoretical efficiencies [3–5]. Intermediate band solar cells are designed with an additional energy band in the band gap between the conduction band and the valence band. The main band gap between the conduction band and the valence band is then split into two sub-band gaps (the smallest E_1 and largest E_2). Electrons can be excited to the conduction band directly from the valence band or through the intermediate band, as shown in Fig. 15.1. The maximum efficiency of an IBSC, having the ideal bandgaps of $E_1 = 0.71$ eV, $E_2 = 1.24$ eV and $E_g = 1.97$ eV, is as high as 63.2 % [6, 7]. As it was mentioned, an optimal IBSC has a total bandgap of about 1.95 eV. 3C-SiC has a relatively wide bandgap ($E_g = 2.2$ eV), which is not suitable to act as a light absorption layer in single junction solar cells. However, 3C-SiC has the closest band gap to the optimal value. So, cubic silicon carbide can be considered a perfect material for intermediate bandgap solar cell thereby making it capable of collecting more than twice the solar energy than silicon based solar cells.

In this work the electronic and optical properties of Fe doped 3C-SiC in different dopant level, are studied with density functional theory (DFT) to survey the candidates of intermediate band (IB) material. Maximum achievable efficiency greatly improved by the introduction of the IB compared to the Silicon host in suitable position.

15.2 Calculation Method

Quantum calculations based on density-functional theory are carried out with the aim of discovering the origin of the electronic properties with substitution of transition metal by Si in crystal structure of 3C-SiC host semiconductor. The simulation is based on Ab initio calculation, using DFT method [8, 9]. The Kohn-Sham equation was solved self-consistently to calculate the properties of the proposed intermediate band materials. 3C-SiC crystallizes in space group of F-43m with lattice constant of $a = b = c = 4.348$ Å and $\alpha = \beta = \gamma = 90°$ at 300 K, respectively. The created unit cell for 3C-SiC has only 18 atoms (see Fig. 15.2a) and by replacement Fe atom with one of silicon atoms in the cell the concentration of Fe atoms will be 5.55 %. For decreasing the Fe concentration we

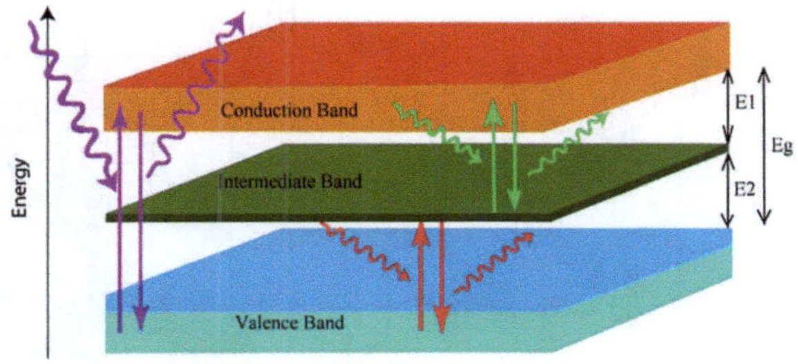

Fig. 15.1 Simplified band diagram showing the possible photogeneration processes. E_1 is the smallest sub-band gap and E_2 the largest, E_g denotes the main band gap

performed a calculation on larger systems. For this purpose we built $2 \times 1 \times 1$, $2 \times 2 \times 1$, and $2 \times 2 \times 2$ supercells (see Fig. 15.2b–d, respectively).

15.3 Simulation Results and Discussion

The band structure along the highly symmetrical direction of BZ and density of states (DOS) of the valence band (VB) and the conduction band (CB) for a 3C-SiC host semiconductor are shown in Fig. 15.3a, b, respectively. The lowest conduction band ege is located at X point and highest valence band energy ege is located at a G point. CB and VB consist s and p orbitals of 3C-SiC but d orbital dos not contributed in DOS of 3C-SiC.

The substitution of Fe with a silicon atom in 3C-SiC introduces intermediate energy levels in the band gap of 3C-SiC host material. The electronic transitions from the valence band to intermediate levels or from the intermediate level to the conduction band can effectively increases the incident light absorption. We calculated the band structure of Fe doped 3C-SiC for four different doping levels. The band structure for different Fe doped materials (1.05, 1.85, 3.22 and 5.55 %) has been depicted in Fig. 15.4a. We observe that, intermediate band levels have been increased by increasing in concentration of Fe atoms in the unit cell. The atomic configurations of Si, C, and Fe are $3s^2$ $3p^2$, $2s^2$ $2p^2$, and $3d^6$ $4s^2$, respectively. Every Si, C, and Fe atom contributes with 2, 2, and 8 valence electrons, respectively. The interaction between Fe orbitals and 3C-SiC orbitals create an isolated IB. In order to identify the orbital composition of these bands the partial density of states (PDOS) of Fe doped 3C-SiC for 1.05, 1.85, 3.22 and 5.55 % Fe concentration have been shown in Fig. 15.4b. This figure shows; the IB made up from contribution of Fe orbitals. DOS in intermediate is formed from a great contribution of Fe d orbitals with a small contribution of Fe p orbitals.

The intermediate band formation in Fe doped 3C-SiC has made it a suitable material for photovoltaic application. For this reason we have calculated the silicon carbide based solar cell power conversion efficiency for this material with the calculated electronic band structures shown in Fig. 15.4a. The calculations are based on the using of the detailed balance theory [9]. Figure 15.5 shows the ideal power conversion efficiency of this solar cell as a function of Fe concentration and

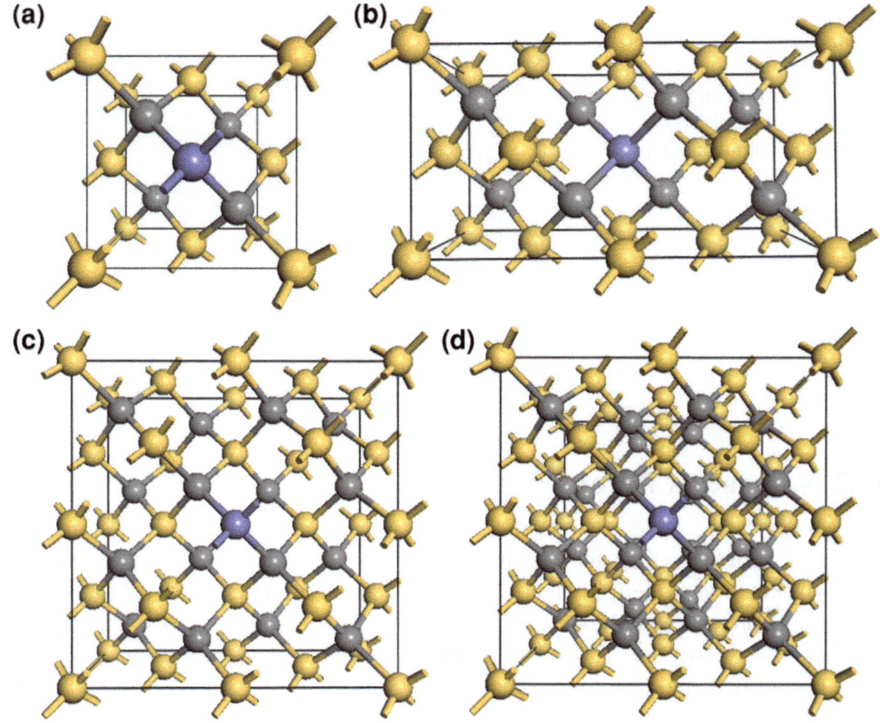

Fig. 15.2 Calculated structure containing **a** 1 × 1 × 1 supercell, **b** 2 × 1 × 1 supercell, **c** 2 × 2×1 supercell, **d** 2 × 2 × 2 supercell

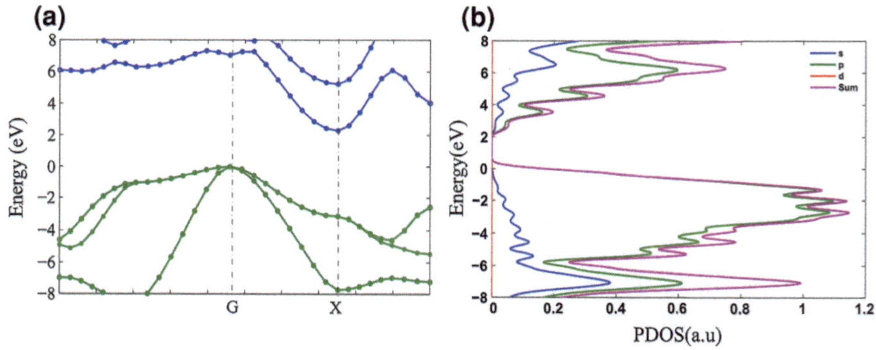

Fig. 15.3 **a** Calculated band structure of the 3C-SiC host semiconductor. **b** The total and partial DOS of the cubic silicon carbide

maximum efficiency obtained for 3.22 % Fe doping, which is higher than the ideal efficiency of any solar cells based on a single junction solar cell and is comparable to the efficiency of double and triple junction cells.

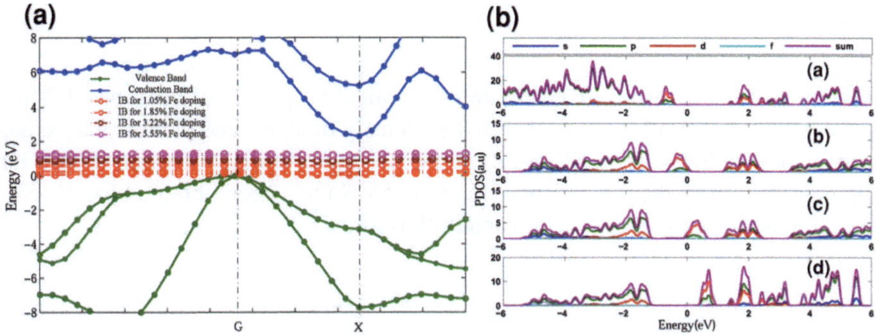

Fig. 15.4 **a** Band structure **b** PDOS for Fe doped 3C-SiC with concentration of 1.05, 1.85, 3.22 and 5.55 %

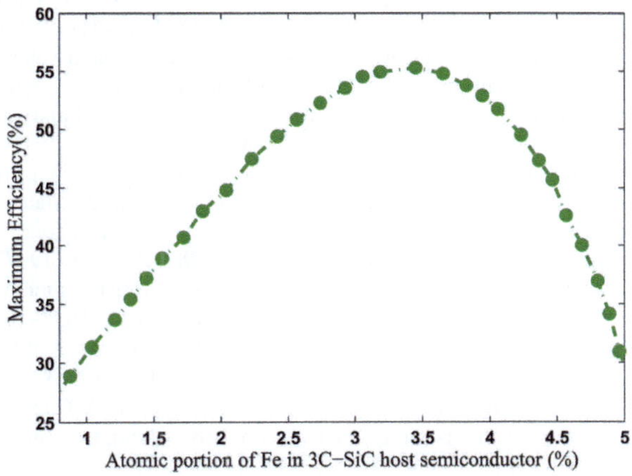

Fig. 15.5 The calculated power conversion efficiency as function of atomic portion of Fe

Table 15.1 The calculated Fe doped 3C-SiC features and performance parameters

Model	N_{Fe}	N_{total}	Fe (%)	E_{IB} (eV)	Eff (%)
$Si_{13}C_4Fe$ (1 × 1 × 1)	1	18	5.55	1.1	28
$Si_{22}C_8Fe$ (2 × 1 × 1)	1	31	3.22	0.95	55
$Si_{37}C_{16}Fe$ (2 × 2 × 1)	1	54	1.85	0.7	41
$Si_{62}C_{32}Fe$ (2 × 2 × 2)	1	95	1.05	0.3	30

In this paper first the band structure and DOS of Fe doped 3C-SiC with first principles calculation is achieved. The compounds having Fe present a narrow half-filled intermediate band isolated from the VB and the CB of the host semiconductor. The influence of the Fe concentration was performed and the maximum achievable efficiency was obtained by detail balance model. The important performed results of this work are collected in Table 15.1.

15.4 Conclusion

In the present study, we have calculated the optical properties of Fe doped 3C-SiC in different concentration of Fe atoms. The density of states and band structure has been achieved for 1.05, 1.85, 3.22, and 5.55 % Fe doping concentration. Our calculation showed the maximum efficiency (55 %) has been obtained for 3C-SiC intermediate band solar cell with substitution of 3.22 % Fe atoms.

References

1. M. Bhatnagar, B.J. Baliga, Comparison of 6H-SiC, 3C-SiC, and Si for power devices. Electron Dev. EEE Trans. **40**(3), 645–655 (1993)
2. R. Arora et al., Charge trapping properties of 3C-and 4H-SiC MOS capacitors with nitrided gate oxides. Nucl. Sci. IEEE Trans. **56**(6), 3185–3191 (2009)
3. A. Luque et al., Operation of the intermediate band solar cell under nonideal space charge region conditions and half filling of the intermediate band. J. Appl. Phys. **99**, 094503 (2006)
4. A. Luque, A. Marti, A metallic intermediate band high efficiency solar cell. Prog. Photovoltaics Res. Appl. **9**(2), 73–86 (2001)
5. T.S. Navruz, M. Saritas, Efficiency variation of the intermediate band solar cell due to the overlap between absorption coefficients. Sol. Energy Mater. Sol. Cells **92**(3), 273–282 (2008)
6. A. Luque, A. Martí, Increasing the efficiency of ideal solar cells by photon induced transitions at intermediate levels. Phys. Rev. Lett. **78**(26), 5014 (1997)
7. A. Martí, L. Cuadra, A. Luque, Quasi-drift diffusion model for the quantum dot intermediate band solar cell. Electron Dev. IEEE Trans. **49**(9), 1632–1639 (2002)
8. M.C. Warren, G.J. Ackland, Ab initio studies of structural instabilities in magnesium silicate perovskite. Phys. Chem. Miner. **23**(2), 107–118 (1996)
9. A.R. Ferreira, M.J. Martins, E. Konstantinova, R.B. Capaz, W.F. Souza, S.S.X. Chiaro, A.A. Leitão, Direct comparison between two γ-alumina structural models by DFT calculations. J. Solid State Chem. **184**(5), 1105–1111 (2011)

Chapter 16
Performance Analysis of Ultra-Thin Silicon Based Tunnel Junctions for Tandem Solar Cell Applications

H. Heidarzadeh, A. Rostami, M. Dolatyari and G. Rostami

Abstract Nowadays silicon solar cells have an efficiency of up to 20 % and in order to increase the efficiency of them, fabrication of multi-junction thin film solar cells with different band gaps is one of the most promising approaches. The silicon based tandem solar cells are third generation new style solar cells with ultra-high efficiency. The sub-cells in a tandem solar cell have different energy band gaps. In order to match the currents between sub-cells, tunnel junctions are used to connect the sub-cells. This work will concentrate on simulating the tunnel junction for application as part of multi-junction solar cell. In this way dopant concentration is changed and the tunnel junction current-voltage characteristics and their Energy band diagram in different dopant levels under equilibrium condition for moderate and usual doping have been calculated. An n++-Si/p++-Si tunnel junction is selected to simulate the overall characteristics of cell by numerical finite element method. We have simulated a symmetric silicon tunnel junction with thickness of 25 nm for n-type and 25 nm for p-type silicon by changing doping value from $1 \times 10e20\ cm^{-3}$ to $2 \times 10e20\ cm^{-3}$. The simulation results show that the doping concentration of $2 \times 10e20\ cm^{-3}$ is suitable for both sides.

16.1 Introduction

Tandem solar cells for concentrator photovoltaics demand excellent tunnel junction interconnects. Tunneling through a highly doped p-n junction is a phenomenon tightly related to quantum theory. Quantum tunneling refers to the quantum

H. Heidarzadeh · A. Rostami (✉)
Photonics and Nanocrystal Research Lab (PNRL), Department of Electrical and Computer Engineering, University of Tabriz, Tabriz, Iran
e-mail: rostami@tabrizu.ac.ir

A. Rostami · M. Dolatyari · G. Rostami
School of Engineering-Emerging Technologies, University of Tabriz, Tabriz, Iran

A. Y. Oral et al. (eds.), *International Congress on Energy Efficiency and Energy Related Materials (ENEFM2013)*, Springer Proceedings in Physics 155,
DOI: 10.1007/978-3-319-05521-3_16,

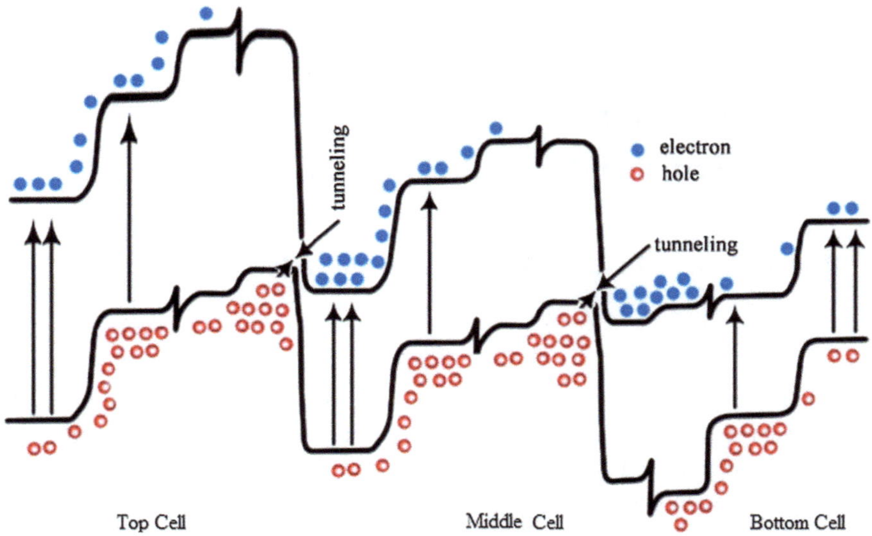

Fig. 16.1 Illustration of transport mechanism in tandem solar cells

mechanical phenomenon where a particle tunnels through a barrier that it classically could not surmount. It has important applications to modern devices such as the tunnel diode and the high speed transistors as well as the tandem solar cells [1–3]. If tandem solar cells are used to attain high efficiencies under high irradiance, one critical aspect of their design is the use of tunnel junctions that interconnect the sub-cells [4]. In this regard, the fundamental requirements for a tunnel junction in a tandem solar cell are (a) low optical absorption, (b) a high peak current, and (c) a low voltage drop. The necessity for a high peak current has special relevance in the ultra-high concentrated photovoltaic systems, in which the photocurrent produced in solar cells under concentrated light is elevated. Therefore, it is important to develop tunnel junctions that can ensure peak currents well above the maximum photocurrent generated by the light spot in the tandem solar cell. So that the tunnel junctions operate in their linear region in which the voltage drop is low.

Traditionally, bulk crystalline silicon solar cells dominate solar energy technology and single crystalline silicon wafers are used to fabricate devices which can achieve conversion efficiency up to 20 % [5]. The fundamental problem with this kind of device is electron (and hole) thermalization in which high energy photons dissipate excess energy as heat [6]. Materials with large band gap alleviate this phenomenon but also lead to lower short circuit current due to no absorption below the fundamental band gap. Tandem cells are proposed as a solution. In this scenario, two or more junctions with different band gaps are connected in series and light incident from the larger band gap side of the device. Carrier tunneling through the common junction is critical part of carrier transport in highly doped p-n junction solar cells. Figure 16.1 illustrates the carrier transport mechanism in the tandem structure solar cells. If the incident light creates an electron-hole pair in the bottom cell, the electron may tunnel from conduction band to the valance band of the toper cell and holes may tunnel from valance band to the conduction band of the bottom cell. This tunneling mechanism from the potential barrier describes the current mismatching solution.

One of the keys to making an effective model of a solar cell is to be able to model all components of the cell. The tunnel junction is a critical component of a tandem solar cell. So, this paper will concentrate on simulating the silicon based tunnel junction for application as part of tandem solar cells.

16.2 Simulation Results and Discussion

To quantify the optical losses in a silicon based tandem cell due to absorption in tunnel junction, we have calculated the absorbance for wavelengths above 500 nm (higher energy photons are considered to be absorbed in the top sub-cell) for different thicknesses of n++ and p++ layers using the transfer-matrix method [7]. The resulting photocurrent density loss is calculated for the constant spectrum from the absorbance assuming an internal quantum efficiency of 100 %. We have simulated a symmetric silicon tunnel junction with thickness of 25, 50 and 75 nm for each side of tunnel junction. The absorption as a function of wavelength is shown in Fig. 16.2 for n++/p++ silicon.

Figure 16.2 shows that decreasing the tunnel junction thicknesses already limit the absorption considerably. For the tandem cell efficiency this amount increases the losses. These calculations show that a significant improvement of tandem cell efficiency can be obtained by reducing the tunnel junction thickness.

When a sufficiently high electric field exists within a p-n junction, local bands may bend sufficiently to allow electrons to tunnel by internal field emission from the conduction band into the valence band; the symmetric behavior occurs for holes. At moderate doping levels, a tunneling effect can be seen in reverse bias, but if the junction doping levels are high enough, then this energy range may also exist in forward bias and tunneling effect can also be appreciated. Tunnel phenomenon in a highly doped p-n junction produces very different current and voltage characteristics than normal p-n junction diode. Figure 16.3a and b shows the current-voltage characteristic and energy band diagram for different dopant levels, respectively. It is clear that high dopant levels are appropriate for tunnel junction in tandem solar cells. When the forward bias on the junction is increased more of the electrons in filled states on the n-side of the junction will be at the same energy level as empty states on the p-side of the junction. This will cause the maximum tunneling effect of electrons moving from the n-side to the p-side. This is when the current peak will occur on I–V curve (this stage is clear in Fig. 16.2a). If the forward bias further increases, the diode will respond as current of the normal p-n junction. The negative resistance portion of the tunnel diode makes it useful in many application including oscillators, amplifiers, switches, and for connecting the layers of a tandem solar cell.

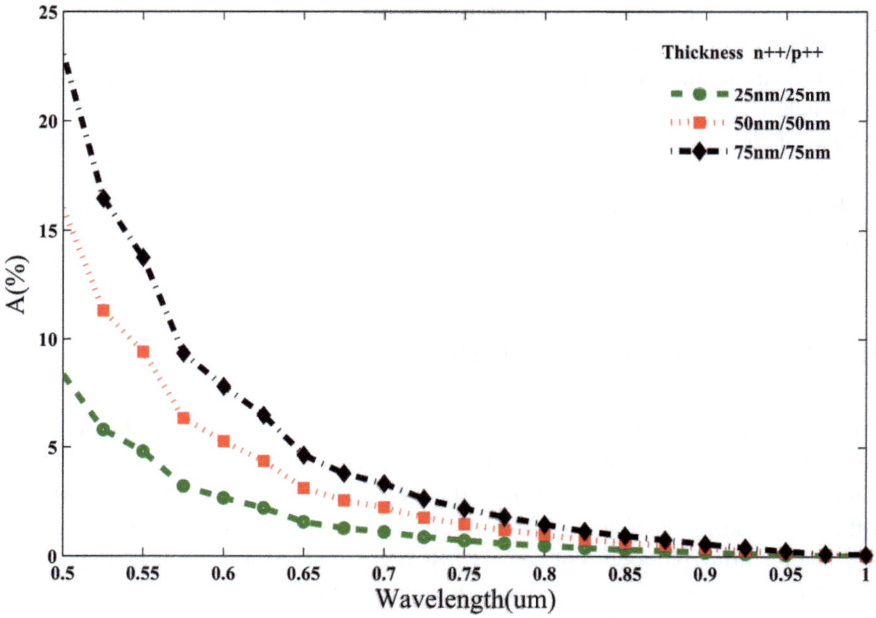

Fig. 16.2 Calculated absorbance in a n++/p++ silicon tunnel junction with different thicknesses

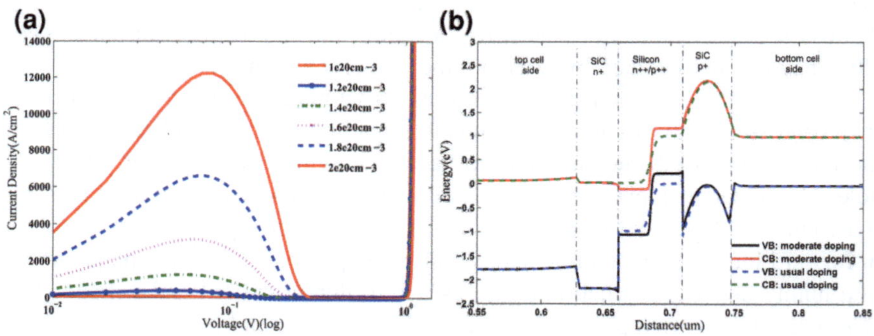

Fig. 16.3 **a** Tunnel junction current-voltage characteristic in different dopant levels. **b** Energy band diagram under equilibrium condition for moderate and usual dopant of tunnel junction

In the tandem cell a BSF (back surface filed) and passivated materials for the top and bottom cells with a band gap higher than the active material (i.e. SiC) is preferred. As can be observed, potential barriers are introduced in the p-side between the SiC barrier and the tunnel junction and bottom cell. However, valence band discontinuity in the p-part of a cell can block the cell current in one direction. Figure 16.3b shows that for the interface between the p++ layer of the tunnel junction and a SiC BSF, this phenomenon is similar. The energy barrier introduces

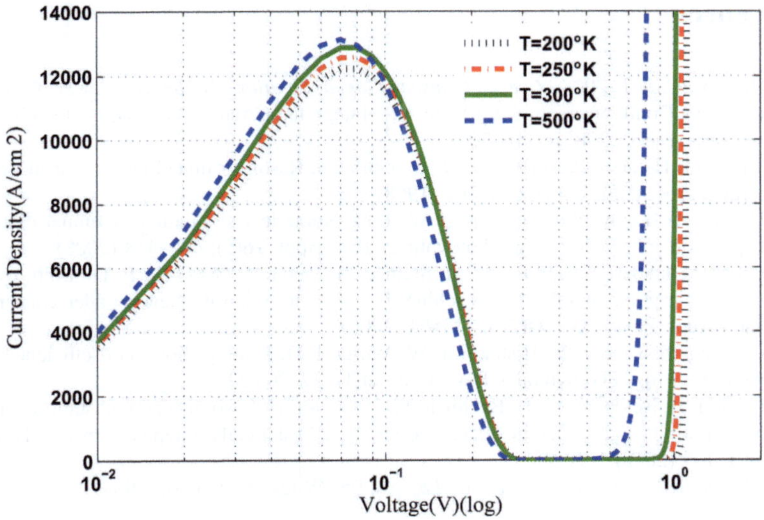

Fig. 16.4 Current-voltage characteristics of silicon tunnel junction at T = 200, 250, 300 and 500 °K

an extra series resistance to the structure. Tunneling of carriers through this barrier will be possible by high doping levels. Creating a perturbation in valence band energy of passivated low doped SiC layer is high that is not desirable for us. However, for n+ SiC the energy barrier do not make large discontinuity in the conduction band.

The temperature dependence of current-voltage characteristics of a n++/p++ silicon tunnel junction has been simulated. Figure 16.4 depicts the current-voltage characteristics of tunnel junction with thickness of 25 nm for n-type and 25 nm for p-type silicon with doping concentration of 2 × 10e20 cm^{-3} for both sides in temperatures of 200, 250, 300 and 500 K.

16.3 Conclusion

In this paper, we evaluated the minimum thickness of high peak current silicon tunnel junction in order to minimize the absorption. We showed that by reducing the tunnel junction thickness, due to absorption in tunnel junction losses decrease. The simulation results show the thickness of 25 nm for n-type and 25 nm for p-type silicon with doping level of 2 × 10e20 cm^{-3} is suitable for high efficient tandem cells. In addition the tunnel junction current-voltage curves in different temperature were calculated.

References

1. C. Hu, in *ICSICT 2008*: *Green Transistor as a Solution to the IC Power Crisis*. 9th International Conference on Solid-State and Integrated-Circuit Technology, October 2008 (IEEE, New York, 2008) pp. 16–20
2. J.P. Sun, G.I. Haddad, P. Mazumder, J.N. Schulman, Resonant tunneling diodes: models and properties. Proc. IEEE **86**(4), 641–660 (1998)
3. M. Hermle, G. Letay, S.P. Philipps, A.W. Bett, Numerical simulation of tunnel diodes for multi-junction solar cells. Prog. Photovoltaics Res. Appl. **16**(5), 409–418 (2008)
4. W. Guter, J. Schone, S.P. Philipps, M. Steiner, G. Siefer, A. Wekkeli, F. Dimroth, Current-matched triple-junction solar cell reaching 41.1 % conversion efficiency under concentrated sunlight. Appl. Phys. Lett. **94**(22), 223504 (2009)
5. M.A. Green, K. Emery, Y. Hishikawa, W. Warta, E.D. Dunlop, Solar cell efficiency tables (version 39). Prog. Photovoltaics Res. Appl. **20**(1), 12–20 (2011)
6. A. Rostami, H. Heidarzadeh, H. Baghban, M. Dolatyari, H. Rasooli, Thermal stability analysis of concentrating single-junction silicon and SiC-based solar cells. J. Optoelectron. Adv. Mater. **15**(1–2), 1–3 (2013)
7. P. Yeh, *Optical Waves in Layered Media*, vol. 95 (Wiley, New York, 1988)

Chapter 17
A Study on Applications of Holography in Solar Energy Installations

Hamed Abbasi, Adeleh Granmayeh Rad, Talie Zarei and Neda Jalali Farahani

Abstract In this paper, applications of holography in the solar energy, photovoltaic concentration, daylighting, illumination, and thermal blocking have been investigated. Holographic elements can be used to concentrate the radiation of the sun onto photovoltaic cells. Moreover the sun radiation is diffracted by the hologram. This has the advantage that proper photovoltaic cells can be installed in different spectral regions. Holographic daylighting systems can diffract sunlight efficiently up to the ceiling in the room. Holograms can be fabricated to reflect or block certain wavelength regions. Holograms are designed for a wavelength in the infrared region and a typical incidence angle of the summer sunlight. In this case the infrared radiation is reflected in the summer and heating of the rooms can be reduced, also, energy for cooling by air conditioners can be reduced. In winter, that the angle of incidence is smaller, a larger wavelength region is reflected, where the infrared intensity is very low.

17.1 Introduction

The well-known scientist, Gabor [1], found basic principles of holography while trying to improve the efficiency of transmission electron microscope in 1948. He performed his first experiments using mercury vapor lamp. After 23 years of

H. Abbasi (✉)
Department of Photonics, Faculty of Physics, University of Kashan, Kashan, Iran
e-mail: hamed.abbasi@grad.kashanu.ac.ir; hamedabbasi@gmail.com

A. Granmayeh Rad
Department of Physics, Faculty of Sciences, Roudehen Branch, Islamic Azad University, Roudehen, Iran

T. Zarei · N. Jalali Farahani
Plasma Physics Research Center, Science and Research Branch, Islamic Azad University, Tehran, Iran

A. Y. Oral et al. (eds.), *International Congress on Energy Efficiency and Energy Related Materials (ENEFM2013)*, Springer Proceedings in Physics 155,
DOI: 10.1007/978-3-319-05521-3_17, © Springer International Publishing Switzerland 2014

experiments, Gabor won the Nobel Prize in 1971. Nowadays, holography has a broad range of applications in various sciences [2]. There are many types of holographic products; one of them is holographic optical elements (HOEs). One of the most applicable holographic optical elements is holographic grating. The production of holographic gratings does not require additional knowledge of holography and its experimental setups. Holographic gratings can be used in solar energy installation. These gratings can be used to concentrate the radiation of the sun onto photovoltaic cells. Also, they can be used for coating solar cells with antireflection thin films. In addition, holographic daylighting systems have been developed to diffract sunlight efficiently up unto the ceilings of buildings. Many analyses and designs have been done for holographic solar systems such as [3–5]. James and Bahaj [6] have predicted the performance of holographic systems for the UK climate. The most common holographic materials developed to be applied in solar energy are dichromated gelatine and photopolymers. Besides holography, there are other optical methods such as using the set of prism [7] and planar micro-optic system [8] to collect and homogenize sunlight in photovoltaic cells. This paper has done a study on applications of holography in the solar energy, photovoltaic concentration, daylighting, illumination, and thermal blocking.

17.2 Solar Concentration and IR Blocking

One interested person can find the history of concentrator development in [9]. Figure 17.1 show how the white light of the sun is divided into different parts after passing the holographic grating. Each part of the sunlight hits the appropriate solar cell. In this way, the infrared part of light, which cannot be transformed into electrical energy and increases the cell temperature, does not hit any of the cells. For instance, one can guide the shorter wavelengths to the AlGaAs/GaAs solar cells and the longer wavelengths to the CulnSe/CdS solar cells. Another option is to guide the short wavelengths to GaAs, middle wavelengths to Si, and long wavelengths to Ge solar cells. In this way, low energy photons hit the low band gap cells and high energy photons hit the high band gap cells.

Other solar concentration systems which are currently less used but are the subject of some study are Fluorescent/Luminescent Concentrators and Quantum Dot Concentrators [10]. Different companies manufacture holographic concentrator; Prism Solar Technologies is a well-known company in this field [11].

Holograms can be fabricated to reflect or block certain wavelength regions. Holograms are designed for a wavelength in the infrared region and a typical incidence angle of the summer sunlight. In this case the infrared radiation is reflected in the summer and heating of the rooms can be reduced, also, energy for cooling by air conditioners can be reduced. In winter, that the angle of incidence is smaller, a larger wavelength region is reflected, where the infrared intensity is very low.

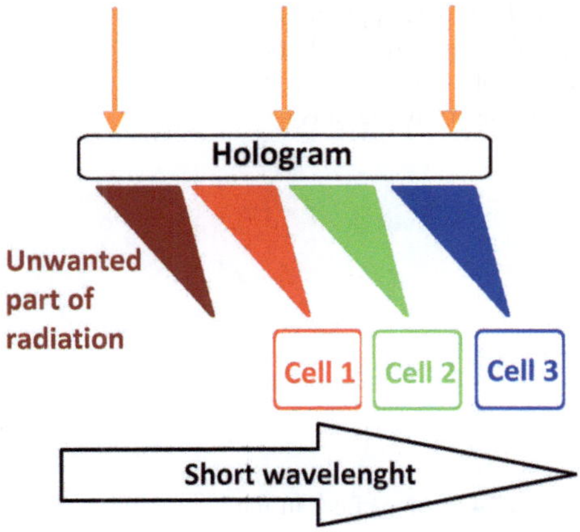

Fig. 17.1 The ability of holographic optical elements to guide the appropriate part of sunlight to the appropriate solar cell

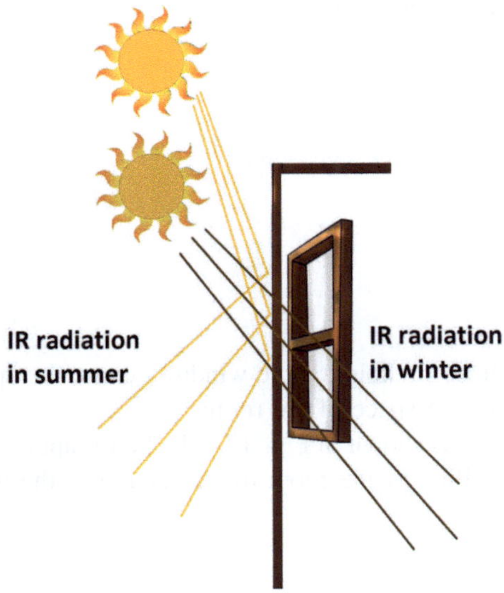

Fig. 17.2 The mechanism of holographic windows in summer and winter

As shown in Fig. 17.2 in the hot weather of summer the IR part of the sunlight which increases the temperature of the rooms does not pass the holographic window, but in winter which rooms need to be heated this part of light passes the window.

17.3 Daylighting

Scientists have always been interested in studying the holographic optical elements for daylighting applications in buildings [12, 13]. Holographic daylighting systems can diffract sunlight efficiently upto the ceiling in the room. Figure 17.3 shows

Fig. 17.3 The appropriate distribution of the sunlight in the whole space of the room by means of holograms

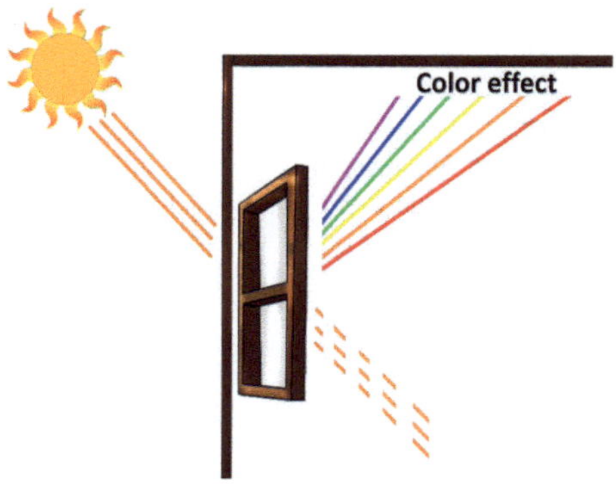

Fig. 17.4 Color effect in the ceiling of the room by means of dispersion of the sunlight using holographic gratings

how the holographic windows cause the appropriate distribution of the light in the whole space of the room.

As shown in Fig. 17.4 the holographic gratings can create the color effect in the ceiling of the room by dispersion of the white light of the sun.

17.4 Making Holographic Gratings

Holographers have always tried to produce large format holograms in different ways [14]. Different types of holograms with various holographic materials have been studied for solar applications such as photopolymers [15–17]. The results show that the photopolymer holographic optical lenses achieved high diffraction efficiency at relatively low spatial frequency and recording intensities; dichromated gelatin holographic layers exhibit high optical quality and diffraction efficiency. The diffraction and the dispersion properties of holographic optical elements are the most important properties for use as solar concentrators for photovoltaic conversion systems. Figure 17.5 shows the experimental setup for producing holographic gratings. The coherent beam of the laser hits the telescopic

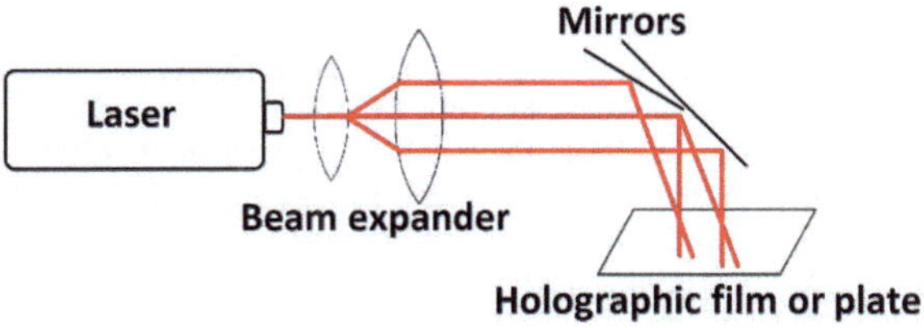

Fig. 17.5 Experimental arrangement for producing holographic gratings

system, beam expander, including a pinhole, a spatial filter and a lens. The expanded beam hits the mirrors which are oriented at different angles, and finally hits the holographic surface.

17.5 Conclusion

In this paper some useful aspects of holographic gratings have been studied for use in solar concentrator systems. Also, a review of the some scientific reports on concentrating systems has been presented. The diffraction and the dispersion properties of holographic optical elements are the most important properties for use as solar concentrators for photovoltaic conversion systems. The results show that the photopolymer holographic optical lenses achieved high diffraction efficiency at relatively low spatial frequency and recording intensities; dichromated gelatin holographic layers exhibit high optical quality and diffraction efficiency. A major proportion of light, is holographic systems, is directed upwards to the ceiling and the rear of the room. Daylighting enhancement systems use different approaches based on the optical phenomena of light propagation, scattering, specular reflection, total internal reflection, optical refraction and optical diffraction. Holographic system uses the principle of diffraction to redirect sunlight. Holograms are produced on films, which are laminated between two glasses, in order to illuminate the darker regions and to reduce glare. Laminated glass with holograms allows a great variety of applications in architecture for solar energy systems, improvement of room comfort as well as design of solar light- and color effects.

References

1. D. Gabor, Nature **161**(4098), 777–778 (1948)
2. A.G. Rad et al., in *Proceeding of the International Conference on Innovation and Collaboration in Engineering Research*, Bucharest, 20–21 June 2013
3. J. Russo et al., in *Proceeding of Photovoltaic Specialists Conference 38th IEEE*, Austin, 3–8 June 2012
4. E.R. Torrey et al., J. Appl. Phys. **109**, 074909 (2011)
5. J.M. Castro et al., Appl. Opt. **49**, 5 (2010)

6. P.A.B. James, A.S. Bahaj, Sol. Energy **78**(3), 441–454 (2005)
7. M. Stefancich et al., Opt. Express **20**, 8 (2012)
8. J.H. Karp et al., Opt. Express **18**, 2 (2010)
9. R.M. Swanson, Prog. Photovolt. Res. Appl. **8**, 93 (2000)
10. D. Chemisana, Renew. Sustain. Energy Rev. **15**, 1 (2011)
11. Prism Solar Technologies, http://www.prismsolar.com. Accessed 17 Oct 2013
12. H.D. Tholl et al., in *SPIE 2017* (1993)
13. J. Ludman et al., in *SPIE 2532* (1995)
14. C.G. Stojanoff, in *SPIE 3011* (1997)
15. S.T. L. Sam, A.P. T. Kumar, in *AIP Conference Proceedings*, vol. 1391 (2011) p. 248
16. H. Akbari et al, in *Proceeding of 2nd International Conference on Sustainable Energy Storage*, Trinity College, Dublin, 19–21 June 2013
17. J.A. Quintana et al., Appl. Opt. **36**, 19 (1997)

Chapter 18
A Process Heat Application Using Parabolic Trough Collector

İbrahim Halil Yılmaz, Mehmet Sait Söylemez, Hakan Hayta and Recep Yumrutaş

Abstract A pilot study has been performed based on a heat process application that is designed, installed and tested at Gaziantep University to establish the technical and economic feasibility of high temperature solar-assisted cooking process. The system has been designed to be satisfying the process conditions integrated with parabolic trough solar collector (PTSC). It is primarily consists of the PTSC array, auxiliary heater, plate type heat exchanger, cooking system and water heating tanks. In the operation of the process heat application, the energy required to cook wheat (used as cooking material) has been supplied from solar energy which is transferred to heat transfer fluid (HTF) by heat exchanging units and finally discharged to water in order to produce bulgur. The performance parameters of the sub-systems and the process compatibility have been accomplished depending on the system operation. In addition that the system performance of the high temperature solar heat process has been presented and the recommendations on its improvement have been evaluated by performing an experimental study. As a result that the use of solar energy in process heat application has been projected and its contribution to economics view with respect to conventional cooking systems has been conducted.

İ. H. Yılmaz (✉) · M. S. Söylemez · H. Hayta · R. Yumrutaş
Department of Mechanical Engineering, University of Gaziantep, 27310 Gaziantep, Turkey
e-mail: iyilmaz@gantep.edu.tr

M. S. Söylemez
e-mail: sait@gantep.edu.tr

H. Hayta
e-mail: hakanhyt@hotmail.com

R. Yumrutaş
e-mail: yumrutas@gantep.edu.tr

A. Y. Oral et al. (eds.), *International Congress on Energy Efficiency and Energy Related Materials (ENEFM2013)*, Springer Proceedings in Physics 155,
DOI: 10.1007/978-3-319-05521-3_18,

18.1 Introduction

Solar energy is free and inexhaustible however it tends to show intermittent and variable character by its nature. The thermal radiation coming from the sun changes instantly during the time. This character may exhibit technical challenges regarding the type of thermal process since most of the thermal systems operate under steady conditions thus need stable energy transfer [1]. Nevertheless, solar systems have been adapted and coupled with many thermal systems for different process applications. These types of applications require a detailed thermal analysis and/or an experimental study for technical feasibility and process compatibility.

Bulgur (in general produced from Triticum durum) is a very famous industrially processed ancient wheat product around the world. Annual bulgur production capacities in Turkey and on the American Continent are approximately 1,125,000 million and 300,000 tonnes, respectively. Due to increase in the market demands, European and Middle East companies have started to construct new bulgur factories and their production capacities are on the level of 80,000 and 120,000 tonnes, respectively [2]. The market enlargement entails cost reduction in order to increase the production capacity.

Bulgur is basically produced using cooking, drying, dehulling, grinding and classification steps [3]. Cooking and drying processes requires remarkable thermal energy which is generally met from fossil fuels in conventional cooking systems. Although the energy need for cooking process is lower than drying, cooking operation requires thermal energy about one-third of drying at moderate temperatures. In order to deliver high temperatures with good efficiency, a high performance solar collector is required [4]. For process-heat applications up to 250 °C could complete the variety of solar thermal collectors however the most feasible can be considered as Parabolic Trough Solar Collector (PTSC) which can effectively produce heat for the concerned process even at higher temperatures up to 400 °C.

In this study, an experimental research has been investigated on solar assisted process heat application through a cooking process using a PTSC array. System design and configuration of the thermal application have been evaluated and its projection and contribution to economics with respect to conventional cooking systems has been examined.

18.2 Experimental Setup

The experimental setup was designed and installed to investigate the thermal performance of cooking system assisted with solar energy. The process layout of the experimental setup and its representation are shown in Figs. 18.1 and 18.2, respectively. The setup consists of two closed loop which are connected each other with a plate type heat exchanger (PHE). In the primary closed loop, the PTSC array composed of a series connected of three collectors was used to receive solar

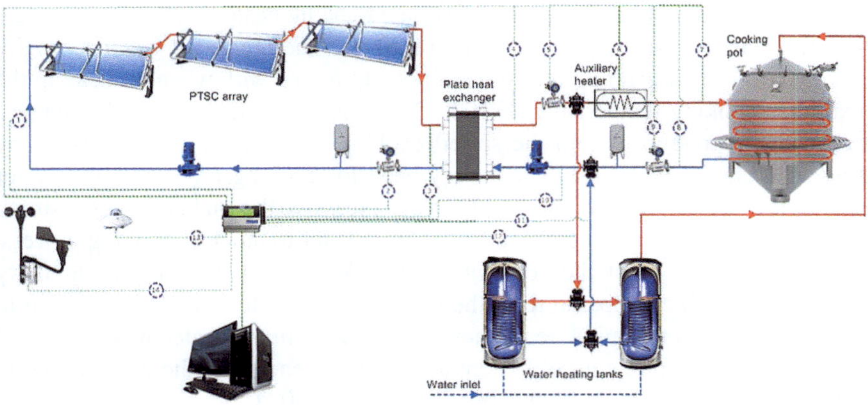

Fig. 18.1 Experimental layout of the process heat application

Fig. 18.2 Experimental setup

radiation and concentrate it onto the receiver element. Thermal oil (Fuchs Renotherm 320) was used in both loops as a working fluid. It was circulated within the absorber tube of the PTSC by means of a frequency-controlled gear pump which keeps flowrate constant. The thermal oil within the secondary loop was heated with the fluid used in the PTSC array via the PHE after the vanes opened. Thermal stability was attained in both loops during heat transfer in the PHE. After steady state conditions and preset temperature were reached in the secondary loop, the water inside the water heating tank was started to being heated for cooking process from ambient to preheating temperature of 90 °C. The preheated water ready for cooking was discharged into the cooking pot and mixed with wheat to heat up the mixture to cooking temperature of 100 °C. During cooking process, the temperature and the pressure within the sealed cooking pot were adjusted constant by using a pressure relief valve. The energy need for cooking process was only for meeting the heat losses from the pot. Thus the excessive thermal energy was stored in the water heating tank to preheat the cooking-water to the next cooking process. The inlet temperature of the cooking pot was kept constant using an auxiliary heating system. The temperature of the thermal oil was fluctuated slightly due to variable structure of coming beam radiation. The auxiliary heater was controlled with a PID temperature control system and supplied energy to the cooking pot to set the inlet temperature at a predefined value, i.e. 150 °C.

18.3 Process Arrangement

Three system arrangements were considered on the process heat application to examine the energy efficiency. The first one is the sizing of the solar system regarding a nominal value due to daily variation of the solar beam radiation. Thus, the time period to make up the nominal value throughout a year is to be identified, for instance, the season between May and September. Such a system design is able to meet the desired conditions at the level of averaged value of solar intensity and provides undersized system selection. Nonetheless this arrangement maintains thermal energy to the heat process on the level of total equivalent solar energy. In other words, in the case of sufficient solar energy i.e. when the solar energy is higher than normalized value, the excessive energy will be stored in water heating tank. When the solar energy is insufficient to cook wheat, the electric heater (auxiliary heating system) will supply energy to the working fluid. In this case, the energy required to cook a specified amount of wheat will be determined according to energy value of the nominal solar radiation. Collector sizing in this condition is to be specified with respect to the nominal value. Thus the beam radiation falling on the aperture of the collector array can be normalized.

The second arrangement is evaluated as the system designed according to a peak beam radiation value that takes place in summer season. In this type of system requires an oversized auxiliary heater and keeps the desired conditions at all times even under the worst conditions but may not be practical. Since an oversized system selection will have a higher initial cost and probably have a higher operating cost because the equipment in this case will run at partial load most of time and thus at a lower efficiency. Thus this arrangement will not store thermal energy for preheating of the cooking-water to the next cooking process and will lower the daily production capacity.

The third arrangement is considered as the system designed to be operating only with solar energy. However this system will not be time and product effective as a result of working under unsteady conditions. Additionally, the product capacity and energy transfer of the system will not be controlled easily. The operating temperatures of both cycles will get lower and fluctuate during the system operation.

18.4 Conclusion

In this study, the system design of the process heat application coupled with solar energy have been investigated and evaluated using an experimental setup. Cooking process has been evaluated as process heat and achieved with the PTSC system. Different system design arrangements have been evaluated and the most feasible system configuration has been proposed from the view of process compatibility and feasibility when coupled with the solar energy system.

Acknowledgments The authors acknowledge the financial support provided by the Scientific Research Projects Governing Unit of Gaziantep University.

References

1. İ.H. Yılmaz, M.S. Söylemez, Design and computer simulation on multi-effect evaporation seawater desalination system using hybrid renewable energy sources in Turkey. Desalination **291**, 23–40 (2012)
2. http://www.investingaziantep.org/
3. M. Bayram, Modelling of cooking of wheat to produce bulgur. J. Food Eng. **71**, 179–186 (2005)
4. S. Kalogirou, The potential of solar industrial process heat applications. Appl. Energy **76**, 337–361 (2003)

Chapter 19
On the Use of Agent-Based Simulation for Efficiency Analysis of Domestic Heating Using Photovoltaic Solar Energy Production Combined with a Heatpump

Jan Treur

Abstract In this paper agent-based simulation is used to analyse the efficiency of domestic heating based on a heatpump together with photovoltaic (PV) solar energy production. A simulation model for the cost (in terms of required kWh per day) of a heating agent based on a heatpump over a year is used, in addition to a simulation model for the yields of a PV production agent estimating the produced solar energy (in kWh per day). In particular, for the heating agent it is analysed how its performance depends on the outdoor temperature, and for the PV-installation agent how the yields depend on irradiation. Based on empirical temperature and irradiation data over a year it is found out which fraction of the energy required per year for heating can be covered by the yields of the PV installation.

19.1 Introduction

In many countries a substantial amount of domestic energy usage per year is used for heating. Due to the negative impacts on the environment of traditional heating systems based on not renewable resources such as gas and oil, often a heatpump is considered as an interesting alternative (e.g., [1, 4, 6, 9, 15]). Most often a heatpump is considered which takes thermal energy from the air (air to water heatpump), as this is most easy to install. The heatpump is driven by electrical energy, which preferably should be produced in a renewable manner as much as possible. Two possibilities for that are by wind energy or by solar energy. In this paper the possibility to use photovoltaic solar energy to produce electricity to drive the pump is analysed. At first sight it may seem that the fact that most solar energy is

J. Treur (✉)
Agent Systems Research Group, VU University Amsterdam, De Boelelaan 1081, 1081 HV Amsterdam, The Netherlands
e-mail: j.treur@vu.nl
URL: http://www.cs.vu.nl/~treur

A. Y. Oral et al. (eds.), *International Congress on Energy Efficiency and Energy Related Materials (ENEFM2013)*, Springer Proceedings in Physics 155,
DOI: 10.1007/978-3-319-05521-3_19,

produced in the warmer season when no or not much heating is needed, will lead to a mismatch between demand and production. This may be true to some extent, but the study reported here was undertaken to find out in more detail which fraction of the required energy still can be covered by solar energy produced by the PV-installation.

The agent-based simulation model consists of a model for the performance of a heating agent depending on the daily outdoor temperature (described in Sect. 19.2) and a model for the performance of a solar energy production agent depending on the amount of irradiation per day (described in Sect. 19.3). In Sect. 19.4 a setup of the simulation experiments combining the two agent models is described and some of the results are discussed. Finally, Sect. 19.5 is a discussion in which further possibilities of the approach are pointed out.

19.2 Modeling Performance of the Heating Agent

In [14] a detailed model for an of an air to water heatpump-based heating agent was introduced. For the current paper, this agent model will be adopted. A central element of the model is the seasonal performance factor (*SPF*) which indicates how much electric energy (in kWh) is needed (to run the heatpump) as input to get a certain amount of heating energy as output for the heatpump over a certain time period:

$$SPF = \frac{energy\ output}{energy\ input}$$

This factor usually varies between 2 and 4; for a given water temperature of the heating system, it strongly depends on the outdoor temperature, and in particular the difference between these two temperatures. Manufacturers often only give indications of these performance factors for just a few water and outdoor temperatures. However, to determine the electricity usage of a heatpump over a year, with all its variations in outdoor temperature, it is needed to have a more systematic estimation of *SPF* for a given water temperature and each possible outdoor temperature, in a realistic context. To obtain a reasonable estimation of how for a given water temperature the performance factor depends on the outdoor temperature, theoretical analyses or lab experiments can be performed. However, such theoretical analyses are often not guaranteed to provide values that occur in realistic situations. A different route is to take empirical data from realistic contexts as a point of departure and make an interpolation and approximation of them by a mathematical function. For example, in [16, p. 2372] this was done based on the manufacturer's catalog data. However, a useful source of more realistic real world data can be found at the website www.liveheatpump.com. This approach was used in [14] and is adopted here as well. In Fig. 19.1 a graph is shown with values from this Website for the average day temperature on the horizontal axis

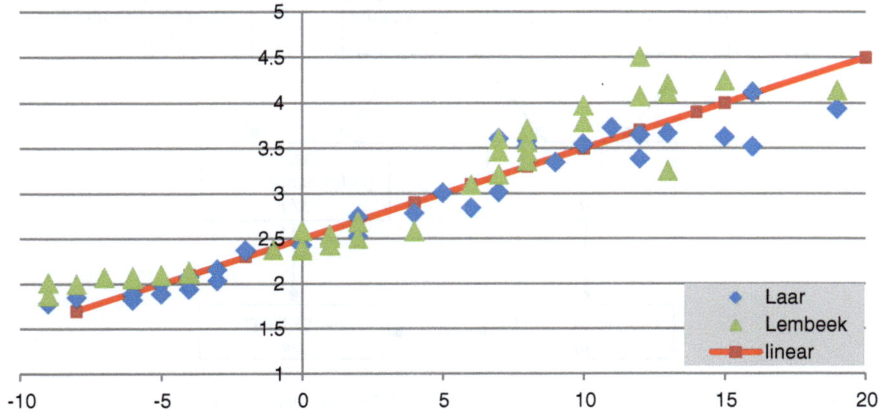

Fig. 19.1 Linear approximation of seasonal performance factors in relation to outdoor temperature compared to empirical data over 2012 for sites in Lembeek and Laar (water temperature 50 °C)

and the performance factor on the vertical axis (for water temperature in the heating system approximately 50 °C).

More specifically, this has been done for the sites at Lembeek and Laar, where the General Waterstage HT heatpump combination WH16/WOH16 is used. Moreover, in Fig. 19.1 a linear approximation of *SPF* for the interval from −10 °C to +20 °C is drawn; this is assumed of the form

$$SPF(T_{od}) = 7.5 - 0.1 * (T_{water} - T_{od}) \quad \text{with } T_{water} = 50$$

Note that in [14] also a quadratic approximation was shown, as was done in [16]. Using the linear approximation (for the sake of simplicity) shown in Fig. 19.1, the seasonal performance factor *SPF* can be estimated on a daily basis throughout a year, when the day temperatures are given. This is one basic ingredient of the model for the heating agent used. A second ingredient of the model concerns how much energy for heating is needed, also depending on the outdoor temperature. A general format to determine how much energy is to be provided for the heating makes use of the concept of *degree day*, denoted by *dd*. This concept is based on the assumption that the amount of energy needed to maintain a difference in temperature (between indoor and outdoor) is proportional to this difference (e.g., see [10]). The number of degree days for a given day linearly relates to the difference between the daily outdoor and the indoor temperature T_{id} (when the latter is higher than the former, else 0):

$$dd(t) = \begin{matrix} \sigma(t)^*(T_{id}(t) - T_{od}(t)) & \text{when} \quad T_{id}(t) > T_{od}(t) \\ 0 & \text{Otherwise} \end{matrix}$$

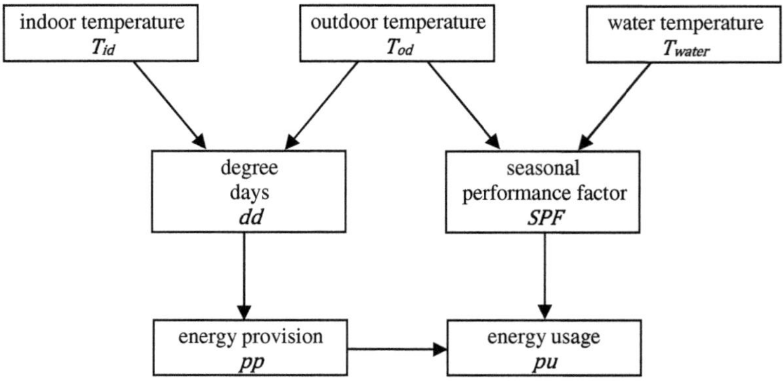

Fig. 19.2 Dependencies of the variables for the model of the heating agent

Here $\sigma(t)$ is as seasonal correction weight factor which is 1.1 for the months November, December, January and February, 1 for the months March and October, and 0.8 for the months April, May, June, July, August and September.

Figure 19.2 shows the variables used in the heating agent model and the dependencies between them, and Table 19.1 summarizes them. The right hand side in Fig. 19.2 describes how the performance factors are determined, and the left hand side describes how the energy demand is determined.

The model results in the electrical energy usage *pu(t)* needed as input for the heating agent. This is determined from the heating energy *pp(t)* provided (as output) to the heating system (*kWh* provided for heating at day *t*) as follows. For each degree day an amount ε of energy (in *kWh*; the value 4.0 kWh/dd is used here) has to be provided as output. Therefore *pp(t)* is determined as (see also Fig. 19.2):

$$pp(t) = \varepsilon \, dd(t)$$

This is the amount provided as output to the heating system, but not the amount *pu(t)* used as input by the heatpump itself, since part of the provided energy *pp(t)* comes from the air in the environment. This is expressed using the seasonal performance factor *SPF*; recall that by definition this is the heatpump's output divided by its input. Therefore the following is obtained:

$$pu(t) = pp(t)/SPF(T_{od}(t)) = \varepsilon \, dd(t)/SPF(T_{od}(t))$$

In addition to the model to determine the energy usage per day a model to describe the cost has been included. Assuming that one *kWh* electricity costs π_{el} euro, the costs *pc(t)* of heating for day *t* by the heatpump is given by

$$pc(t) = pu(t) * \pi_{el} = (\pi_{el}\varepsilon/SPF(T_{od}(t))) \, dd(t)$$

Table 19.1 Main concepts for the heating agent

Notation	Description	Unit
SPF	Seasonal performance factor	
$T_{od}(t)$	Average outdoor temperature at day t	°C
$T_{id}(t)$	Average indoor temperature at day t	°C
$T_{water}(t)$	Average water temperature of the heating system at day t	°C
$\sigma(t)$	Seasonal weight factor	
dd(t)	Degree days at day t	dd
pp(t)	Heat pump heating energy provision for day t	kWh
pu(t)	Heat pump electrical energy use for day t	kWh
pc(t)	Heat pump electrical energy cost for day t	euro
ε	Energy needed per degree day	kWh/dd
π_{el}	Price of electrical energy	euro/kWh

Note that this cost parameter π_{el} depends on the source of the electrical energy. If it is taken from an external production company it may be higher than when it is taken from the own PV-installation. This is discussed further in Sect. 19.4.

19.3 Modeling the Photovoltaic Solar Energy Production Agent

In [13] a detailed agent-based model of a photovoltaic solar energy production system has been described. The agent model used here will be an abstracted form of the agent-based model from [13] in two ways. A main difference is that in [13] the PV installation is modelled as a system composed of multiple agents, where each agent consists of one solar panel and one connected micro-inverter. In contrast, here the PV installation as a whole is modelled as one solar energy production agent. This is a form of abstraction in the agent cluster dimension (cf. [3]). A second type of abstraction with respect to the model in [13] takes place in the temporal dimension (cf. [3]). While [13] describes a dynamical model for the pattern over time in detail at a grain size of the hours or minutes of a day (e.g., with updated states per half an hour), in the current approach these processes are aggregated into a dynamical model based on states per day as follows. For a summary of the symbols used, see (Table 19.2). In general, the power $P_{out}(u)$ generated as output at time u can be described as a function of:

- used irradiation at u, which itself depends on
  - the available irradiation $irr(u)$ at time u
  - the efficiency ρ_a due to angle and orientation of the panel
  - the efficiency ρ_s due to shadow

- the maximal power $P_{panelpeak}$ of the panels (Watt peak)
- the efficiency ρ_{panel} of the panels
- the maximal power $P_{inverterpeak}$ of the inverters
- the efficiency ρ_{inv} of the inverters

Note that as another form of temporal abstraction all efficiency factors ρ_i are assumed to be aggregated over a year, so that they are taken constant within the year. The following relations are assumed for each point in time *t*:

- Provided power $P_{panel}(u)$ by panel: $\min(P_{panelpeak}, \rho_a\ \rho_s\ irr(u))\ \rho_{panel}$
- Provided power $P_{out}(u)$ by inverter: $\min(P_{inverterpeak}, P_{panel}(u))\ \rho_{inv}$

Given that $\rho_{panel} < 1$, when it is assumed that $P_{inverterpeak} > P_{panelpeak}$ and $P_{panelpeak} > \rho_a\ \rho_s irr(u))\ \rho_{panel}$, this can be simplified into:

$$P_{out}(u) = \rho_a \rho_s \rho_{panel} \rho_{inv}\ irr(u)$$

Such a relation can also be aggregated to the energy (in kWh) obtained for a day *t*, where the integration is taken over times *u* within day *t*:

$$\begin{aligned} E_{day}(t) &= \int P_{out}(u)du \\ &= \int \rho_a \rho_s \rho_{panel} \rho_{inv} irr(u)du \\ &= \rho_a \rho_s \rho_{panel} \rho_{inv} \int irr(u)du \\ &= \rho_a \rho_s \rho_{panel} \rho_{inv} irr_{day}(t) \end{aligned}$$

For the case of no shadow $\rho_s = 1$. For cases that a most optimal angle is used, also the value for ρ_a can be set to 1; for example, for mid-Europe often 30 or 35 is assumed for this optimal angle. In such a case it simplifies into:

$$E_{day}(t) = \rho_{panel} \rho_{inv} irr_{day}(t)$$

In the general case as the model addresses in particular how the energy production is distributed over the days in a year, the energy fraction $EF_{day}(t)$ of a day relative with respect to the year production is a relevant notion:

$$EF_{day}(t) = E_{day}(t)/E_{year}$$

with

$$E_{year} = \Sigma E_{day}(t) = \rho_a \rho_s \rho_{panel} \rho_{inv}\ irr_{year}$$

Table 19.2 Main concepts for the energy production agent

Notation	Description	Unit
$P_{panelpeak}$	Watt peak of panel (W)	W
ρ_{panel}	Efficiency of the panel	
$P_{invpeak}$	Max power of inverter (W)	W
ρ_{inv}	Efficiency of the inverter	
$P_{out}(u)$	Outgoing power at time u	W
$irr(u)$	Irradiation at time u	W/m^2
ρ_a	Efficiency due to angle and orientation	
ρ_s	Efficiency due to shadow	
$E_{day}(t)$	Produced energy on day t	kWh
$EF_{day}(t)$	Fraction of the year production of produced energy on day t	kWh

where

$$irr_{year} = \Sigma irr_{day}(t)$$

When this division $E_{day}(t)\ /E_{year}$ is done, the constant efficiency factors $\rho_{a,}\ \rho_{s},$ $\rho_{panel},\ \rho_{inv}$ are divided out, and a dependence of $EF_{day}(t)$ on the day irradiation remains:

$$EF_{day}(t) = irr_{day}(t)/irr_{year}$$

This provides the form of temporal abstraction (cf. [3]) for the solar energy production agent model used in the simulation experiments described in Sect. 19.4.

19.4 Simulation Experiments

In the previous two sections it was described how models can be made for the two agents. These agent models can be used in particular to describe consumption of energy by a heatpump for the heating agent over days in a year, and production of energy by a PV-installation agent over days in a year. In general the distribution of the consumption over days in a year (high during winter, low during summer) has a far from perfect match with the distribution of the production over days in a year (low during winter, high during summer). In the simulation experiments discussed here empirical data have been used for the year 2012 in the Netherlands, covering both the irradiation per day and the average day temperature per day. As a first example, in Fig. 19.3 the day distributions of the day fractions (of a year) of production and consumption are shown for a case in which the year production is assumed exactly equal to the year consumption. It can be seen that in November, December, January and February (from about day 305 to day 60) the consumption strongly exceeds the production, whereas from May until August (from about day

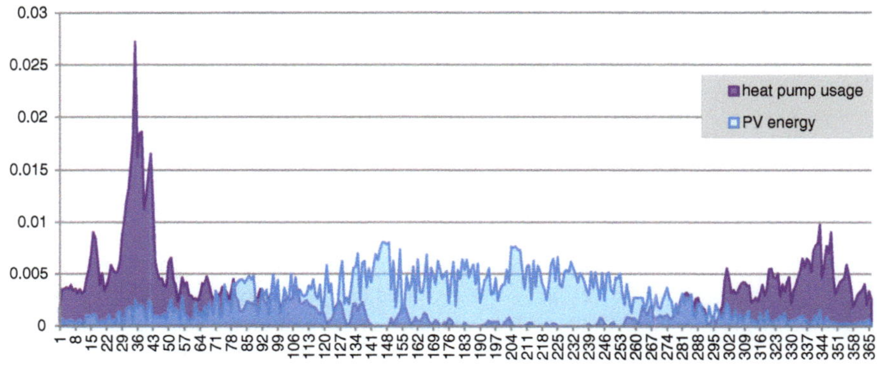

Fig. 19.3 Distributions and overlap for equal overall year production and consumption over 2012 (0.33 of consumption covered)

120 to day 250) it is the other way around. Nevertheless, an overlap can be found that in total covers a non-neglectable fraction 0.33 of the year consumption amount, mostly concentrated in the periods March to May and September to October.

In a general case the overall year production and consumption will not be exactly the same. It might also be wise to go for a higher production level per year in order to increase the 33 % coverage of the consumption. It has been analysed in how far this would increase the coverage. In Fig. 19.4 two of such (a bit extreme) cases are depicted, respectively for production 5 times the consumption (upper graph, resulting in 72 % coverage of consumption) and for 10 times the consumption (lower graph, resulting in 91 % coverage of consumption); note that the vertical scales are shown more condense here. This shows that increasing the production level indeed can increase the coverage.

Similar simulations have been done for a number of different proportions between production and consumption, for a factor 0 to 20 of production with respect to consumption. The results are as shown in Fig. 19.5. For very high production levels per year with respect to consumption, an asymptote of 1 is reached.

As a further step assumptions have been made for costs. It is assumed that per kWh the own produced energy is much cheaper than the energy that has to be bought externally. For values 0.05 resp. 0.25 euro per kWh for own resp. externally produced energy, the graph shown in Fig. 19.6 (with derivative in Fig. 19.7) is found. Here an asymptote is found with value 0.20 (consumption fully based on own produced energy). The derivative can be used to assess how much investment in additional capacity of the PV installation is reasonable.

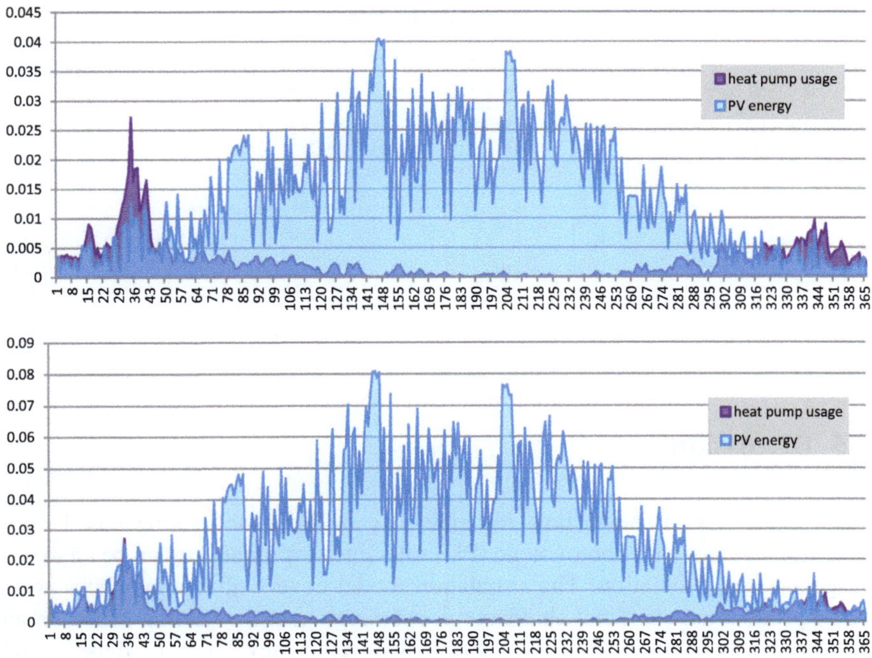

Fig. 19.4 Unequal distributions of overall year production and consumption over 2012: **a** *upper* graph: year production 5 times year consumption (0.72 of consumption covered), **b** *lower* graph: year production 10 times year consumption (0.91 of consumption covered)

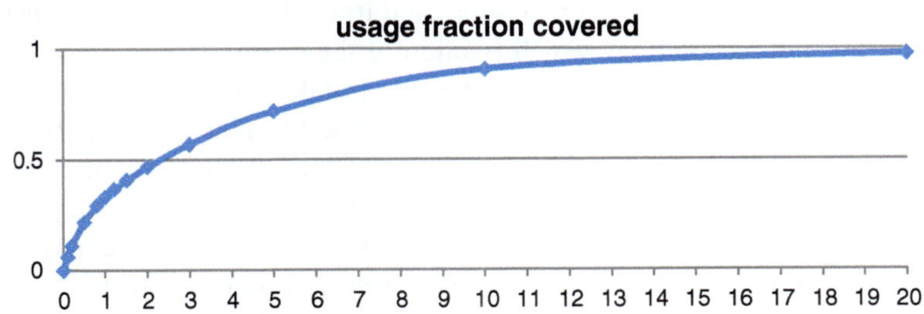

Fig. 19.5 Consumption coverage fraction for different factors for overall year production w.r.t. overall consumption over 2012

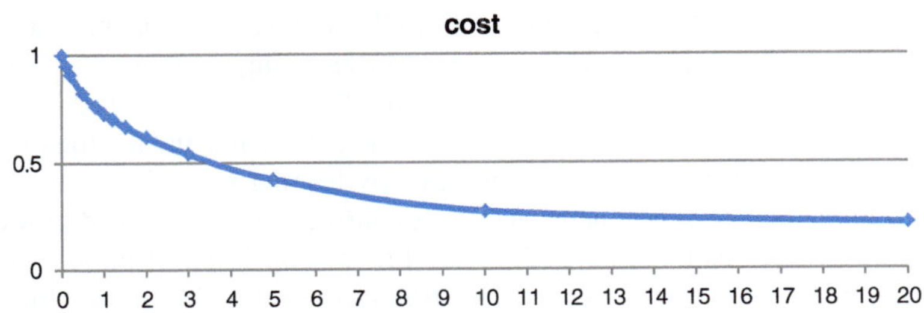

Fig. 19.6 Cost for different factors of overall year production w.r.t. overall consumption over 2012

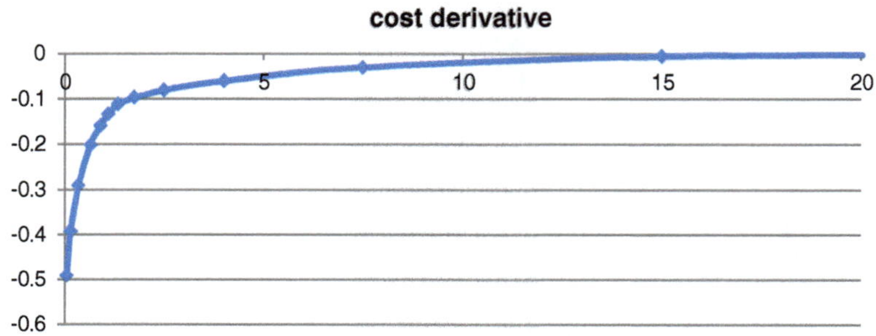

Fig. 19.7 Derivative of cost for different factors of overall year production w.r.t. overall consumption over 2012

19.5 Discussion

In this paper it was shown how the efficiency of domestic heating based on a heatpump together with photovoltaic (PV) solar energy production can be analysed by agent-based simulation. The simulation model covers of two agents: a solar energy production agent and an energy consuming agent for heating based on an air to water heatpump. For both agents their efficiency strongly depends on dynamically varying environmental circumstances, in particular on the amount of irradiation (production agent) and on the outdoor temperature (heating agent). For the heating agent it was analysed how exactly its daily performance depends on the outdoor temperature, and for the PV-installation how the daily yields depend on irradiation. Based on empirical temperature and irradiation data over a year it was found out for different settings which fraction of the energy required per year for heating can be covered by the yields of the PV installation.

Within a day the heatpump usage was assumed to fit to the time the PV energy production takes place. During this period of sunshine the temperatures may be higher than the average day temperature, so the performance factor may be a bit better in reality than estimated in the model used here. However, there may well be as well a need to heat when there is just little or no sunshine, for example early in the morning. This will count in the opposite direction. To inverstiage this variation, the simulations also have been performed for the minimun and maximum day temperatures instead of the average day temperature. For example, for the case of equal overall production and usage, this provides a variation around the 33 % coverage (for average day temperatures) from 30 % coverage (for minimum day temperatures) to 36 % coverage (for maximum day temperatures). To obtain a more detailed approach for such types of intraday effects, a refined model is needed in which hours or even minutes per day are considered and the lifestyle and choices of the customer concerning when exactly heating is needed.

The combination of solar energy and heatpumps have been studied as well in other literature such as (e.g., [2, 5, 7, 8, 11, 12, 15, 16]. In all of these cases the source of the heatpump was not air but ground or water. Moreover, the solar energy installation used was in the form of solar thermal collectors and added to the heatpump loop to heat the water. In these respects they address situations quite different from what was addressed in the current paper, in which the source was air, and the solar energy production was based on photovoltaic panels to obtain electrical energy, not thermal energy.

References

1. E. Acikkalp, H. Aras, Comparing Geothermal Heat Pump System with Natural Gas Heating System. in *Proceedings of the World Renewable Energy Congress*, Sweden, 2011, ed. by B. Moshfegh, pp. 1337–1344
2. K. Bakirci, O. Ozyurt, K. Comakli, O. Comakli, Energy analysis of a solar-ground source heat pump system with vertical closed-loop for heating applications. Energy **36**(2011), 3224–3232 (2011)
3. T. Bosse, M. Hoogendoorn, M.C.A. Klein, J. Treur, A Three-Dimensional Abstraction Framework to Compare Multi-Agent System Models. in *Computational Collective Intelligence: Technologies and Applications*, eds. by J.-S. Pan, S.-M. Chen, N.T. Nguyen. Proceedings of the 2nd International Conference on Computational Collective Intelligence, ICCCI'10, Part I. Lecture notes in artificial intelligence, vol. 6421 (Springer, 2010), pp. 306–319
4. F. Busato, R.M. Lazzarin, M. Noro, Two years of recorded data for a multisource heat pump system: a performance analysis. Appl. Therm. Eng. **57**(2013), 39–47 (2013)
5. D. Carli, M. Ruggeri, M. Bottarelli, M. Mazzer, Grid-assisted photovoltaic power supply to improve self-sustainability of ground-source heat pump systems, in *Proceedings of the 2013 IEEE International Conference on Industrial Technology (ICIT)*. IEEE (2013), pp. 1579–1584
6. K.J. Chua, S.K. Chou, W.M. Yang, Advances in heat pump systems: a review. Appl. Energy **87**(2010), 3611–3624 (2010)
7. S. Deng, Y.J. Dai, R.Z. Wang, T. Matsuura, Y. Yasui, Comparison study on performance of a hybrid solar-assisted CO_2 heat pump. Appl. Therm. Eng. **31**, 3696–3705 (2011)
8. S. Deng, Y.J. Dai, R.Z. Wang, Performance study on hybrid solar-assisted CO_2 heat pump system based on the energy balance of net zero energy apartment. Energ. Buildings **54**, 337–349 (2012)
9. A. Hepbasli, Y. Kalinci, A review of heat pump water heating systems. Renew. Sustain. Energy Rev. **13**(2009), 1211–1229 (2009)
10. J. Jeong, C.S. Kim, J. Lee, Household electricity and gas consumption for heating homes. Energy Policy **39**, 2679–2687 (2011)
11. O. Ozgener, A. Hepbasli, Performance analysis of a solar-assisted ground-source heat pump system for greenhouse heating: an experimental study. Build. Environ. **40**(2005), 1040–1050 (2005)
12. O. Ozgener, A. Hepbasli, 2007. A review on the energy and exergy analysis of solar assisted heat pump systems renewable and sustainable energy reviews **11**(3), 482–496 (2007)
13. J. Treur, Analysis of configurations for photovoltaic solar energy production using agent-based simulation, in *Proceedings of the Second International Workshop on Smart Energy Networks and Multi-Agent Systems, SEN-MAS'13*. Lecture notes in artificial intelligence, eds by M. Klusch, M. Thimm, M. Paprzycki, vol. 8076 (Springer, 2013a), pp. 372–385
14. J. Treur, Dynamic allocation of a domestic heating task to gas-based and heatpump-based heating agents, in *Proceedings of the Second International Workshop on Smart Energy Networks and Multi-Agent Systems, SEN-MAS'13*. Lecture notes in artificial intelligence, eds by M. Klusch, M. Thimm, M. Paprzycki, vol. 8076 (Springer, 2013b), pp. 386–399

15. B. Xu, Y. Zhang, J. An, K. Han, J. Yi, Feasibility study of the winter heating about solar-water source heat pump system in hot summer and cold winter zone, in *Proceedings of the 4th Electronic System-Integration Technology Conference (ESTC'12)*. IEEE (2013), pp. 289–292
16. W.B. Yang, M.H. Shi, H. Dong, Numerical simulation of the performance of a solar-earth source heat pump system. Appl. Therm. Eng. **26**(2006), 2367–2376 (2006)

Chapter 20
The Influence of the Construction of the Cooling System of Semiconductor Devices on the Watt-Hour Efficiency of DC–DC Converters

Janusz Zarębski and Krzysztof Górecki

Abstract In the paper the influence of cooling conditions of semiconductor devices on the characteristics of a boost converter is considered. The form of the thermal model of semiconductor devices is proposed and some results of calculations and measurements of the characteristics of this converter are shown. The investigations were performed for the selected types of power MOSFETs operating at different cooling conditions.

20.1 Introduction

The power supply systems converting the electrical energy obtained from wind power systems or solar cells contain DC–DC converters [1]. The aim of this circuit is to change the value of the parameters of voltage obtained from renewable power sources. DC–DC converters contain semiconductor devices operating as switches.

As it's commonly known, the watt-hour efficiency of DC–DC converters strongly depends on the parameters of semiconductor devices [2, 3], particularly on a voltage drop on the transistors and diodes operating in the on-state. These parameters, in turn, strongly depend on the device internal temperature [4]. This temperature results from self-heating, that means, it depends on the power dissipated in the considered device and on the construction of the device cooling system (free or forced) [5–7].

The paper presents the results of investigations concerning the influence of the construction of the cooling system of the power MOSFET on the watt-hour efficiency of DC–DC converters containing this transistor and designed to be used in the power systems with renewable power sources. By means of electrothermal

J. Zarębski · K. Górecki (✉)
Department of Marine Electronics, Gdynia Maritime University, Gdynia, Poland
e-mail: gorecki@am.gdynia.pl

A. Y. Oral et al. (eds.), *International Congress on Energy Efficiency and Energy Related Materials (ENEFM2013)*, Springer Proceedings in Physics 155,
DOI: 10.1007/978-3-319-05521-3_20,

models of the power MOSFET and the diode, the characteristics of the selected DC–DC converters at different cooling conditions of the power MOSFET are calculated. The results of calculations are compared with the results of measurements.

20.2 Device Electrothermal Models

In the paper [8] the non-linear thermal model of the semiconductor device taking into account, among other things, the influence of the construction of the cooling system on thermal resistance of this device is proposed. This model is described by means of the classical equation of the form

$$T_j = T_a + R_{th} \cdot p_{th} \tag{20.1}$$

where T_j denotes the device internal temperature, T_a—the ambient temperature, R_{th}—the thermal resistance, and p_{th}—the thermal power dissipated in the device.

The paper [8] proposes the analytical description of the dependence of thermal resistance on such factors as: geometrical dimensions of the elements of the heat flow path, the length of metal leads, the area of solder pads, the length of the heat-sink, the power dissipated in the investigated device and the ambient temperature.

The thermal model is a component of the electrothermal models, taking into account self-heating phenomena. Electrothermal models of diodes and power MOSFETs dedicated to the analysis of DC–DC converters was proposed in the paper [3]. In these models, called hybrid electrothermal models, the following things are used: SPICE built-in models of the diode and the MOSFET, controlled voltage sources representing the additional voltage drop between the terminals of the investigated devices, due to an increase in the value of their internal temperature, and the RC Foster network representing the transient thermal impedance of the device. The RC network is excited by the current source representing the thermal power dissipated in the device, and the voltage on this source corresponds to the device internal temperature.

20.3 Investigations Results

The investigations were performed for the boost converter of the diagram shown in Fig. 20.1. The voltage source U_{ster} produces the rectangular pulses train of the period equal to 9 μs, which controls the transistor T_1. The resistor R_G of the resistance equal to 100 Ω limits the value of the gate current. The inductance of the choking-coil L_1 is 650 μH, and the capacity of the capacitor C = 470 μF. Transistors IRF640 and IRF840, as well as diodes 1N5822 and BY229 are used. The considered devices operate at different cooling conditions.

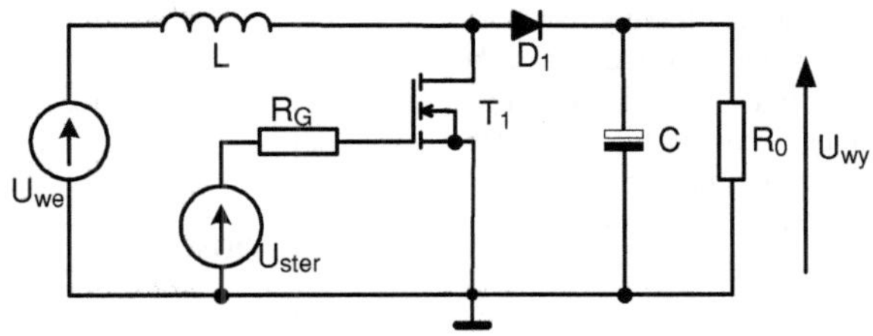

Fig. 20.1 Diagram of the investigated converter

The considered transistors differ from one other by the value of the on-resistance R_{ON} which is equal to 0.67 Ω for the transistor IRF840 and 0.13 Ω for the transistor IRF640. In turn, the voltage drop on the forward biased diode BY229 is equal to about 0.9 V, and on the Schottky diode 1N5822—to about 0.4 V. The research was conducted for the considered transistors situated on the heat-sink made from the shaped piece A-4463 of the length 70 mm or without any heat-sink. The diode BY229 was placed on the heat-sink made from the shaped piece A-5409 of the length 38 mm.

The measured value of thermal resistance of the diode 1N5822 is 50 K/W. The thermal resistance of each of the considered transistor situated on the heat-sink is equal to 5.7 K/W, and for the transistors operating without any heat-sink—55 K/W. The thermal resistance of the diode BY229 situated on the heat-sink is equal to 20 K/W.

In Figs. 20.2, 20.3, 20.4 and 20.5 the calculated and measured dependences of the watt-hour efficiency of the considered converter on the duty of the control signal d and on the load resistance R_0 obtained for different cooling conditions of the considered semiconductor devices are presented. In these figures, points mark the results of measurements, and lines—the results of calculations.

In Fig. 20.2 the dependence of the watt-hour efficiency η of the considered converter from the duty d of the signal controlling the transistor for the converter containing the transistor IRF640 and the diode 1N5822 is presented, whereas in Fig. 20.3—for the converter containing the transistor IRF840 and the diode BY229.

As it can be noticed, for both sets of semiconductor devices the dependence η(d) is a decreasing function in the whole considered range of changes d. For the transistor situated on the heat-sink higher values of the watt-hour efficiency are obtained than in the case is which this device operates without any heat-sink. This is due to the almost tenfold difference in the value of thermal resistance of the transistor, which in turn results in the considerably greater value of the internal temperature of the transistor operating without any heat-sink. This temperature rise causes an increase in the value of the transistor on-resistance R_{ON} resulting in additional energy losses in the transistor. These losses are bigger when the value of the duty d is higher which corresponds to the greater value of the output voltage of the converter [1, 2, 9]. The values of the watt-hour efficiency obtained for the converter containing the transistor situated on the heat-sink are even 30 % higher than the value of this efficiency for the converter with the transistor operating

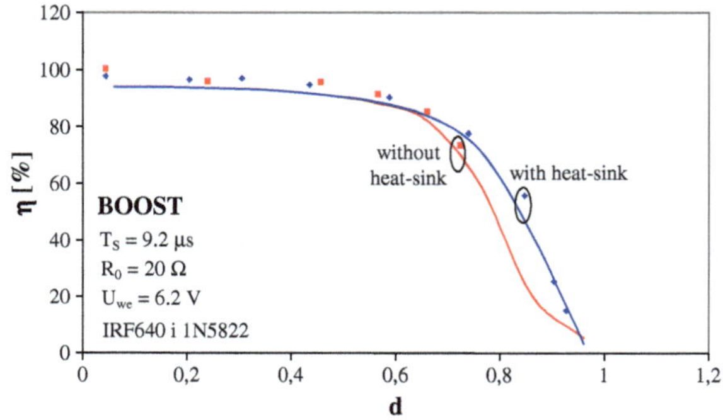

Fig. 20.2 Measured and calculated dependences of the watt-hour efficiency on the duty factor for the converter including the transistor IRF640 and the diode 1N5822

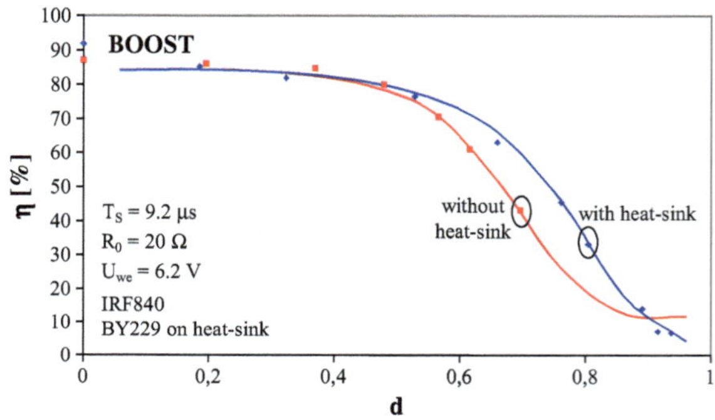

Fig. 20.3 Measured and calculated dependences of the watt-hour efficiency on the duty factor for the converter including the transistor IRF840 and the diode BY229

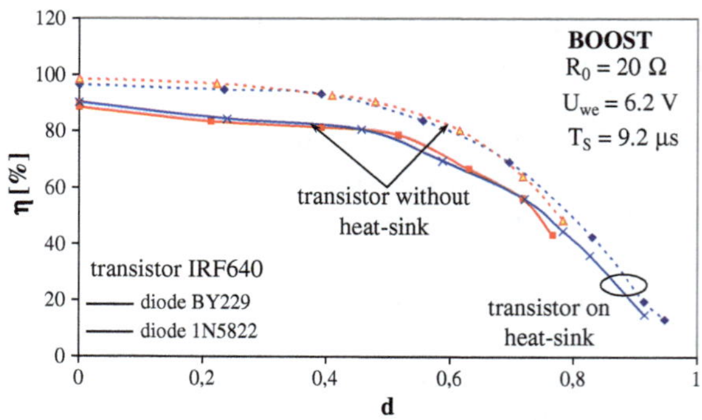

Fig. 20.4 Measured dependences of the watt-hour efficiency on the duty factor for the converter including the transistor IRF640 and different diodes

without any heat-sink. The influence of the cooling system of the transistor on the dependence η(d) is most visible in the range of high values of the duty factor.

Comparing the characteristics presented in Figs. 20.2 and 20.3 one can state that the considerably higher values of the watt-hour efficiency are obtained for the converter containing the transistor IRF640 characterized by the smaller value of the transistor on-resistance. Differences in the obtained values of the watt-hour efficiency for the considered sets of semiconductor devices reach even 10 %.

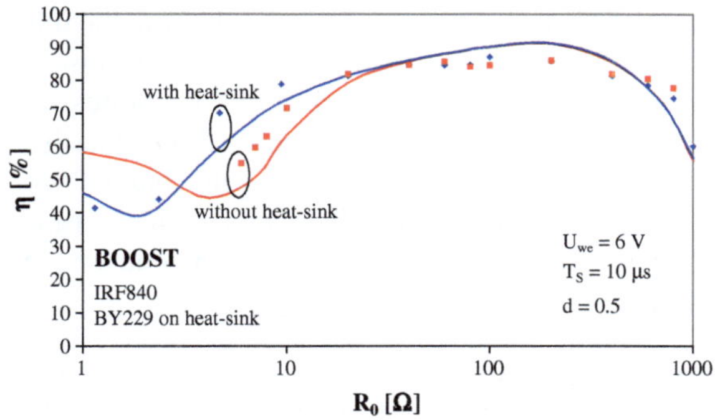

Fig. 20.5 Measured and calculated dependences of the watt-hour efficiency on the load resistance for the converter including the transistor IRF840 and the diode BY229

In turn, in Fig. 20.4 the dependences of the watt-hour efficiency of the converter with the transistor IRF640 on the value of the duty d obtained with the use of different diodes are presented.

As one can notice, in the presented figures, the replacement of the Schottky diode 1N5822 with the diode BY229 is followed by visible worsening of the watt-hour efficiency of the investigated converter. Both for the transistor situated on the heat-sink, and for the transistor operating without any heat-sink a decrease in the value of the watt-hour efficiency above 10 % is observed. The observed changes of the value of this parameter of the converter result first of all, from the differences in the value of the forward voltage of the considered diodes.

The presented above results were obtained at the load resistance of the converter $R_0 = 20\ \Omega$. In Fig. 20.5 the dependences of the efficiency of the considered converter on the load resistance at $d = 0.5$, are presented.

As it is visible, over a wide range of changes of the resistance R_0 the constant value of the watt-hour efficiency of the converter is observed. At the values of the resistance R_0 exceeding $20\ \Omega$ practically the lack of the influence of the cooling conditions of the transistor on the course of the dependence $\eta(R_0)$ is observed. This is due to the small value of the power lost in the transistor. In turn, within the range of small values of the resistance R_0 the visible differences in the values of the considered parameters for the converter with the transistor situated on the heat-sink and without any heat-sink are obtained, because in this range the transistor on-resistance (strongly increasing in the function of the temperature) has the essential influence on the watt-hour efficiency of the converter.

20.4 Conclusion

As it results from the presented investigations, the properties of the applied cooling system of the switching transistor influence the watt-hour efficiency of DC–DC converters in an essential way. Particularly, very distinct changes in the course of the dependences $\eta(d)$ and $\eta(R_0)$ appear for transistors characterized by the high value of the on-resistance. As it results from the information presented in the paper [10] and from the catalogue data analysis of power MOS transistors, the value of this resistance increases together with the admissible value of the voltage between the drain and the source of the power MOS transistor. Therefore, in the converters used in the power supply systems with solar cells or with turbines of wind, where output voltage exceeds 200 V, this problem bears great significance. Ensuring efficient cooling of the transistors in the converter will allow obtaining the higher value of the watt-hour efficiency.

Acknowledgments This project is financed from the funds of National Science Centre which were awarded on the basis of the decision number DEC-2011/01/B/ST7/06740.

References

1. M.H. Rashid, *Power Electronic Handbook* (Academic Press, Elsevier, 2007)
2. K. Górecki, A New electrothermal average model of the diode-transistor switch. Microelectron. Reliab. **48**(1), 51–58 (2008)
3. K. Górecki, J. Zarębski, Modeling nonisothermal characteristics of switch-mode voltage regulators. IEEE Trans. Power Electron. **23**(4), 1848–1858 (2008)
4. K. Górecki, J. Zarębski, Influence of MOSFET model form on characteristics of the boost converter. Informacije MIDEM **41**(1), 1–7 (2011)
5. K. Górecki, J. Zarębski, Nonlinear compact thermal model of power semiconductor devices. IEEE Trans. Compon. Packag. Technol. **33**(3), 643–647 (2010)
6. V. Szekely, A new evaluation method of thermal transient measurement results. Microelectron. J. **28**(3), 277–292 (1997)
7. P.E. Bagnoli, C. Casarosa, M. Ciampi, E. Dallago, Thermal resistance analysis by induced transient (TRAIT) method for power electronic devices thermal characterization. I. fundamentals and theory. IEEE Trans. Power Electron. **13**(6), 1208–1219 (1998)
8. K. Górecki, J. Zarębski, Modeling the influence of selected factors on thermal resistance of semiconductor devices. IEEE Trans. Compon. Packag. Manuf. Technol. **4**(3), 421–428 (2014)
9. R. Ericson, D. Maksimovic, *Fundamentals of power electronics* (Kluwer Academic Publisher, Norwell, 2001)
10. J. Zarębski, ON-resistance of power MOSFETs. Informacije MIDEM **35**(1), 1–4 (2007)

Chapter 21
New Method for Analytical Photovoltaic Parameters Identification: Meeting Manufacturer's Datasheet for Different Ambient Conditions

Javier Cubas, Santiago Pindado and Carlos de Manuel

Abstract At present, photovoltaic energy is one of the most important renewable energy sources. The demand for solar panels has been continuously growing, both in the industrial electric sector and in the private sector. In both cases the analysis of the solar panel efficiency is extremely important in order to maximize the energy production. In order to have a more efficient photovoltaic system, the most accurate understanding of this system is required. However, in most of the cases the only information available in this matter is reduced, the experimental testing of the photovoltaic device being out of consideration, normally for budget reasons. Several methods, normally based on an equivalent circuit model, have been developed to extract the *I-V* curve of a photovoltaic device from the small amount of data provided by the manufacturer. The aim of this paper is to present a fast, easy, and accurate analytical method, developed to calculate the equivalent circuit parameters of a solar panel from the only data that manufacturers usually provide. The calculated circuit accurately reproduces the solar panel behavior, that is, the *I-V* curve. This fact being extremely important for practical reasons such as selecting the best solar panel in the market for a particular purpose, or maximize the energy extraction with MPPT (Maximum Peak Power Tracking) methods.

J. Cubas (✉) · S. Pindado · C. de Manuel
Instituto Universitario de Microgravedad "Ignacio da Riva", ETSI Aeronáuticos, Universidad Politécnica de Madrid, Madrid, Spain
e-mail: j.cubas@upm.es

S. Pindado
e-mail: santiago.pindado@upm.es

C. de Manuel
e-mail: c.manuelnavio@upm.es

A. Y. Oral et al. (eds.), *International Congress on Energy Efficiency and Energy Related Materials (ENEFM2013)*, Springer Proceedings in Physics 155,
DOI: 10.1007/978-3-319-05521-3_21,

21.1 Introduction

It is a fact that energy efficiency of electric systems requires a good understanding of each one of its parts. In case of systems that involve photovoltaic panels, the analysis of the system must take into account the behavior of the mentioned panels depending on, at least, sun radiation levels, temperature, and connected loads. In the case of space applications, the cumulative ionizing radiation is another important factor. The optimization of the system requires then a mathematical modeling of the photovoltaic devices (solar panels or solar cells). One of the most popular photovoltaic systems models is based on an equivalent circuit. In the present work the mentioned equivalent circuit model is analyzed in order to get the best approximation to the solar panel/cells behavior. This work was carried out in the framework of the UPMSat-2 satellite project at the IDR/UPM Institute (Madrid, Spain), nevertheless, the calculation process can be applied to every photovoltaic system no matter if it is designed for space or terrestrial applications.

21.2 Solar Cell/Panel Modeling

The simpler equivalent circuit of a solar cell/panel involves a current source and one or several diodes connected in parallel. A series resistance and a shunt resistance, R_s, and R_{sh}, are normally added to the circuit to take respectively into account losses in cell solder bonds, interconnection, junction box, etc., and the current leakage through the high conductivity shunts across the p-n junction. The most popular equivalent circuit is shown in Fig. 21.1, it includes a current source, one diode and the mentioned resistances. It reproduces quite well the behavior of the solar cell/panel, that is, the *I*-*V* curve, see Fig. 21.2. In Fig. 21.2 open circuit, maximum power, and short circuit points are indicated. These points together with their variation as a function of the temperature and the radiation exposure (also known as fluence) are the normal information included in manufacturers' datasheets see Table 21.1.

The circuit model of Fig. 21.1 is defined by the following expression [1]:

$$I = I_{pv} - I_0\left[\exp\left(\frac{V + IR_s}{NaV_T}\right) - 1\right] - \frac{V + IR_s}{R_{sh}}, \tag{21.1}$$

where I_{pv} is the photocurrent delivered by the constant current source, I_0 is the reverse saturation current corresponding to the diode, N is the number of cells in series, and a is the ideality factor that takes into account the deviation of the diodes from the Shockley diffusion theory (the value of this factor, is assumed to be constant and between 1 and 1.5 [2] for one-junction cells. V_T is the thermal voltage of the diode and depends of the charge of the electron, q, the Boltzmann constant, k, and the temperature, T, with (21.2).

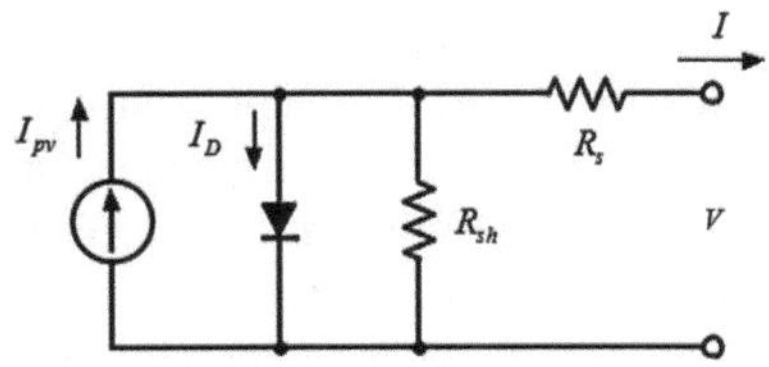

Fig. 21.1 Equivalent circuit for a solar cell

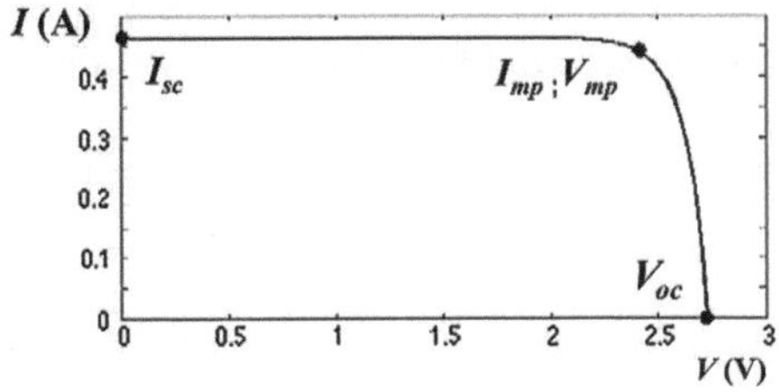

Fig. 21.2 Example of the I-V curve of a solar cell

Table 21.1 Characteristic points and its variation with temperature and fluence of Emcore ZTJ photovoltaic cell from manufacturer datasheet

Fluence (e/cm^2)	V_{oc} (V)	I_{sc} (A)	V_{mp} (V)	I_{mp} (A)	ΔV_{oc} (mV/°C)	ΔI_{sc} (mA/°C)	ΔV_{mp} (mV/°C)	ΔI_{mp} (mA/°C)
0	2.726	0.463	2.410	0.439	−6.3	0.31	−6.7	0.24
3.00×10^{13}	2.617	0.458	2.362	0.435	–	–	–	–
1.00×10^{14}	2.590	0.454	2.338	0.435	−6.6	0.30	−7.0	0.24
5.00×10^{14}	2.481	0.449	2.241	0.421	–	–	–	–
1.00×10^{15}	2.426	0.435	2.193	0.413	−6.9	0.30	−7.3	0.28
3.00×10^{15}	2.344	0.412	2.097	0.377	–	–	–	–
1.00×10^{16}	2.235	0.380	2.000	0.325	−7.4	0.31	−6.6	0.36

AM0 (1,353 W m^{-2}) T_r = 28 °C [3]

$$V_T = \frac{q}{kT}. \tag{21.2}$$

These parameters, together with R_s and R_{sh}, are adjusted to fit the circuit behavior to the solar cell/panel testing results. As testing results depend on the mentioned temperature, radiation and fluence, the aforementioned parameters will change from one case to another.

21.3 Parameter Calculation

The parameters of the equivalent electric circuit, I_{pv}, I_0, a, R_s, and R_{sh}, can be numerically [4] or analytically [5] extracted from *I-V* curves experimentally measured, or can be calculated from the data included by the manufacturer in the datasheet (normally, the mentioned three characteristic points: open circuit, maximum power, and short circuit). In this second case the procedure use to be numerical [2, 6]. In the present work, an analytical procedure [7] has been selected to extract the circuit parameters based on the characteristic points. This method is direct, simple, non-iterative, and has been proved to be as accurate as other more sophisticated numerical methods [8, 9]. The first step consists in the estimation of parameter a, whose value will be in the bracket 1–1.5. This first estimation of one of the 5 parameters of the circuit is necessary, as the number of equations resulting from the characteristic points is only 4. The other 4 parameters can be derived, for specific temperature, irradiation and fluence conditions from equations:

$$\begin{aligned} &\frac{aV_T V_{mp}\left(2I_{mp}-I_{sc}\right)}{\left(V_{mp}I_{sc}+V_{oc}\left(I_{mp}-I_{sc}\right)\right)\left(V_{mp}-I_{mp}R_s\right)-aV_T\left(V_{mp}I_{sc}-V_{oc}I_{mp}\right)} \\ &\quad = \exp\left(\frac{V_{mp}+I_{mp}R_s-V_{oc}}{aV_T}\right), \\ R_{sh} &= \frac{\left(V_{mp}-I_{mp}R_s\right)\left(V_{mp}-R_s\left(I_{sc}-I_{mp}\right)-aV_T\right)}{\left(V_{mp}-I_{mp}R_s\right)\left(I_{sc}-I_{mp}\right)-aV_T I_{mp}}, \\ I_0 &= \frac{(R_{sh}+R_s)I_{sc}-V_{oc}}{R_{sh}\exp\left(\frac{V_{oc}}{aV_T}\right)}, \\ I_{pv} &= \frac{R_{sh}+R_s}{R_{sh}}I_{sc}. \end{aligned} \tag{21.3}$$

As said, temperature variations will modify the behavior of the photovoltaic device, that is, its I-V curve. More specifically, the characteristic points of the curve will also be modified, the variations being lineal with the temperature as it can be appreciated in the example of Table 21.1. An increase of temperature will increase the current level but decrease the voltage level, and as a result the maximum power point will also decrease. The variations of current and voltage levels of open circuit, maximum power, and short circuit points as a function of the temperature can be expressed as:

$$\begin{aligned} V_{oc,T} &= V_{oc,T_r} + \Delta V_{oc}(T-T_r); \quad V_{mp,T} = V_{mp,T_r} + \Delta V_{mp}(T-T_r); \\ I_{sc,T} &= I_{sc,T_r} + \Delta I_{sc}(T-T_r); \quad I_{mp,T} = I_{mp,T_r} + \Delta I_{mp}(T-T_r); \end{aligned} \tag{21.4}$$

where T_r is the reference temperature. There are several methods in the existing literature that relates directly the equivalent circuit parameters with the temperature [2, 10]. Nevertheless, a different procedure has been followed in the present work: (21.3) were solved for a specific temperature with the characteristic points values calculated previously with (21.4). This procedure implies much more precision than other methods that estimate parameter variation with temperature, as the calculated circuits are directly forced to meet the manufacturers' data.

Usually, manufacturers do not include in the datasheets any information regarding the behavior of the solar cell/panel at different irradiation levels.

However, it is known that the *I-V* curve is essentially invariant with intensity in ranges around one solar constant. On the other hand, temperature affects I_{sc} linearly and V_{oc} exponentially, whereas R_s remains not affected for temperature variations [11]. Those conditions, once simulated using an equivalent circuit, lead to the following expression [2]:

$$I_{pv,G} = I_{pv,G_r} \frac{G}{G_r}, \tag{21.5}$$

where G is the irradiance on the cell/solar panel, $I_{pv,G}$ is the photocurrent delivered by the current source of the equivalent circuit, and G_r and $I_{pv,Gr}$ are the reference values.

The radiation exposure or fluence radiation is translated into a degradation of the photovoltaic system performances. Obviously, the fluence modifies the characteristic points of the *I-V* curve, as it can be observed in the example of Table 21.1. However, in this case the variation is not linear, and normally is not easy to interpolate using the manufacturers' data. Additionally, fluence radiation degradation is a slow process that will take months or years to affect visibly the operation of the cell. For this reason it seems reasonable to calculate separately the circuit parameters for each one of the fluence levels specified in the manufacturers' datasheets (see Table 21.1), and apply the proper one for each stage of the mission.

21.4 Example of Application, LTSpice Model

In this Section, the method explained above is applied in the performance analysis of a commercial solar cell, Emcore ZTJ, for every irradiance and temperature condition, using LTSpice models fitted to the experimental data from the manufacturer (see Table 21.1). Although, the mentioned analysis is carried out for zero fluence, it can be easily reproduced for different levels of radiation exposure (see other fluence levels in Table 21.1).

As said in Sect. 21.1, the value of parameter a is firstly estimated as $a = 1.1$. Taking also into account that the studied cell is triple junction $N = 3$. Solving (21.3) at AM0 (1,353 W m^{-2}) and $T_r = 28$ °C, it is possible to calculate the rest of the equivalent circuit parameters, see Table 21.2. Repeating this process for temperatures in the bracket 0–100 °C, the variation of the mentioned parameters as a function of the temperature can be obtained, see Figs. 21.3, 21.4, 21.5, and 21.6.

Taking into account that, with the exception of I_{pv}, all parameters are not dependent of the irradiance, and fitting the temperature using the least squares method, the following 21.6 can be then derived.

Table 21.2 Parameters of equivalent circuit of Emcore ZTJ

N	a	I_{pv} (A)	I_0 (A)	R_s (Ω)	R_{sh} (Ω)
3	1.1	0.463	6.80×10^{-15}	6.09×10^{-2}	284.4

AM0 (1,353 W m^{-2}) $T_r = 28$ °C

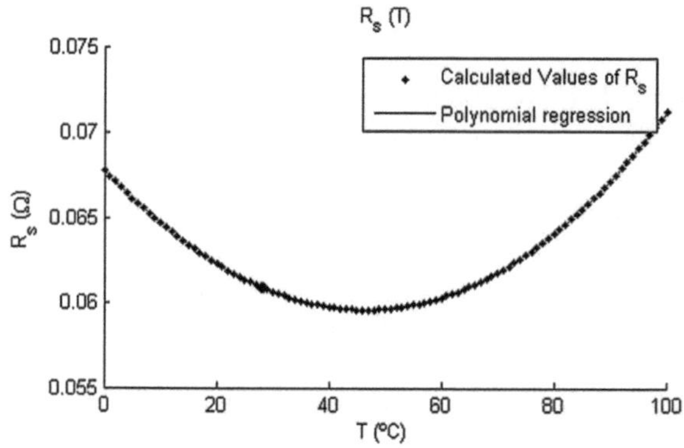

Fig. 21.3 Calculated $R_s(T)$ and polynomial regression

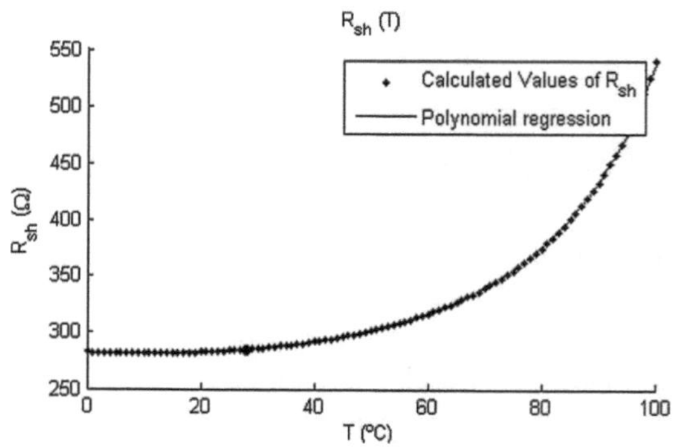

Fig. 21.4 Calculated $R_{sh}(T)$ and polynomial regression

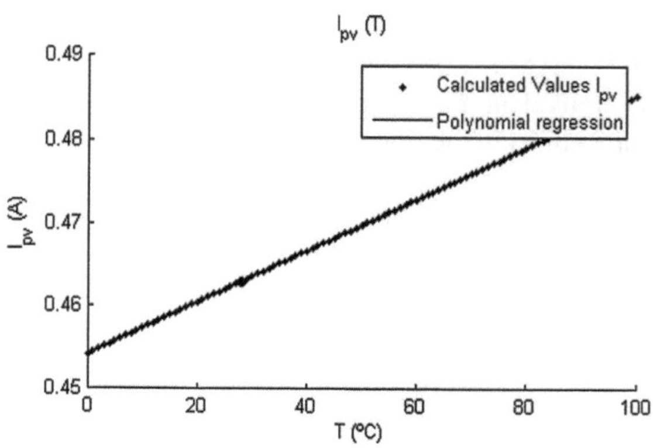

Fig. 21.5 Calculated $I_{pv}(T)$ and polynomial regression

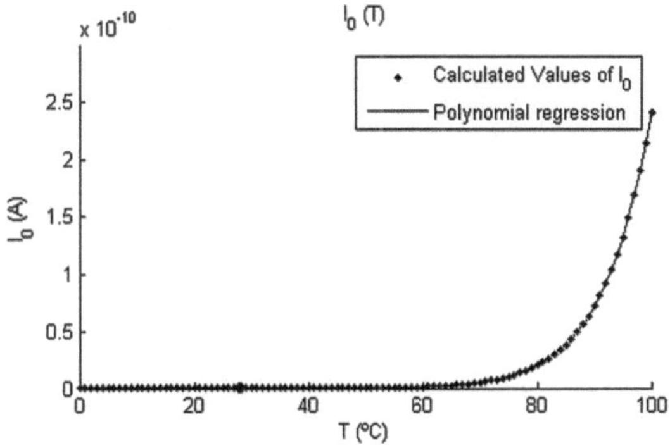

Fig. 21.6 Calculated $I_0(T)$ and polynomial regression

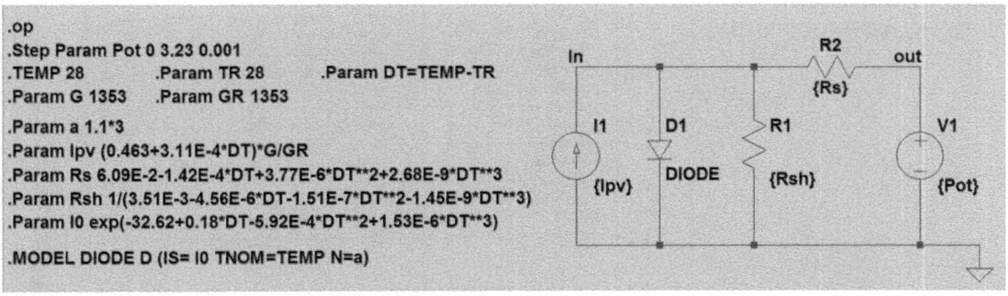

Fig. 21.7 LTSpice model for Emcore ZTJ solar cell

$$
\begin{aligned}
a &= 1.1 \\
I_{pv}(T,G) &= \left(0.463 + 3.11 \times 10^{-4}\Delta T\right)\frac{G}{G_r}, \\
R_s(T) &= 6.09 \times 10^{-2} - 1.42 \times 10^{-4}\Delta T + 3.77 \times 10^{-6}\Delta T^2 + 2.68 \times 10^{-9}\Delta T^3, \\
R_{sh}(T) &= 1/\left(3.51 \times 10^{-3} - 4.56 \times 10^{-6}\Delta T - 1.51 \times 10^{-7}\Delta T^2 - 1.45 \times 10^{-9}\Delta T^3\right), \\
I_0(T) &= \exp\left(-32.62 + 0.18\Delta T - 5.92 \times 10^{-4}\Delta T^2 + 1.53 \times 10^{-6}\Delta T^3\right).
\end{aligned}
\tag{21.6}
$$

Finally, programming (21.6) in a LTSpice model (see Fig. 21.7), it is possible to obtain a quiet accurate estimation of the studied photovoltaic device behavior for every irradiation, temperature and fluence condition, see Figs. 21.8 and 21.9.

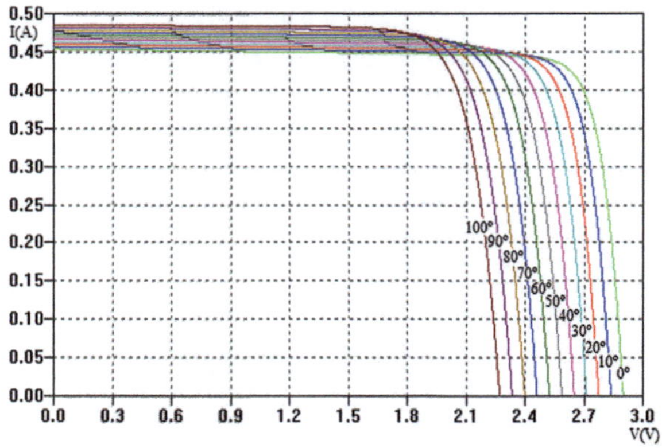

Fig. 21.8 Curves IV for Emcore ZTJ solar cell, $G = 1{,}353\ \mathrm{W/m^{-2}}$; $T(0{:}100\ °\mathrm{C})$

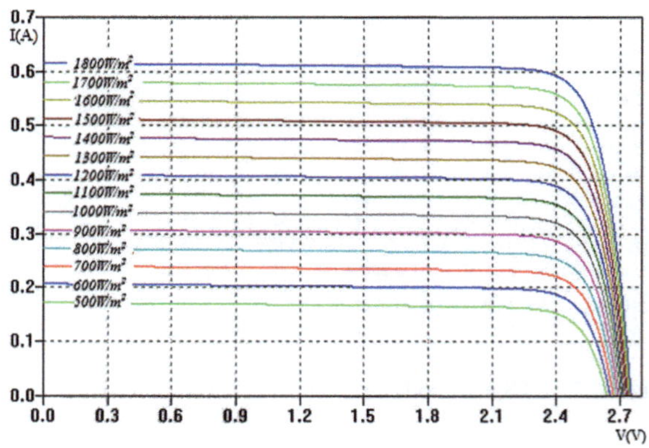

Fig. 21.9 Curves IV for Emcore ZTJ solar cell, $T = 28\ °\mathrm{C}$; $G(500{:}1{,}800\ \mathrm{W/m^{-2}})$

21.5 Conclusions

In the present work a simple but accurate method to analyze the performances of a photovoltaic device for different working conditions is described. This method is based on manufacturers' data. The present work is part the research framework associated to the design and construction of the UPMSat-2 satellite at the Universidad Politécnica de Madrid (Polytechnic University of Madrid).

References

1. K.L. Kennerud, Analysis of performance degradation in CdS solar cells. IEEE Trans. Aerosp. Electron. Syst, 912–917 (1969)
2. M.G. Villalva, J.R. Gazoli, E.R. Filho, comprehensive approach to modeling and simulation of photovoltaic arrays. IEEE Trans. Power Electron. **24**, 1198–1208 (2009)
3. T. Easwarakhanthan, J. Bottin, I. Bouhouch, C. Boutrit, Nonlinear minimization algorithm for determining the solar cell parameters with microcomputers. Int. J. Solar Energy **4**, 1–12 (1986)

4. J.C.H. Phang, D.S.H. Chan, J.R. Phillips, Accurate analytical method for the extraction of solar cell model parameters. Electron. Lett. **20**, 406–408 (1984)
5. W. Xiao, W.G. Dunford, A. Capel, in *A Novel Modeling Method for Photovoltaic Cells*. 2004 IEEE 35th Annual Power Electronics Specialists Conference (IEEE Cat. No.04CH37551) (2004), pp. 1950–1956)
6. J. Cubas, S. Pindado, M. Victoria, On the analytical approach for modeling photovoltaic systems behavior. J. Power Sources **247**, 467–474 (2014)
7. W. Gong, Z. Cai, Parameter extraction of solar cell models using repaired adaptive differential evolution. Sol. Energy **94**, 209–220 (2013)
8. A. Askarzadeh, A. Rezazadeh, Parameter identification for solar cell models using harmony search-based algorithms. Sol. Energy **86**, 3241–3249 (2012)
9. S. Lineykin, in *Five-Parameter Model of Photovoltaic Cell Based on STC Data And Dimensionless*. 2012 IEEE 27th Convention of Electronical and Electronics Engineers in Israel (2012), pp. 1–5
10. H.S. Rauschenbach, *Solar Cell Array Design Handbook, The Principles and Technology of Photovoltaic Energy Conversion*. (Van Nostrand Reinhold Co, New York, 1980)
11. Emcore: ZTJ photovoltaic cell (2011), www.emcore.com

Chapter 22
Design of Biomass Gasification and Combined Heat and Power Plant Based on Laboratory Experiments

Juma Haydary and Ľudovít Jelemenský

Abstract Three types of wooden biomass were characterized by calorimetric measurements, proximate and elemental analysis, thermogravimetry, kinetics of thermal decomposition and gas composition. Using the Aspen steady state simulation, a plant with the processing capacity of 18 ton/h of biomass was modelled based on the experimental data obtained under laboratory conditions. The gasification process has been modelled in two steps. The first step of the model describes the thermal decomposition of the biomass based on a kinetic model and in the second step, the equilibrium composition of syngas is calculated by the Gibbs free energy of the expected components. The computer model of the plant besides the reactor model includes also a simulation of other plant facilities such as: feed drying employing the energy from the process, ash and tar separation, gas-steam cycle, and hot water production heat exchangers. The effect of the steam to air ratio on the conversion, syngas composition, and reactor temperature was analyzed. Employment of oxygen and air for partial combustion was compared. The designed computer model using all Aspen simulation facilities can be applied to study different aspects of biomass gasification in a Combined Heat and Power plant.

22.1 Introduction

In the past 25 years, biomass has gained renewed interest as a long term renewable energy source. Due to its widely abundant feedstock, the energy liberated from biomass can significantly contribute to the future of energy generation. Lignocellulosic materials are a major constituent of most biomass and they are

J. Haydary (✉) · Ľ. Jelemenský
Institute of Chemical and Environmental Engineering, Faculty of Chemical and Food Technology, Slovak University of Technology, Radlinského 9, 812 37 Bratislava, Slovakia
e-mail: juma.haydary@stuba.sk

A. Y. Oral et al. (eds.), *International Congress on Energy Efficiency and Energy Related Materials (ENEFM2013)*, Springer Proceedings in Physics 155,
DOI: 10.1007/978-3-319-05521-3_22,

composed of cellulose, hemi-cellulose and lignin. Wood chips consisted cellulose and lignin are the best example of this group of materials. One of the most effective methods for conversion Ligno-cellulosic biomass is gasification.

A complete computer model of gasification describing all aspects of the process integrated in a Combined Heat and Power (CHP) plant can be a very useful tool for process development and conceptual process design. Much work is currently focused on biomass gasification modelling. The ASPEN PLUS process simulator has been used by different investigators to simulate coal, biomass and waste gasification. Doherty et al. [1] used ASPEN PLUS for the computer simulation of a biomass gasification-solid oxide fuel cell power system. They assumed zero-dimensional and equilibrium composition of gas. Nikoo and Mahinpey [2] simulated the biomass gasification in a fluidised bed reactor applying ASPEN PLUS. Using a user subroutine, hydrodynamics and reaction kinetics of combustion and steam gasification reactions were included in the model. They used a Yield model for the decomposition of the biomass. An energy assessment of a CHP plant with integrated biomass gasification using ASPEN PLUS was made by Damartizis et al. [3]. They assumed that biomass devolatilisation takes place instantaneously and the volatile products consist of H_2, CO, CO_2, CH_4, and H_2O. Zheng et al. [4] used ASPEN PLUS to simulate biomass integrated gasification combined cycle systems at corn ethanol plants. They found the ASPEN PLUS to be suitable software for the simulation of this technology. Simulation of the biomass gasification using ASPEN PLUS was studied also by other authors [5–7]. Puig-Arnawat et al. [8] presented a Review and analysis of biomass gasification models. A significant part of this review is devoted to the models that use ASPEN PLUS. Practically no work analysed in this review deals with the devolatilisation kinetics and particle sizes.

The model developed in this work is based on the experimental data obtained under laboratory conditions. The gasification process was modelled in two steps. The first step describes thermal decomposition of the biomass based on a kinetic model and the equilibrium composition of syngas is calculated by the Gibbs free energy of the expected components in the second step.

22.2 Mathematical Model

Basic processes of a gasification CHP plant are shown in Fig. 22.1. Mathematical models of individual processes include material and energy balance equations and phase equilibrium calculations. Besides a gasifier in other cases the models integrated in ASPEN PLUS were used. The gasification process was modelled in two steps. A kinetic model including heat transfer inside biomass particles for thermal decomposition of biomass was proposed. The following rate equation was considered:

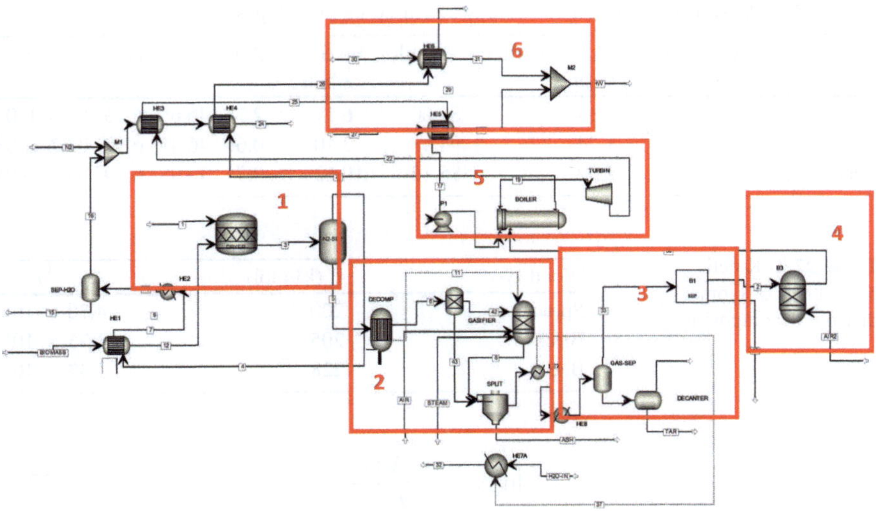

Fig. 22.1 Scheme of gasification and CHP plant. *1*-drying, *2*-gasification, *3*-gas cleaning, *4*-gas combustion, *5*-power generation, *6*-heat recovery

$$\frac{d\alpha}{dt} = \sum_{j}^{M}\sum_{i}^{N}\left(A_i \cdot e^{-\frac{E_{a,i}}{RT}} \cdot (1-\alpha_i)^{n_i}\cdot\right)w_j \tag{22.1}$$

where A is the apparent pre-exponential factor, E_a is the apparent activation energy, T is the temperature, t is time, R is the gas constant, α is the conversion, n the reaction order and w is the mass fraction of individual types of biomass. Index i represents the reaction step during the thermal decomposition (some types of biomass are decomposed in more than one step) and index j represents the type of biomass in the feed.

Assuming a spherical homogenous particle with radius r and taking into account the assumptions presented in [9], the enthalpy balance equation for a defined volume element can be written in the following form:

$$\lambda\left(\frac{\partial^2 T}{\partial r^2} + \frac{2}{r}\frac{\partial T}{\partial r}\right) + \Delta_r H\rho\frac{\partial\alpha}{\partial t} = \rho\, c_p\frac{\partial T}{\partial t} \tag{22.2}$$

where λ is the effective thermal conductivity, ρ is the average density, c_p the average specific heat capacity, and $\Delta_r H$ is the reaction enthalpy in J kg^{-1}. Details on the thermal decomposition model are published in our previous work [9]. The second step of the gasification process is the reaction of decomposed volatiles and char with oxygen and steam (partial oxidation, steam reforming, shift reaction). These reactions were modelled assuming that the chemical equilibrium is reached. The equilibrium constants were calculated using the free Gibbs energy:

Table 22.1 Proximate and elemental analyses of used biomass types (wt%)

Wood type	Moisture, when received	Moisture after drying	Volatiles	Fixed carbon	Ash	C	H	N	S
Spruce	20	2.14	89.00	6.11	2.75	46.65	6.23	0.06	0.07
Birch	19	1.74	94.48	3.10	0.68	46.45	6.10	0.07	0.03
Beech	21	2.52	86.12	10.62	0.74	46.98	6.19	0.01	0.01

Table 22.2 Kinetic parameters of biomass thermal decomposition

Wood type	E (kJ/mol)	A (s^{-1})
Spruce	200	4.78×10^{14}
Birch	205	1.75×10^{15}
Beech	228	1.43×10^{17}

$$\ln K_e = \frac{-\Delta_r G^o}{RT} \tag{22.3}$$

where K_e is equilibrium constant. R is the gas constant and $-\Delta_r G^o$ is the standard Gibbs energy of reaction.

22.3 Experimental Results

Three different types of wooden biomass were analysed using an elemental analyser (Vario Macro Cube, ELEMENTAR Analysensysteme GmbH) and a simultaneous thermogravimetric/differential scanning calorimetric device (Netzsch STA 409 PC Luxx). Table 22.1 shows the results of elemental and proximate analyses. For a mixed biomass, the content of individual components was calculated based on the mass fraction of the biomass' type in the mixture. The same method was applied for the determination of kinetic parameters of the mixed biomass devolatilisation. Kinetic parameters of individual biomass types were determined using thermogravimetric measurements and a multi-curve iso-conversion approach described in our previous works [9, 10]. Table 22.2 shows the average values of activation energies and pre-exponential factors for the reaction order n = 1.

Tars in the produced gas present a major problem for the electricity generating equipment downstream from the gasifier. In this work, the tar content of the produced gas was studied in a laboratory unit described in Fig. 22.2. The standard tar measurement method [11] was integrated into the experimental apparatus. The influence of temperature and catalyst type in a secondary reactor on gas tar content were estimated (Fig. 22.3). Dolomite and a mineral mixture termed as AFRC were used as catalysts. For details see our previous work [12]. The results of experimental measurements were used in the Aspen simulation model of an industrial scale biomass gasification plant.

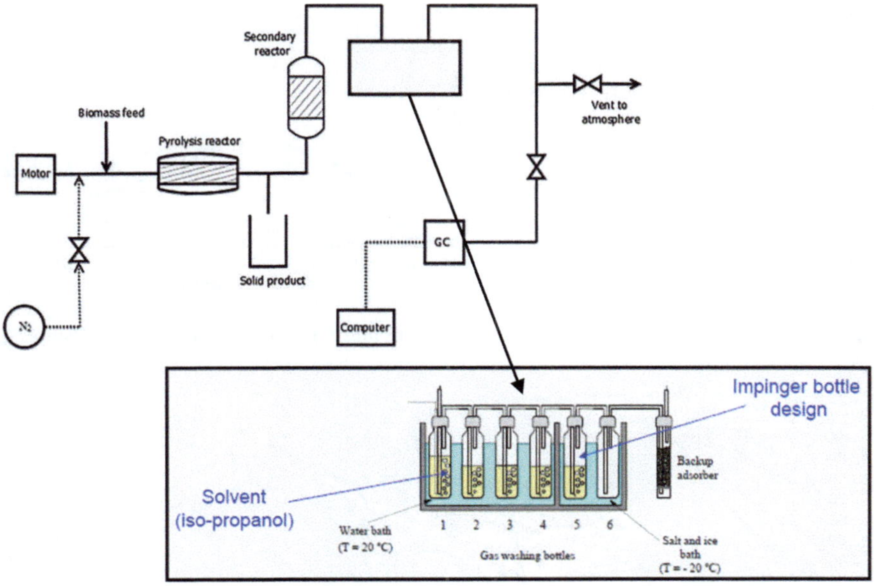

Fig. 22.2 Scheme of experimental apparatus for the tar content determination

Fig. 22.3 Tar content of the produced gas at different temperatures

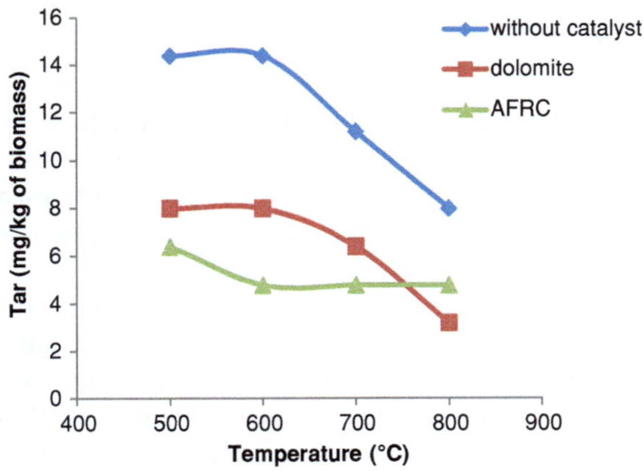

22.4 Simulation Results and Discussions

The Aspen simulation scheme is shown in Fig. 22.1. A feed of 18 tons/h of mixed wooden chips (33 wt% of spruce, 27 wt% of birch and 40 wt% of beech) with the proximate and elemental compositions shown in Table 22.1 and atmospheric pressure were considered. Before entering the gasifier, the biomass was subjected to a drying process where the content of moisture was reduced to 2 wt%. The gasifier model consisted of a user modelled decomposition reactor, a tar separator

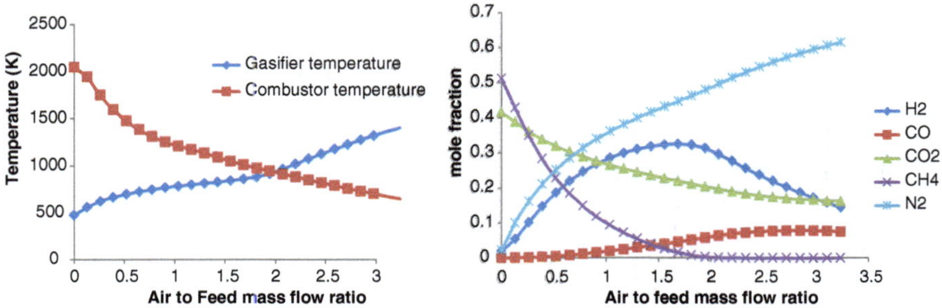

Fig. 22.4 Effect of air to feed mass ratio on temperature in gasifier and on gas composition

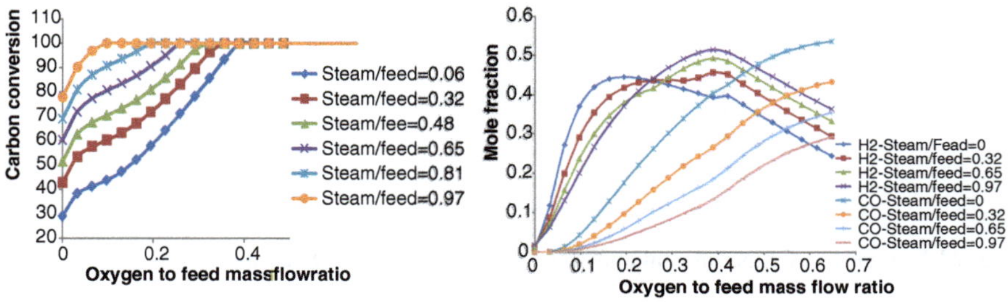

Fig. 22.5 Effect of oxygen and steam mass to feed mass ratio on carbon conversion and gas composition

and a Gibbs equilibrium reactor model. The produced gas was cooled and cleaned from ash and tar. The heat recovered was used in a steam boiler. The gas was combusted after its cleaning. Flue gases were used in a steam cycle for power generation. The remaining heat of the flue gases was used to produce hot water of 80 °C.

Describing the complete results of this simulation is beyond the scope of this paper. In this work we focus only on some results of the gasification step. Two different gasification agents were studied; gasification with air and steam and gasification with oxygen and steam. Figure 22.4 shows the temperature of gases at the outlet of the gasifier, temperature of gases at the outlet of the combustor and composition of the gases from the gasifier at different air to biomass mass flow ratios. For this simulation, the ratio of steam to biomass mass flow was 0.65 and air flow to the combustor was controlled to obtain the mole fraction of oxygen in the flue gases of 11 vol.%. The results shown in Fig. 22.4 indicate that an optimum ratio of air to biomass feed mass is around 2. At this rate, the carbon conversion is completed, the concentration of hydrogen shows a maximum and the temperature in the gasifier is high enough for the decomposition reactions. However, if a higher temperature in the gasifier is required, air is not a suitable gasification agent. Even though higher temperatures can be achieved when using air, the produced gas could not be used in a steam cycle due to its very low heating value.

In case of gasification with pure oxygen and steam, the produced gas has much higher heating value and higher temperatures in the gasifier and combustor are possible. Figure 22.5 shows the carbon conversion and concentration of the most important components in the produced gas (H_2 and CO) at different oxygen and steam mass flow to biomass mass flow ratios. The optimum ratio of oxygen to biomass mass flow moves between 0.12 and 0.4 according the steam mass flow.

22.5 Conclusion

Combining experimental results and user developed models with the ASPEN simulation environment enables the simulation and design of the gasification process to be integrated in a CHP plant. The optimum air to biomass mass flow ratio was found to be 2 for the steam to biomass mass flow ratio of 0.6. Gasification of biomass with air results in low heating value of the gas produced. At the air to biomass mass flow ratio lower than 2, the gasifier temperature is not high enough for complete conversion and at a value higher than 2, the content of combustible gases is too low to be used in a steam boiler. Optimum oxygen to biomass mass flow ratio was found to be around 0.4 for the steam to biomass mass flow ratio of 0.3. By increasing the steam mass flow, the equilibrium content of hydrogen in the produced gas increased; however, the gas heating value and the gasifier temperature decreased.

Acknowledgments This work was supported by the Grant VEGA No. 1/0757/13 from the Slovak Scientific Grant Agency.

References

1. W. Doherty, A. Reynolds, D. Kennedy, Computer simulation of a biomass gasification-solid oxide fuel cell power system using Aspen Plus. Energy **35**(12), 4545–4555 (2010)
2. M.B. Nikoo, N. Mahinpey, Simulation of biomass gasification in fluidized bed reactor using ASPEN PLUS. Biomass Bioenergy **32**(12), 1245–1254 (2008)
3. Th. Damartzis, S. Michailos, A. Zabaniotou, Energetic assessment of a combined heat and power integrated biomass, gasification–internal combustion engine system by using Aspen Plus. Fuel Process. Technol. **95**, 37–44 (2012)
4. H. Zheng, N. Kaliyan, R.V. Morey, Aspen Plus simulation of biomass integrated gasification combined cycle systems at corn ethanol plants. Biomass Bioenergy **56**, 197–210 (2013)
5. C. Chen, Y.Q. Jin, Y.H. Yan, Y. Chi, Simulation of municipal solid waste gasification in two different types of fixed bed reactors. Fuel **103**, 58–63 (2013)
6. C.R. Vitasari, M. Jurascik, K.J. Ptasinski, Exergy analysis of biomass-to-synthetic natural gas process via indirect gasification of various biomass feedstock. Energy **36**, 3825–3837 (2011)
7. L. Shen, Y. Gao, J. Xiao, Simulation of hydrogen production from biomass gasification in interconnected fluidized beds. Biomass Bioenergy **32**, 120–127 (2008)
8. M. Puig-Arnavat, J.C. Bruno, A. Coronas, Review and analysis of biomass gasification models. Renew. Sustain. Energy Rev. **14**(9), 2841–2851 (2010)

9. J. Haydary, D. Susa, Kinetics of thermal decomposition of aseptic packages. Chem. Pap. **67**(12), 1514–1520 (2013)
10. J. Haydary, Ľ. Jelemenský, L. Gašparovič, J. Markoš, Influence of particle size and kinetic parameters on tire pyrolysis. J. Anal. Appl. Pyrol. **97**, 73–79 (2012)
11. W.L. van de Kamp, P.J. de Wild, H.A.M. Knoef, J.P.A. Neeft, J.H.A. Kiel, Tar measurement in biomass gasification, standardisation and supporting R&D. ECN-C–06-046, July 2006
12. J. Haydary, D. Susa, J. Dudáš, Pyrolysis of aseptic packages (tetrapak) in a laboratory screw type reactor and secondary thermal/catalytic tar decomposition. Waste Manag **33**, 1136–1141 (2013)

Chapter 23
Thermomechanical Properties of Quartz Intended for Carbothermic Process for Silicon Production

Aissa Kefaifi, Tahar Sahraoui, Abdelkrim Kheloufi, Yacine Berbar and Nadjib Drouiche

Abstract The technology of solar grade silicon production intended for solar cells manufacturing consists in three successive stages: silica raw material ore enrichment, carbothermic reduction of silica to obtain metallurgical silicon grade (MG-Si) and purification of metallurgical silicon grade for obtaining the silicon solar grade (SoG-Si). Our work was focused on the preparation of the charge (SiO_2 and C) to obtain the metallurgical silicon grade, which aims to study the silica thermomechanical properties as raw material for the carbothermic process. Various experiments at laboratory scale were developed on quartz samples as the explosion tests, heat tests, and mechanical tests in order to derive friability indices, thermal resistance and heat index. The results have allowed us to make a preliminary conclusion on silica intended for carbothermic process based on its thermomechanical characteristics as well as its better performance in the muffle furnace.

23.1 Introduction

Today, many advanced materials depend on the silica used as feedstock for the production of silicon. We cite a few examples of the use of these materials, in the electronic industry and the photovoltaic (PV) industries. The (PV) one is growing in excess of 40 % [1].

Also silicon is used as raw material in the chemical and steel industries. Therefore the demand on silicon is experiencing unprecedented growth in recent

A. Kefaifi (✉) · T. Sahraoui · A. Kheloufi · Y. Berbar · N. Drouiche
Centre de Recherche en Technologie des Semi-conducteurs pour l'Energétique (CRTSE), 2, Bd Frantz Fanon BP140, Alger-7 Merveilles 16027 Algiers, Algeria
e-mail: kefaifi_22@yahoo.fr

T. Sahraoui
e-mail: Tahar.sahraoui@nasr-dz.org

A. Y. Oral et al. (eds.), *International Congress on Energy Efficiency and Energy Related Materials (ENEFM2013)*, Springer Proceedings in Physics 155,
DOI: 10.1007/978-3-319-05521-3_23,

Table 23.1 Type of silica studied

Type of silica	Location of the deposit	
Quartz	DL5	EL HOGGAR
	DL7	EL HOGGAR
Sandstone	DL1	EL TAREF
	DL2	EL TAREF

years. The difference between the supply and consumption has led to a shortage prompted scientists and industrialists in order to get better value for money have conducted studies and research to produce a low cost silicon and high purity.

The silicon solar grade production technology consists of three successive stages which are: the ore processing and the reduction of silica by carbon to obtain metallurgical grade silicon MG-Si purification of metallurgical grade silicon for the obtaining solar grade silicon SoG-Si [2–4].

In this work we perform burst tests on silica (sandstone, quartz) in a muffle furnace to determine and optimize the technological parameters: mechanical resistance, thermal resistance, charge friability in the furnace for studying the silica behaviour at high temperature [5].

23.2 Experimental Techniques

This test is intended to determine the ability of silica (quartz sandstone) to explode and produce fine particles. These are a qualitative and comparative testing, which allows us to classify the silica according to its qualities.

We have taken two types of slice (sandstone and quartz) as samples to study and each type was studied by two different nuances. Table 23.1 summarizes the deferent types of grades used in this study.

23.2.1 Heat Index

An amount of 400–500 g of silica (quartz sandstone) is introduced in the muffle furnace at 1,300 °C for 1 h. After cooling to room temperature, the samples were carefully placed on a series of screens (20, 10, 4 and 2 mm) [6, 7].

23.2.2 Thermal Strength Index

The same amount of silica (quartz, sandstone) in the muffle furnace at 1,300 ° C is introduced for 1 h, and then introduced in a quantity such HANNOVER drum

Table 23.2 Thermomechanical parameters calculated silica

Sample	Heat index HI (%)	Friability index FI (%)	Thermal strength index STI (%)
DL1	73.5	7.75	23.31
DL2	67.5	6.5	23.93
DL5	52.75	11.25	19.25
DL7	56.75	8.75	21.87

which rotates about its axis with a rotational velocity of 40 rpm for 2.5 min [8, 9]. The TSI is given by the following relationship TSI % = (+20 mm) % (−20 + 10 mm) % (−20 + 4 mm) % (−20 + 2 mm).

23.2.3 *Friability Index*

The same method was used to calculate the index of the thermal strength only silica is used without heat treatment (heating). The FI is the percentage by weight of silica (quartz sandstone) which is less than +2 mm. It is noted that the HI, FI and TSI indices are expressed in percentage, and this method of characterization (burst test) is standard, as described by Alnæs (1986), Johannesen (1998) and Birkeland (2004) [10].

23.3 Results and Discussion

After conducting tests on burst silica (sandstone, quartz), different indices were calculated (Heat Index, friability and thermal resistance indices) and were placed on a table summarizing (Table 23.2, Figs. 23.1, 23.2).

Most of the samples placed in the muffle furnace are fragmented into small parts due to the phenomenon of thermal expansion. It is noted that the hang burst test it was noted that in addition to fracturing silica, sound effects (popping sound) occurs due to burst of the silica. The initiation of the rupture of the silica (quartz, sandstone) begins at a temperature of 573 °C [7, 8] (Fig. 23.3).

The most important cause of failure is the existence of the inclusion or opening of micro cracks or grain boundaries. All samples sudden transformation to 573 °C (α phase transformation to β).

Extreme fracturing observed for sample DL5 (Quartz) and DL7 (Quartz) as we noticed a low degree of fracturing samples DL1et DL2 (sandstone). These observations are in agreement with the calculated parameter (Index friability FI) for the four samples. Moreover, after cooling to ambient air samples, fragmentation and disintegration continues, due to thermal fatigue [11]. The latter interpreted by Birkeland and Carstens (1989) thermal fatigue could cause the opening of grain

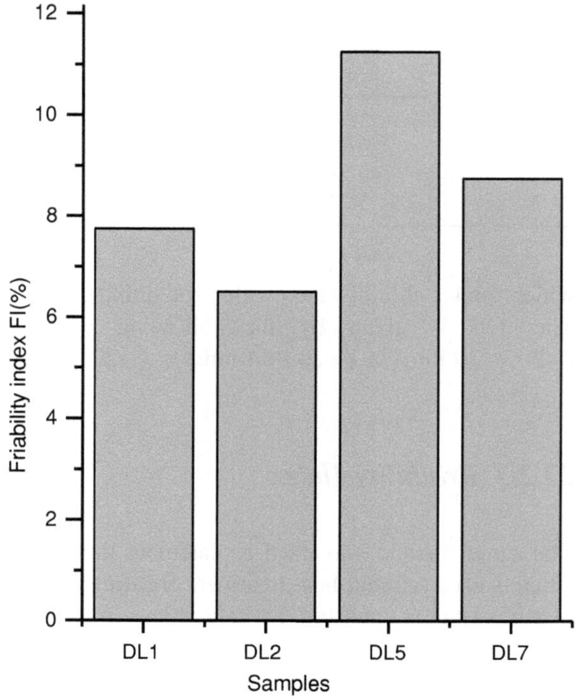

Fig. 23.1 Variation the heat index in silica

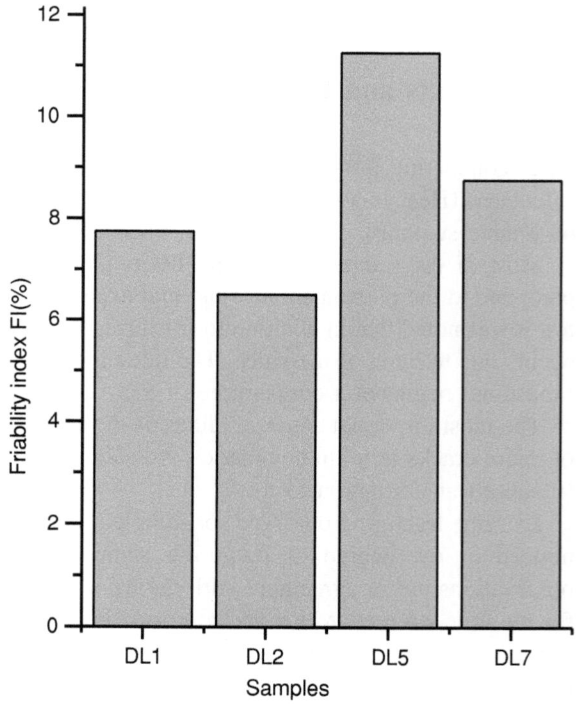

Fig. 23.2 Variation friability index in the silica

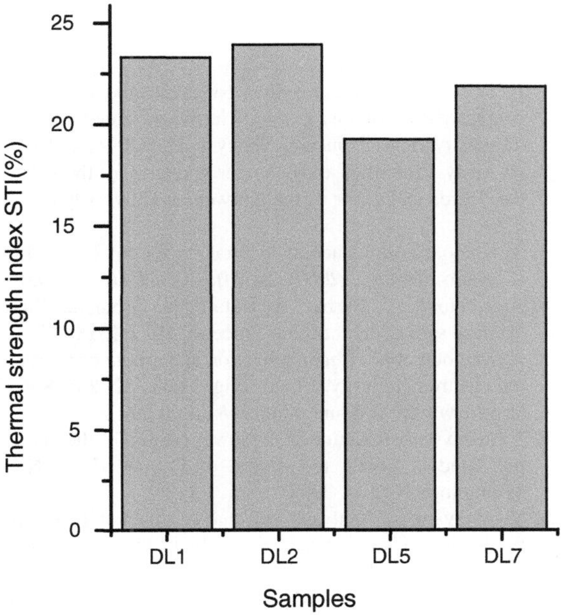

Fig. 23.3 Variation Index TSI thermal resistance in silica

boundaries in the silica accompanied by the development of micro cracks [12]. We also note that the value of the heat index is maximum for the DL1 sample. This implies that this sample has a better resistance to disintegration compared to other samples under the effect of temperature. The histogram above shows the sample DL5 has the highest friability index calculated therefore value, the latter is the more friable relative to other samples [13–15] it also notes that the same sample has the lowest among all the samples studied index thermal resistance.

23.4 Conclusion

The silica carbothermic technology to obtain the metallurgical grade silicon is a long and complex process. It depends on various factors, starting with the preparation of the charge until obtaining of a final product (metallurgical silicon grade). This process takes place in stable conditions in an electric arc furnace; we must first optimize all operating parameters of the process. The fundamental objective of this work is first, to characterize the raw material (silica) intended for the carbothermic process for obtaining metallurgical silicon grade, to optimize the charge ($SiO2 + C$) and finally thus valorize national deposits may be used in the electrometallurgical field. The bursting tests of the silica (sandstone, quartz) showed that the sample (DL5) is the most suitable to be used for the operation of carbothermic process as well as its better mechanical and thermal characteristics, which leads to a better performance in the muffle furnace.

References

1. L. Désindes, Ultra-pure silica for electrometallurgy: Geology and physical and chemical characteristics of quartz ore. Thesis presented for obtaining the grade of Doctor of the university Henri Poincaré, Nancy 1, Dec 2004, 433 p
2. K. Aasly, Properties and behavior of quartz for the silicon process. Thesis presented to obtain the degree of Doctor of the Norwegian University of Science and Technology, NTNU, Aug 2008
3. A. Kheloufi, Acid leaching technology for obtaining a high-purity silica for photovoltaic area. D. Sauro Pierucci. (2009). doi:10.3303/CET0917119.pp.197-202
4. A. Kheloufi, Y. Berbar, A. Kefaifi, N. Drouiche, Silica sand etching behaviour during a primary step of its leaching process. AIP J **1315**, 271–276 (2010)
5. A. Kheloufi et al., Characterization and quartz enrichment of the Hoggar deposit intended for the electrometallurgy. Chem. Eng. Trans. **32**, 889–894 (2013)
6. http://www.sfc.fr/donnees/mines/si/texsi.htm
7. J. Gustavsson, Reaction in the lower portion of the blast furnace with focus on silicon. Thesis presented to obtain the degree of Doctor of the Norwegian University of Science and Technology NTNU, 2004
8. V. Myrvågnes, Analysis and characterization of fossil carbonaceous materials for silicon production. Thesis presented to obtain the degree of Doctor of Norwegian University of Science and Technology NTNU, Jan 2008
9. A. Çiftja, T.A. Engh, M. Tangstad, *Refining and Recycling of Silicon: A Review*. NTNU report, Trondheim, Feb 2008. ISBN: 978-82-997357-2-8
10. H. Nordmark, Microstructure studies of silicon for solar cells defects, impurities and surface morphology. Thesis presented. To obtain the degree of Doctor of Norwegian University of Science and Technology, 2009 NTNU.janvier
11. Y.Sakaguchi, M. Ishisaki et al., Production of high purity silicon silica using AC-arc furnace with by carbothermic reduction of heated shaft. ISIJ Int. **32**(5), 643–649, (1992), Japan
12. S.M. Karabanov et al., *Study of solar grade silicon by the method of carbothermic reduction; Ryazan metal ceramics instrumentation plant JSC*. In 17th EPVSEC, Germany (2001), pp. 77–79
13. P. Maycock, *Silicon supply and word PV industry; PV energy systems; INC pveenergy.com*. In 14th Workshop on Crystalline Silicon Solar Cells & Modules: Materials and Processes Août (2004), pp. 51–65
14. M. Abdoukhanov, Development of technological bases for the production of metallurgical grade silicon of high quality photovoltaic application. Pos Khim J. M.XLV N°5–6,107–111, Russie (2001)
15. C.P. Khattak, D.B. Joyce, F. Schmid, A simple process to remove boron from metallurgical grade silicon. Sol. Ener. Mater. Sol. Cells **74**(2002), 77–89

Chapter 24
The Influence of the Material of the Transformer Core on Characteristics of the Selected DC–DC Converters

Krzysztof Górecki and Janusz Zarębski

Abstract In the paper the influence of the ferromagnetic material used for the construction of the core of the impulse-transformer on the characteristics of a half-bridge converter is considered. The investigated network is described and some results of measurements are shown. On the basis of the obtained characteristics of the investigated converters some suggestions for the designers of such circuits are formulated.

24.1 Introduction

Power supply systems converting the electrical energy generated in systems with renewable energy sources contain DC–DC converters [1]. One of the typical requirements for these circuits is to assure a galvanic separation between their input and output [2]. Accordingly, transformer DC–DC converters, whose the important component of which is the impulse-transformer, are typically used.

The impulse-transformer contains at least two windings and the core made of ferromagnetic material [3]. Nowadays producers of ferromagnetic cores offer cores made from various kinds of ferromagnetic materials. In each of this group one can find cores of different shapes, geometrical sizes and different proportional participation of each component in alloys or mixtures.

As it results from the previous papers of the authors [4, 5], properties of the core and its model influence in an essential manner the characteristics of isolated and non-isolated converters [6].

K. Górecki (✉) · J. Zarębski
Department of Marine Electronics, Gdynia Maritime University, Gdynia, Poland
e-mail: gorecki@am.gdynia.pl

A. Y. Oral et al. (eds.), *International Congress on Energy Efficiency and Energy Related Materials (ENEFM2013)*, Springer Proceedings in Physics 155,
DOI: 10.1007/978-3-319-05521-3_24,

In the paper some results of measurements of the characteristics and exploitive parameters of the half-bridge DC–DC converter containing the impulse-transformer with the core made from different ferromagnetic materials are presented. This converter commonly collaborates with renewable power sources. The research was conducted in a wide range of changes of load resistance, frequency and the duty factor of the control signal. Particularly, the influence of the mentioned factors on the output voltage and the watt-hour efficiency of the examined converter and on the temperature of transistors and diodes operated in the converter and on the temperature of the core and windings of the impulse-transformer are examined. On the basis of the obtained results of measurements the operating conditions of the converter in which each group of cores of ferromagnetic assures the optimum values of the exploitive parameters of the considered converter, are shown.

24.2 Investigated Converter

In Fig. 24.1 the diagram of the investigated converter is presented. As it is visible, it consists of three functional blocks: the PWM controller of the type of SG2525A (A), the driver with IR2110 (B) and the power circuit with the considered half-bridge converter (C).

The PWM controller SG2525A existing in the block A generates the rectangular pulses train controlling signal whose frequency is regulated by means of the resistor R_6, whereas the value of the duty factor is set with the use of the potentiometer R_1. The driver is supplied from the voltage source U_C of the value equal to 15 V. The driver SG2525A makes a possible to obtain the control signal of the frequency non-resident to 400 kHz [7]. Together with an increase in the value of the frequency the range of change of the duty factor gets narrower.

The main component of the block B is the driver IR2110, which plays the role of the power amplifier of the control signal. In this block, the constant voltages U_{DD} and U_{CC} of the values equal to 15 V, assure the suitable polarization of the driver terminals. Additionally, the Schottky diode D_1 of the type 1N5822 and the capacitor C_8 operate in the bootstrap configuration.

The block C includes the half-bridge converter with the constant input voltage equal to $U_{we} = 25$ V and the load resistance R_0. The considered converter contains the following components: the power MOS transistor IRF 540, the Schottky diode 1N5822, capacitors of capacity 47 μF and the choking-coil of inductance 220 μH. In the considered converter the pulse transformers, containing the primary winding and secondary winding containing 20 turns of the copper wire in the enamel of the diameter 0.8 mm, wound on the following cores: ferrite RTF-25x15x10 (F-867), powder RTP-26.9x14.5x11 (T106-26) and nanocrystalline RTN-26x16x12 (M-070), are applied successively.

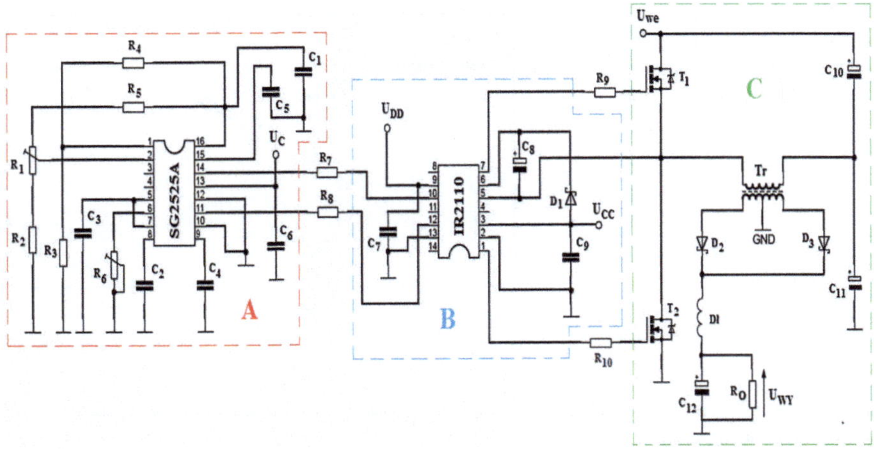

Fig. 24.1 Diagram of the considered half-bridge converter with the controller and the driver

24.3 Results of Measurements

For the converter from Fig. 24.1 the characteristics at the steady-state over a wide range of changes of the frequency f and the duty factor d of the control signal and the load resistance R_0 are presented. Figures 24.2, 24.3, 24.4, 24.5, 24.6, and 24.7 show the influence of the selection of the transformer core on the characteristics of the considered converter.

In Figs. 24.2 and 24.3 the measured dependences of the output voltage of the converter U_{wy} (Fig. 24.2) and the watt-hour efficiency η (Fig. 24.3) on the duty factor d obtained at the frequency f = 50.5 kHz and the load resistance $R_0 = 5\ \Omega$ are shown.

As one can observe, at the low frequency of switching, the dependence $U_{wy}(d)$ obtained for the converters with all the considered cores are almost linear and one cannot observe essential influence of the core material on their course. Only for $d > 0.35$ it is visible that in the converter with the ferrite core the higher values over 15 % of the output voltage are obtained. The watt-hour efficiency is an increasing function of the factor d and attains the greatest values for the converter with the ferrite core. The use of the powder core causes a fall of the efficiency by about 10 %.

In turn, in Figs. 24.4 and 24.5 the measured dependences of the converter output voltage U_{wy} (Fig. 24.4) and the watt-hour efficiency η (Fig. 24.5) on the duty factor d obtained at the frequency f = 204 kHz and the load resistance $R_0 = 5\ \Omega$ are presented.

Comparing the results of measurements obtained at the frequency f = 50.5 kHz (Figs. 24.2, 24.3) and f = 204 kHz (Figs. 24.4, 24.5) one can observe that an increase of the frequency caused a visible decrease in the value of the watt-hour

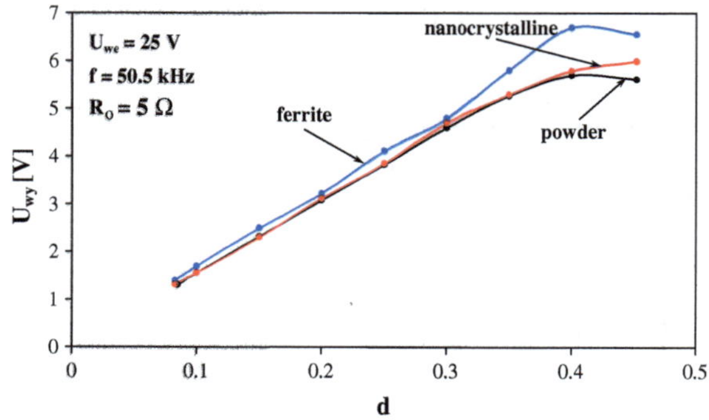

Fig. 24.2 Measured dependences of the output voltage on the duty factor at f = 50.5 kHz and $R_0 = 5\ \Omega$

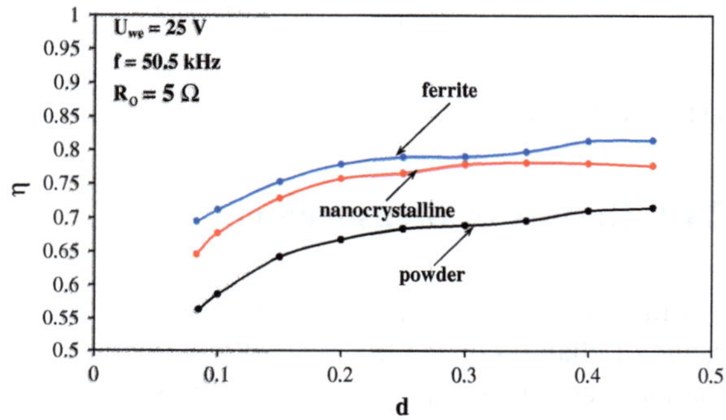

Fig. 24.3 Measured dependences of the watt-hour efficiency on the duty factor at f = 50.5 kHz and $R_0 = 5\ \Omega$

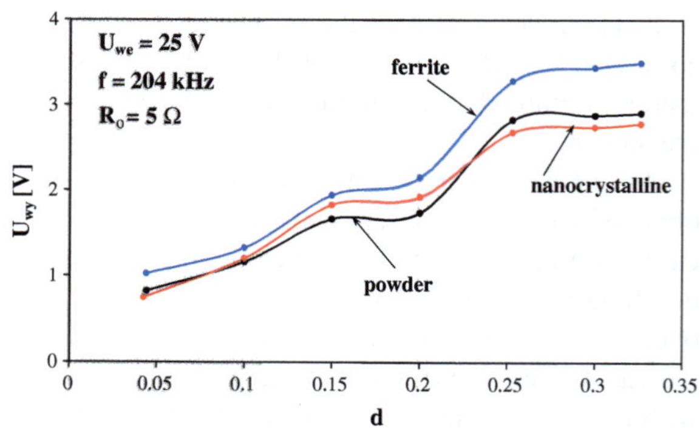

Fig. 24.4 Measured dependences of the output voltage on the duty factor at f = 204 kHz and $R_0 = 5\ \Omega$

efficiency and the output voltage for all the considered cores. The change of the shape of the characteristics $U_{wy}(d)$ is also worth noticing.

In Figs. 24.6 and 24.7 the influence of the load resistance on the converter output voltage U_{wy} (Fig. 24.6) and the watt-hour efficiency η (Fig. 24.7) at the frequency f = 50.5 kHz and the duty factor d = 0.3, is illustrated.

The voltage U_{wy} is for all the cores an increasing function of the load resistance, whereas within the range $R_0 < 1\ \text{k}\Omega$ the greatest values of the output voltage are

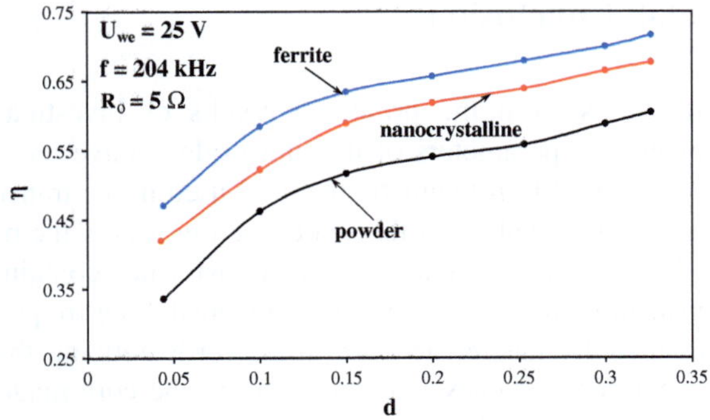

Fig. 24.5 Measured dependences of the watt-hour efficiency on the duty factor at f = 204 kHz and $R_0 = 5\ \Omega$

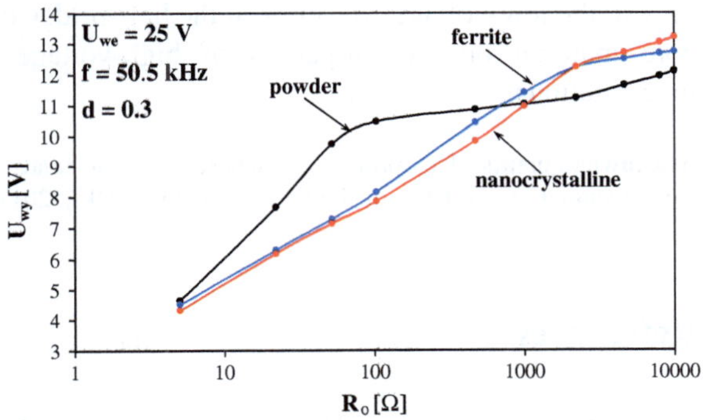

Fig. 24.6 Measured dependences of the output voltage on the load resistance at f = 50.5 kHz and d = 0.3

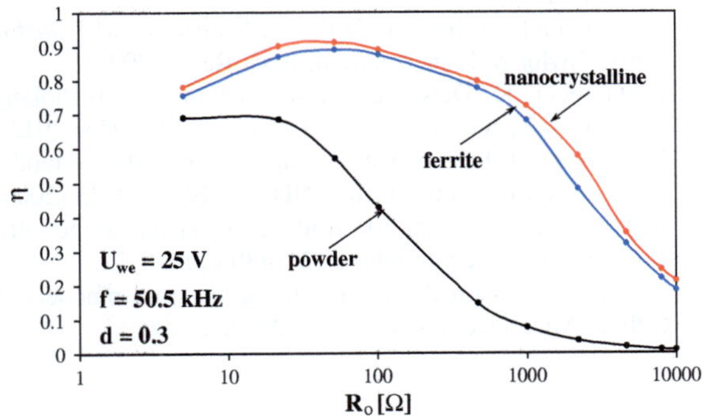

Fig. 24.7 Measured dependences of the watt-hour efficiency on the load resistance at f = 50.5 kHz and d = 0.3

obtained for the converter with the powder core, on the other hand for higher R_0—the greatest values of U_{wy} are reached for the converter with the nanocrystalline core. The highest efficiency was obtained for the nanocrystalline core, and the least for the powder core. It is worth noticing that the dependence $\eta(R_0)$ possesses the maximum at R_0 equal to about 40 Ω.

24.4 Conclusions

It results from the presented results of investigations that the values of the exploitive parameters of the half-bridge converter are strongly dependent on the selection of ferromagnetic material used in the transformer core. Within the range of the small values of the switching frequency, the highest values of the watt-hour efficiency are obtained for the transformer containing the nanocrystalline core, instead within the range of the high switching frequency—the highest efficiency is assured by the ferrite core. It is worth noticing that essential worsening of the converter efficiency is obtained using the core made from powdered iron.

The presented results of measurements can be useful to designers of switching-mode power supplies. One ought however to notice that the presented results refer only to the low-voltage converter of the half-bridge converter. The influence of the core material on the properties of high-voltage converters demands further investigations.

Acknowledgments This project is financed from the funds of National Science Centre which were awarded on the basis of the decision number DEC-2011/01/B/ST7/06738.

References

1. M.H. Rashid, *Power Electronic Handbook* (Academic Press, Elsevier, 2007)
2. R. Ericson, D. Maksimovic, *Fundamentals of Power Electronics* (Kluwer Academic Publisher, Norwell, 2001)
3. A. Van den Bossche, V.C. Valchev, *Inductors and Transformers for Power Electronics* (CRC Press, Taylor & Francis Group, Boca Raton, 2005)
4. K. Górecki, K. Detka, Electrothermal model of choking-coils for the analysis of dc–dc converters. Mater. Sci. Eng. B **177**(15), 1248–1253 (2012)
5. K. Górecki, W.J. Stepowicz, Comparison of inductor models used in analysis of the buck and boost converters. Informacije MIDEM **38**(1), 20–25 (2008)
6. K. Górecki, J. Zarębski, Electrothermal analysis of the self-excited push-pull dc–dc converter. Microelectron. Reliab. **49**(4), 424–430 (2009)
7. J. Zarębski, K. Górecki, The electrothermal macromodel of voltage mode PWM controllers for SPICE. Microelectron. J. **37**(8), 728–734 (2006)

Chapter 25
A Novel Zinc Diffusion Process for Fabrication of High Performance GaSb Thermophotovoltaic Cells

Liangliang Tang and Hong Ye

Abstract This paper reports a novel zinc diffusion method for forming emitters in GaSb thermophotovoltaic cells. A closed quartz-tube diffusion system using Zn–Ga alloys as the diffusion sources is designed to realize p-type doping in N-GaSb wafers, the surface high-concentration zinc diffusion region is suppressed by this diffusion method, the GaSb cells fabricated using this method shows good quantum efficiency in the near-infrared bands. Compared to the conventional pseudo-closed-box (or semi-closed box) diffusion method using Zn–Sb alloys, the zinc profiles obtained from the novel diffusion method have no surface high-concentration zinc diffusion region which could have shortened the lifetime of photo-generated carriers, thus the controllability of the etch-back process after front-side metallization is significantly improved. The cost of the cell fabrication is reduced since no protective gas is required during the diffusion process.

25.1 Introduction

The GaSb cells are ideal candidates for thermophotovoltaic (TPV) energy conversion systems, and they have already been successfully used in micro-scale [1, 2], meso-scale [3] and home used large-scale [4, 5] TPV generators. The GaSb cells can be fabricated using a zinc diffusion method [6–14] or epitaxial methods adopting MOVPE (metal organic vapor phase epitaxy) and MOCVD (metal organic chemical vapor deposition) techniques [15–17]. The zinc diffusion method is the most attractive one among the above techniques for fabricating low-cost GaSb cells.

L. Tang · H. Ye (✉)
Department of Thermal Science and Energy Engineering, University of Science and Technology of China, 230027 Hefei, People's Republic of China
e-mail: hye@ustc.edu.cn

L. Tang
e-mail: tangll@mail.ustc.edu.cn

A. Y. Oral et al. (eds.), *International Congress on Energy Efficiency and Energy Related Materials (ENEFM2013)*, Springer Proceedings in Physics 155,
DOI: 10.1007/978-3-319-05521-3_25,

Zn–Sb alloys are commonly used as the diffusion source during the diffusion process, in which a kink-and-tail shaped diffusion profile with a high-concentration surface region is formed. The heavily doped surface region is beneficial for ohmic contact with the front electrode, however, heavily doping could shorten the lifetime of photo-generated carriers, and is harmful to the collection of carriers. Precisely etching after front-side metallization was used for enhancing the quantum efficiency, which could result in the short circuit current rising by a factor of about 2 [6]. However, the frontside emitter etch was time consuming and could not be readily automated. The wafer was etched, rinsed, tested for many times before the electrical current reached its optimum value. Each etching process should be precisely controlled because the voltage would drop quickly beyond the optimum etching depth, therefore, no standard etch time could be reliably established [6, 9].

In order to solve the difficulty of the etching process, Fraas et al. [9] and Gruenbaum et al. [10] invented a two-step diffusion process to optimize the fronside doping profile. In their invention, a light zinc diffusion initially creates the active region of the cell, followed by a heavy zinc diffusion under the grid lines to produce a low ohmic contact resistance. The frontside emitter etch was eliminated in this method, while extra diffusion and photolithography process were added, thus the total cost could have increased.

In our previous investigation of zinc diffusion in N-GaSb, we found that Ga atoms from the diffusion sources can suppress the formation of the high-concentration surface region in zinc profiles [18]. This phenomenon plays a direct guiding role for the fabrication of GaSb cells. We fabricated GaSb cells using this new diffusion method, details about our experiments and analyses are presented below.

25.2 Device Fabrication Using a Novel Diffusion Process

Tellurium-doped ($n = 2{\sim}7 \times 10^{17}\ cm^{-3}$), $\langle 100 \rangle$-oriented GaSb substrates from the Institute of Semiconductors (Chinese Academy of Sciences) were used in the experiments. The N-GaSb substrates were washed in xylene, acetone, and methanol, followed by etching the oxide layer in diluted HCl solution. An insulating layer of SiO_2 with a thickness of approximately 0.11 μm was deposited as a coating on the upper surface of GaSb substrates using PECVD (Plasma Enhanced Chemical Vapor Deposition), and the central area of the wafers was exposed using the standard photolithography technique (Fig. 25.1a).

The p/n junctions were realized by zinc diffusion into the N-GaSb wafers. In previous investigations, the diffusion sources were usually Zn–Sb alloys and a pseudo-closed-box diffusion method was used, the zinc diffusion profiles showed kink-and-tail shaped under the above conditions [19–21]. Phosphorus diffusion profile in silicon also shows kink-and-tail shaped [22] as that of zinc in GaSb, nevertheless, the surface high-concentration phosphorus diffusion region before the kink point is much shorter than that of zinc profiles, and it doesn't affect the collection of photo-generated carriers seriously. In order to fabricate high

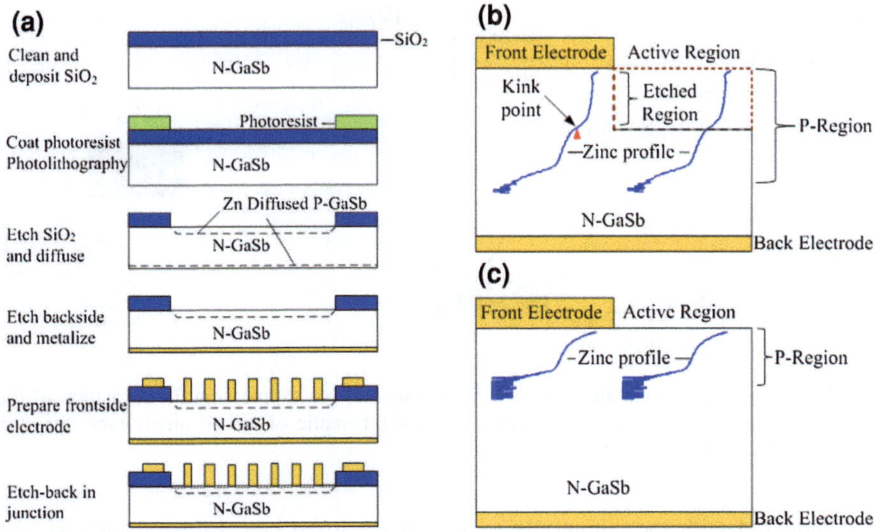

Fig. 25.1 Fabrication process of GaSb cells (**a**) and cell structures using different diffusion process (**b**) cells fabricated using conventional Zn–Sb alloys as the diffusion sources and pseudo-closed-box diffusion method; **c** cell structures fabricated using Zn–Ga alloys as the diffusion sources and closed quartz-tube diffusion method

performance GaSb cells, the region before the kink point should be precisely removed (Fig. 25.1b). This process is difficult to control (details were explained in the introduction) and the electrode needs to withstand the etching solution for long time. Here we designed a closed quartz-tube diffusion system (Fig. 25.2b) for forming zinc profiles which have only the tail regions of the kink-and-tail shaped profiles (Fig. 25.1c).

Some researchers considered that the zinc diffused samples with kink-and-tail profiles were related to extended defects and not suitable for modeling [23, 24]. In our recent investigations, we have found out that both the surface and tail region of the kink-and-tail profiles show perfect regularities and if Ga atoms are added into the diffusion sources, the high-concentration surface diffusion regions could be completely suppressed (Fig. 25.2a) [18]. Based on the above principle, the etching process in fabricating GaSb cells may be simplified if we use Zn–Ga alloys as the diffusion sources. Figure 25.2b shows the schematic of the diffusion systems, a quartz boat with N-GaSb wafers and Zn–Ga sources is placed in the bottom of the quartz tube I, a slightly smaller tube IIs is inserted into the tube I. The operation process is as follows: fix the tube I with the above parts using a flange joint firstly; close the valve I, open the valve II and vacuum pump, evacuate the air; and then close the valve II, open the valve I and allow some argon to flow into the tube until the atmospheric pressure is reached; afterwards, close the valve I and evacuate the argon, burn the middle of tube II using a flame gun under the evacuating state and form a vacuum seal between the two tubes. The sealed quartz tube is removed from the flange joint and kept at 500 °C for 2 h in a resistance-heated furnace.

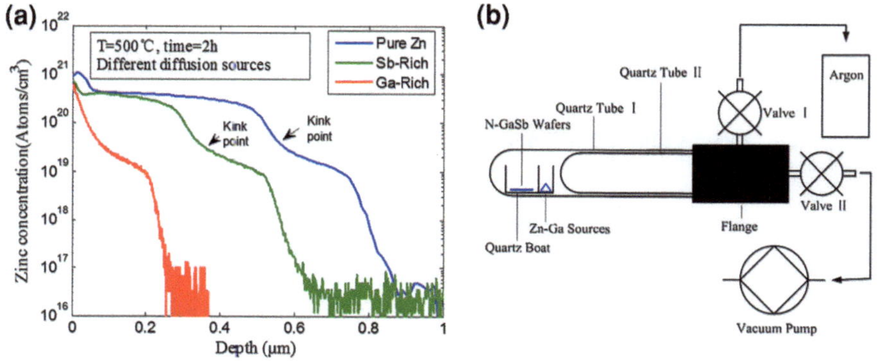

Fig. 25.2 The effect of Ga atoms from diffusion sources for suppressing the formation of the high-concentration surface diffusion region (**a**) and schematic of closed quartz-tube diffusion method using Zn–Ga alloys (**b**)

Compared with the commonly used pseudo-closed-box diffusion method, the closed quartz-tube diffusion method has two main advantages. Firstly, if Zn–Ga alloys are used as the diffusion sources, the zinc diffusion profiles will show box shaped without the high-concentration diffusion region (Fig. 25.2a), which is beneficial for cell manufacturing, the box shaped profiles can't be formed using pseudo-closed-box diffusion method due to that the system is not hermetic completely, Zn–Ga alloys couldn't keep a saturated pressure because that Zn–Ga alloys show liquid phase even at the room temperature and can be readily volatile. Secondly, protective gas (argon) is not required during the diffusion process when using the new diffusion method, only a little argon is required as the flushing gas during the evacuation process, while the protective gas is needed during the whole diffusion process when using the pseudo-closed-box diffusion method. The sealed quartz tube should be cracked after diffusion, the reason for this design is to facilitate the preparation of small experimental samples, and a diffusion system without the need to break tubes should be redesigned for industrial production. In general, the new diffusion method is beneficial for both the simplification of cell manufacturing and cost reduction.

After the diffusion process, the front of the wafers was protected by a photosensitive resist layer and the p-doping region of the backside was etched. The metals of electrode was deposited using magnetron sputtering, Ti(50 nm)/Pt(50 nm)/Ag(250 nm) layers were deposited for back-side electrode and Pt(50 nm)/Ag(250 nm) layers for front-side electrode. Our selection of materials referred to Fraas's design [8] except that the Au layer was changed to Ag layer for lower cost. The pattern of the front-side electrode was defined through a hollowed-out metal plate during the sputtering, the standard lift-off photolithography was not used because that the metal mask method is more convenient for preparation the electrode of the cell samples.

25.3 Device Performance and Analysis

The quantum efficiency of the GaSb cell without anti-reflection (AR) layers was measured using Newport IPCE/QE 200 tester. Figure 25.3 showed the external quantum efficiency (EQE) of three kinds of cells: A1 represents the cell which was not etched after metallization; A2 and A3 represent the cell etched in an ammonium sulfide [$(NH_4)_2Sx$] solution for 5 and 10 min after metallization, the etching depth was 0.06 and 0.12 μm respectively.

The Ga atoms from the diffusion sources can suppress the forming of the high-concentration regions. However, zinc concentration in the extremely shallow region is still high due to the escaping of small amounts of Ga atoms in the GaSb wafers (Fig. 25.3a); if the shallow region was etched, the EQE would increase; this phenomenon was reflected in the performance of cells A1 and A2. The cell performance only increased a little instead of rising by about a factor of 2 as that using conventional diffusion method, indicating that the new diffusion method is beneficial for cell manufacturing. If the cell A2 was etched in $(NH_4)2Sx$ solution for another 5-min, the EQE of the visible band will increase while the EQE of near-infrared bands will decrease (see cell A3), which is detrimental to its use in TPV systems.

Before comparing the performance of our new diffusion method made cells and that from JX Crystals Inc., it should be noted that the EQE data obtained from the tester we used is lower than the JXC' testing data (Fig. 25.4a), the following data were all obtained from our tester. The Internal quantum efficiency (IQE) of cell A2 was computed using the following equations:

$$\mathrm{IQE} = \frac{\mathrm{EQE}}{1 - R} \tag{25.1}$$

$$\mathrm{IQE} = \frac{\mathrm{EQE}}{(1 - R)(1 - A_{\mathrm{grid}})} \tag{25.2}$$

Here R is the reflectance of the cell and Agrid is the fraction of the cell covered with the metal grid (in this case: grid width = 0.1 mm; grid spacing = 0.9 mm; Agrid = 0.1). (25.1) represents the IQE calculated when taking account of the impact of the front electrode shielding. (25.2) represents the IQE calculated when neglecting this impact.

As shown in Fig. 25.4, the IQE of our cell is about 10–15 % lower than that of JXC's cell. It can be explained as follows: the front electrode of JXC's cell was made of the standard lift-off photolithography and the gird design is narrower and denser (gird = 0.0163 mm; grid spacing = 0.102 mm) than ours, thus the current collection ability is better than our design; the Au layer of electrode was changed to the Ag layer for lower cost, the conductivity of Au is better than Ag; our cells were fabricated in several laboratories due to the dispersion of the devices, all the

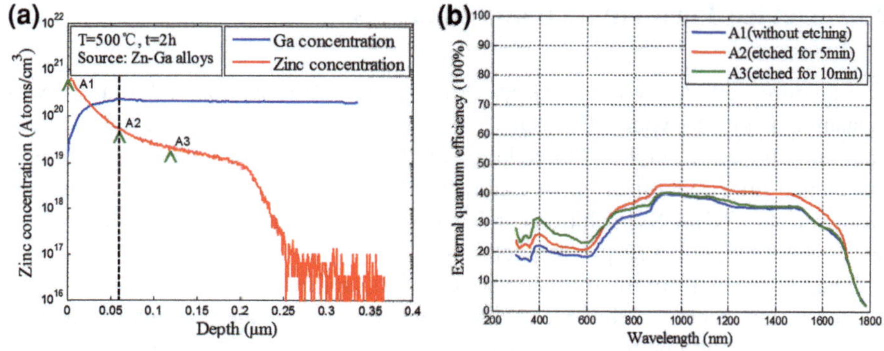

Fig. 25.3 Different etching depth from the surface of GaSb cells (**a**) and the corresponding EQE (**b**)

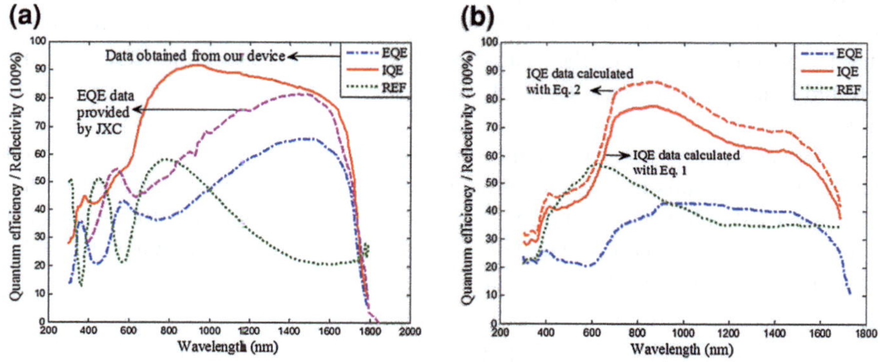

Fig. 25.4 The performance of GaSb cells. **a** EQE, IQE and REF data of the commercial GaSb cell obtained from Newport IPCE/QE 200 tester and the EQE data provided by JXC; **b** EQE, IQE and REF data of cell A2, the IQE data were calculated using 25.1 and 25.2

steps were not completed in a same clean room. The above reasons may account for the lower performance of our cells, we believe that the cell performance could be improved if a full set of excellent manufacturing devices were provided.

25.4 Conclusions

In this paper, a closed quartz-tube zinc diffusion system was designed for fabricating GaSb cells. Compared to the conventional pseudo-closed-box diffusion method, the novel zinc diffusion method can not only suppress the formation of the high-concentration diffusion front, but also reduce the cost of cell fabrication in that only a little protective gas was used. After the metallization, the cell performance would reach its optimal value if it was etched in $(NH_4)_2S_x$ solution for only 5 min. Although the cell performance is still about 10–15 % lower than that of JXC's, it is believed that it could be increased if the manufacturing device condition is improved.

Acknowledgments This work is funded by the National Basic Research Program of China (grant no 2009CB939904), the Fundamental Research Funds for the Central Universities.

References

1. S.K. Chou, W.M. Yang, K.J. Chua, J. Li, K.L. Zhang, Development of micro power generators–A review. Appl. Energ **88**, 1–16 (2011)
2. L.C. Chia, B. Feng, The development of a micropower (micro-thermophotovoltaic) device. J. Power Sources **165**, 455–480 (2007)
3. Y.H. Li, Y.S. Lien, Y.C. Chao, D. Dunn-Rankin, Performance of a mesoscale liquid fuel-film combustion-driven TPV power system. Prog. Photovoltaics Res. Appl. **17**, 327–336 (2009)
4. L.M. Fraas, J.E. Avery, H.X. Huang, Thermophotovoltaic furnace-generator for the home using low bandgap GaSb cells. Semicond. Sci. Technol. **18**, 247–253 (2003)
5. A.W. Bett, O.V. Sulima, GaSb photovoltaic cells for applications in TPV generators. Semicond. Sci. Technol. **18**, 184–190 (2003)
6. L.M. Fraas, G.R. Girard, J.E. Avery, B.A. Arau, V.S. Sundaram, GaSb booster cells for over 30 % efficient solar-cell stacks. J. Appl. Phys. **66**(8), 3866–3870 (1989)
7. L.M. Fraas, J.E. Avery, P.E. Gruenbaum, V.S. Sundaram, K.Emery, R.Matson, *Fundamental characterization studies of GaSb solar cells.* in Process of 22nd IEEE Photovoltaic Specialists Conference (New York, 1991), pp. 80–83
8. L.M. Fraas, J.E. Avery, R.G. Gerald, Tandem photovoltaic solar cell with III-V diffused junction booster cell. US Patent 5091018, 1992
9. L.M. Fraas, V.S. Sundaram, J.E. Avery, P.E. Gruenhaum, E. Malocsay, III-V solar cells and doping processes. US Patent 5217539, 1993
10. P.E. Gruenhaum, V.T. Dinh, V.S. Sundaram, Gallium antimonide infrared solar cells with improved efficiency and manufacturability. Sol. Energ Mater. Sol. Cells **32**, 61–69 (1994)
11. O.V. Sulima, A.W. Bett, Fabrication and simulation of GaSb thermophotovoltaic cells. Sol. Energ Mater. Sol. Cells **66**, 533–540 (2001)
12. G. Rajagopalan, N.S. Reddy, H. Ehsani, I.B. Bhat, P.S. Dutta, R.J. Gutmann, G. Nicols, O.V. Sulima, A simple single-step diffusion and emitter etching process for high-efficiency gallium-antimonide thermophotovoltaic devices. J. Electron. Mater. **32**, 1317–1321 (2003)
13. V.P. Khvostikov, M.G. Rastegaeva, O.A. Khvostikova, S.V. Sorokina, A.V. Malevskaya, M.Z. Shvarts, A.N. Andreev, D.V. Davydov, V.M. Andreev, High-efficiency (49 %) and high-power photovoltaic cells based on gallium antimonide. Phys. Semicond. Devices **40**, 1242–1246 (2006)
14. A.S. Vlasov, E.P. Rakova, V.P. Khvostikov, S.V. Sorokina, V.S. Kalinovsky, M.Z. Shvarts, V.M. Andreev, Native defect concentration in Czochralski-grown Te-doped GaSb by photoluminescence. Sol. Energ Mater. Sol. Cells **94**, 1113–1117 (2010)
15. D.Martin, C.Algora, *Theoretical comparison between diffused and epitaxial GaSb TPV cells*, in Thermophotovoltaic Generation of Electricity, Sixth Conference, (Rome, 2004), pp. 311–319
16. T.Schlegl, F.Dimroth, A.Ohm, A.W. Bett, *TPV modules based on GaSb structures*, in Thermophotovoltaic Generation of Electricity, Sixth Conference, (Rome, 2004), pp. 285–293
17. M.G. Mauk, V.M. Andreev, GaSb-related materials for TPV cells. Semicond. Sci. Technol. **18**, 191–201 (2003)

18. H.Ye, L.L. Tang, K.J. Li, The intrinsic relationship between the kink-and-tail and box-shaped zinc diffusion profiles in n-GaSb, *Semicond. Sci. Technol.* **28,** 1–6 (2013). 015001. doi:10.1088/0268-1242/28/1/015001
19. V.S. Sundaram, Gruenbaum. Zinc diffusion in GaSb. J. Appl. Phys. **73**(8), 3787–3789 (1993)
20. A.W. Bett, S. Keser, O.V. Sulima, Study of Zn diffusion into GaSb from the vapour and liquid phase. J. Cryst. Growth **181**, 9–16 (1997)
21. O.V. Sulima, A.W. Bett, M.G. Mauk, B.Y. Ber, P.S. Dutta. *Diffusion of Zn in TPV materials: GaSb, InGaSb, InGaAsSb and InAsSbP.* in Thermophotovoltaic Generation of Electricity, Fifth Conference, (Rome, 2002), pp. 402–413
22. A. Bentzen, G. Schubert, J.S. Christensen, B.G. Svensson, A. Holt, Influence of temperature during phosphorus emitter diffusion from a spray-on source in multicrystalline silicon solar cell processing. Prog. Photovoltaics Res. Appl. **15**, 281–289 (2007). doi:10.1002/pip.731
23. K.Sunder, H.Bracht. Zinc and gallium diffusion in gallium antimonide. Phys. Rev. B 2007; **75**: 245210(1–9)
24. S.P. Nicols, H. Bracht, M. Benamara, Z. Liliental-Weber, E.E. Haller, Mechanism of zinc diffusion in gallium antimonide. Phys. B **308–310**, 854–857 (2001)

Chapter 26
Impact of Barium on Surface and Reactivity of TiO_2

Madani Ghelamallah, Radia Imane Fertout and Soufi Kacimi

Abstract A series of samples noted TiBaX (where X = at.% of Ba) have been prepared by hydrolysis, in neutral medium from TiO_2 and γ-$BaCO_3$. These samples were calcined under air at 450 and 1150 °C then characterized by specific surface area (BET), XPS experiments, thermogravimetry (TG) and differential thermal analysis (DTA). Obtained results show that after hydrolysis, the conversion of the anatase did not take place. For the samples without barium, after calcination at 450 °C, the specific surface of TiBa0 is not modified. At 1150 °C the effect of sintering is significant; the surface of TiO_2 is lower than 1 $m^2\ g^{-1}$ and anatase structure is transformed to rutile. The addition of barium decrease surfaces of the calcined samples with 450 °C. After calcination at 1150 °C, the surface of the oxide increases with the rate of barium added. Indeed, the surfaces of TiO_2 triple when 15 % of barium is added. XPS results shows presence of oxides, carbonates and oxy-carbonates. The TG analysis under H_2/Ar, show that volatile specie as water of hydratation is eliminated at 100 °C.

M. Ghelamallah (✉)
Laboratoire de Matériaux, Applications et Environnement,
Faculté des Sciences et de la Technologie, Université de Mascara,
BP 305 Route de Mamounia, 29000 Mascara, Algeria
e-mail: gh_madani@hotmail.com

R. I. Fertout
Laboratoire de Matériaux et Catalyse, Faculté des Sciences, Université Djillali Liabes,
BP 89 Cité Larbi Ben M'hidi, 22000 Sidi Bel-Abbes, Algeria

S. Kacimi
Laboratoire de Chimie Appliquée, Institut des Sciences et de la Technologie,
Centre Universitaire d'Ain Temouchent, BP 284 Route de Sidi Bel-Abbes,
46000 Ain Temouchent, Algeria

A. Y. Oral et al. (eds.), *International Congress on Energy Efficiency and Energy Related Materials (ENEFM2013)*, Springer Proceedings in Physics 155,
DOI: 10.1007/978-3-319-05521-3_26,

26.1 Introduction

The properties of TiO_2 draw more and more attention of the researchers and industrialists. The reasons are economic, environmental and of energetic nature. The examples of utilization of TiO_2 are numerous:

- In electronics luminescent materials [1], ceramic dielectrics [2] for integrated circuits for microwave, isolators, filters, ceramics, superconductors [3], etc.
- In catalysis reactions of oxidative coupling of methane (COM) [4–6], oxidation, hydrogenation, and hydrogenolysis of hydrocarbons, automotive post-combustion [7–11] etc. One of essential parameter is the stability of materials surface [7]. In order to limit some damage under the reactions conditions: lower activities, reduction in surfaces, doping of titania by group II_A elements increases the activity and selectivity during the COM [12–14]. This efficiency is attributed to a decrease of density and a high basicity of these catalysts [15]. In this paper, we are interested to the effect of barium on surface and reactivity of TiO_2.

26.2 Experimental

26.2.1 Sample Preparation

TiO_2 used in this study, from Merck, purity 99.9 % in mass; their surface area is about 10.3 $m^2\ g^{-1}$. $BaCO_3$ (from BDH Chemicals Ltd Poole) is at 99 %, it surface is 2.3 $m^2\ g^{-1}$ environ. TiO_2 was hydrolysed in excess of distilled, deionised (D.D) water for 16 h at 80 °C [19]. $BaCO_3$ was added to this solution, the mixture was homogenized at the same temperature during 5 h. The obtained products were filtered and dried overnight in air at 120 °C. Calcinations were led under air flow of (3 l/h), in dynamic reactor: with heating rate of 10°/min and isothermal at 450 and 1150 °C in order to obtain TiBaX (X = at.% of Ba).

26.2.2 Samples Characterisation

Nitrogen physisorption measurements were performed in a Micromeritics ASAP 2000 apparatus at −196 °C. Helium was used as the carrier gas. The samples were degassed in vacuum for at least 2 h at 120 °C before analysis. X-ray diffraction (XRD) measurements were performed on a Philips PW1800 diffractometer fitted with an anti-cathode CuKα ($\lambda_{Cu} = 1.5406$ Å). XPS experiments were performed using the RIBER spectrometer equipped with a monochromatized MgK_α source for excitation. Binding energy (B.E.) values were referenced to the binding energy of the C_{1S} core level (284.9 eV). A mixed Gaussian/Lorentzian peak fit procedure was used to simulate the experimental photopeaks according to the software supplied by VG Scientific. The thermobalance used is a SETARAM TG-DTA92-16 including an electric balance with continuous signal, a furnace equipped with a regulator-programmer of temperature and a recorder. Measurement is made under oxidizing atmosphere (air: 3 Lh^{-1}) with a speed of heating of 10° min^{-1} from ambient to 1200 °C.

26.3 Results and Discussion

26.3.1 Hydrolysis and Calcination of Alone Oxide

At 450 °C, the results of surfaces measurements represented in (Table 26.1) show clearly that the hydrolysis treatment is without effect on surface of titanium oxide. One notes a very light increase of 3 % which could be due to the molecules of water, physically absorbed, after calcination and degasification during measurement [12].

No conversion was announced, this oxide would be only hydrated. However, the sintering effect is more significant at 1150 °C, since surface value of this oxide loses 93 % of its initial value [13]. XRD patterns recorded on TiBa0 shows that TiO_2 is in anatase form before and after hydrolysis and calcinations at 450 °C. It is transformed to rutile at 1150 °C Fig. 26.1 [14].

26.3.2 Effect of Barium on the Surface of TiO_2

The surfaces results of samples calcined at 450 °C are reported in Table 26.1. The surfaces of TiBaX decreases when barium increases. This decrease of the surface can be explain by algebra amount of surfaces of TiO_2 (10.3 m^2 g^{-1}) and $BaCO_3$ (2.3 m^2 g^{-1}) according $S_{TiBaX} = S_{TiBa0} + S_{BaCO3}$. This indicates an enlargement of the particles of oxide when $BaCO_3$ is added [15].

After calcination at 1150 °C, the surfaces of oxides increases, in general, when the rate of barium increase and the resistance to the sintering is reinforced. Indeed, the surface triple when one added 15 % of barium. This increase could be due to the fine homogenization reinforced by extra granular electrics forces between titanium oxide and barium oxide. The results show that 10 % of barium presents a limit for the conservation of the surface of TiO_2 [16, 17]. The decreases of its surface compared to calcined at 450 °C, is due to the sintering phenomena which makes increase of the particles size.

The presence of carbonates and oxy-carbonates are highlighted, for all samples, by XPS analysis. Indeed results shows in Table 26.2.

Three peaks C_{1S} at: 284.9 eV, one due to the contamination by carbon, 286.2 eV peak associated to carbonate linked to an oxygen C–O and 288.4 eV peak due to carbon linked to two oxygen O–C=O.

Table 26.1 Specific surface area of TiBaX

Samples	After calcination at 450 °C	After calcination at 1150 °C
TiBa0	10.6	0.8
TiBa10	8.7	0.9
TiBa15	7.2	3

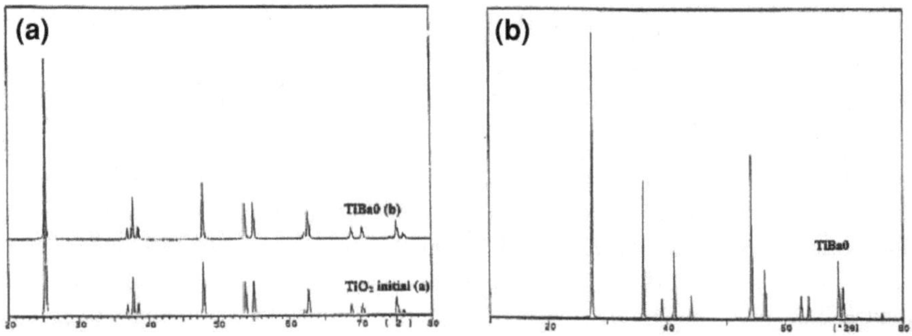

Fig. 26.1 XRD measurements on TiBa0. **a** Initial sample (*a*) and calcined at 450 °C (*b*)—TiO_2 in anatase form. **b** Sample calcined at 1150 °C—TiO_2 rutile form

Table 26.2 Results of XPS analysis, binding energies (eV) of core electrons of TiBaX samples calcined at 450 °C

Samples	C_{1s}	O_{1s}	$Ti_{2p3/2}$	$Ba_{3d5/2}$
TiBa0	284.9 (69)	530.0 (72)	458.5	–
	286.4 (23)	531.6 (15)		
	288.4 (8)	532.5 (13)		
TiBa10	284.9 (69)	530.0 (70)	458.5	779.8
	286.4 (23)	531.5 (16)		
	288.5 (8)	532.4 (14)		
TiBa15	284.9 (69)	530.1 (68)	458.5	779.8
	286.4 (23)	531.6 (12)		
	288.5 (8)	532.5 (20)		

Values between parentheses are relative peak intensities

Three peaks O_{1S} at: 530 eV corresponds to O^{2-}, 531.5 eV peak due to carbonates and 533 eV peak associated to adsorbed H_2O.

It was show that samples contain an important quantity of water, and the addition of barium makes increase the quantity of H_2O adsorbed [18]. The displacement of peak 533 towards 532.5 eV indicates that the adsorption of water is reinforced. These results show also peaks associated to the cation:

Peak of $Ti2p_{3/2}$: at 458.5 eV which characterizes the ion Ti^{4+} [19].

Peak $Ba3d_{5/2}$: at 779.5 eV peak associated to Ba^{2+}.

Table 26.3 ATG results of reduction under H_2/Ar

Sample	$\Delta_1(\%)$	H_2O Ads. (mol/g)	Majority products		Δ_2 (%)	$\delta = \Delta_1 - \Delta_2$ (%)
			Before	After		
TiBa0	0.85	0.025	TiO_2	$TiO_{1.985}$	0.29	−0.04

Δ_1 is $\Delta m/m$ measured and Δ_2 is calculated on the basis of majority product formulas

26.3.3 Thermogravimetric Analysis

Followed by thermogravimetry (Table 26.3), from the ambient to 1000 °C under H_2/Ar, the reduction of sample calcined at 450 °C show volatile species that leave at 100 °C; this is essentially the water of hydratation.

26.4 Conclusion

A series of samples, TiBaX where X is the atomic percentage in barium was prepared by hydrolysis of TiO_2 and $BaCO_3$ in neutral medium. These samples was calcined under air at 450 and 1150 °C, and then characterized by BET specific surface area, XPS experiments and thermogravimetric analysis (TG). After calcination at 450 °C, the specific surface of TiO_2 remains unchanged and it's in anatase form. XPS experiments confirm that TiO_2 is reduced on surface to $TiO_{1.985}$. The addition of barium decreases the surface of TiO_2. XPS results shows presence of oxides, carbonates and oxy-carbonates in TiBaX. After calcination at 1150 °C, sintering is more significant for alone oxide. The XRD pattern of TiBa0 shows that TiO_2 is in the rutile form. At same temperature, the BET measurements show that surface of titania increase with rate of barium added. Indeed, with 15 % of barium the surfaces of TiO_2 triple. TG analysis under H_2/Ar shows volatile species that leave at 100 °C; this is essentially the water of hydratation.

References

1. N.R. Khalid, E. Ahmed, Z. Hong et al., Ceram. Int. **39**(4), 3569–3575 (2013)
2. C.C. Wu, C.F. Yang, Y.T. Hsieh et al., Ceram. Int. **38**(1), 223–227 (2012)
3. V. Iliev, D. Tomova, S. Rakovsky, Desalisation **260**(1–3), 101–106 (2010)
4. W.W. Roczniki, Chemistry **30**, 733 (1956)
5. M.C.J. Bradford, M.A. Vannice, Catal. Today **50**(1), 87–96 (1999)
6. K.I. Shimizu, T. Kaneto, T. Fujishima et al., Appl. Catal. A **255**(1–2), 185 (2002)
7. C. Cantau, T. Pigot, J.C. Dupin et al., J. Photochem. Photobiol. A: Chem. **216**(1–3), 201–208 (2010)
8. D.L. Trimm, Appl. Catal. **7**, 249 (1983)
9. H. Sreemoolanadhan, M.T. Sebastian, P. Mohanan, Br. Ceram. Trans. **34**, 581 (1991)
10. V. Harlé, M. Vrinat, J.P. Scharff et al., Appl. Catal. A **196**(2), 261–269 (2000)

11. M.D. Mitchell, A. Vannice, Ind. Eng. Chem. Fundam. **23**(1), 88 (1984)
12. Z.B. Wei, W. Yan, H. Zhang et al., Appl. Catal. A **167**(1), 39–48 (1998)
13. S.H. Song, X. Wang, P. Xiao, Mat. Sc. Eng. B **94**(1), 40–47 (2002)
14. S. Music, M. Gotic, M. Ivanda et al., Mat. Sc. Eng. B **47**(1), 33–40 (1997)
15. J. Ramirez, A. Gutiérrez-Alejandre, Catal. Today **43**(1–2), 123–133 (1998)
16. N. Ikemiya, J. Yoshitomi, S. Hara et al., Nippon Kinzoku Gakkaishi **57**(5), 527 (1993)
17. B. Bloch, B.G. Ravi, R. Chaim, Mat. Lett. **42**(1), 61 (2000)
18. B.M. Reddy, B. Chowdhury, P.G. Smirniotis, Appl. Catal. A **211**(1), 19–30 (2001)
19. S. Bourgeois, P. Le Seigneur, M. Perdereau, Surf. Sc. **328**(1–2), 105–110 (1995)

Chapter 27
Structural Study of Rare Earth Oxides Doped by Barium

Madani Ghelamallah, Soufi Kacimi and Radia Imane Fertout

Abstract Two series of samples noted LnBaX (where Ln = Nd or Sm, X = at. % of Ba) were dispersed in an excess of bi-distilled and demineralised water during 16 h at 80° from Ln_2O_3 and γ-$BaCO_3$. These samples were calcined under air at 450 and 1150 °C then characterized by X-ray diffraction, scanning electron microscopy (SEM) and X-ray photoelectron spectroscopy. After calcination at 450 °C, XRD and XPS measurement shows the presence of Ln_2O_3, barium carbonates, oxy-carbonates and oxy-hydroxides LnO(OH) in LnBaX. At 1150 °C, results shows that NdBaX and SmBaX are formed by the BaO and Ln_2O_3. However neodymium and samarium oxides do not incorporate barium.

27.1 Introduction

The examples of utilization of rare earth oxides are numerous: electronics luminescent materials, ceramic dielectrics, integrated circuits for microwave, isolators, filters, ceramics, superconductors and in catalysis. However, most of these studies were resolved by structural, conductive properties and chemical behavior. In catalysis: reactions of oxidative coupling of methane [1–3], oxidation, hydrogenation and hydrogenolysis of hydocarbons, automotive post-combustion [4–8] etc.

M. Ghelamallah (✉)
Laboratoire de Matériaux, Applications et Environnement, Faculté des Sciences et de la Technologie, Université de Mascara, BP 305 Route de Mamounia, 29000 Mascara, Algeria
e-mail: gh_madani@hotmail.com

S. Kacimi
Laboratoire de Chimie Appliquée, Institut des Sciences et de la Technologie, Centre Universitaire d'Ain Temouchent, BP 284 Route de Sidi Bel-Abbes, 46000 Ain Temouchent, Algeria

R. I. Fertout
Laboratoire de Matériaux et Catalyse, Faculté des Sciences, Université Djillali Liabes, BP 89 Cité Larbi Ben M'hidi, 22000 Sidi Bel-Abbes, Algeria

A. Y. Oral et al. (eds.), *International Congress on Energy Efficiency and Energy Related Materials (ENEFM2013)*, Springer Proceedings in Physics 155,
DOI: 10.1007/978-3-319-05521-3_27,

The doping of rare earth oxides by alkaline earth increases the activity and selectivity in COM [9–11], this efficiency is attributed to lower density and high basicity of these catalysts [12]. In this paper we investigate the effect of barium on the structures and microstructures of Nd_2O_3 and Sm_2O_3.

27.2 Experimental

27.2.1 Sample Preparation

Ln_2O_3 precursors (Ln = Nd and Sm) used in this study, are from Merck, purity 99.9 % in mass. $BaCO_3$ (from BDH Chemicals Ltd Poole) is at 99 %. Nd_2O_3 and Sm_2O_3 were hydrolysed in excess of distilled, deionized (D.D) water for 16 h at 80 °C [8]. $BaCO_3$ was added to these solutions, the mixture was homogenized at the same temperature during 5 h. The obtained products were filtered and dried overnight in air at 120 °C. Calcinations were led under air flow of (3 l/h), in dynamic reactor: with heating rate of 10°/min and isothermal at 450 and 1150 °C in order to obtain NdBaX and SmBaX (X = at.% of Ba).

27.2.2 Samples Characterisation

X-ray diffraction (XRD) measurements were performed on a Philips PW1800 diffractometer fitted with an anti-cathode CuKα ($\lambda_{Cu} = 1.5406$ Å). Scanning electron microscopy (SEM) studies were performed on Philips SEM 505 microscope operating at 25 kV and magnification values ×5000–10000. This instrument is equipped with an energy dispersive X-ray microanalyzer EDAX, with permitted analytical electron microscopy measurements. The samples were sputter coated with gold. XPS experiments were performed using the RIBER spectrometer equipped with a monochromatized MgK_α source for excitation. Binding energy (B.E.) values were referenced to the binding energy of the C_{1S} core level (284.9 eV).

27.3 Results and Discussion

27.3.1 Samples Analysis by XRD and SEM

27.3.1.1 Samples Calcined at 450 °C

The XRD patterns of LnBX calcined under air at 450 °C are reported on Fig. 27.1a, b respectively for (NdBaX) and (SmBaX). The analysis of these spectra shows that samples represents mixtures of oxides, oxy-hydroxides, hydroxides and

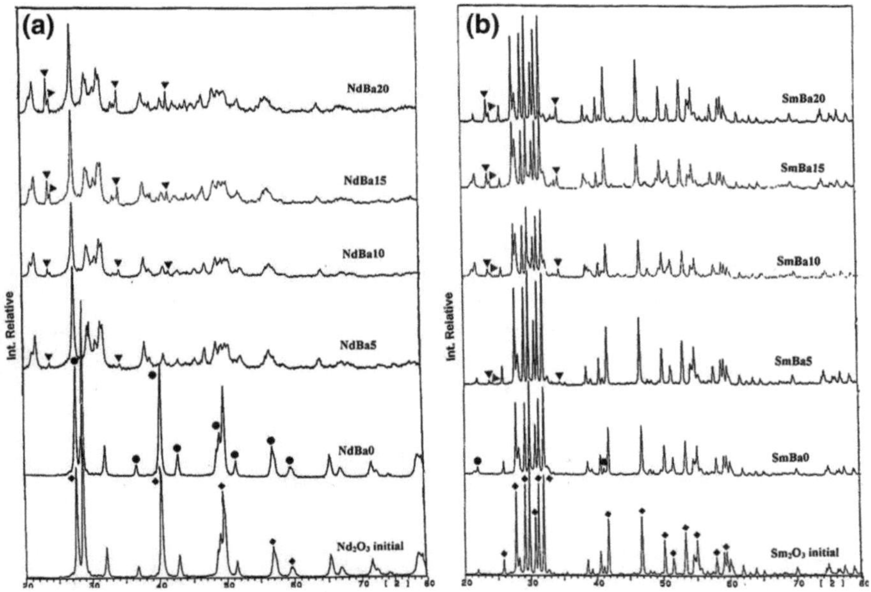

Fig. 27.1 XRD measurements on LnBaX calcined at 450 °C. **a** *Left* Mixture of Nd(OH) (*black circle*), Nd_2O_3 (*black diamond suite*) and $BaCO_3$ (*black down-pointing triangle*). **b** *Right* Mixture of Sm(OH) (*black circle*), Sm_2O_3 (*black diamond suit*) and $BaCO_3$ (*black down-pointing triangle*)

carbonates. No interaction between rare earth oxides and barium carbonate has been found.

27.3.1.2 Samples Calcined at 1150 °C

After calcination at 1150 °C, XRD patterns (Fig. 27.2a, b) of samples shows:

- No interaction between neodymium and barium, the sample is a mixture of Nd_2O_3 and BaO.
- A slight displacement of the lines of Sm_2O_3 and BaO in SmBa15. These results could be due to an early interaction between these two oxides [13].

The study by scanning microscopy SEM show the absence of well-defined crystalline phase for NdBa15 (Fig. 27.3a), the compound is amorphous [14]. For Sm_2O_3 there is a beginning of crystallization shows a slight interaction as detected by XRD (Fig. 27.3b). For BaO–Ln_2O_3 systems, the ability to form a solid solution depends on the ionic size of Ln, the method of preparation and temperature. The compatibility of the sizes of Ba^{2+} and Ln^{3+} ions present a predominant role in the formation of a solid solution [15]. So the probabilities of formation of a solid solution decrease with increasing size difference between Ln^{3+} and Ba^{2+}. This difference is about 0.34, 0.30 respectively for Sm^{3+}, Nd^{3+}).

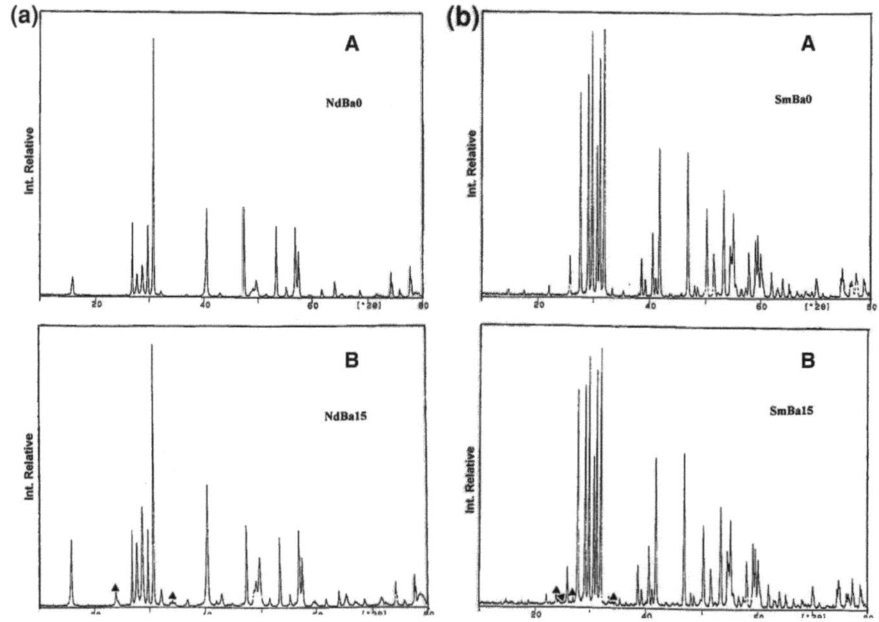

Fig. 27.2 XRD measurements of LnBa0 and LnBa15 calcined at 1150 °C. **a** *Left* (*A*) raies of Nd_2O_3 (*B*) Mixture of Nd_2O_3 and BaO. **b** *Right* (*A*) raies of Sm_2O_3 (*B*) mixture of Sm_2O_3 and BaO

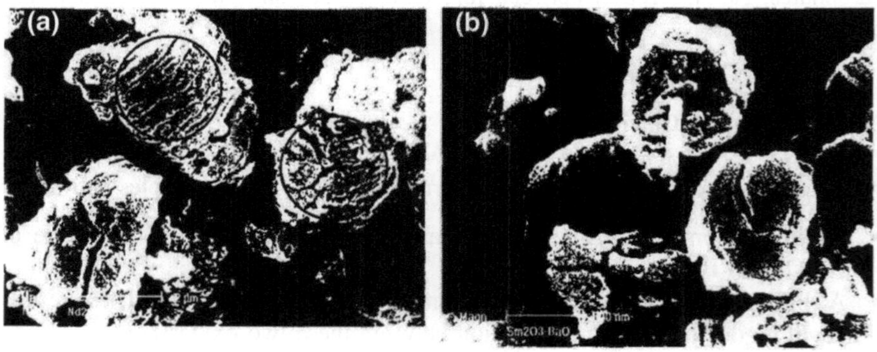

Fig. 27.3 SEM Micrographs on Ln_2O_3–BaO systems. **a** *Left* BaO–Nd_2O_3 system. **b** *Right* Sm_2O_3–BaO system

27.3.2 XPS Analysis for Samples Calcined at 450 °C

The Table 27.1 below shows the results of XPS of LnBaX samples calcined at 450 °C. This technique reveals the presence of carbonates and oxy-carbonates for all samples. Indeed, the results show:

Table 27.1 Results of XPS analysis, binding energies (eV) of core electrons of LnBaX samples calcined at 450 °C

Samples	C1s	O1s	Ln3d5/2	Ba3d5/2
	284.9 (64)	530.2 (28)		
NdBa0	286.3 (14)	532.0 (57)	981.7	
	288.6 (22)	533.6 (15)		
	284.9 (63)	530.3 (34)		
NdBa15	286.3 (14)	531.9 (41)	981.7	780
	288.3 (23)	533.1 (25)		
	284.9 (62)	530.1 (36)		
SmBa0	286.3 (21)	531.7 (45)	132.7	
	288.8 (17)	533.6 (19)		
	284.9 (65)	530.0 (35)		
SmBa15	286.4 (25)	531.7 (49)	132.7	779.9
	288.5 (10)	533.4 (16)		

Values between parentheses are relative peak intensities

The presence of carbonates and oxy-carbonates are highlighted, for all samples, by XPS analysis. Indeed results shows in (Table 27.1).

Three peaks C_{1S} at: 284.9 eV, one due to the contamination by carbon, 286.4 eV peak associated to carbonate linked to an oxygen C–O and 288.5 eV peak due to carbon linked to two oxygen O–C=O.

Three peaks O_{1S} at: 530.2 eV corresponds to O^{2-}, 531.5 eV peak due to carbonates and 533 eV peak associated to adsorbed H_2O.

It was shown that samples contain an important quantity of water, and the addition of barium makes decrease the quantity of H_2O adsorbed [13]. These results show also peak associated to the cations:

Two peaks of $Ln3d_{5/2}$ at 981.7 eV peak due to Nd^{3+} and other at 132.7 eV which characterizes Sm^{3+} [14]. Peak $Ba3d_{5/2}$: at 779.5 eV peak associated to Ba^{2+}.

27.4 Conclusion

Two series of samples, NdBaX and SmBaX where X is the atomic percentage in barium, were prepared by hydrolysis of Nd_2O_3, Sm_2O_3 and $BaCO_3$ in neutral medium. These samples were calcined under air at 450 and 1150 °C, and then characterized by X-ray diffraction (XRD), XPS experiments and scanning electron microscopy. For samples without barium (NdBa0 and SmBa0) calcined at 450 °C, they contain Nd_2O_3, $Nd(OH)_3$, NdO(OH), Sm_2O_3 and SmO(OH) respectively. XRD and XPS measurements of samples with barium have shown the presence of hydroxides $Ln(OH)_3$, oxy-hydroxydes LnO(OH), oxides (Ln_2O_3) and barium carbonate ($BaCO_3$). No insertion of barium into structures of Nd_2O_3 or Sm_2O_3 was detected. After calcination at 1150 °C, the XRD patterns show that oxides are in

Ln_2O_3 form. For samples with barium, XRD and SEM measurements show that LnBaX are formed only the Ln_2O_3 and BaO. A beginning of crystallization was detected by XRD and shows a slight interaction between Sm_2O_3 and BaO.

References

1. W.W. Roczniki, Chemistry **30**, 733 (1956)
2. M.C.J. Bradford, M.A. Vannice, Catal. Today **50**(1), 87–96 (1999)
3. K.I. Shimizu, T. Kaneto, T. Fujishima et al., Appl. Catal. A **255**(1–2), 185 (2002)
4. C. Cantau, T. Pigot, J.C. Dupin et al., J. Photochem. Photobiol. A: Chem. **216**(3), 201–208 (2010)
5. D.L. Trimm, Appl. Catal. **7**, 249 (1983)
6. H. Sreemoolanadhan, M.T. Sebastian, P. Mohanan, Bristish. Ceram. Trans. **34**, 581 (1991)
7. V. Harlé, M. Vrinat, J.P. Scharff et al., Appl. Catal. A **196**(2), 261–269 (2000)
8. M.D. Mitchell, A. Vannice, Ind. Eng. Chem. Fundam. **23**(1), 88 (1984)
9. Z.B. Wei, W. Yan, H. Zhang et al., Appl. Catal. A **167**(1), 39–48 (1998)
10. S.H. Song, X. Wang, P. Xiao, Mat. Sci. Eng. B **94**(1), 40–47 (2002)
11. S. Music, M. Gotic, M. Ivanda et al., Mat. Sci. Eng. B **47**(1), 33–40 (1997)
12. J. Ramirez, A. Gutiérrez-Alejandre, Catal. Today **43**(1–2), 123–133 (1998)
13. Q. Zhiyru, X. Xianran, Y. Vencia, W. Soukun, C. Xiaolong, L. Jingkui, X. Sishen, J. Alloy. Compd. **202**, 77 (1993)
14. J.G. Pepin, E.R. Vance, G.J. McCarthy, J. Solid State Chem. **38**, 360 (1981)
15. W. Wong-ng, B. Paretzkin Jr, E.R. Fuller, J. Solid State Chem. **85**, 117 (1990)

Chapter 28
Object-Oriented Modeling of an Energy Harvesting System Based on Thermoelectric Generators

Marco Nesarajah and Georg Frey

Abstract This paper deals with the modeling of an energy harvesting system based on thermoelectric generators (TEG), and the validation of the model by means of a test bench. TEGs are capable to improve the overall energy efficiency of energy systems, e.g. combustion engines or heating systems, by using the remaining waste heat to generate electrical power. Previously, a component-oriented model of the TEG itself was developed in Modelica® language. With this model any TEG can be described and simulated given the material properties and the physical dimension. Now, this model was extended by the surrounding components to a complete model of a thermoelectric energy harvesting system. In addition to the TEG, the model contains the cooling system, the heat source, and the power electronics. To validate the simulation model, a test bench was built and installed on an oil-fired household heating system. The paper reports results of the measurements and discusses the validity of the developed simulation models. Furthermore, the efficiency of the proposed energy harvesting system is derived and possible improvements based on design variations tested in the simulation model are proposed.

28.1 Introduction

Nowadays, energy harvesting systems become more important with regard to the shortage of resources. There is a need to use the inserted energy more efficient and for such a case, TEGs are very practical. They are capable to generate electrical

M. Nesarajah (✉) · G. Frey
Chair of Automation, Saarland University, 66123 Saarbrücken, Germany
e-mail: marco.nesarajah@aut.uni-saarland.de

G. Frey
e-mail: georg.frey@aut.uni-saarland.de

A. Y. Oral et al. (eds.), *International Congress on Energy Efficiency and Energy Related Materials (ENEFM2013)*, Springer Proceedings in Physics 155,
DOI: 10.1007/978-3-319-05521-3_28,

power by using the waste heat of e.g. heating systems. Already in the 60s, and still in these days, they are used in spacecraft [1]. Their advantages are the simple scalability as well as the low maintenance, due to the lack of mechanical components. That is why an application in the car industry is imaginable (see [2–4]). The Seebeck-effect describes the phenomenon that a temperature difference between the both TEG surfaces produces electrical power.

A component-oriented modeling of the thermoelectric (TE) material itself was already described by the authors in [5]. There, a library for dynamic simulation of TE devices in 1D spatial resolution was developed in the open modeling language Modelica® [6].

Modelica® is an open component-oriented modeling language. The structure-preserving way of modeling and the reusability of components allow the modeling of complex physical systems with an acausal concept. As the TEG library uses the standard interfaces of Modelica®, the ease of integration in other existing Modelica® libraries is assured.

The model from [5] respects the temperature dependencies of material properties (Seebeck coefficient, thermal conductivity, and electrical resistivity) and some geometrical parameters, e.g. the thickness of TE legs. In [7], the library from [5] was expanded to an overall library for complete TEGs. With the new model not only the TE material itself can be modeled, but also the complete TEG, including the surrounding ceramic plates and the internal contact resistances.

This contribution describes the extension of the model from [5, 7] to an overall thermoelectric energy harvesting system in Modelica®. It contains, beside of the model for an entire TEG, the cooling system, the heat source as well as the power electronics. This simulated energy harvesting system will be compared with the real test bench on an oil-fired household heating system. In Sect. 28.2, the setting of the test bench is described. The Modeling of an Energy Harvesting System, containing the particular components as well as the complete system is given in Sect. 28.3. The simulation results compared with the test reading is presented in Sect. 28.4 and in Sect. 28.5, a conclusion follows.

28.2 Test Bench

On a real oil-fired household heating system, a test bench is built to validate the simulation model. Related test benches on heating systems are already described in [8, 9], pure test benches in [10, 11]. The purpose is to harvest energy from the waste heat by TEGs for electrical power supply. For installing the TEGs, the exhaust pipe of the heating system is favored to have no disturbance on the regular house-heating process. Therefore, this concept is also appropriate for any other common oil-fired household heating system.

The Test bench consists of four, identical *energy-harvesting-modules*. Each module has two similar sides (*TEG-modules*) and one TEG-module involves an aluminum profile adequate to the form of the exhaust pipe, two temperature

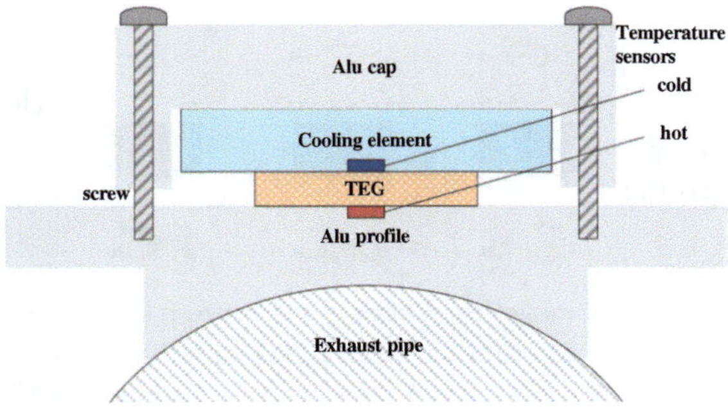

Fig. 28.1 Schematic diagram of a TEG-module; obviously are the sensors, the aluminum profiles, the TEG and the cooling element

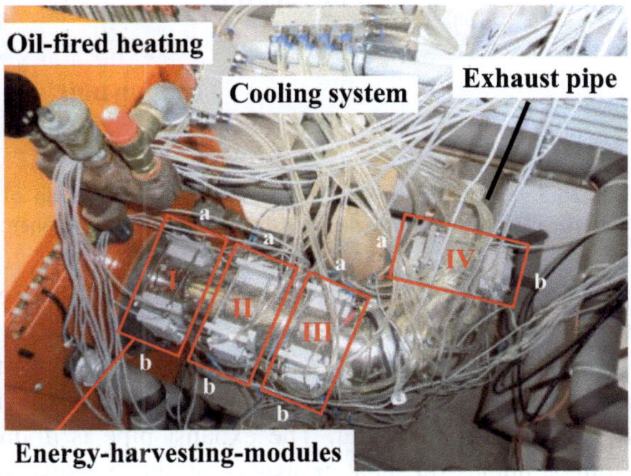

Fig. 28.2 Test bench on an exhaust pipe of an oil-fired household heating system consisting of four energy-harvesting-modules (marked with *roman numerals*), each includes two TEG-modules (marked with '*a*' and '*b*')

sensors, one on each surface of a TEG as well as the cooling element and an aluminum cap. Figure 28.1 shows the schematic diagram of such a TEG-module, Fig. 28.2 the complete composition. The data are logged with LabView.

28.3 Modeling of an Energy Harvesting System

As aforementioned, the modeling of the energy harvesting system takes place in Modelica®. The simulation model for the TEG itself comes from [5, 7]. Additionally to the TEG model, the Modelica® standard fluid, electrical and thermal library are used to model the other components.

First, a TEG-module is modeled, compared to Fig. 28.1. Therefore, the pure TEG model from [5, 7] is extended by the heat transfer paste on each side. The aluminum components are implemented by using thermal conductors and heat capacitors and additionally include the free convection of the heat to the ambient air. With the help of the standard fluid library, the cooling element is modeled. Consequently, the used Modelica® connectors for the TEG-module are electrical

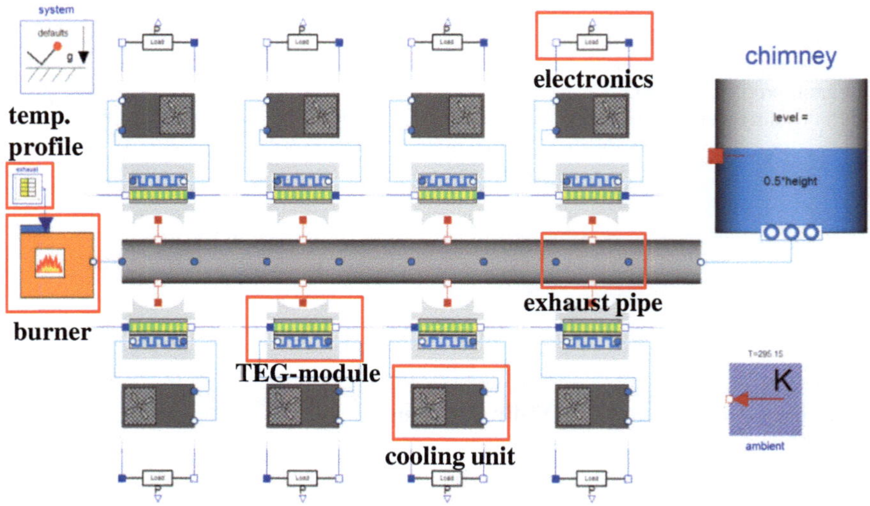

Fig. 28.3 Model of the energy harvesting system on an oil-fired heating based on TEGs in Modelica®/Dymola; *indicated* are the different subcomponents

pins (voltage and current), fluid ports (mass flow rate and pressure) and heat ports (temperature and heat flow rate).

In the simulation model, like in the real component, the cooling unit consists of a pump and a tank. Thereby, the cooling medium is simplified for the model and is described by pure water. The exhaust pipe is divided into two subcomponents. Both simulate the heat transport of the exhaust air through the pipe wall. In addition, one component describes the heat flow to the TEG-modules and the other the heat flow to the ambient air. With the help of a prescribed temperature profile and a mass flow source, the burner is modeled. The electronics involves a load resistor, a voltage and a current sensor. Figure 28.3 shows the complete model of the energy harvesting system, consisting of the different subcomponents for burner, exhaust pipe, TEG-module, cooling unit and the electronics.

28.4 Simulation Versus Measurement

For the test reading, the oil-fired heating begins to work and after three and a half minutes, the cooling system is switched on, off after 23 min. Figure 28.4 shows exemplary the data for TEG Ia and TEG Ib, whereat the open-circuit voltage of TEG Ia, the load voltage of TEG Ib—for a load resistor of 2.4 Ω—as well as the temperature of the exhaust pipe and the temperatures on the TEGs surfaces are recorded. Conspicuously are the fall and the rise of the temperatures on the cold side of the TEGs, depending on the cooling system, and the approximately proportional behavior of the voltages to the temperature differences. Moreover it is obviously that

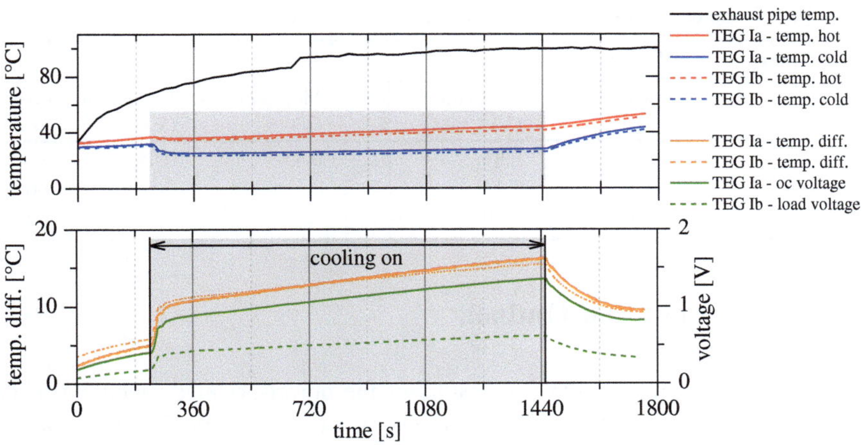

Fig. 28.4 Test reading; *top* exhaust pipe temperature as well as hot and cold side temperatures of TEG Ia and TEG Ib; *bottom* temperature differences for TEG Ia and TEG Ib as well as the open circuit (*oc*) voltage of TEG Ia and the load voltage of TEG Ib

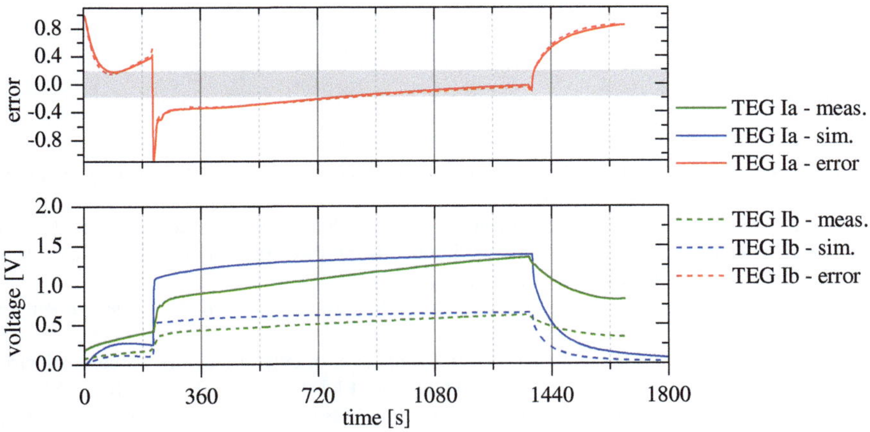

Fig. 28.5 Simulation versus measurement; *top* error curves; *bottom* comparison between the voltage test reading of Fig. 28.4 and the simulation results

the behavior of TEG Ia and TEG Ib is identically, comparing the temperature differences and dissenting, comparing the voltages (open circuit vs. load).

Figure 28.5 compares the simulation results with the voltage test reading of Fig. 28.4 and shows the error curves. Evidently is that the trend of both error curves is identically, but due to the dynamic, the error alternates enormously. Probably, some heat capacity is not taken into account precisely enough (specific heat capacity of water is 4.18 kJ/(kg K) and of the real used cooling medium—ethylene glycol—2.40 kJ/(kg K)). In static cases the simulation provides good results, compare Table 28.1.

Table 28.1 Static comparison between simulation and measurement

Exhaust pipe temperature (°C)	TEG Ia (oc)			TEG Ib (load)		
	Measurement (V)	Simulation (V)	Error (%)	Measurement (V)	Simulation (V)	Error (%)
80	0.95	0.97	+2	0.45	0.45	0
90	1.06	1.13	+7	0.5	0.53	+6
100	1.34	1.29	−4	0.61	0.6	−2

28.5 Conclusions and Outlook

This contribution presents the first results of the object-oriented modeling of an energy harvesting systems based on thermoelectric generators. The test reading was compared with the simulation results and it was shown that the energy harvesting system model reflects the reality basically. Within the next months, the test bench will be modified and the simulation will get more precisely, mainly through the integration of the correct medium parameters. Moreover, the power electronics will be extended.

References

1. J. Yang, T. Caillat, Thermoelectric materials for space and automotive power generation. MRS Bull. **31**(03), 224–229 (2006)
2. G.J. Snyder, Small thermoelectric generators. Electrochem. Soc. Interface **17**(3), 54 (2008)
3. J. LaGrandeur, Vehicle fuel economy improvement through thermoelectric waste heat recovery, in DEER Conference, Chicago, IL, 2005
4. J. Fairbanks, Thermoelectric applications in vehicles—status 2008. U.S. Department of Energy, 2008
5. F. Felgner, L. Exel, M. Nesarajah, G. Frey, Component-oriented modeling of thermoelectric devices for energy system design. IEEE Trans. Ind. Electron. **61**(3), 1301–1310 (2014)
6. Modelica and the Modelica Association—Modelica Association. https://www.modelica.org/
7. M. Nesarajah, L. Exel, G. Frey, Modelica library for dynamic simulation of thermoelectric generators, in *Proceedings 11th European Conference on Thermoelectrics (accept.)* (2013)
8. K. Qiu, A. Hayden, A natural-gas-fired thermoelectric power generation system. J. Electron. Mater. **38**(7), 1315–1319 (2009)
9. K. Qiu, A. Hayden, Development of thermoelectric self-powered heating equipment. J. Electron. Mater. **40**(5), 606–610 (2011)
10. J. Vázquez, R. Palacios, M.A. Sanz-Bobi, A. Arenas, Test bench for measuring the electrical properties of commercial thermoelectric modules, in *22th International Conference on Thermoelectrics—ICT* (2003)
11. F.A. Leavitt, N.B. Elsner, J.C. Bass, *Use, Application and Testing of Hi-Z Thermoelectric Modules* (Hi-Z Technology Inc., 2007)

Chapter 29
Extensible Wind Towers

Marco Sinagra and Tullio Tucciarelli

Abstract The diffusion of wind energy generators is restricted by their strong landscape impact. The PERIMA project is about the development of an extensible wind tower able to support a wind machine for several hundred kW at its optimal working height, up to more than 50 m. The wind tower has a telescopic structure, made by several tubes located inside each other with their axis in vertical direction. The lifting force is given by a jack-up system confined inside a shaft, drilled below the ground level. In the retracted tower configuration, at rest, tower tubes are hidden in the foundation of the telescopic structure, located below the ground surface, and the wind machine is the only emerging part of the system. The lifting system is based on a couple of oleodynamic cylinders that jack-up a central tube connected to the top of the tower by a spring, with a diameter smaller than the minimum tower diameter and with a length a bit greater than the length of the extended telescopic structure. The central tube works as plunger and lifts all telescopic elements. The constraint between the telescopic elements is ensured by special parts, which are kept in traction by the force of the spring and provide the resisting moment. The most evident benefit of the proposed system is attained with the use of a two-blade propeller, which can be kept horizontal in the retracted tower configuration.

M. Sinagra (✉) · T. Tucciarelli
Dipartimento di Ingegneria Civile, Ambientale, Aerospaziale, dei Materiali, Università degli Studi di Palermo, viale delle Scienze, edificio 8, Palermo, Italy
e-mail: marco.sinagra@unipa.it

T. Tucciarelli
e-mail: tullio.tucciarelli@unipa.it

T. Tucciarelli
Sezione di Idraulica e efficienza delle risorse, Istituto Euro-Mediterraneo di Scienza e Tecnologia, Via Emerico Amari, 123 90139 Palermo, Italy

A. Y. Oral et al. (eds.), *International Congress on Energy Efficiency and Energy Related Materials (ENEFM2013)*, Springer Proceedings in Physics 155,
DOI: 10.1007/978-3-319-05521-3_29,

29.1 Introduction

The occurring strong increment of the wind velocity along with the elevation requires the wind turbines to be located in exposed places, so that wind turbines and their towers have often a significant impact on the country landscape. This impact, along with the turbine noise and the possible inference of the blades with the bird life is the major issue against further construction of wind energy plants.

A partnership of Italian companies and research institutes are starting the PERIMA (Produzione Eolica con Ridotto IMpatto Ambientale) project, co-funded by 2007–2013 Regional Operational Programme of the European Regional Development Fund, which aims to develop a telescopic tower for medium power wind turbines, able to move up and down in a short time and without the use of any external machinery. The project starts from the observation that in many locations wind turbines produce significant power only during part of the total yearly time. In the remaining time a cheap and fast removal of both the turbine and the tower would not provide a significant reduction of the total produced energy. For example, in several Mediterranean locations along the coast, most of the energy production occurs in winter, when most of the tourists are missing. In other cases most of the energy could be produced in short periods of time, when weather conditions make the relative impact of the plant on the surrounding environment much less severe.

A telescopic structure, made of several tubular elements with different diameters moving inside each other, has been envisioned as the best solution for the addressed problem [1]. An alternative choice is the tilt-up tower [2]. This technique has been already adopted for towers with small height, maximum 20 m, but has the inconvenient of (1) leaving a significant impact inside the tower area after his reposition, (2) being difficult to apply for high towers.

The main challenges to be faced by the project are (1) the need of an efficient lifting system, able to rise a load of the order of 500 kN for a displacement of tens of meters, (2) the construction of a junction system automatically linking the tubular elements when the tower is fully extended. The use of a simple oleodynamic actuator is prevented by its cost [3], while the junctions of the elements are required to balance all the momentum generated by the horizontal forces acting on the turbine, but must be armed and disarmed easily, without the need of a direct human action.

In this paper the authors describe the proposed lifting system and linking device with reference to a wind tower of 30 m, supporting a 60 kW wind generator (Fig. 29.1).

29.2 The Telescopic Tower

The length of each tubular element can be set equal to the maximum length allowed for trunk transportation. This length is usually about 10 m. In the example of Fig. 29.2 the tower is split in four elements, one of them kept fixed below the ground level. Each tubular element has two junctions at the ends, which are better

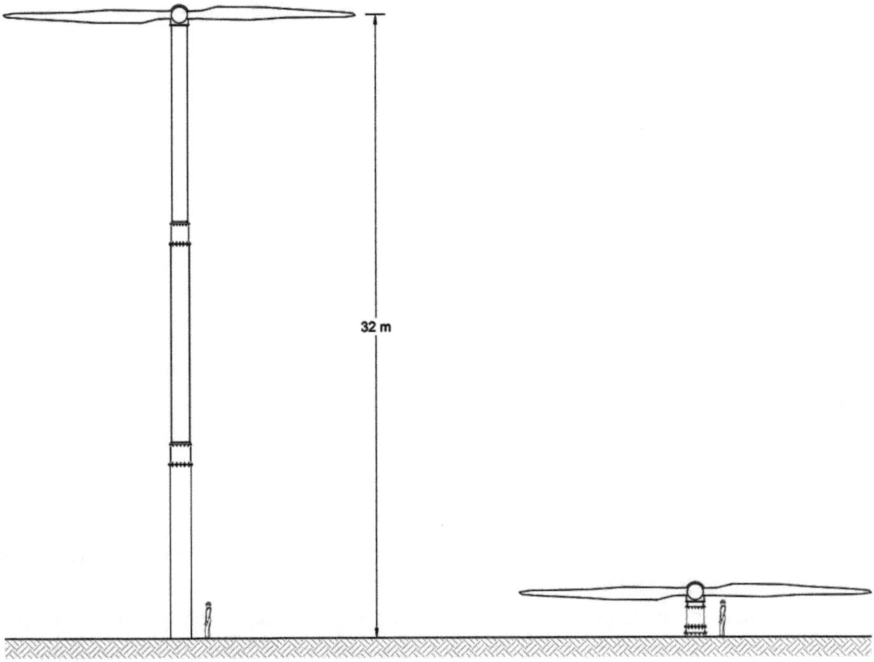

Fig. 29.1 Telescopic tower: **a** tower extended; **b** tower retracted

Fig. 29.2 Section of telescopic tower

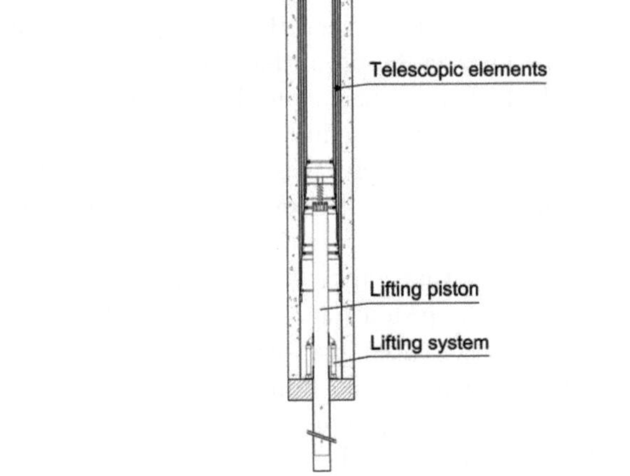

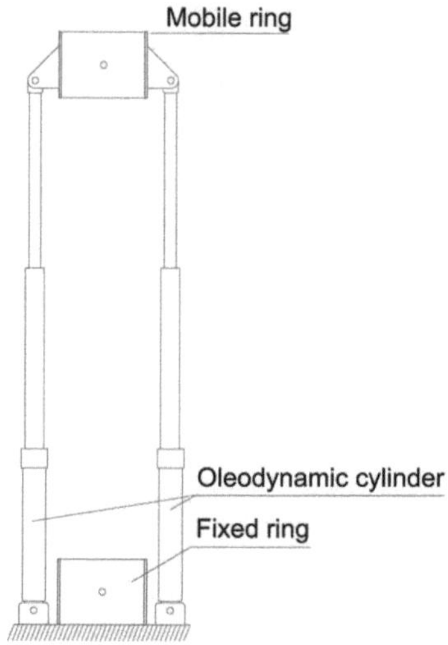

Fig. 29.3 Lifting system

described in the following section, and slides on the larger element during the rising time through two sets of small carts. One set of carts is fixed on the external side of the lower junction of the moving element, the other one is fixed on the internal side of the upper junction of the larger element. The carts are designed to provide only a small momentum during the rising time, when the blades of the wind generator are kept fixed. A shaft is drilled to host all the tubular elements. All the elements, at rest, are laid on a metallic ring located 1–2 m above the shaft foundation. In the example of Fig. 29.2 the outer diameters of the telescopic elements, from top to bottom, are 0.8, 0.94, 1.08 and 1.22 m (Fig. 29.2). The thickness for all tubes is equal to 10 mm.

The tower lifting takes place via a piston, which is a pipe with a diameter smaller than the minimum wind tower diameter and a length a bit greater than the length of the extended telescopic structure, which runs along a borehole drilled below the tower foundation. The movement is transferred by the piston to the innermost tubular element, which drags the larger one when the lower junction of the innermost element meets the upper junction of the larger one. Observe that, at work, the load on the tip of the piston is equal at least to the total weight of the tower. Unless horizontal control is given to the piston displacement along its extension, structural stability condition has to be verified.

The lifting of the piston is guaranteed for a length of 1–2 m by a couple of oleodynamic actuators. The main case of each actuator and the end of its arm are bound to a pair of rings (Fig. 29.3). Both rings can be either bound or free from the piston by the extension or the retraction of small cylinders entering holes excavated inside the piston (jack-up system). The distance between two holes in the vertical direction along the piston is equal to the maximum displacement of the arm of the actuators. An automatic system guarantees that one of the two rings is always bound to the piston, either to drop the arm after each lifting step or to lift the tower during the lifting (or the dropping) step.

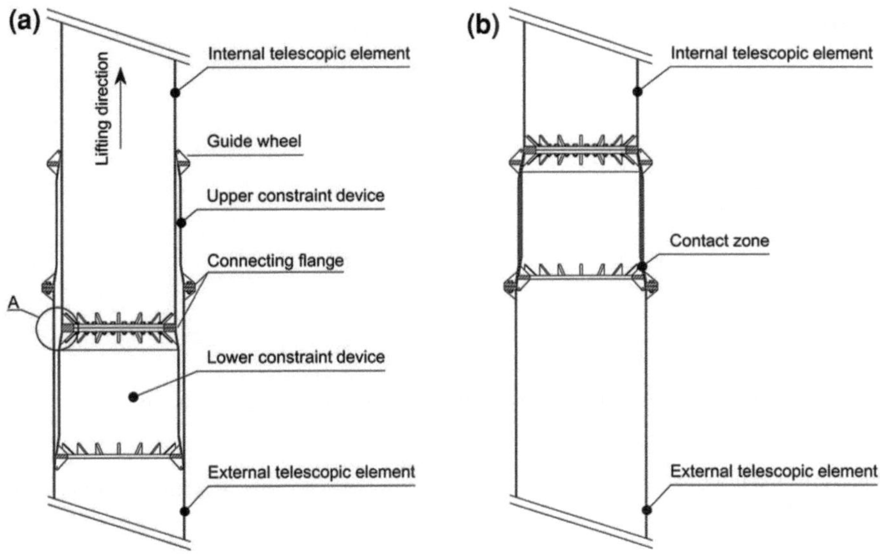

Fig. 29.4 Joint system and guidance system; **a** during lifting, **b** after lifting

29.3 The Tubular Element Junctions

The two cart sets fixed on each element junction balance the momentum given by the wind force only during the lifting (or dropping) time, but the same momentum has to be balanced during the working time, or even when the wind generator is at rest, but strong wind forces act on the tower. This requires a robust junction system and a self-locking device. Each proposed junction has two steel elements with a truncated cone shape [4] and a relatively small height (0.15 m in the example). The larger radius of the smaller element is equal to the smaller radius of larger one and the two elements are spaced out by a cylinder. The cylinders of the mating junctions have the same height, but a bit different diameter, such that the smaller one can move inside the larger one. After the inner tubular element is raised up, the smaller and the larger cone shaped elements of the two junctions mate (Fig. 29.4). The reaction forces will have two components; the vertical one will transfer the lifting force to the larger element, the horizontal one will provide the momentum required to balance the momentum of the external forces (Fig. 29.5). Observe that, to avoid the instability of the piston, an elastic element like a spring must be placed between its tip and the base of the innermost tubular element (Fig. 29.6). This is because part of the vertical component of the junction reaction, due to the external momentum, could be otherwise balanced by the piston itself. In the proposed example the cone shaped elements are disposed at a distance equal to 0.85 m. The slope of the cone must large enough to guarantee a reaction component tangent to the surface small enough (with respect to the normal one) in order to avoid the scroll of the two junctions. On the other hand, the slope must be small enough to

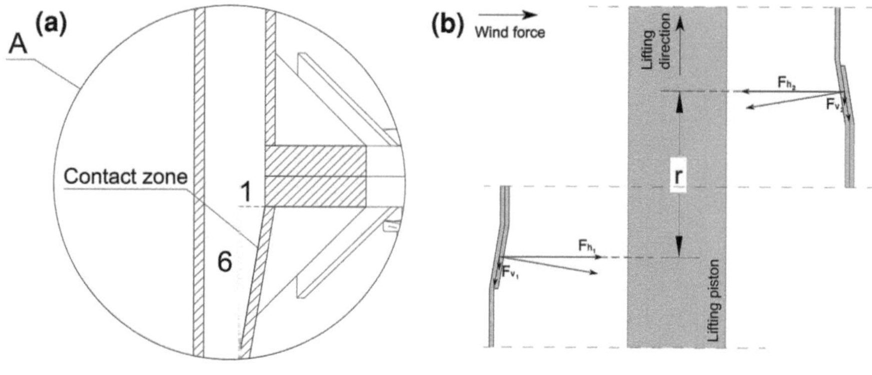

Fig. 29.5 Junction details **a** slope of the contact area; **b** reaction forces scheme

Fig. 29.6 Detail of the lifting spring

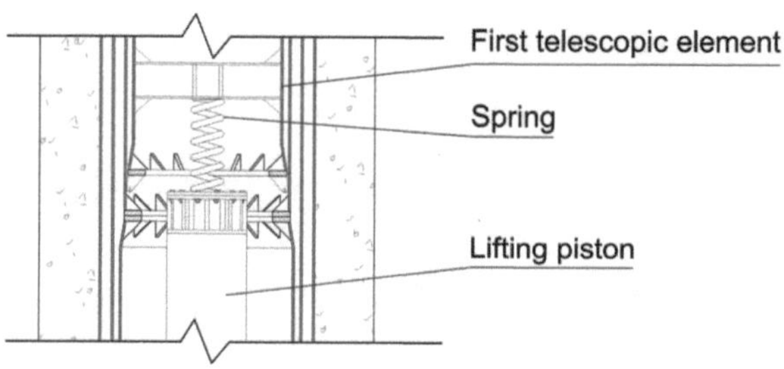

allow a large tolerance between the cylinder diameters, as well as to avoid the block of the cone mating during the tower drop. A ratio 1:6 seems to be a good compromise.

The force provided by the elastic element must guarantee a tensile strength in all the junctions. Using standard steel with a standard friction coefficient, even a force almost equal to the tower weight should guarantee the adhesion of the mated junctions. On the other hand, the variability of the wind forces could provide compressive forces on the top of the tower and a security margin is required.

29.4 Conclusions

This work has shown the innovative proposal of a telescopic wind tower, which is well located inside the international technological research oriented to renewable sources. The main strength of this system is its construction with existing components commercially available, a feature that makes it potentially competitive on a commercial level. The cost increment deriving from a larger tower complexity is compensated by a significant reduction of transport and mounting costs and civil works for road construction, as well as by a significant increment in the number of possible installation sites. Finally, observe that the drilling works necessary for the lifting system can be used also to put a filter and a water pump immediately below the shaft foundation. For a plant that needs electricity and water, such as a farm, the cost of drilling could be afforded only once for both needs.

Acknowledgments We gratefully acknowledge funding from the Italian Ministry of Education, University and Research (MIUR) through the PRIN project "Hydroelectric energy by osmosis in coastal areas".

References

1. A.P. Henderson, U.S. Patent US 6782667 B2, 31 Aug 2004
2. ARE—American Resource and Energy, Motorized Self-Raising Wind Tower System, http://www.arewindtowers.com/raising-systems/motorized. Accessed 04 Oct 2013
3. B. Rexroth, Jacking systems, http://www.boschrexroth.com/en/xc/industries/machinery_applications_engineering/offshore_1/products___solutions_2/jacking_systems. Accessed 07 Oct 2013
4. Consiglio Nazionale delle Ricerche (CNR), Norma sul calcolo di verifica del dimensionamento dei collegamenti forzati mozzo-albero. Milano, 1999 (In italian)

Chapter 30
Zirfon® as Separator Material for Water Electrolysis Under Specific Conditions

María José Lavorante, Juan Isidro Franco, Pablo Bonelli, Gerardo Martín Imbrioscia and Héctor José Fasoli

Abstract Hydrogen production through alkaline water electrolysis requires improvements to use renewable energy more efficiently. In the process of converting electrical to chemical energy, efforts are focused on reducing energy loss. Electrolysers play two important roles in this process: one of them is as a hydrogen producer and the other is as a storage mechanism. A storage mechanism occurs when there is an excess of renewable energy that can be stored in the form of hydrogen (chemical energy), which is the fuel for the following step, turning chemical into electrical energy again. Electrolysers research is focused on: separators and electrodes materials, electrolytic solutions and cell design. The ideal situation for a separator in an electrolyser is to possess low electric resistance. For that purpose, we compared Zirfon®, silicone and the system without separator. This work studied the behaviour of Zirfon® under specific working conditions: room temperature, an electrolytic solution of potassium hydroxide 35 % w/w and five different distances between electrodes. In order to carry out this experiment, we designed and constructed a special electrolytic cell. The experimental results showed that Zirfon® separator increases the system resistance approximately 15 % when compared to the same system without separator, but it reduces resistance

M. J. Lavorante (✉) · J. I. Franco · G. M. Imbrioscia
Departamento de Investigación y Desarrollo de Energías Renovables (CITEDEF-EST), San Juan Bautista de La Salle 4397 B1603ALO Provincia de Buenos Aires, Argentina
e-mail: mjlavorante@gmail.com; mlavorante@citedef.gob.ar

P. Bonelli
Programa de Investigación y Desarrollo de Fuentes Alternativas de Materia Prima y Energía (PINMATE), Departamento de Industrias, Facultad de Ciencias Exactas y Naturales, Universidad de Buenos Aires Ciudad Universitaria, Pabellón I C1428EHA Buenos Aires, Argentina

H. J. Fasoli
Pontificia Universidad Católica Argentina de los Buenos Aires, Alicia Moreau de Justo 1500 C1107AAZ Ciudad Autónoma de Buenos Aires, Argentina

A. Y. Oral et al. (eds.), *International Congress on Energy Efficiency and Energy Related Materials (ENEFM2013)*, Springer Proceedings in Physics 155,
DOI: 10.1007/978-3-319-05521-3_30,

when compared to silicone (excellent insulator). Another result proved that the distances between electrodes proposed in this work did not show bubbles resistance because the system performance improved as the distance became shorter.

30.1 Introduction

Hydrogen production using alkaline water electrolysis has been a well-known technology for around 200 years, and it offers the advantages of simplicity, availability and flexibility. In spite of those benefits, only 4 % of the world production is obtained by this method. The most important reason for this is its high electricity consumption of about 4.5–5 kWh/m_N^3 H_2 in the majority of industrial electrolysers [1–3]. On the other hand, alkaline water electrolysis can store, in the form of hydrogen, the excess of renewable energy, for example electricity from photovoltaic, wind and other sources. That permits the electrolysers to be an important part of the so-called "hydrogen economy", as transformers and storage devices of primary energy [2, 3]. The use of a renewable energy source may be considered the most sustainable method for hydrogen production.

Alkaline water electrolysis has long been used in different industries such as chemical, pharmaceutical, electronic, metallurgic and the production of food. In all these applications, electrolysers operate for long periods, which would require testing if they are to be used intermittently, when the source of renewable energy is available. Therefore, the challenge that this technology has to overcome is the capacity to work intermittently. In order to make this method of production more efficient the efforts were focused on the design of cells with zero-gap geometry; the reduction of the overpotential with new electrocatalytic materials and development of new material for diaphragms [1–3]. Another attempt is to try to lower, as much as possible, the cost of the electricity used to reduce its operational cost.

Focusing on diaphragm materials, this work investigates the behaviour of Zirfon® as separator material in alkaline water electrolysis.

Initially, alkaline water electrolysers used asbestos as separator material, but it was proved to be carcinogenic and its use was forbidden [4]. The industry started to look for a new material and with that purpose Zirfon® was developed. Zirfon® is a porous composite material composed of ZrO_2 particles, in the form of powder, embedded in a polysulfone matrix and hydrophilic in nature. The film-casting technique is the basis of the manufacturing procedure [5, 6]. Zirfon® offers good chemical and physical properties: high thermal and chemical stability, low resistance towards ionic migration and excellent wettability. This last characteristic, wettability, is very important in systems where hydrogen and oxygen are being evolved, because it avoids the generation of dry islands, in which it does not allow the electrolytic solution to be in contact with the separator [6]. This characteristic is improved with the addition of ZrO_2 to the polysulfone matrix, which results in a denser structure. Zirfon® has applicability in the field of alkaline water electrolysis, $Ni–H_2$ battery systems and as ultrafiltration membranes [7, 8].

In this research the behaviour of Zirfon® under specific working conditions was studied: room temperature, an electrolytic solution of potassium hydroxide 35 % w/w and five different distances between electrodes. In order to carry out this experiment, a special electrolytic cell was designed and constructed. The results were compared with the same system using silicone as separator material and without a separator.

30.2 Experimental

The behaviour of Zirfon® as separator material for alkaline water electrolysis was tested in a special electrolytic cell designed and constructed with the additional purpose of studying the effect of the gap between electrodes in this specific system.

As a general description, the special electrolytic cell consists of a cubic container, a set of five different pairs of gauges, two guide brackets, two electrodes, four screws and two mobile locks. The material used for the electrodes was stainless steel 316L, the screws were made of steel and the rest of the pieces were made of crystal acrylic. The isometric view of the electrolytic cell is presented in Fig. 30.1. As a detailed description, the electrolytic cell consists of a base having in its center a channel that position parallel to the electrodes. This channel extends through the walls of the cell to ensure not only the position of the separator but also the watertightness of the system. The gauges allow setting the distance between electrodes. The guide brackets are the pieces that support the electrodes secured by two screws, which act as electrical connectors. The gauges are placed between the wall and the electrode, to position the electrodes, with respect to one another according to the necessity or the requirement only by changing the selected gauge. The isometric exploded view of the parts of the electrolytic cell is shown in Fig. 30.2.

Once the desired distance between electrodes is reached, the system is secured by the mobile locks. Those pieces block the movement of the electrodes during the experiences maintaining the gap constant between electrodes. The electrodes received a cleaning treatment before their use in the different determinations. The treatment consisted of these steps [9]: wash the electrodes with distilled water and allow to dry. Soak up a filter paper in acetone and use it to clean the surface of the electrode; allow solvent to evaporate. Soak up a filter paper in ethylic alcohol and use it to clean the surface of the electrode; allow solvent to evaporate.

When all the pieces are in the right position, an electrolytic solution of KOH 35 % w/w (J. T. Baker 87.0 %) prepared with bidistilled water, is added to the electrolytic cell in addition to a drop of sodium dodecyl sulphate. Sodium dodecyl sulphate acts as a surfactant reducing the superficial tension of the solution.

The electrical connectors are switched to a power source Agilent N5743A System DC Power Supply (12.5 V/60 A, 750 W). Given a certain potential

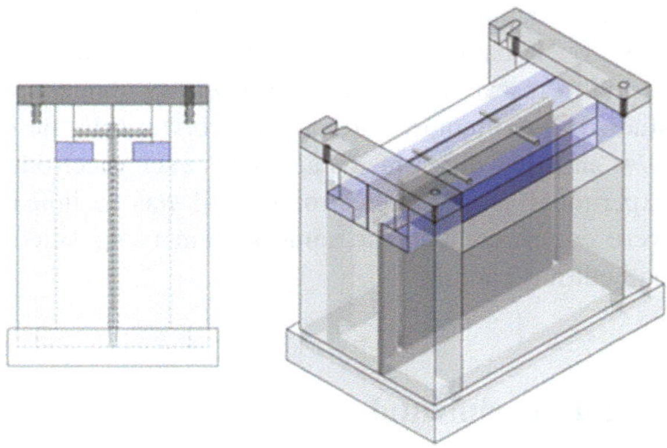

Fig. 30.1 Isometric view of the electrolytic cell

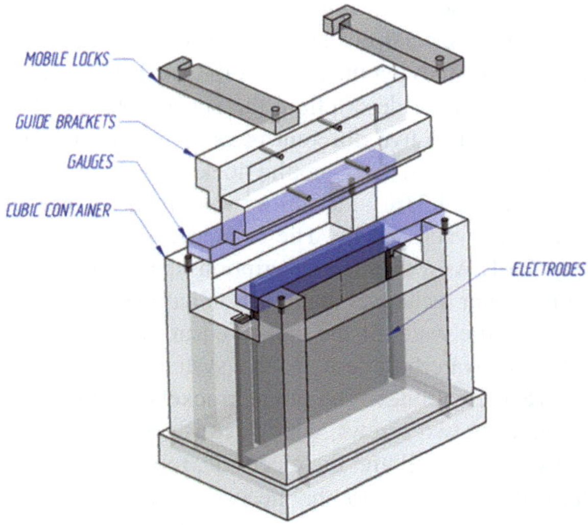

Fig. 30.2 Isometric exploded view of the electrolytic cell

ranging from 0.0 to 4.0 V, current measurements are made. The monitor parameters controlled are voltage, current and temperature. The working conditions are atmospheric pressure and room temperature. All the experiences were carried out twice. The monitor parameters were used to calculate current density, decomposition potential and resistance.

The distances between electrodes studied in this work were: 1.65; 1.45; 1.35; 1.00 and 0.75 cm.

30.3 Discussion and Results

The results were analysed in order to study and compare the effect of the gap between electrodes using Zirfon® as separator and the system without it. From the graphic representation of current density as a function of applied voltage

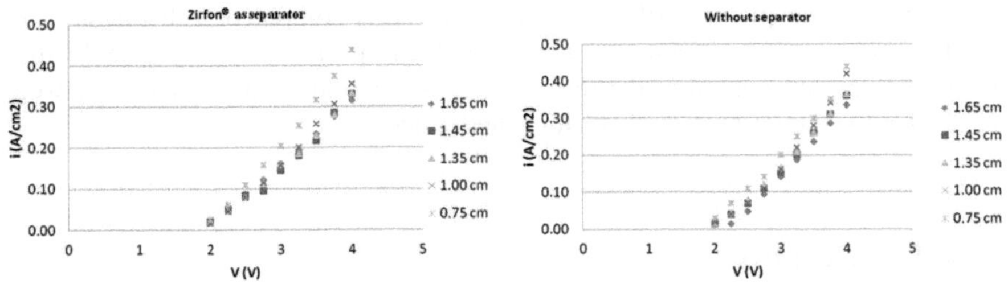

Fig. 30.3 Graphic representation of the current density as a function of the applied voltage differences in a system with Zirfon® as separator and without separator

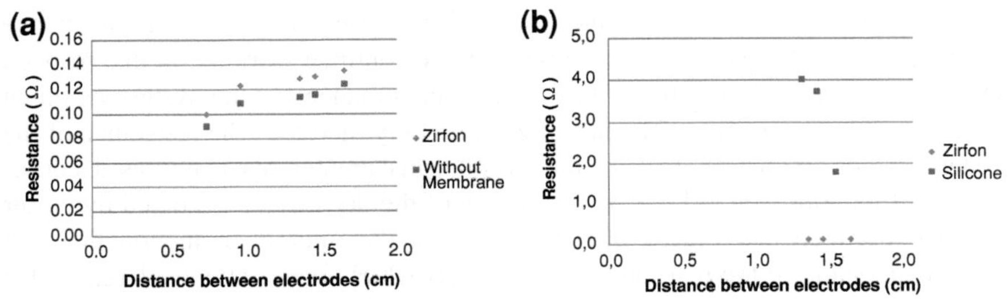

Fig. 30.4 Graphic representation of the resistance as a function of the distances between electrodes for a system with: **a** Zirfon and without separator; **b** Zirfon and silicone

differences for the distances studied, it is observed that as distance grows, the performance of the system decreases. Figure 30.3 shows the behaviour of the same system using Zirfon® as separator and without a separator. Same results were obtained for other researchers that work with different materials for electrodes and separators [10, 11].

From the graphic representation of the current as a function of applied voltages differences for all the experiences, the linear equation was obtained with the objective of determining the current density, the resistivity and the resistance of the system.

The values calculated for the resistance in these two systems (with and without Zirfon® as separator) were presented in Fig. 30.4a. This graphic representation exhibited a similar behaviour between them and the differences were established by the resistance of the Zirfon® separator itself, approximately 15 %, if it is compared with the system without separator.

Silicone was also tested as alternative material to be used as separator. It was chosen precisely because of its hydrophobicity, its poor wettability in electrochemical systems in which there is gases production. In the experiments where silicone was used, the silicone separator was placed 1 mm over the channel to allow electrolytic solution to be in contact in cathode and anode compartments. Only three distances between electrodes were studied for this purpose: 1.65; 1.45

and 1.35 cm. The silicone behaves differently if it is compared with Zirfon®: the resistance of the system increases when the electrodes are closer. Figure 30.4b is the graphic representation of the resistance as a function of the distances between electrodes for the systems with Zirfon® and silicone.

30.4 Conclusions

In this experimental work the behaviour of Zirfon® as separator material for alkaline water electrolysis was studied. Five different distances were tested and as it was expected, the resistance increased with longer distances. Bubble resistance was not observed in the distance studied because the system performance improved as the distance became shorter, but we continue working on that respect. The distance of 0.75 cm presented the best performance for the system as much with and without separator. The use of Zirfon® only increases the resistance of the system in approximately 15 %. Silicone presents greater resistance as a consequence of its poor wettability. The formation of the dry islands is helped by closer distance between electrodes. Wettability is not the only phenomenon that can modify the value of the resistance; other effects which were not considered in this work can also affect the ionic conductivity of the electrolytic system.

Acknowledgments Authors want to thank the Defence Ministry for the PIDDEF 21/11 and 22/11 and the Army Project Direction (DIEJ) for the given support. Authors are grateful to the Prototype Department of CITEDEF for the construction of the device used in the experiments.

References

1. V.M. Nikolic, G.S. Tasic, A.D. Maksic, D.P. Saponjic, S.M. Miulovic, M.P.M. Kaninski, Raising efficiency of hydrogen generation from alkaline water electrolysis–energy saving. Int. J. Hydrogen Energy **35**, 12369–12373 (2010)
2. K. Zeng, D. Zhang, Recent progress in alkaline water electrolysis for hydrogen production and applications. Prog. Energy Combust. Sci. **36**, 307–326 (2010)
3. D. Stojic, M.P. Marceta, S.P. Sovilj, S.S. Miljanic, Hydrogen generation from water electrolysis-possibilities of energy saving. J. Power Sources **118**, 315–319 (2003)
4. Ph Vermeiren, W. Adriansens, J.P. Moreels, R. Leysen, Evaluation of the Zirfon® separator for use in alkaline water electrolysis and Ni–H_2 batteries. Int. J. Hydrogen Energy **23**, 321–324 (1998)
5. Ph Vermeiren, J.P. Moreels, A. Claes, H. Beckers, Electrode diaphragm assembly for alkaline water electrolysers. Int. J. Hydrogen Energy **34**, 9305–9315 (2009)
6. Ph Vermieren, W. Adriansens, R. Leysen, Zirfon®: a new separator for Ni–H_2 batteries and alkaline fuel cells. Int. J. Hydrogen Energy **21**, 679–684 (1996)
7. S. Kuypers, I. Genné, R. Leysen, Surface characteristics of Zirfon® composite ultrafiltration membranes. J. Microscopy **177**, 313–319 (1994)
8. J. Schaep, C. Vandecasteele, R. Leysen, W. Doyen, Salt retention of Zirfon® membranes. Sep. Purif. Technol. **14**, 127–131 (1998)

9. M.J. Lavorante, R. Munaro, J.I. Franco, H.J. Fasoli, A.R. Sanguinetti, Estudio sistemático del comportamiento de electrodes de acero inoxidable 316L picados con HCl para su uso en electrolizadores bipolares alcalinos. Avances en Energías Renovables y Medio Ambiente **15**, 9.13–9.18 (2011)
10. N. Nagai, M. Takeuchi, T. Kimurab, T. Okaa, Existence of optimum space between electrodes on hydrogen production by water electrolysis. Int. J. Hydrogen Energy **28**, 35–41 (2003)
11. A.F.M. Mahrous, I.M. Sakr, A. Balabel, K. Ibrahim, Experimental investigation of the operation parameters affecting hydrogen production process through alkaline water electrolysis. Int. J. Therm. Environ. Eng. **2**, 113–116 (2011)

Chapter 31
Onshore Wind Farms: Value Creation for Stakeholders in Lithuania

Marija Burinskienė, Paulius Rudzkis and Adomas Kanopka

Abstract With the costs of fossil fuel consistently rising worldwide over the last decade, the development of green technologies has become a major goal in many countries. Therefore the evaluation of wind power projects becomes a very important task. To estimate the value of the technologies based on renewable resources also means taking into consideration social, economic, environmental, and scientific value of such projects. This article deals with economic evaluation of electricity generation costs of onshore wind farms in Lithuania and the key factors that have influence on wind power projects and offer a better understanding of social-economic context behind wind power projects. To achieve these goals, this article makes use of empirical data of Lithuania's wind power farms as well as data about the investment environment of the country.Based on empirical data of wind power parks, the research investigates the average wind farm generation efficiency in Lithuania. Employing statistical methods the return on investments of wind farms in Lithuania is calculated. The value created for every party involved and the total value of the wind farm is estimated according to Stakeholder theory.

31.1 Introduction

Continual population and economic growth has meant increased energy consumption worldwide. From 2004 to 2008, the world population grew by 5 %, whereas the gross energy production increased by 10 %, and the yearly CO_2

M. Burinskienė (✉) · P. Rudzkis
Vilnius Gediminas Technical University, Saulėtekio Avenue 11, 10223 Vilnius, Lithuania
e-mail: marija.burinskiene@vgtu.lt

P. Rudzkis
e-mail: paulius.rudzkis@gmail.com

A. Kanopka
Mykolas Romeris University, Ateities Street 20, 08303 Vilnius, Lithuania
e-mail: adomas.kanopka@gmail.com

A. Y. Oral et al. (eds.), *International Congress on Energy Efficiency and Energy Related Materials (ENEFM2013)*, Springer Proceedings in Physics 155,
DOI: 10.1007/978-3-319-05521-3_31,

emission increased by 10 % [9]. If such tendencies continue, the IEA predicts that, by 2030, energy demand worldwide will have increased by approximately 60 %, and carbon dioxide emission will have increased by 62 %.

Wind energy is expected to play an important role in the future. By 2020, approximately 180 GW of both onshore and offshore wind power, corresponding to 10–15 % of all energy produced in EU power installations, could be generated in the European Union alone. As estimated by the Global Wind Energy Council [7], in 2020, approximately 16 % of electricity consumed worldwide will be generated by wind energy.

In Lithuania wind power is one of the most rapidly growing technologies of renewable energy. As opposed to 2009, in 2010, the generation of electricity in wind power plants in Lithuania grew by 43 % which corresponded to 3.9 % of all electricity generated in the country. The amount of electricity generated in hydro power plants in Lithuania is small in comparison, and, moreover, the usage of hydro resources is limited due to the flat landscape.

In the last decade, a variety of models have been developed for analyzing long-term trends in renewable and wind energy costs and payback periods necessary for the return of the investments for wind farms [1, 8, 20, 22]. A usual way to look at the long-term cost trend is by applying the concept of the experience curve, which analyses the cost development of a product or a technology as a function of cumulative production, based on recorded data. The experience curve is not a forecasting tool based on estimated relationships; it merely points out that if specific trends continue to prevail in the future, then we may see the proposed decrease. However, some of these models use incompatible specifications, not all of which can be compared directly. Experience curves for wind energy have been presented in scientific papers [10, 17]. As a rule, the price for wind energy is determined by calculating the direct value. However, wind farms create not only direct, but also indirect value, which refers to their social value and the price and properties of inferred resources. Consequently, it is necessary to ensure that both the evaluation and the development of wind power projects are performed in the economically most optimal fashion.

This paper attempts to contribute to a better understanding of how to substantiate wind power projects economically.

The objective of the article is: to evaluate the implementation costs and the created value of onshore wind farms in Lithuania, to identify key factors that have an influence on wind power projects, and to contribute a better understanding of the social-economic context behind wind power projects.

The expected outcomes of this study are:

- improving the evaluation of wind power plant investment projects in Lithuania by analyzing their investment environment and prospects
- identification of the total economic value and distribution of value shares to stakeholders.

To achieve these goals, empirical data about Lithuania's wind power farms as well as data about their investment environment have been collected.

Technological development plays a major role in decreasing the overall cost of wind energy; the growing demand raises the price of power generation units in the short run. The wind energy generation costs have increased by 20 % in the past 3 years, due to the rising prices of key raw materials and an unexpected surge in the demand for wind turbines, following the approval of favorable support policies in large markets, such as those of the USA, China, and the second round of European Member States [1]. For the reasons mentioned, in this article, we refer to the financial and technical characteristics of the already existing wind farms.

The empirical analysis presented in this paper is based on the data provided by the operator of the largest wind farm (30 MW) in the Baltic countries and owner of the closed joint stock company "Vėjųspektras," which has an extensive experience in developing wind energy projects. Its wind power farm consists of 15 E-70 model power turbines, manufactured by German Enercon, GmbH; the installed capacity of each turbine is 2 MW. The wind power farm is located in the seaside area, which is recognized as a location, commercially viable for the development of wind power farms. Therefore, the data about this farm is a suitable basis for an analysis of the value and costs of wind energy in the seaside area.

Methods: This article carries out a financial analysis, a Cost-Benefit analysis, and a Discounted Cash Flow analysis. The Cost-Benefit analysis (CBA) is an analytical method applied by businesses and governments in order to determine whether the economic net benefit, brought by a project or a public sector program, outweighs the costs [2, 18]. CBA makes use of a set of widely adopted methods for evaluating investments, which include the Discounted Cash Flow (DCF) method. The DCF method helps estimate the current value of an investment by discounting projected future revenues and costs.

31.2 The Efficiency of the Wind Resources of the Lithuanian Onshore Area in Generating Electricity

The most important factor affecting the profitability of investments in wind energy is the local wind resource. Differences in wind speed explain most of the differences in the costs per kWh between specific countries and projects. The development of wind energy in Lithuania has less favorable natural conditions than in Latvia and Estonia [24]. This is determined by a short stretch of the coastal line in Lithuania. The deeper into the continent goes to measure wind speed and the effectiveness of wind power plants, the more they tend to decrease [14].

A number of UNDP supported studies have been conducted in order to identify wind energy resources in Lithuania [19]. The wind speed measurement data (m/s) obtained at 10 m above ground level by Hydro meteorological stations has led to a conclusion that the wind speed in Lithuania is insufficient for the development of industrial wind energy production. Measurements of wind speed, conducted in 1996–1997, showed that in the seaside area, at 50 m above ground, wind speed reaches 7.4 m/s [23] while Lithuania average is 2.52–4.55 m/s. The regional nature of wind energy is thus obvious; however, adequate technological improvements as well as simply increasing the height of wind turbine towers allow for exploiting the entire territory of Lithuania.

Multi-year wind measurements conducted in the coastal area of Lithuania between 1995 and 2003 revealed that, in Klaipeda's region, near Giruliai district, the average wind speed was 6.4 m/s. During different seasons, the wind speed can vary by up to 50 %, but the average annual speed measurements differ only slightly.

The most favorable conditions for developing wind energy in Lithuania can be found along a 50 km stretch of the coastal line [11, 15]. To determining a correct micro location of each individual wind turbine successfully is thus crucial for the economic success of any wind energy project. The most attractive territories for installing power plants are the Baltic Sea and the Curronian lagoon, where the average wind speed is 8 m/s.

One of the biggest technical problems posed by wind energy development, compared to traditional forms of energy generation, is the dependence of wind power on atmospheric stability. Consequently, the exact amount of wind produced electricity cannot be predicted for every particular time period, which causes problems in terms of system control and the balance between the demand and supply of electricity. Unlike with traditional power generating technologies, electricity production in wind farms cannot be unclearly calculated, and efficiency rates vary according to a geographic location and to winds prevailing there at any particular period of time. However, the production of electricity by a wind generator installed in a specific location is a stationary random process. Therefore, the average efficiency of a wind generator and its potential errors can be statistically evaluated, referring to historical data.

As can be seen in Fig. 31.1, the monthly fluctuations in the efficiency of the wind farm are sufficiently large. During the investigation period, the average farm efficiency was about 0.238. To evaluate possible limitations of calculating average efficiency and to construct a confidence interval, the standard error of estimate should be taken into account.

Having evaluated the confidence interval for the average efficiency of the wind farm, it can be said with 95 % probability that, in the long run, the average efficiency of a wind turbine should be between 0.208 and 0.269. In this case, 1 kW of installed capacity should produce from 1.82 to 2.36 MWh of electricity on the average.

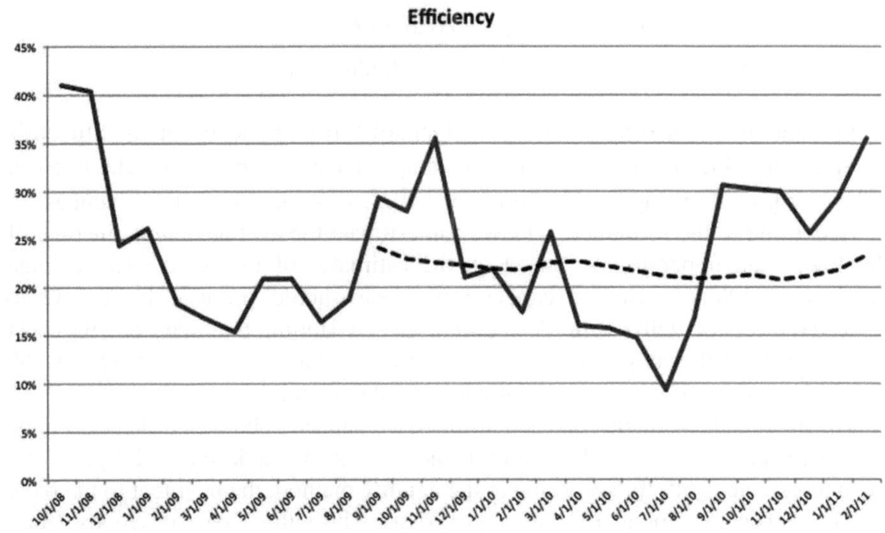

Fig. 31.1 The average efficiency dynamics of a wind farm. *Note* the data is obtained from a wind farm consisting of 15 wind generators, where the installed capacity of one power generator is 2 MW

31.3 The Components of Investments in Onshore Wind Energy in Lithuania

There are four groups of parameters that influence wind power costs: 1—capital costs, 2—operation and maintenance costs, 3—the amount of electricity produced, 4—the discount rate and economic lifetime of the investment.

Investments in electricity generation are directly related to electricity production; therefore, they are quite clearly defined [12]. On the other hand, investments in infrastructure, which involves environmental impact assessments, roads, and so on, are unique in each case; such investments are typically smaller when developing a larger wind power farm. In this article, the assessment is made, taking a 30 MW (15 × 2 MW) wind farm as an example.

The major part of the investment (approximately 90 %) in developing a wind farm is allocated to the power generator, whereas the infrastructure is given only approximately 10 % of the total amount of costs.

Taking into consideration the average estimate of the efficiency of wind energy, we can predict the average income of a wind farm. Wind power produced in Lithuania is subsidized: wind produced electricity is purchased for 81.1–107 euros/MWh; moreover, pollution-reducing projects receive Emission Reduction Units (ERUs), which can be sold on the market. Thus the revenue of a wind farm consists of two parts, both of which are related to the performance ratio, and ERUs also depend on market prices. Empirical assessment of the wind farm data can be constructed in the lower and upper limits of the average annual income:

$$P((\overline{E} - t * SEM) * 8.76 * (p + 0.626 * \overline{ERU}) \\ < R < (\overline{E} + t * SEM) * 8.76 * (p + 0.626 * \overline{ERU})) = 0.95 \quad (31.1)$$

P—random probability, *SEM*—the standard error of the mean, t—Student's distribution value, $\overline{E}$—efficiency of the sample mean, *p*—power purchase price, $\overline{ERU}$—projected average market price of ERUs, *R*—average annual revenues.

According to the formula (31.1), we can estimate the average annual income of the lower and upper limits. Based on the estimates of the wind farm average efficiency 1 kW of installed capacity per year should produce from 1.82 to 2.36 MWh of electricity, with 95 % probability. Assuming that one ERU will cost about 6 euro in the long run, the average gross revenues should be between 165 and 214 euros, approximately 177.4 euros on the average.

The average investment for 1 kW of installed capacity should be expected to be approximately 1561 euros. Therefore, if the term of the bank loan is 20 years and the interest rate is 5.5 %, according to the annuity method, the payment to the bank should consist of about 131 euros per year, while the Cost of Goods (rent, insurance, maintenance, etc.) should be about 23.7 euros and management costs (salaries and other expenses)—about 5 euros. Looking at the total costs imposed on 1 kW installed capacity, the largest part of the cost is bank payments, which account for approximately 83 % of the total sum, meanwhile the Cost of Goods, together with operating expenditure, constitute only about 17 %. Consequently, changes in interest rates have the biggest impact on the general cost and thus can substantially influence the profitability of the project itself. Assuming the interest rate to be 5.5 %, and the loan period to be 20 years, the total annual cost is estimated to be approximately 160 euros.

The current value of the project depends on the cash flows that will be generated in the future. Referring to the above calculations, over the next 20 years, the average annual income of the wind farm should be between 165 and 214 euros, while the average annual expenses would be approximately 160 euros. Consequently, the average annual cash flow should be within the range between 5 and 54 euros. The total present value of the project can be evaluated, using the discounted cash flow method.

In calculating the current value, estimated cash flows and the 20-year period of time as indicated above will be taken into account, standard 5 % discount rate is used, which is slightly lower than the aforementioned interest rate. The residual value of the project should be divided into two parts: that of the infrastructure and of the power generator. In this article, the residual value of the infrastructure after 20 years is equaled to 161 euros, the initial amount of the investment. It is assumed that the average value of the infrastructure will increase the size of the discount rate, and the investment in the generator is measured, taking into consideration the receipt of benefits; therefore, the amortization is not depreciated. The residual value of the generator should be viewed, assuming the probability that the generator does not wear out completely in 20 years and can be used longer.

Taking into consideration the aforementioned assumptions and the formula (31.1), the average value of the project at the moment can be calculated to fall within the range of 92.2 and 663.5 euros/kWh or approximately 377.8 euros/kWh on the average.

However, if a wind farm is evaluated as a commercial project not subsidized by the state, one which is expected to produce electricity for sale on the market, the situation changes significantly. In 2011, the average price for electricity on Lithuania's power market was about 48 euros/MWh, whereas wind energy was purchased for around 81 euros/MWh. Generating electricity in the wind power-station costs from 68 to 88 euros/MWh (from 160 euros/2.36 MWh to 160 euros/1.82 MWh). Thus wind energy is purchased for a price approximately 68 % higher than the market price. Using the DCF method, the interval of the net present value of state subsidies (NPVs) can be calculated. Therefore, the net present value of the state's subsidies for new wind farms should be between 751 and 973 euros per 1 kW of installed power. Consequently, the value of state subsidies is significantly higher than the benefits of the project developers. From the point of view of a consumer, when the value estimation is based only on the cost-benefit basis, wind power becomes commercially unattractive. However, it is also clear that the development of wind energy creates indirect benefits, such as encouraging production, reducing pollution, etc. Therefore, it is necessary to assess the impact of wind power more generally, taking into consideration not only its prices and cash flows, but also its created value for all the parties involved.

31.4 Indirect Wind Energy Costs and Value: Stakeholders' Perspective

When evaluating the projects for the development of wind farms, the overall impact of the development on all the parties involved should be taken into consideration. No detailed analysis or assessment of societal and environmental costs and benefits of renewable energy has been conducted yet. The theoretical and conceptual basis for evaluating the impact of wind power on the involved parties is provided by the stakeholder theory (SH). The concept of stakeholders was formulated by Freeman in 1984: "any group or individual who can affect or is affected by the achievement of the organization's objectives" [6]. This concept is ambiguous and has been debated.

SH theory sees an organization as a complex system and analyses it according to descriptive, instrumental, and normative aspects. The descriptive approach mainly focuses on the interests of the stakeholder group. The instrumental approach allows for an investigation of stakeholders' connections within and with an organization and their consistency, while the normative approach considers stakeholders' legally defined interests [3]. Thus, the SH theory helps not only to define the organizational goals, but to take into account the interests of the SH group [4].

The SH theory distinguishes between five groups of stakeholders: the owners (shareholders), employees, customers, suppliers, and a society at large. All these groups are important for an organization, its survival and development. Producing electricity differs from conventional businesses, but interested parties are similar overall. The electricity sector can be divided into the following groups: the country's economy, producers, consumers, society, and suppliers. These groups have very different roles and weight in developing wind generated energy value, and for this reason, these should be assessed for each group separately at the beginning of a new project.

Producers—in this article, producers are seen as project developers, who organize the construction of a wind farm and operate it. The value of the producers can be defined very simply, through future cash slurry. According to the DCF method, the producers can be said to be creating value which ranges between 92.2 and 663.5 euros, depending on the actual efficiency of a wind farm.

Consumers—the largest consumers of electricity are legal entities, whose main interest is in lowering the price. The direct benefits of providing electricity to the customers are measured by their willingness to pay for the power. Increasing the production of wind energy has negative impact on consumers, and the impact of new wind energy projects can be estimated to range between 751 and 973 euros/kW.

Society—the public benefits of wind energy can be defined as environmental friendliness and creation of work places. Wind energy is green energy, which reduces environmental pollution. Public benefits of wind energy (apart from consumption) can be seen through emissions trading scheme (ETS), the environmental value of a new wind power plant (using the DCF method) ranges from 142 to 184 euros per 1 kW of installed capacity.

The assessment of society benefits in terms of work places created during the development of a project is a complex undertaking. There is a widespread opinion in scholarly literature that operational costs of fossil fuel technologies, along with societal and environmental costs, are less competitive economically than those of wind power technology [5, 13, 16, 21, 25].

Suppliers—in this article, suppliers are seen as a balance of power system controllers and the guarantor of the power reserve. The production of wind energy is volatile, so wind energy has negative impact on the electricity system and requires additional reserves. However, this effect is difficult to measure because wind energy is but a small part of all electricity production and requires almost no supplementary costs, but even a small increase in them makes it more difficult to balance the system and ensure the reserve.

The country's economy—one of the interested parties in this case is the state, whose value is perceived based on the country's economy. The development of wind energy has systematic impact on any country's economic development, because it affects foreign trade, creates jobs, develops technologies, encourages the production, etc. The benefits on the country's economy are difficult to assess because they depend on a number of complex factors. In each country, the impact of the development of wind farms on the national economy should be assessed and modeled separately, taking into account such factors as unemployment, GDP structure, etc. This requires a complex analysis, which will not be carried out in this article.

Taking into consideration to the interested parties listed above, it is possible to estimate the total created value of a new wind farm. To simplify the process, let us assume that all the interested parties are equally important. In addition, only the value of only two of the interested parties overlaps directly. This is because producers receive the benefits from ERU, and the public gets the benefits of reduced pollution. General value can be calculated:

$$GV = G + V + VP + T + E - v^* = 378 - 862 + 163 + T + E - 98 \quad (31.2)$$

Here G—the value of producers, V—the value of consumers, VP—the value of the public, T—the value of suppliers, and E—the value of the country's economy.

As the estimate (31.2) shows, according to the SH theory, a new farm with 1 kW installed capacity depends on two non-assessed values, that of the suppliers and that of the country's economy. According to the estimate, it can be stated that the total value will be positive if the created value for the country's economy is larger than the harm done to suppliers by at least 419 euros.

31.5 Conclusions

1. The most efficient fair wind resources in Lithuania are found along the coastal line of the Baltic Sea. Measurements show that the most favorable conditions for developing onshore wind energy in Lithuania exist in the stretch of about 50 km along the coastal line.
2. Having evaluated the confidence interval of the average efficiency of a wind farm, with the 95 % probability in the long run, the average efficiency of a wind turbine should be between 0.208 and 0.269, and 1 kW of installed capacity should produce from 1.82 to 2.36 MWh of electricity on the average.
3. A financial and economic analysis of investment projects reveals that the return on investments in wind farms in Lithuania is estimated to range between 92.2 and 663.5 euros (per 1 kW of installed capacity), depending on the efficiency of wind turbines.
4. While determining the value of wind energy generated for different SH groups, it was revealed that the state subsidies are higher than the producers' value and that the total value of wind energy will be positive if the created value for the country's economy will be at least 419 euros higher than the harm done to the suppliers.
5. Most of the arguments presented in scholarly literature support an opinion that operational costs of fossil fuel technologies, along with societal and environmental costs, are less competitive economically than those of wind power technology.

References

1. M.I. Blanco, The economics of wind energy. Renew. Sustain. Energy Rev. **13**, 1372–1382 (2009)
2. K.A. Brekke, The numéraire matters in cost-benefit analysis. J. Public Econ. **64**, 117–123 (1997)
3. T. Donaldson, L.E. Preston, The stakeholder theory of the modern corporation: concepts, evidence and implications. Acad. Manage. Rev. **20**, 65–91 (1995)
4. W.M. Evan, R.E. Freeman, A stakeholder theory of the modern corporation: kantian capitalism, in *Contemporary Issues in Business Ethics,* 5th edn (2005), pp. 76–84
5. J. Firestone, W. Kempton, Public opinion about large offshore wind power: *underlying factors*. Energy Policy **35**, 1584–1598 (2007)
6. R.E. Freeman, *Strategic Management: A Stakeholder Approach* (Pitman, Boston, 1984)
7. Global Wind Energy Council, GWEC, Global Wind Energy Outlook Report (2006)
8. J.B. Greenblatt, S. Succar, D.C. Denkenberger, R.H. Williams, R.H. Socolow, Baseload wind energy: modeling the competition between gas turbines and compressed air energy storage for supplemental generation. Energy Policy **35**(3), 1474–1492 (2007)
9. IEA Key World Energy Statistics (2006, 2010)
10. H.M. Junginger, A. Faaij, W.C. Turkenburg, Global experience curves for wind farms. Energy Policy **33**, 133–150 (2005)
11. V. Katinas, A. Markevičius, M. Tamašauskienė, J.Z. Vilemienė, Vėjo srauto energetinių parametrų Lietuvos pajūrio regione tyrimas. Energetika **56**(3–4), 193–201 (2010)
12. A. Klevienė, Formation of knowledge structure in field of renewable energy sources implementation. Ekonomikairvadyba **16**, 766–772 (2011). ISSN 1822-6515
13. M. Lenzen, J. Munksgaard, Energy and CO_2 life-cycle analyses of wind turbines—review and applications. Renew. Energy **26**, 339–362 (2002)
14. M. Marčiukaitis, Wind farms power prediction in Lithuania. Kaunas LEI 1–10 (2011)
15. A. Markevicius, V. Katinas, M. Marciukaitis, Wind energy development policy and prospects in Lithuania. Energy Policy **35**(4), 893–901 (2007)
16. Y. Maruyama, M. Nishikido, T. Iida, The rise of community wind power in Japan: enhanced acceptance through social innovation. Energy Policy **35**, 2761–2769 (2007)
17. L. Neij et al., Experience curves: a tool for energy policy assessment. Lund University, Lund (SE). IMES/EESS Report, 40 (2003)
18. D. Pearce, G. Atkinson, S. Mourato, *Cost-Benefit Analysis and the Environment: Recent Developments* (OECD Publishing, Paris, 2006)
19. O. Rathman, *The UNDP/GEF regional Baltic Wind Energy Programme* (Riso National Laboratory, Roskilde, 2003)
20. J. Šliogerienė, A. Kaklauskas, E. Zavadskas, J. Bivainis, M. Seniut, Environment factors of energy companies and their effect on value: analysis model and applied method. Technol. Econ. Dev. Econ. **15**(3), 490–521 (2009)
21. D. Štreimikienė, External costs of energy security and climate change. Int. Energy J. **11**, 225–233 (2010)
22. D.J. Swider, L. Beurskens, S. Davidson, J. Twidell, J. Pyrko, W. Prüggler, H. Auer, K. Vertin, R. Skema, Conditions and costs for renewables electricity grid connection: examples in Europe. Renew. Energy **33**(8), 1832–1842 (2008)
23. *The wind regime at Butinge*. Lithuania (Institute for Energy Technology, Kjeller, 1996)
24. T. Tomson, M. Kallis, Assessment on development of wind energy in offshore regions of Estonia. J. Energy Power Eng. **5**, 794–801 (2011)
25. S.V. Valentine, Understanding the variability of wind power costs. Renew. Sustain. Energy Rev. **15**, 3632–3639 (2011)

Chapter 32
Evaluation of New Thermally Conductive Geopolymer in Thermal Energy Storage

Matěj Černý, Jan Uhlík, Jaroslav Nosek, Vladimír Lachman, Radim Hladký, Jan Franěk and Milan Brož

Abstract This paper describes an evaluation of a newly developed thermally conductive geopolymer (TCG), consisting of a mixture of sodium silicate and carbon micro-particles. The TCG is intended to be used as a component of high temperature energy storage (HTTES) to improve its thermal diffusivity. Energy storage is crucial for both ecological and economical sustainability. HTTES plays a vital role in solar energy technologies and in waste heat recovery. The most advanced HTTES technologies are based on phase change materials or molten

M. Černý (✉) · J. Uhlík
PROGEO s.r.o., Tiché údolí 113 252 63 Roztoky u Prahy, Czech Republic
e-mail: progeo@1progeo.cz; cernym@fzp.czu.cz

J. Uhlík
e-mail: progeo@1progeo.cz

M. Černý
Faculty of Environmental Sciences, Czech University of Life Sciences Prague, Kamýcká 129, Praha 6 165 21 Suchdol, Czech Republic

J. Nosek
Technical university of Liberec, Studentská 2, 461 17 Liberec, Czech Republic
e-mail: jaroslav.nosek1@tul.cz

V. Lachman
ISATech s.r.o., Osadní 26 170 00 Prague 7, Czech Republic
e-mail: vlachman@isatech.cz

R. Hladký
Division Geotechnika, ARCADIS CZ a.s., Geologická 4 152 00 Praha 5, Czech Republic
e-mail: radim.hladky@arcadis.cz

J. Franěk
Czech Geological Survey, Klárov 3 118 21 Praha 1, Czech Republic
e-mail: jan.franek@geology.cz

M. Brož
Institute of Rock Structure and Mechanics, Academy of Science Czech Republic, v.v.i., V Holesovickach 41 182 09 Praha 8, Czech Republic
e-mail: mbroz@irsm.cas.cz

A. Y. Oral et al. (eds.), *International Congress on Energy Efficiency and Energy Related Materials (ENEFM2013)*, Springer Proceedings in Physics 155,
DOI: 10.1007/978-3-319-05521-3_32,

salts, but suffer with economic and technological limitations. Rock or concrete HTTES are cheaper, but they have low thermal conductivity without incorporation of TCG. It was observed that TCG is stable up to 400 °C. The thermal conductivity was measured in range of 20–23 W m^{-1} K^{-1}. The effect of TCG was tested by heating a granite block with an artificial fissure. One half of the fissure was filled with TCG and the other with ballotini. 28 thermometers, 5 dilatometers and strain sensors were installed on the block. The heat transport experiment was evaluated with COMSOL Multiphysics software.

32.1 Introduction

At present times, under the threat of global climate changes and environmental pollution, energy management is becoming crucial. The main trends in the energy industries are based on energy saving and renewable resources. The greatest disadvantage of renewable energy resources, primary wind and solar, is the mismatch between the supply and demand of energy. It can be minimized with thermal energy storage (TES) [7]. Energy savings can also be achieved by storing waste heat from industrial processes.

TES can be classified in terms of by their operating temperatures. Low temperature TES operates below 120 °C [7] or 200 °C [5]. Applications can be found in building heating and cooling [3]. High temperature thermal energy storage (HTTES) operates above these temperatures. HTTES are mainly used at solar thermal power plants, where the solar energy is transformed into high-temperature steam to drive turbines producing electricity [7].

Heat storing techniques are based on three basic principles. The first type uses sensible heat. When a phase change appears, the energy is stored in the form of latent heat, corresponding to the second type. The third type, called thermochemical storage, occurs when endothermic and exothermic reversible reactions are used [5]. The first two are more common, while the third is being studied [1, 4].

Solid sensible heat storage materials do not undergo phase change in the temperature range of the storage process. The ability to store sensible heat for a particular material greatly depends on the value of its energy density, i.e. the heat capacity per unit volume. For a material to be useful in a TES application, it must be inexpensive and have good thermal conductivity [6]. Solid sensible heat storage has to include some kind of heat exchanger to transfer the thermal energy to or from the heat transfer fluid. If the storage media has lower conductivity, the input costs increase for more complex heat exchangers. Cast ceramics and concrete were tested as TES at the Plataforma Solar de Almeria in a research project of the German Aerospace Centre [11]. Recycled industrial ceramics made by vitrification of asbestos containing wastes were also studied [8] as candidates to be used as sensible TES material.

The concept of the research project “Reversible Storage of Energy” (RESEN) is based on the use of high heat capacity of rock for energy accumulation and its conductivity enhancement. A newly developed thermally conductive geopolymer (TCG) is added to improve the charging rates and ensure less extensive heat exchanges.

This study describes granite rock and TCG sandwich heating experiment. The evaluation of the experiment is based on a numerical heat transfer model assembled in COMSOL Multiphysics [2]. The software was previously used for modelling of rock heating experiment [9] or geothermal system study [10].

32.2 Materials and Methods

A survey of common rocks in the Czech Republic was performed at the beginning of the RESEN project. The basic geomechanical properties were measured for samples from 41 locations. 4 sedimentary, 12 volcanic, 9 plutonic and 16 metamorphic rocks were examined. The ten best candidates were chosen for further investigation of their thermal properties. A sample of common concrete was added for comparison. Finally, the granite rock from Panské Dubénky was chosen as the best rock fill of designed HTTES. This granite has the highest heat capacity and lowest thermal conductivity decrease with increasing temperature. It is also stable and does not exhibit degradation after cyclic heating up to a temperature of 380 °C.

Thermally conductive material was developed in an attempt to increase heat transfer in HTTES. This material is based on a geopolymer mixture of graphite, specifically, a mixture of sodium-silica water glass, metakaolin and flake graphite of various grain sizes. It has 6–10× higher thermal conductivity compared to rocks and high thermal endurance. The cyclic heating of TCG, demonstrated endurance up to 380 °C under conditions in the presence of atmospheric oxygen. The thermal endurance of the base geopolymer matrix is higher than 800 °C and the maximum operational temperature is thus limited by the thermally conductive additive (graphite), which is oxidized by oxygen at temperatures above 400 °C. The thermal conductivity and heat capacity of TCG were measured using a HotDisk Constant Thermal Analyser, yielding values of 19.7–23.3 $W\ m^{-1}\ K^{-1}$ and 1.05–1.51 $MJ\ m^{-3}\ K^{-1}$—in dependence on the grain-size distribution of the conductive matter. TCG also has following parameters: density 1,150 $kg\ m^{-3}$, porosity 40 %, compressive strength 4 MPa and saturated hydraulic conductivity $7 \times 10^{-9}\ m\ s^{-1}$.

The concept of described experiment was based on heating a rock block with an artificial fissure and comparing temperature changes with and without the use of TCG. To eliminate external influences, the fissure was divided into two symmetrical halves, so that comparison could be made on the basis of one set of measurements. One half of the fissure was filled with TCG and the other with

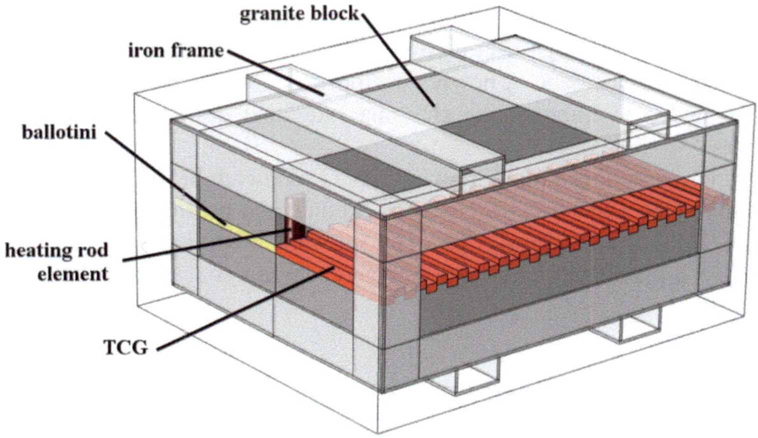

Fig. 32.1 Experimental setup and its model geometry representation in COMSOL. Front half of top block plate is removed

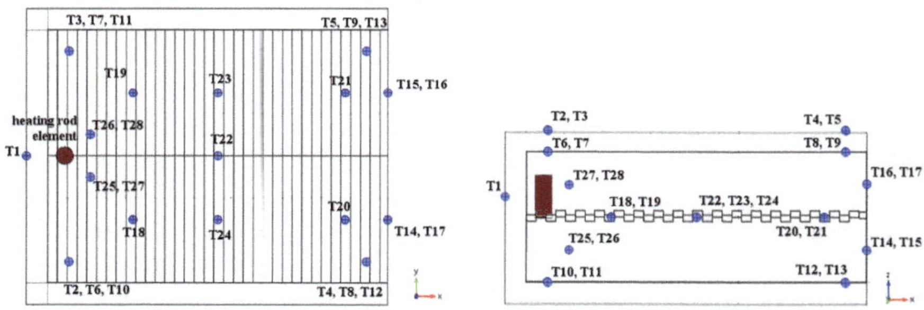

Fig. 32.2 Placement of thermometers, *on the left* is the top view; *on the right* is a side view

ballotini (originally granite gravel was planned). The effect of TCG was evaluated by a numerical model on the basis of measured temperature field.

A scheme of the experimental setup is depicted in Fig. 32.1. A granite block with dimensions 800 × 600 × 300 mm was used. It was horizontally divided into two plates, which were notched on the cut plane. Each notch was 10 mm deep and 23 mm wide on an average. The space with notches between the plates was filled with a 15 mm thick layer of TCG and ballotini. The resulting block is 305 mm high with a 5 mm continuous layer of TCG and 10 mm TCG in the filled notches. A hole was drilled by the side in the upper plate for the heating rod element (Fig. 32.2). Other holes were drilled for a total of 28 thermometers. The block was reinforced by a metal frame and dilatometers were installed before the experiment. The whole block was encased in 50 mm thick FoamGlas panels as thermal insulation except for the side farthest form the heating rod to promote the heat flux in the direction of the fissure.

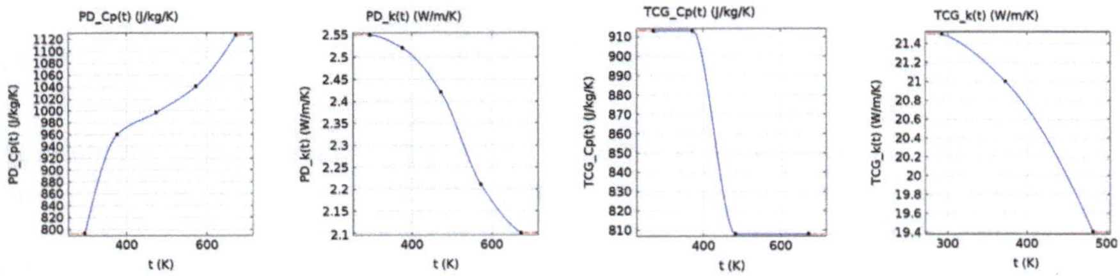

Fig. 32.3 Shapes of the temperature dependent functions of the specific heat capacity (Cp) and the thermal conductivity (k) of Panské Dubénky granite (PD) and TCG

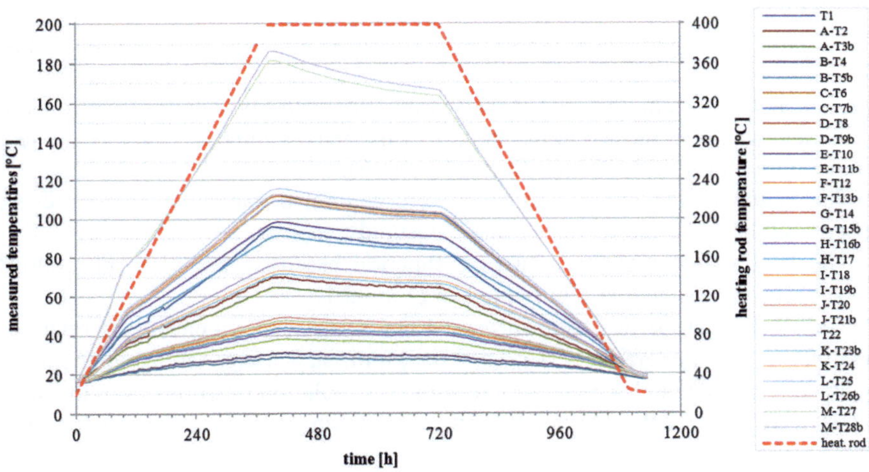

Fig. 32.4 Measured temperatures (*solid lines*) and controlled temperature at the heating rod element (*dashed line*)

The experiment started by heating the rod element. The maximal possible power load was 630 W. The space in the borehole between the heating rod and the block was filled with TCG. 2 thermometers were inserted in each of granite plates near the heating rod, 7 in the fissure, 9 between granite surface and the insulation, 4 on the top of the insulation and 4 on the side farthest from the heating element. Their position is depicted in Fig. 32.2.

For the first period, the thermal gradient at the heating rod was set at 1 °C per hour because of hardening the TCG layer. The final temperature 400 °C was reached after 16.6 days. This temperature was fixed for another 14 days (2nd period). In 3rd period a declining temperature gradient 1 °C per hour was set at the rod, as can be seen in Fig. 32.4.

The numerical model was created with the Heat Transfer Module in the COMSOL Multiphysics finite element software package. According to the experiment, the modelled process corresponds to unsteady heat conduction in solid. The governing equation is:

$$\rho C_p \frac{\partial T}{\partial t} = \nabla \cdot (k \nabla T) + Q \tag{32.1}$$

where T is the temperature, t is time, Q is the heat source, and the remaining coefficients are the material properties: the density ρ, heat capacity Cp, and thermal conductivity k.

A combination of two boundary conditions was used on the surfaces of the thermal insulation and the metal frame. These are called convective cooling (32.2) and surface-to-ambient radiation (32.3) in COMSOL.

$$-\boldsymbol{n} \cdot (-k \nabla T) = h \cdot (T_{ext} - T) \tag{32.2}$$

$$-\boldsymbol{n} \cdot (-k \nabla T) = \varepsilon \sigma (T_{ext}^4 - T^4) \tag{32.3}$$

where $\boldsymbol{n}$ is the normal vector, k is the thermal conductivity, h is the heat transfer coefficient, ε is the surface emissivity, T is the temperature, T_{ext} is the ambient temperature and σ is the Stefan-Boltzmann constant.

The values of the input parameters were calibrated manually by minimising differences between the modelled and measured temperatures.

The heat source was set as a time function based on recorded power load values. The heat source is represented as Q in 32.1.

The thermal properties were measured before the experiment in the range of 20–400 °C for the granite, TCG and ballotini, and were set as temperature dependent in the model. The piecewise cubic interpolation functions from the table values were used for the thermal dependency. The function curves are displayed in Fig. 32.3.

The computational mesh was automatically generated in COMSOL with the "fine resolution" settings. The entire mesh consists of 944,300 tetrahedral elements and 158,746 triangular surface elements.

At first stage of simulations we neglected iron frame, but modelled temperatures have not fit the measurements. The model results were closer to measurements after adding representation of metal frame.

32.3 Results

The total duration of the experiment was 1,134 h (47.25 days). The temperatures and power input were recorded every 30 s. The measured temperatures are depicted in Fig. 32.4, the measured power input is shown in Fig. 32.7.

There is visible temperature trend change after 94 h from beginning of experiment. The anomaly is visible for every thermometer except for the heating rod. Before this event the temperature gradient at the nearest thermometer was 1 °C/h, than it changed to 0.2 °C/h. During more than three days the gradient was increasing to 0.4 °C/h. The temperature at the heating rod was 88 °C at the

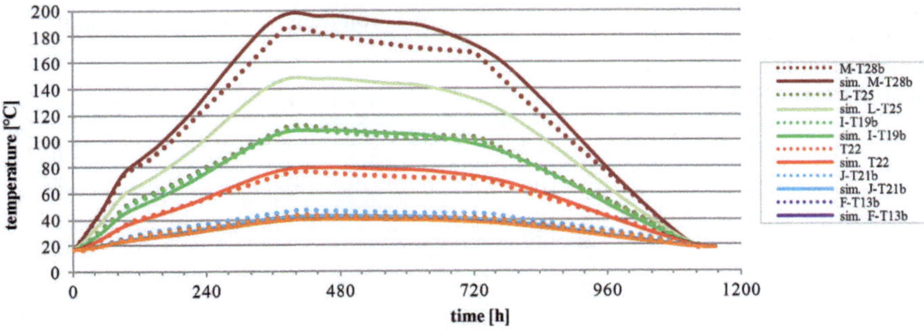

Fig. 32.5 Measured (*dotted lines*) and simulated (*solid lines*) temperatures of selected thermometers

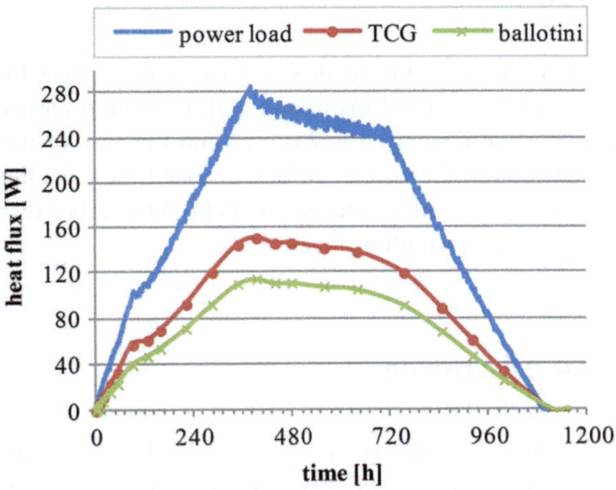

Fig. 32.6 Heating element power load (measured) and heat fluxes from the heating rod element (modelled) to the half of granite block filled with TCG and the half with ballotini

beginning of the event. The lowest gradient was reached at the temperature of 111 °C at the heating rod. At the same time, there are visible changes in the trend of the power input curve. During 2nd period with fixed temperature on the rod, the measured power input and the temperatures in the granite block were decreasing.

It was impossible to reproduce observed changes in temperature slopes without adding thermally resistive layer at the surface of the heating rod with $R_s = 0.0025\ \mathrm{K\ m^{-2}\ K^{-1}}$ from 94th hour. With a focus on the symmetrically placed thermometers, the TCG conductivity function (*PD_k*(*t*) in Fig. 32.4) values had to be reduced by 30 % to achieve measured temperature differences. The results for the final calibrated model compared to the measured values are depicted in Fig. 32.5.

The contribution of TCG was evaluated by comparing the normal heat flux from the heating element to the half of the granite block filled with ballotini and the half filled with TCG. At the maximum, the heat flux from the heating rod to the TCG filled half was 38 % higher and on the average, it was 24 % higher than the flux to the half with ballotini. The plots of simulated heat fluxes are depicted in Fig. 32.6.

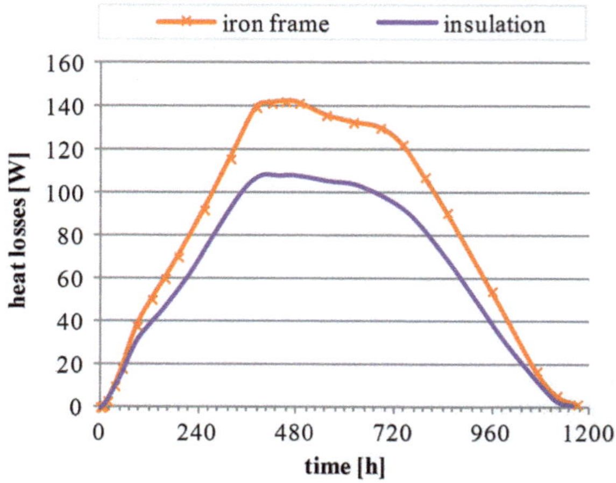

Fig. 32.7 Modelled heat losses through the surface of the insulation (2.013 m^2) and the iron frame (0.5198 m^2)

The other evaluated fluxes correspond to heat losses (Fig. 32.7) from the surfaces of the thermal insulation and uncovered parts of the iron frame. The area of the thermal insulation is 2.01 m^2 and the area of the uncovered frame is 0.52 m^2. Even though the frame surface is four times smaller, the heat loss is higher. The frame significantly affects the heat fluxes, because it represents a heat bridge through the insulation layer.

32.4 Discussion

The observed temperatures 88–111 °C during the anomaly occurrence can suggest problems related to the residual water. Another explanation could be based on heat contact destruction between the rod, TCG and the granite, caused by different thermal expansion of these materials. Prior to this experiment, TCG alone was tested with a cyclic thermal load and no structural changes were observed. The granite block has not been disassembled yet, because more cyclic heating tests will be performed.

The thermometers consist in a 35 mm long metal rod, but model probes are dimensionless points. If the thermometer is placed in a steep temperature gradient, its size can lead to averaging of the surrounding temperatures. This could explain remaining differences between the simulation and measured data.

Adding a thermally resistive layer to the surface of the heating rod in the model corresponds to a possible loss of thermal contact, which is assumed to occur because of the constant temperature rise while the power input was decreasing. The need of the 30 % reduction of the TCG conductivity function could be assigned to the contact resistance between granite block and TCG or other uncertainties.

32.5 Conclusion

The heating experiment was performed on a granite block with an artificial fissure filled with TCG and ballotini. Unexpected temperature development was observed at the installed thermometers. When we neglected some of the measurements, it was possible to reproduce the experiment with finite element simulation and evaluate the heat fluxes in each half. Adding a 15 mm TCG layer to 300 mm granite block increased the heat flux by 24 % on average compared to the ballotini-filled half, although the model estimations indicate conductivity decline in the application to cca 15 W m^{-1} K^{-1}, compared to 22 W m^{-1} K^{-1} measured at the TCG samples. More experiments should be performed on TCG application to identify the cause of anomaly at 94th hour of the experiment.

Acknowledgments This work was done within research project "Reversible Storage of Energy" funded by Technology Agency of the Czech Republic (Project No. TA01020348).

References

1. A.H. Abedin, M.A. Rosen, Assessment of a closed thermochemical energy storage using energy and exergy methods. Appl. Energy **93**, 18–23 (2012)
2. COMSOL—Multiphysics Modeling and Simulation Software, Multiphysics Modeling and Simulation Software (2013). http://www.comsol.com
3. L.F. Cabeza, C. Castellón, M. Nogués, M. Medrano, R. Leppers, O. Zubillaga, Use of microencapsulated PCM in concrete walls for energy savings. Energy Build. **39**(2), 113–119 (2007)
4. R. Dunn, K. Lovegrove, G. Burgess, A review of ammonia-based thermochemical energy storage for concentrating solar power. Proc. IEEE **100**(2), 391–400 (2012)
5. D. Fernandes, F. Pitié, G. Cáceres, J. Baeyens, Thermal energy storage: "How previous findings determine current research priorities". Energy **39**(1), 246–257 (2012)
6. A.I. Fernandez, M. Martínez, M. Segarra, I. Martorell, L.F. Cabeza, Selection of materials with potential in sensible thermal energy storage. Sol. Energy Mater. Sol. Cells **94**(10), 1723–1729 (2010)
7. A. Gil, M. Medrano, I. Martorell, A. Lázaro, P. Dolado, B. Zalba, L.F. Cabeza, State of art on high temperature thermal energy storage for power generation. Part 1—Concepts, materials and modellization. Renew. Sustain. Energy Rev. **14**(1), 31–55 (2010)
8. X. Py, N. Calvet, P. Echegut, C. Bassada, R. Olives, A. Meffre, E. Veron, S. Ory, Recycled material for sensible heat based thermal energy storage to be used in concentrated solar thermal power plants. J. Solar Energy Eng. **133**(3) (2011)
9. P. Rálek, H. Hokr, Numerical simulation of temperature and stress fields in the rock heating experiment, in *COMSOL Conference 2012 Milan Proceedings* (COMSOL Inc., USA, 2012)
10. S. Saeid, R. Al-Khoury, F. Barends, An efficient computational model for deep low-enthalpy geothermal systems. Comput. Geosci. **51**, 400–409 (2013)
11. R. Tamme, D. Laing, W.-D. Steinmann, Advanced thermal energy storage technology for parabolic trough. J. Sol. Energy Eng. **126**(2), 794–800 (2004)

Chapter 33
Transient Response of Different Highly Conductive PCM Composites

Mohammed H. Mahmoud Alhamdo, Bashar A. Bdaiwi and Ali H. Hasan

Abstract In this work, the thermal conductivity of paraffin wax has been enhanced by employing four different high conductivity additives infiltrated within the PCM. These include the use of Graphite Powder (GP), Copper Particles (CP), Aluminum oxide Particles (AP), and Copper Network (CN). Four different types of pure waxes were selected, tested and compared. Twelve samples of wax/additives composites were prepared by adding different mass ratios of 3, 6 and 9 % of additives by weight in each type of wax. The storage system contains spherical capsules filled with composite PCMs that are packed in an insulated cylindrical storage. Transient temperature based governing equations have been developed and solved numerically by both ANSYS FLUENT 14 code and by numerical implicit time marching model. With progress of time, results showed that the numerical predictions of ANSYS software start to deviate from the experimental observations. The grade-B paraffin was found to be the best one. Results indicate that all the enhancement methods have significant effect on the thermal response of the system. However, the utilization of 6 % additives by weight has been found to provide the best enhancement effect. The developed new-sort CN composite is found to produce the best thermal response due to its good homogeneity with wax and its high conductivity. Results showed that for CN composite with 6 % additives, the charging and discharging time decreased by 26.4 and 30.3 % respectively than that of pure wax and the thermal conductivity enhanced by 2.57 times that of pure wax.

M. H. Mahmoud Alhamdo (✉) · B. A. Bdaiwi · A. H. Hasan
Department of Mechanical Engineering, College of Engineering, Al-Mustansirya University, Baghdad, Iraq
e-mail: profmohamedh@yahoo.com

B. A. Bdaiwi
e-mail: phdbashar@yahoo.com

A. H. Hasan
e-mail: mscalih@yahoo.com

A. Y. Oral et al. (eds.), *International Congress on Energy Efficiency and Energy Related Materials (ENEFM2013)*, Springer Proceedings in Physics 155,
DOI: 10.1007/978-3-319-05521-3_33,

Nomenclature

C	Specific heat
CA	PCM capsules
H	Enthalpy
K	Thermal conductivity
L	Storage length
LH	Latent heat
P	Pressure
r	Radial coordinate inside PCM capsule
R	Radius of PCM capsule
r_0	Radius of the storage
t	Time
T	Temperature
u	Fluid axial velocity
$\vec{V}$	Velocity vector
WF	Working fluid
x	Axial direction

33.1 Introduction

The use of phase change materials (PCMs) for Latent Heat Energy Storage (LHES) has received a great interest in recent years. Among the PCMs proposed, paraffin wax attracts considerable attention. However, in spite of desirable properties of paraffin wax, the low thermal conductivity is its major drawback decreasing the rate of heat stored and released during melting and solidification processes.

Various techniques have been used to increase the thermal response of PCMs. These have included the use of pin and plate fin heat sinks in the PCM system [1–3]. Other methods to improve thermal response have included the use of metalled foams with PCM [4–6]. Ukrainczyk et al. [7] reported that large difference exist between paraffin wax thermo physical properties values available in open literatures. For example, the thermal conductivity of wax is reported of even about 100 % variations in different studies. Recently, Maheswari and Reddy [8] reported that the reason for difference between the theoretical and experimental results is the material thermal properties of the PCM. Thus the material thermal properties of the PCM should be well known in order to obtain accurate results with the numerical methods [8].

So, while there have been many studies of various techniques to improve paraffin PCM thermal response, there has been no comprehensive study which tested each of the common LHES styles under identical operating conditions. Therefore, the objectives of this research are to:

1. Establish a self-adopted thermo-physical properties of available waxes and composites used as PCMs.
2. Investigate the thermal effects of various types of additives that infiltrated within PCM with different mass fraction ratios.
3. Make series of transient numerical simulations by using both ANSYS 14.0 code and implicit-model to account for the thermal gradient inside spherical PCM capsules packed in a cylindrical storage with air as the working fluid.
4. Build an experimental facility to validate the results obtained from the numerical simulations during charging and discharging cycles.

33.2 Numerical Modeling

Figure 33.1 shows the physical representation of the LHES. A cylindrical storage is randomly packed with PCM spherical capsules. The working fluid (air) flows through the porous spacing between capsules.

The FLUENT CFD software by ANSYS is used to simulate and analyze the phase change processes of LHES. The governing conservation equations are:

The continuity equation:

$$\nabla \overrightarrow{V} = 0 \tag{33.1}$$

The momentum equation:

$$\rho_{WF} \cdot \frac{D\overrightarrow{V}}{Dt} = -\nabla P + \mu \nabla^2 \overrightarrow{V} \tag{33.2}$$

where ∇P is the pressure gradient within the velocity field.

The energy equation of the working fluid is:

$$\rho_{WF} \cdot \frac{DH_{WF}}{Dt} = K_{WF} \nabla^2 T_{WF} + S \tag{33.3}$$

where S is the dissipation function that includes the energy transformed into heat due to the fluid shear stress. The energy equation of the composite PCM is:

$$\rho_{CA} \frac{DH_{CA}}{Dt} = K_{CA} \nabla^2 T \tag{33.4}$$

where the enthalpy of the PCM:

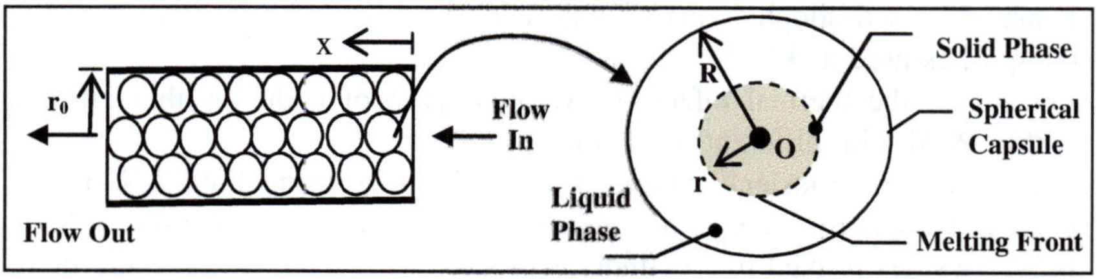

Fig. 33.1 Physical representation of the numerical modeling

$$H_{CA} = H_{sensible} + H_{latent} \tag{33.5}$$

And

$$H_{sensible} = H_0 + C_{CA}(T - T_0) \tag{33.6}$$

H_0 and T_0 are arbitrary reference values, since only enthalpy change is required for analysis.

$$H_{latent} = \beta \cdot LH_{CA} \tag{33.7}$$

where β is the liquid fraction ($0 \leq \beta \leq 1$). In order to fully solve for the above system of equations, there will be a need to initial and boundary conditions:

At $t = 0$; $T_{WF} = T_{initial}$, $T_{CA} = T_{initial}$ and $V = 0$
At $t > 0$; $V_{inlet} = u_{x,inlet}$

For the exit flow, there is no need to define any condition, since ANSYS extrapolates the data from interior domain. After creating the geometry, mesh generation tool that included in the FLUENT software workbench is used to generate the model mesh. The SIMPLE (Simi-Implicit Method for Pressure-Linked Equations) algorithm was utilized for solving the governing equations.

For time marching implicit method, the energy equations, and initial and boundary conditions are;

$$\varepsilon \cdot \rho_{WF} \cdot C_{WF}\left(\frac{\partial T_{WF}}{\partial t} + u\frac{\partial T_{WF}}{\partial x}\right) = Kx_{WF} \cdot \frac{\partial^2 T_{WF}}{\partial x^2} + Kr_{WF}\left[\frac{\partial^2 T_{WF}}{\partial r_0^2} + \frac{1}{r_0}\frac{\partial T_{WF}}{\partial r_0}\right] + h \cdot a_C\left(T_{CA} - T_{WF}\right) \tag{33.8}$$

$$\rho_{CA} \cdot C_{CA} \cdot \frac{\partial T_{CA}}{\partial t} = K_{CA}\left[\frac{\partial^2 T_{CA}}{\partial r^2} + \frac{2}{r}\frac{\partial T_{CA}}{\partial r}\right] + h \cdot a_C\left(T_{WF} - T_{CA}\right) \tag{33.9}$$

At $t = 0$; $T_{WF} = T_{WF,\ initial}$ [for $0 \le x \le L$] and $T_{CA} = T_{CA,\ initial}$ [for $0 \le r \le R$; $0 \le x \le L$]

At $t > 0$; $T_{WF} = T_{WF,\ initial}$ [for x=0] and $\frac{\partial T_{WF}}{\partial x} = 0$ [for x=L]

$\frac{\partial T_{CA}}{\partial r} = 0$ [for r=0] and $k_{CA} \frac{\partial T_{CA}}{\partial r} = h \cdot (T_{WF} - T_{CA}|_{r=R})$ [for r=R ; $0 \le x \le L$]

Where (ε) is porosity, (a) is surface area and (h) is heat transfer coefficient. The solution of the set of above equations is accomplished with a time marching implicit method. The iteration accuracy was selected to be;

$$| T_{CA(x,t+1)} - T_{CA(x,t)} | = 0.0001$$

33.3 Experimental Work

The experimental investigation is divided into three major parts; obtaining the thermo physical properties of PCMs, investigating the melting front inside single capsule and investigating the transient thermal response during charging and discharging processes for packed capsules system.

The thermal conductivity of paraffin wax has been enhanced by employing four different high conductivity additives infiltrated within the PCM. These include the use of Graphite Powder (GP), Copper Powder (CP), Aluminum oxide Particles (AP) and fabricated Copper Network (CN).

Four different types of pure waxes were selected, tested and compared. Two of them produced by major Iraqi oil company (Al-Durra refinery) named as grade-A and grade-B, the third one is a commercial Iranian product (KCC, Co.) and the fourth is natural bee wax (Haman Co., Germany).

Twelve samples of wax/additives composites were prepared by dispersion different mass ratios of 3, 6 and 9 % of additives by weight in each type of liquid wax. To ensure stability and homogeneity of the composite, it's mixed by hand for approximately five minutes and then sonicated for three hours using ultrasonic processor. For the Copper Network (CN), the capsule is filled with CN and then poured the liquid wax to it directly.

A self-adopted detailed experimental study is presented to determine thermo physical properties of each type of pure wax and composite under investigation. The melting point of waxes has been obtained by using (Stuart) melting point device; the supply heating to the sample is at a rate (0.1 °C/s). During density measurements of liquid and solid phases, a significant change occurs during the phase change process. This leads to a significant contraction upon solidification. Therefore, during encapsulation the PCM occupies only 88 % of the capsule and the upper segment of 12 % contains air. The viscosity measurements were done by using BROOKFILLD Viscometer with a spindle rotate speed of 100 rpm. A calorimeter is used to measure the specific heat and latent heat of fusion. The measured data for pure waxes are shown in Table 33.1.

Table 33.1 Measured properties of pure waxes

Wax type	Grade-A	Grade-B	Commercial	Natural
Melting point (°C)	60.4	63.7	61.5	69.5
Solid density (kg/m^3)	787.9	796	797.8	831.3
Liquid density (kg/m^3)	765.4	766.5	767.1	789.8
Viscosity (kg/m s)	0.0116	0.0127	0.0123	0.0165

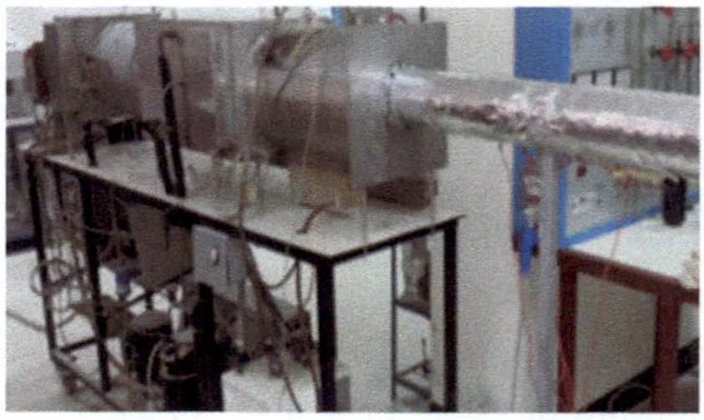

Fig. 33.2 Photo of the experimental setup

The thermal conductivity of pure and composite PCM was done based on the ASTM standards. The steady state test method is taken primarily from ASTM C177-04 and ASTM C518-02 standard methods. All samples were cut and polished to dimension of 2.5 cm thick and 10 cm radius. After around 25 min, the steady state is achieved. The validity of the measurement obtained from the steady state manufactured device has been checked by repeating the measurement using computerized thermal conductivity measuring system (TK-04, Germany). This system is based on the transient standard ASTM D5930 and ASTM D5334. Results obtained from the two devices were nearly convergence to each other (except for the CN/PCM composite).

Figure 33.2 represent a photo of the experimental facility that consist of 3-phase variable speed air blower, 1,000 W pre-heater, vapour compression refrigeration system, four 500 W electrical heaters, ducting system with mixing baffles.

The PCMs are encapsulated in plastic capsules of an inner diameter of 39 mm with a wall thickness of 0.5 mm. Totally 80 spherical capsules are filled with (25 g) of PCM have been packed inside the cylindrical test section. Eighteen thermocouples of (K type) with data logger were used to record the temperatures of composite PCM capsules and air at five different locations along the axial and radial direction in the test section. The phenomenon of wax melting is captured by high resolution digital camera that fixed in a suitable position in front of the test rig. A glass capsule that had an outer diameter of 50 mm and wall thickness of 1 mm was placed in a controlled heated air flow stream. In addition to providing visualization photographs, temperature readings were obtained using seven thermocouples that were placed on a horizontal and vertical planes passing through sphere center. To check the reputability of the data, all experiments were performed at least twice (on different days). The error associated with primary experimental measurements of the temperature of PCM, working fluid and the mass flow rates were done using the root sum square method. The estimated values for uncertainties a temperature is 1.072 % and for mass flow rate is 0.53 %.

33.4 Results

The propagation of melting fronts of grade-A pure wax are shown in Fig. 33.3. It can be seen that the top half melts at a faster rate than the bottom half for pure wax and the solid PCM has sunk to the bottom of the capsule due to heavier density of the solid PCM. It is concluded that free convection dominates in the top side of capsule, while heat conduction occurs only at the bottom half.

Figure 33.4 shows the transient melting fronts of grade-B/CN composite. It is observed that the melting occurs equally around the solid PCM inwards the center of the sphere. Also, it is clear that the CN holds the solid PCM and restricted it from sinking to the capsule bottom. For other types of additives (GP, CP and AP), no visualization photos were possible, since the color of composite has been changed to dark color. As compared to pure wax a greater amount of wax has been melted with the same boundary conditions due to larger amount of heat conduction. It is found that due to presence of 6 % CN by weight, the observed melting period was reduced by 19.23 % for grade-A wax, 21.05 % for grade-B wax, 17.02 % for commercial wax and 19.11 % for natural wax.

Figure 33.5 show the ANSYS numerical results for grade-A wax with 6 % additives by weight. Figure 33.6 shows a comparison between the numerical and experimental results obtained for temperature history of grade-A wax/6 % GP composite. Also, Fig. 33.7 shows comparison between experimental and numerical simulations inside capsule contains grade-A wax/9 % AP composite. The results indicate that the ANSYS numerical predictions start to deviate from the experimental observation with the progress of time. However, it should be noted that the

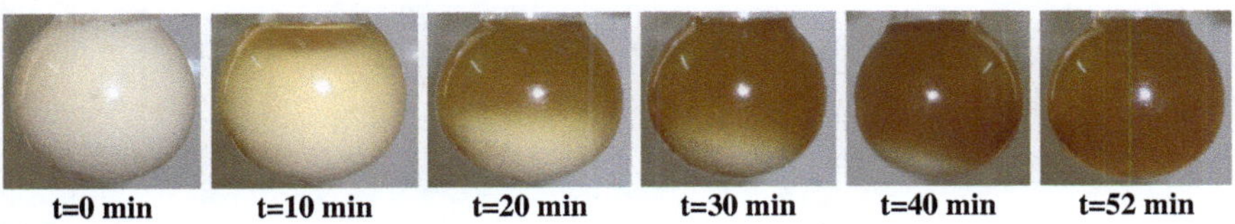

Fig. 33.3 Visualization of the transient melting front of pure grade-A wax

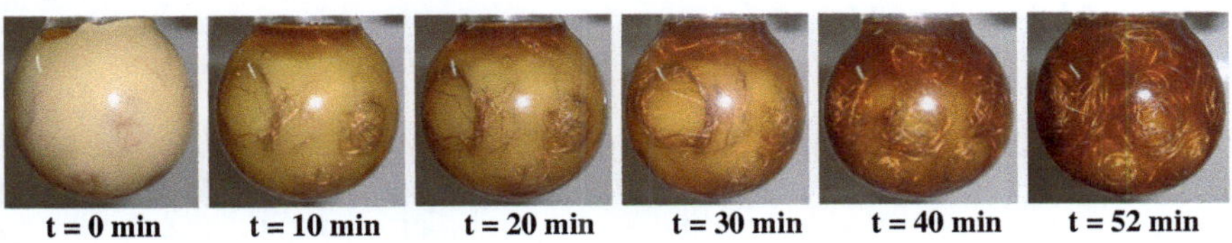

Fig. 33.4 Visualization of transient melting front of grade-B wax/CN composites

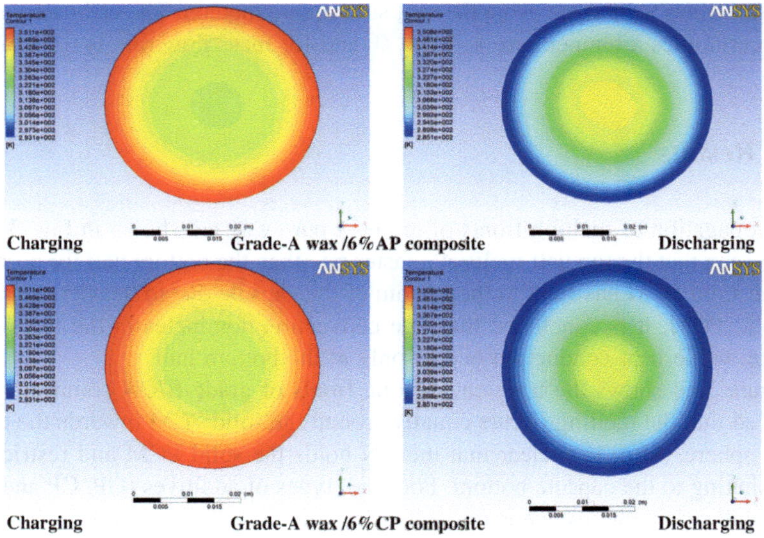

Fig. 33.5 ANSYS results for composite grade-A wax during melting and solidification

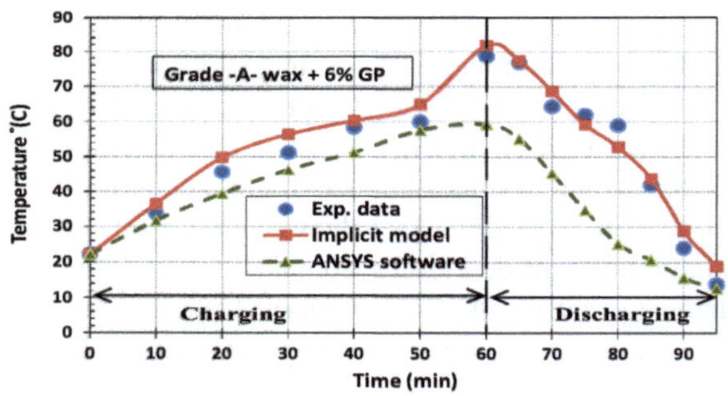

Fig. 33.6 Comparison between experimental and numerical results of melting and solidification processes

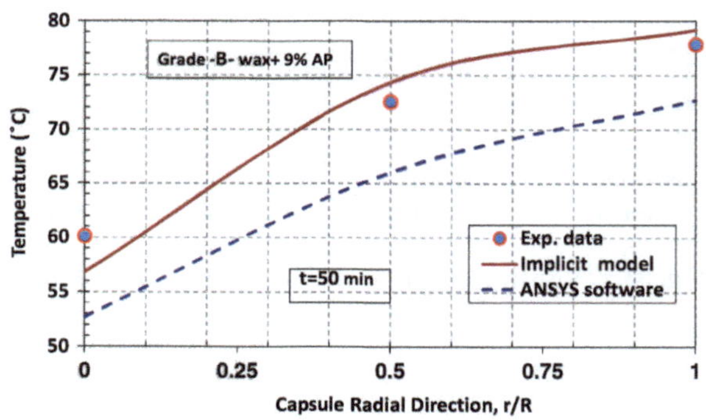

Fig. 33.7 Comparison between experimental and numerical results of melting process inside single capsule

PCM solid shape predicated in this work by ANSYS software (Fig. 33.5) were agree with the results just obtained recently from ANSYS software by Maheswari and Reddy [8]. However, it is clear that the developed implicit model is able to predict the thermal response more accurately than ANSYS code.

The thermo-physical properties of paraffin wax dispersed with 0, 3, 6 and 9 % of additives by weight are plotted as a function of mass fraction in Figs. 33.8 and 33.9. Results show that the specific heat and latent heat of composite PCM is lower than the pure wax and its value will reduced with the increase of the mass fraction of additives. As can be seen from results, the specific heat has drastic drop when the CN mass fraction reaches 9 %.

Figure 33.10a–d shows the variation of thermal conductivity of pure and composite PCMs with respect to mass fraction of additives. For all types of additives under investigation, results revealed that composite thermal conductivity increases with increasing of additives rate in the PCM. The utilization of 6 % additives by weight has been found to provide the best enhancement effect. This mass fraction is found to significantly increase the composite thermal conductivity with little decrease in specific and latent heats and good homogeneity with waxes to form composite PCMs. Grade-B paraffin wax has been found as the most preferred type due to its compatibility with all additives under investigation and due to its high latent heat and high thermal conductivity.

Figure 33.11 shows the thermal effect of adding different infiltrated high conductivity materials, with various mass ratios within grade-B wax. Results indicate that all the enhancement method have significant effect on the thermal response of the system. However, it is found that the CN composite improve the rate of heat transfer during both charging and discharging modes even if the mass fraction of the CN is just as low as 3 % with the PCM. The developed new-sort CN composite, that using combined sensible and latent heat storage concept is found to produce the best thermal response due to its good homogeneity with wax and its high conductivity due to their continuous inter-connected structure. The AP composite is found to exert little effect on melting/solidification processes and slight enhancement in thermal conductivity of the waxes under investigation.

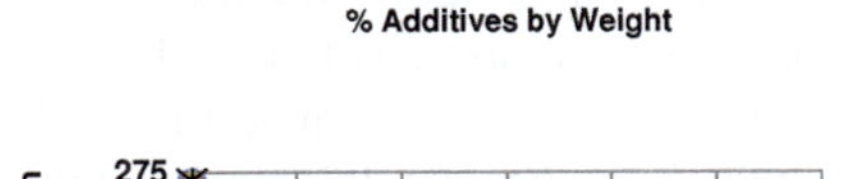

Fig. 33.8 Liquid specific heat of different grade-B wax composites

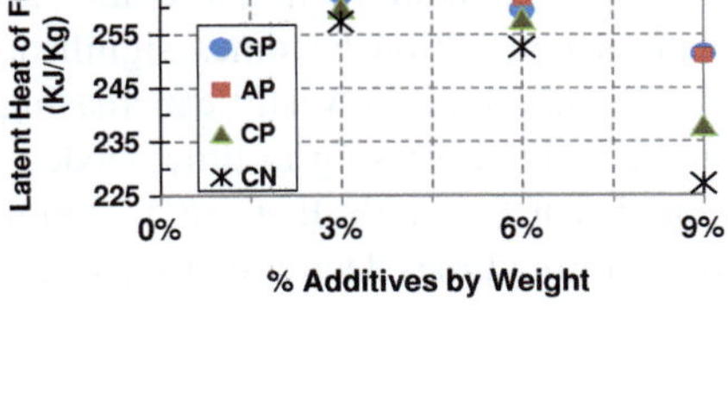

Fig. 33.9 Latent heat of different grade-B wax composites

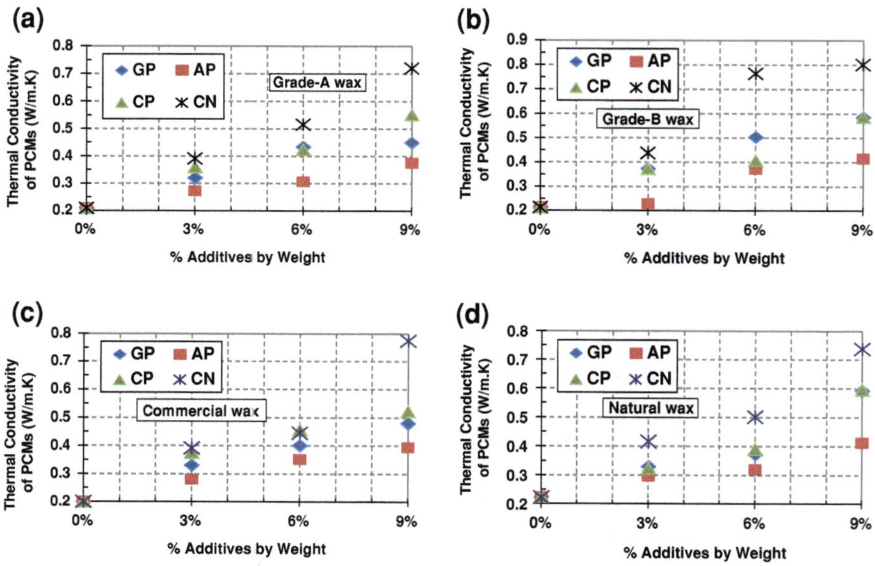

Fig. 33.10 **a** Thermal conductivity for various composites of grade-A wax. **b** Thermal conductivity for various composites of grade-B wax. **c** Thermal conductivity for various composites of commercial wax. **d** Thermal conductivity for various composites of natural wax

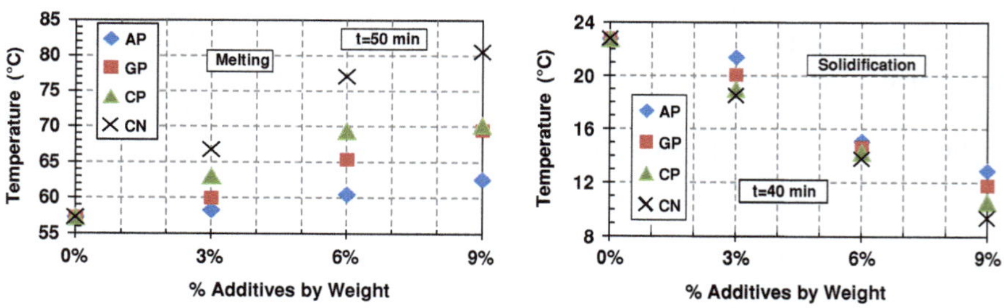

Fig. 33.11 Thermal effect of adding various mass ratios of different materials within grade-B wax

Figure 33.12 shows the melting and solidification of packed capsules contain composites composed of grade-B wax with CN of 3, 6 and 9 % by weight. It can be noted that the phase change time reduces with the increase of the CN mass ratio within the composite. It is clear that thermal response enhancement is greater during solidification than melting as conduction plays a greater role in solidification while natural convection becomes significant during melting. The increase to 9 % by weight additives show a little increasing of thermal conductivity than of those of 6 % and a little decreasing in time needed to complete melting and solidification. It is found that for grade-B wax/CN composite with 6 % additives, the charging and discharging times decreased by 26.4 and 30.2 % respectively than that of pure wax.

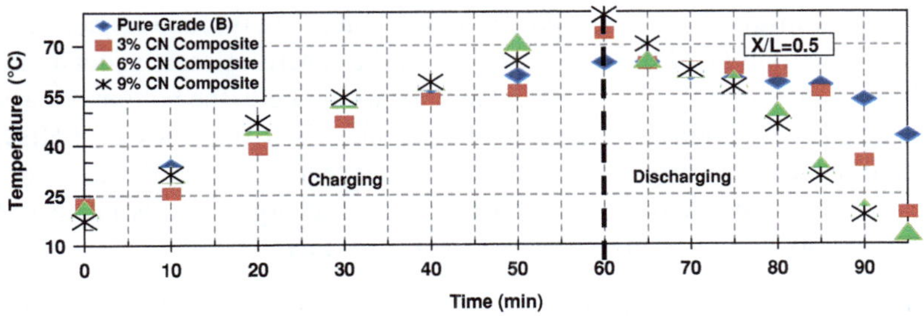

Fig. 33.12 Melting and solidification profiles for different CN composite packed capsules

33.5 Conclusions

Experimental and numerical investigation has been carried out to find the thermal conductivity enhancement of paraffin different waxes composites by employing four different high conductivity additives infiltrated within waxes. Based on the findings of the present work, the following conclusions were drawn:

1. Among the four types of waxes under investigation, grade-B paraffin wax has been found as the most preferred type due to its compatibility with all additives under investigation and due to its high latent heat and high thermal conductivity.
2. Compared to the results of pure PCMs, the thermal response of all composite PCMs is enhanced by embedding additives into it. It has been found that the selection of 6 % additives by weight has been found to provide the most suitable enhancement effect. This optimal value is found to utilize maximum advantage of both conduction and heat capacity effects in the composite.
3. The comparisons between experimental and numerical results indicate that the predictions of ANSYS start to deviate from experimental data with the progress of time. The implicit model was able to predict the response more accurately.
4. To eliminate errors of wax thermal properties that previously addressed in literature, a self-adopted experimental investigation has been conducted on each type of pure and composite waxes under investigation. It is found that increasing the mass fraction of CN additives from 3, 6 and 9 % by wt increase the thermal conductivity of grade-B wax by 105.14, 257.47 and 274.76 % respectively.
5. The developed new-sort CN composite is found to produce the best thermal response due to its good homogeneity with wax and its high conductivity. The AP composite is found to exert little effect on melting/solidification processes and slight enhancement in thermal conductivity.
6. For melting process of pure waxes inside single spherical capsule, it is found that conductive heat transfer dominates during early periods only. After that, and because of molten wax, the free convection is dominated in the top half of the

spherical capsule. The melting process of CN composite is found to occur equally around the solid PCM inwards the center of the sphere. Also, it's found that the CN holds the solid PCM and restricted it from sinking to the bottom of capsule.
7. Cyclic analysis indicates that the charge and discharge rates of LHES are greatly enhanced using composites. It's found that for grade-B wax/CN composite with 6 % additives, the charging and discharging times decreased by 26.4 and 30.2 % respectively than that of pure wax.

This work shows a great potential of utilizing the new-sort of CN composite as a PCM in LHES systems.

References

1. N. Zheng, R.A. Wirtz, A hybrid thermal energy storage device, Part 2: thermal performance figures of merit. J. Electron. Packag. **126**, 8–13 (2004)
2. U. Stritih, An experimental study of enhancement heat transfer in rectangular PCM thermal storage. Int. J. Heat Mass Transf. **47**, 2841–2847 (2004)
3. M.J. Huang, P.C. Eames, B. Norton, Phase change materials for limiting temperature rise in building integrated photovoltaic. Sol. Energy **80**, 1121–1130 (2006)
4. K. Nayak, S. Saha, K. Srinivasan, P.Dutta, A numerical model for heats sinks with phase change materials and thermal conductivity enhancers. Int. J. Heat Mass Transfer **49** (2006)
5. S. Krishnan, J.Y. Murthy, S.V. Garimella, Analysis of solid-liquid phase change under pulsed heating. ASME J. Heat Transfer **129**, 395–400 (2007)
6. A. Siahpush, J. O'Brien, J. Crepeau, Phase change heat transfer enhancement using copper porous foam. ASME J. Heat Transfer **130** (2008)
7. N. Ukrainczyk, S. Kurajica, J. Sipusie, Thermophysical comparison of five commercial paraffin waxes as latent heat storage materials. Chem. Biochem. Eng. **24**, 129–137 (2010)
8. C. Maheswari, M. Reddy, Thermal analysis of thermal energy storage system with phase change material. Int. J. Eng. Res. Appl. (IJERA) **3**, 617–622 (2013)

Chapter 34
Energy Harvesting in the Microwaves Spectrum Using Electrically Small Resonators

Mohammed R. AlShareef and Omar Ramahi

Abstract This paper presents energy harvesting improvements utilizing electrically small resonators. A single bow-tie antenna and a split ring resonator (SRR) working at the same frequency of 5.8 GHz are designed. The bow-tie antenna achieves about 12 % power efficiency, whereas the SRR achieves more than 37 %. A new efficiency term is also proposed to take into account the footprint reduction and efficiency advancement resulting from SRR elements. Two arrays of SRRs and bow-tie antennas are designed to occupy identical footprints. Microwave power harvesting efficiency of 60 and 15.2 % for the SRRs and bow-tie arrays, respectively, are achieved.

34.1 Introduction

Recently, microwave energy harvesting has received much attention, especially in Wireless Power Transmission (WPT) applications. In WPT, previous works utilized rectenna systems that consist of three major elements: the antenna, the rectifying circuit and the load to trap and harvest microwave radiation [1–7] and convert it to DC power. Generally, any advancement has been focused on improving the conversion efficiency rather than energy harvesting. Moreover, the potential radiation that can be harvested or received by means of a classical antenna is very small owing to the harvesting capability of the antenna. Hence, rectenna arrays with different configurations have been introduced [2, 8–10] to increase rectenna system power transfer capability and gain better power harvesting efficiency. Although one rectenna system using 20 × 20 array rectenna elements demonstrated about 75 % DC conversion efficiency, the entire efficiency

M. R. AlShareef (✉) · O. Ramahi
Department of Electrical and Computer Engineering, University of Waterloo,
Waterloo, ON N2L 3G1, Canada
e-mail: mohammed.alshareef@uwaterloo.ca

A. Y. Oral et al. (eds.), *International Congress on Energy Efficiency and Energy Related Materials (ENEFM2013)*, Springer Proceedings in Physics 155,
DOI: 10.1007/978-3-319-05521-3_34,

from the transmitter to the receiver was only about 33 % [8]. Employing an effective collector will, therefore, be extremely advantageous in enhancing the overall efficiency of the WPT systems. In [11], electrically small resonators, which are much smaller (in size) than their free-space wavelength, were recently reported showing an improvement of power trapping efficiency. In this paper, a comparison between a bow-tie antenna and SRR in terms of power harvesting is presented. Furthermore, a new efficiency definition is proposed, taking into account the footprint miniaturization.

34.2 Power Harvesting Using Classical Antenna and SRR

Since the most effective element in power harvesting systems is the collector or harvester, this work is devoted to enhancing the power collecting efficiency of the harvester. Thus, a bow-tie antenna and an SRR structure are compared as harvester elements. The SRR is designed on a Rogers substrate with a dielectric constant ($\varepsilon_r = 2.2$) and thickness ($h = 0.787$ mm), and simulated using HFSS Ansoft [12]. The geometric dimensions of the resonator are optimized to obtain its resonance frequency at 5.8 GHz, where the arm length $l = 6.3$ mm, arm width $w = 1.1$ mm, and the gap $g = 0.8$ mm as shown in Fig. 34.1.

The microstrip bow-tie antenna has been selected to compare with the SRR due to its low profile. This antenna is designed to operate at the next higher ISM band (5.8 GHz) to miniaturize the aperture size without compromising the element efficiency as claimed in [4]. In [13], a closed-form formula to calculate the resonance frequency, which is based on the bow-tie antenna geometry, is developed as follows:

$$f_r = \frac{c \times k_1}{2\pi\sqrt{\varepsilon_e}}$$
$$\varepsilon_e = \left(\frac{\varepsilon_r + 1}{2}\right) + \left(\frac{\varepsilon_r - 1}{2}\right)\left(1 + \frac{12h}{W_e}\right)^{1/2} \tag{34.1}$$

$$W_e = \frac{W + W_c}{2} \tag{34.2}$$

where ε_e is the effective permittivity, W_e is the effective bow-tie side width, and k_1 is the eigenvalue and obtained by using the Rayleigh-Ritz method:

$$k_1 = 2\sqrt{\delta_1 (W + W_c)/(L^2(3W + W_c))} \tag{34.3}$$

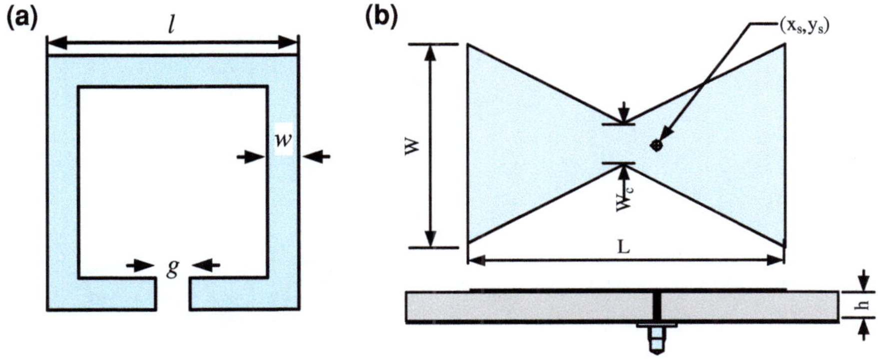

Fig. 34.1 Schematic representations of the **a** split ring resonator, and **b** bow-tie microstrip antenna. Note that schematic is not drawn to scale

where δ_1 is a coefficient and chosen as $\delta_1 = 3.45$. The calculated dimensions of the bow-tie antenna are as follows: $W = 9$ mm, $W_c = 3$ mm, and $L = 13.81$ mm, a Rogers 5,880 substrate with a dielectric constant (ε_r) of 2.2 and thickness (h) of 0.787 mm is used in this design. A 50 Ω coaxial line probe feed located at $x_s =$ 0 mm and $y_s = 0.75$ mm is used to excite the antenna. The structure geometry is shown in Fig. 34.1 The reflection coefficient S_{11} of the designed bow-tie antenna is illustrated in Fig. 34.2a. The antenna operates at 5.8 GHz and its bandwidth, which is determined at the 10 dB return loss, is 1.1 %. This narrow bandwidth may be attributed to the thin substrate and compact size of the antenna.

A comparison of the power efficiency of the two structures is conducted to show the power harvesting improvement using the SRR inclusion, where a 50 Ω resistive load is placed over the coaxial line feeding port of the bow-tie antenna. A plane-wave perpendicular to the mictrostrip plane is then shone to excite the antenna. By applying the Poynting vector, the time-average power is expressed as

$$P = \frac{1}{2}\oint \mathbf{S_{av}} \cdot ds = \frac{1}{2}\oint Re[\mathbf{E} \times \mathbf{H}^*] \cdot ds \tag{34.4}$$

where $\mathbf{S_{av}}$ is the average power density, and for an electric field of a uniform plane wave **E**, the average power density can be calculated as

$$P_D = \frac{|\mathbf{E}^2|}{2\eta}. \tag{34.5}$$

The power harvesting efficiency of the bow tie antenna loaded by a 50 Ω resistance is plotted in Fig. 34.2b. The power efficiency is computed using the HFSS calculator by dividing the power dissipated on the resistive load surface over

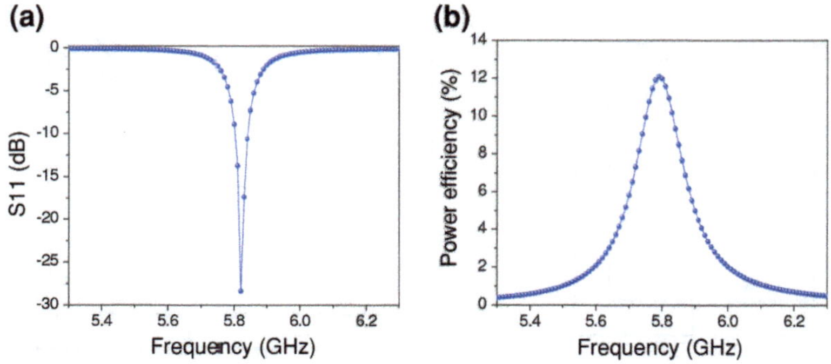

Fig. 34.2 Simulated **a** reflection coefficient (S_{11}), and **b** power efficiency of the bow-tie microstrip antenna loaded by 50 Ω

the power density (P_D) multiplied by the cross-sectional area (A) of the air box, thus:

$$\eta_{eff} = \frac{P_{diss}}{P_D \times A} \tag{34.6}$$

Although the maximum power efficiency occurs at the resonance frequency (5.8 GHz) as expected, this antenna, which has been used in rectenna systems, shows a low power harvesting capability. In other words, the highest efficiency that can be achieved by this bow-tie antenna is only about 12 %.

The same previous steps are applied to the SRR structure to prove the power harvesting enhancement; however, the SRR has a different characteristic impedance. This impedance is much higher than 50 Ω since it depends largely on the SRR geometry, such as the arm length and the gap width. Thus, to achieve the highest possible power efficiency, the resonator should be loaded with a resistance that matches the input impedance. Hence, it is found that the optimal resistance value of the SRR designed at 5.8 GHz is 2,000 Ω. A resistive sheet is placed across the SRR gap, where the highest electric field intensity is observed. A substantial power efficiency improvement is realized using the SRR, in which more than 37 % power efficiency is achieved. In other words, the resonator harvesting efficiency improvement is more than double that of the bow-tie antenna system. Figure 34.3 shows the harvesting power efficiency when using the SRR loaded by different resistance values.

34.3 Power Harvesting Efficiency Per Unit Area

In addition to enhancing harvesting efficiency through use of electrically small resonators, SRR also allows footprint miniaturization, because the coupling that occurs between classical antennas does not arise. Therefore, a new efficiency term

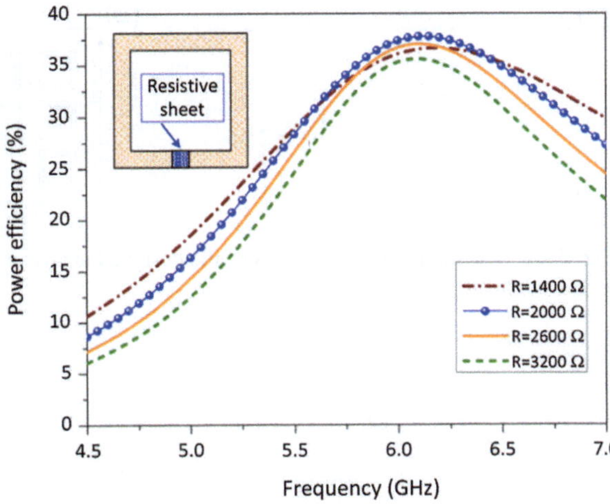

Fig. 34.3 Power efficiency of the SRR loaded by different resistances

is proposed to take into account the size reduction as well. For that reason, two comparable arrays of SRRs and bow-tie antennas are made to show SRR's miniaturization and efficiency enhancement.

For the first case, an array composed of two bow-tie antennas is designed at ISM band 5.8 GHz. In order to prevent any coupling or cross talking between the two antennas, the separation between them is varied until the coupling becomes suppressed. It is found that the minimum separation is about $\lambda/2$, at which point the coupling effect becomes negligible. The geometry and dielectric material of the two bow-tie antennas used in the array are similar to that used in the individual one shown in the previous section. A second array, having a footprint identical to the bow-tie antenna array is made of 21 SRRs. Figure 34.4 depicts the two arrays of the bow-tie and SRRs occupying the same area A of $L \times W = 80$ mm $\times$ 30 mm.

The two arrays shown in Fig. 34.4 are excited by two different orientations of a plane-wave, depending on the array that is excited. For the bow-tie array, an orthogonal plane-wave (direction of propagation k perpendicular to the antenna plane (xy-plane) is shone; then, a power dissipated on the loaded resistance (50 Ω) for the two antennas is calculated. As illustrated in Fig. 34.5, about 15 % power efficiency is achieved, which is approximately identical to that of the single bow-tie antenna, at the resonance frequency (5.8 GHz). With the same area of the previous array, 21 SRRs occupy this array, and each resonator is loaded by a 2,000 Ω, which is an optimal resistance as introduced before. Instead of calculating the power efficiency, the harvested power amount is computed since it is difficult to estimate the exact amount of incident power on each SRR due to the

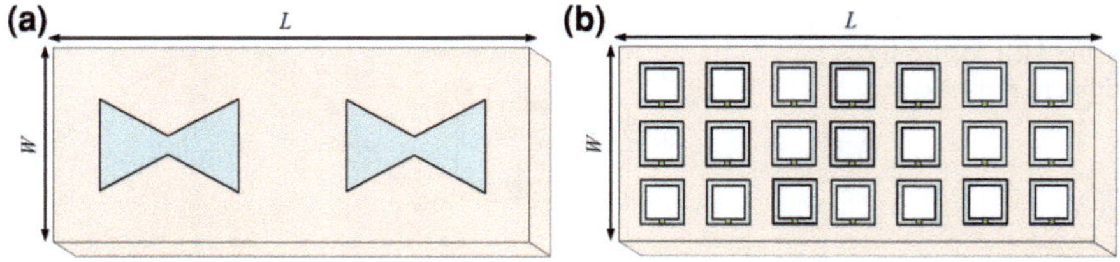

Fig. 34.4 Schematic representations of the designed arrays of **a** bow-tie microstrip antenna, and **b** SRR element

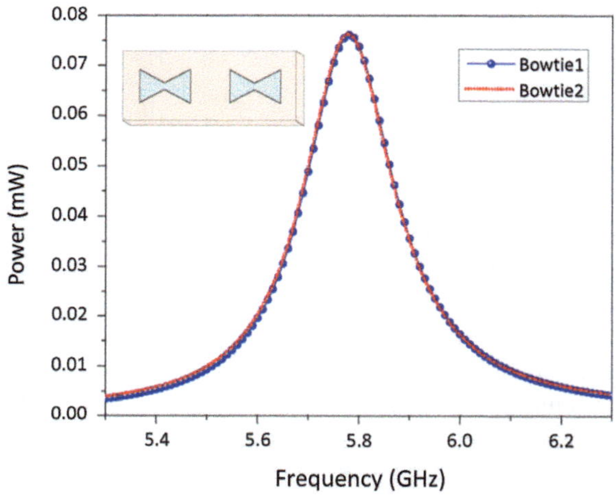

Fig. 34.5 Power collected by the array of bow-tie antennas loaded by 50 Ω

mutual coupling among them. Figure 34.6 shows the power that can be collected by a single SRR if 1 mW is shone over the array. Figure 34.6a shows the power collected by the first row in the array, while Fig. 34.6b and c show the power collected by the second and the third rows, respectively.

Table 34.1 summarizes the potential power that can be harvested for the two arrays. If a plane-wave carrying 1 mW is applied on each structure, the total amount of power collected at resonance frequency is 0.597 mW and 0.153 mW for SRRs and bow-tie arrays, respectively.

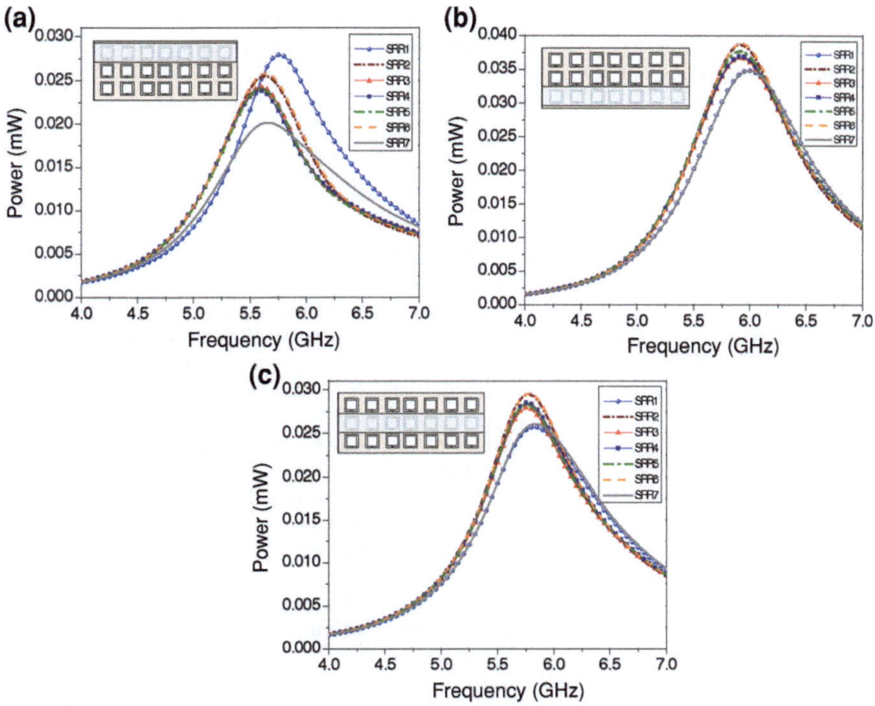

Fig. 34.6 Power collected by the array of SRRs loaded by 2,000 Ω. **a** Frist row, **b** second row, **c** third row

Table 34.1 The amount of power harvested for bow-tie and SRR arrays

Array type	Average power (mW)	No. of structures	Total power (mW)
Bow-tie	0.0761	2	0.152
SRR	0.0284	21	0.597

34.4 Conclusions

A comparison between a bow-tie antenna and split ring resonator (SRR) was made to show the efficiency improvement in power harvesting. SRR showed a substantial power efficiency improvement over that of the classical antenna. The bow-tie antenna achieved about 12 % power efficiency, whereas the SRR achieved more than 37 %. Moreover, a new efficiency definition was proposed to take into account the size reduction achieved by utilizing the resonator coupling concept, which in turn, increases the total power efficiency. Therefore, two arrays having the same area, both designed at 5.8 GHz, produce 0.152 and 0.597 mW, as power harvested for the bow-tie antennas and SRRs, respectively.

Acknowledgments The authors would like to thank King Abdulaziz City for Science and Technology (KACST) and the Natural Sciences and Engineering Research Council of Canada (NSERC) for supporting this work.

References

1. Y.-H. Suh, K. Chang, A high-efficiency dual-frequency rectenna for 2.45- and 5.8-GHz wireless power transmission. IEEE Trans. Microw. Theory Tech. **50**, 1784–1789 (2002)
2. Y.-J. Ren, K. Chang, New 5.8-GHz circularly polarized retrodirective rectenna arrays for wireless power transmission. IEEE Trans. Microw. Theory Tech. **54**, 2970–2976 (2006)
3. J. McSpadden, T. Yoo, K. Chang, Theoretical and experimental investigation of a rectenna element for microwave power transmission. IEEE Trans. Microw. Theory Tech. **40**, 2359–2366 (1992)
4. J. McSpadden, L. Fan, K. Chang, Design and experiments of a high-conversion-efficiency 5.8-ghz rectenna. IEEE Trans. Microw. Theory Tech. **46**, 2053–2060 (1998)
5. Y.-J. Ren, K. Chang, Bow-tie retrodirective rectenna. Electron. Lett. **42**, 191–192 (2006)
6. B. Strassner, K. Chang, 5.8-GHz circularly polarized dual-rhombic-loop traveling-wave rectifying antenna for low power-density wireless power transmission applications. IEEE Trans. Microw. Theory Tech. **51**, 1548–1553 (2003)
7. Y.-H. Suh, K. Chang, IN *A Novel Dual Frequency Rectenna for High Efficiency Wireless Power Transmission at 2.45 and 5.8* GHz. Microwave Symposium Digest, 2002 IEEE MTT-S International, vol. 2 (2002), pp. 1297 –1300
8. D.-G. Youn, Y.-H. Park, K.-H. Kim, Y.-C. *Rhee,* in *A Study on the Fundamental Transmission Experiment for Wireless Power Transmission System.* TENCON 99. Proceedings of the IEEE Region 10 Conference, vol. 2 (1999), pp. 1419–1422
9. N. Shinohara, H. Matsumoto, Experimental study of large rectenna array for microwave energy transmission. IEEE Trans. Microw. Theory Tech. **46**, 261–268 (1998)
10. Y.-H. Park, D.-G. Youn, K.-H. Kim, Y.-C. Rhee, in *A Study on the Analysis of Rectenna Efficiency for Wireless Power Transmission,* TENCON 99. Proceedings of the IEEE Region 10 Conference, vol. 2 (1999), pp. 1423–1426
11. O.M. Ramahi, T.S. Almoneef, M. AlShareef, M.S. Boybay, Metamaterial particles for electromagnetic energy harvesting. Appl. Phys. Lett. **101**(17), 173903 (2012)
12. A. Inc., http://www.ansys.com
13. X.-P. Zhang, S.-S. Zhong, in *Resonant Frequencies and Dual-Frequency Operation of a Bow-tie Microstrip Antenna.* Antennas and Propagation Society International Symposium, IEEE, vol. 2 (2002), pp. 60–63

Chapter 35
3D Blade Vibration Measurements on an 80 m Diameter Wind Turbine by Using Non-contact Remote Measurement Systems

Muammer Ozbek and Daniel J. Rixen

Abstract Non-contact optical measurement systems photogrammetry and laser interferometry are introduced as cost efficient alternatives to the conventional wind turbine/farm monitoring systems that are currently in use. The proposed techniques are proven to provide an accurate measurement of the dynamic behavior of a 2.5 MW—80 m diameter—wind turbine. Several measurements are taken on the test turbine by using 4 CCD cameras and 1 laser vibrometer and the response of the turbine is monitored from a distance of 220 m. The results of the infield tests and the corresponding analyses show that photogrammetry (also can be called as videogrammetry or computer vision technique) enable the 3D deformations of the rotor to be measured at 33 different points simultaneously with an average accuracy of ±25 mm, while the turbine is rotating. Several important turbine modes can also be extracted from the recorded data. Similarly, laser interferometry (used for the parked turbine only) provides very valuable information on the dynamic properties of the turbine structure. Twelve different turbine modes can be identified from the obtained response data.

35.1 Introduction

Contemporary testing and monitoring strategies require accelerometers or strain gauges to be used inside the blade or tower for dynamic measurements performed on wind turbines [1]. However, these measurement systems are sensitive to

---

M. Ozbek (✉)
Faculty of Mechanical, Maritime and Materials Engineering, Engineering Dynamics Section, Delft University of Technology, Mekelweg 2, 2628 CD Delft, The Netherlands
e-mail: ozbek.muammer@yahoo.com; m.ozbek@tudelft.nl

D. J. Rixen
Institute of Applied Mechanics, Technische Universität München, Boltzmannstr. 15, 85748 Garching, Germany
e-mail: rixen@tum.de

A. Y. Oral et al. (eds.), *International Congress on Energy Efficiency and Energy Related Materials (ENEFM2013)*, Springer Proceedings in Physics 155,
DOI: 10.1007/978-3-319-05521-3_35,

lightning and electro-magnetic fields. Besides, some extra installations inside the blades such as placement of cables for power supply and data transfer are required for these applications. The signals from rotating sensors on the blades are transferred to stationary computer via slip rings or by radio/wireless transmission. For large commercial turbines the required installations and preparations (sensor calibration) may be very expensive and time consuming [1–3].

Due to the technical limitations in sensor installations, the conventional systems mentioned above can only be applied at certain locations on the turbine. Unfortunately, the last 10–15 m of the blades (close to tip) are not accessible and therefore cannot be instrumented. In practice, sensors are usually placed at the root regions of the blades. However, the response measured at the root region only may not provide useful information about the location and extent of a possible damage close to the tip of the blade. The reliability of the extracted information is directly related to the number of measurement points.

Fiber-optic strain gauges are proposed to be a promising alternative to accelerometers and conventional strain gauges since optical sensors are not prone to electro-magnetic fields or lightning. However, it is reported that some feasibility tests are still needed to ensure the effective and cost efficient use of this measurement system. The factors affecting the performance of fiber-optic sensors such as sensitivity to humidity and temperature variations and the corresponding error compensation methods should also be investigated further [4].

In this work, two non-contact optical measurement systems (photogrammetry and laser interferometry), which do not require any sensors to be installed in the structure, are proposed to be promising and cost efficient alternatives for measuring the vibration response of wind turbines. Unlike conventional measurement systems (accelerometers, piezo-electric or fiber-optic strain gauges), optical measurement techniques do not require any sensors to be placed on the turbine. Therefore, no additional preparations such as cable installations for power or data transfer are needed inside the blade or tower. However, some reflective markers should be placed (or painted) on the structure. These markers are made up of a retro-reflective material, which is 1,000 times more reflective than the background blade material. Since the markers are in the form of very thin stickers (with a diameter of 400 mm) they do not have any effect on aerodynamic performance of the blades. During the tests, a total of 55 markers were placed on the turbine (11 markers for each blade and 22 markers on the tower). Placement of the markers on the blade and their final distribution throughout the structure can be seen in Fig. 35.1a, b respectively. In photogrammetry, markers are used as targets to be tracked by the camera systems and all the targets can be tracked simultaneously.

During in-field tests, dynamic behavior of the turbine was monitored from a measurement distance of 220 m by using a modified PONTOS system consisting of four CCD cameras. Photogrammetric measurements were performed by GOM mbH (GOM Optical Measuring Techniques). Although photogrammetry is efficiently used in smaller scales by a wide variety of disciplines, this method was applied for the very first time to a MW scale wind turbine within the scope of this research project.

Fig. 35.1 a–b Placement of markers and their final layout on the turbine [1]

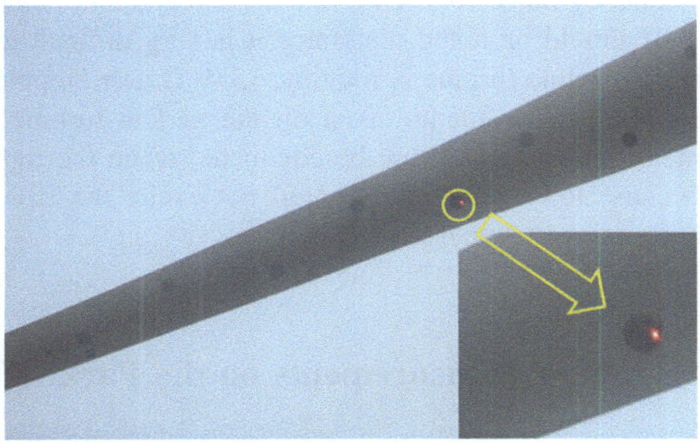

Fig. 35.2 Reflection of laser beam from the marker on the blade

In laser interferometry, a laser vibrometer continuously sends a laser beam to the target and receives the beam reflected from its surface. If the object is moving, this causes a frequency change and phase shift between the sent and reflected beams. By detecting this frequency change (Doppler principle), velocity of the moving object can be found. If the object itself has a reflective surface, no extra retro-reflective markers are needed. However, since the blade material was not reflective enough and the distance between the laser source and the turbine was very long, high quality laser signals could only be acquired if the laser was targeted to the markers. Once the quality of reflected laser beam is assured, laser vibrometer can measure the vibration of the blade with very high accuracy (even in micron level).

During the tests, laser interferometry measurements were taken by using a Polytec OFV 505 laser head and OFV 5,000 controller with VD06 velocity decoder. These systems were located in the field at a distance of 200 m from the turbine. An SLR (Super Long Range) lens which enables an increased measurement range up to 300 m was also required to take measurements from this distance. Figure 35.2 shows the reflection of laser beam from the marker on the blade. Different from photogrammetry, laser vibrometer can only measure the motion of a single point at a time. However, it is still possible to measure all the markers distributed throughout the blade successively. Although this approach increases test duration, it enables vibration characteristics of the structure to be acquired with the same spatial resolution as provided by photogrammetry.

A third system (used as a reference), which has already been installed in the turbine as a part of a long term wind load monitoring campaign, consists of 6 strain gauges placed at the root region of the three blades (2 gauges per blade) and 2 strain gauges located at the tower base. These strain gauges are used to measure flapwise and edgewise vibration of the blades and fore aft and side to side vibration of the tower at a sampling frequency of 32 Hz. All the data recorded by 3 different systems were then synchronized by using a GPS clock whose absolute time accuracy is approximately 10 ms. Considering the fact that frequencies that are expected to dominate the response of the wind turbine are mostly in low frequency range (0–5 Hz), this accuracy can be considered as quite sufficient.

It should be noted that since it is very difficult to keep the laser on the same marker while turbine is rotating, LDV (Laser Doppler vibrometer) was only used for the measurements taken on the parked turbine. Similarly, photogrammetric measurements could not be conducted when the turbine was at parked condition because low wind speeds could not excite the structure sufficiently resulting in high noise to signal ratios.

35.2 LDV Measurements on the Parked Turbine

Table 35.1 summarizes the modal parameters (frequencies) calculated by using strain gauge and LDV measurements. Considering the strain gauge signals, it can be easily seen that some frequencies can be identified either from in-plane strains or from out of plane strains only but not from both and that some other frequencies can only be identified from tower signals. It can also be seen that all the frequencies can be identified by analyzing LDV measurements. However, these frequencies cannot be detected from a single data block only. LDV measurement contains a single channel data recorded on a specific marker at a time. The targeted marker may not be at a suitable location to detect some of the modes (or frequencies). Therefore, it is required to try several locations (markers) and time series to identify all the modes. However, the frequencies identified by using two different systems (strain gauge and laser) are always in good coherence.

35.3 Photogrammetry Measurements on the Rotating Turbine

During in-field tests, dynamic behavior of the turbine was monitored from a measurement distance of 220 m by using four CCD cameras. The vibration characteristics of the rotor were measured at 33 different points simultaneously.

Table 35.1 Modal parameters calculated for the parked turbine

Mode	Blade strain out of plane	Blade strain in plane	Tower strain	LDV measurements
1st longitudinal tower	X		X	X
1st lateral tower			X	X
1st BW flapwise (yaw)	X			X
1st FW flapwise (tilt)	X			X
1st symmetric flapwise	X			X
1st BW edgewise (Vertical)		X	X	X
1st FW edgewise (Horizontal)		X		X
2nd BW flapwise (yaw)	X	X		X
2nd FW flapwise (tilt)	X	X		X
2nd symmetric flapwise	X			X
Tower torsion	X	X		X

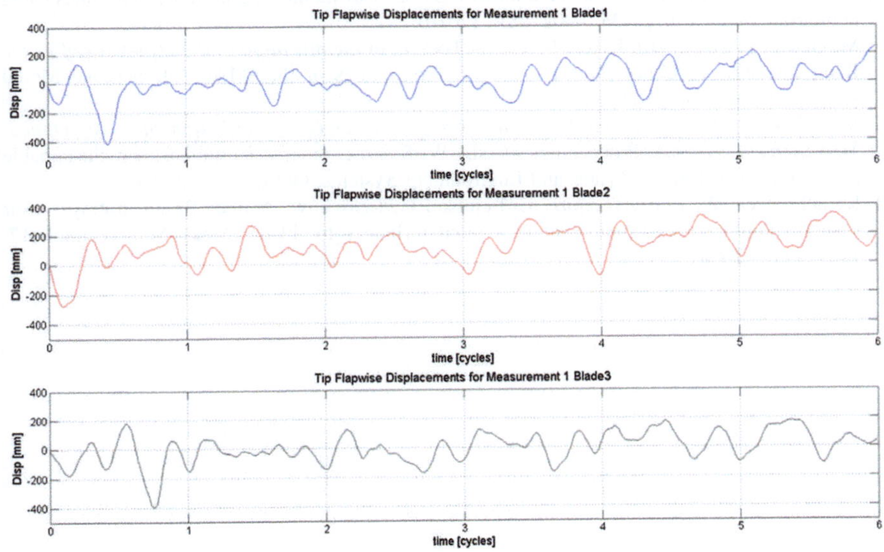

Fig. 35.3 Time normalized tip flapwise displacements records

Although photogrammetry is efficiently used in smaller scales by a wide variety of disciplines, this method was applied for the very first time to a MW scale wind turbine within the scope of this research project.

A typical displacement time history measured in flapwise direction for the tip markers of 3 blades can be seen in Fig. 35.3. In the figure horizontal axis represents the number of rotation cycles in the recorded data. It can be seen that the tip of the blade can experience a relative displacement up to 102.4 cm during rotation.

Analysis results showed that the deformations of the turbine can be measured with an average accuracy of ±25 mm from a measurement distance of 220 m [1].

35.4 Conclusions

Wind turbines have very specific characteristics and challenging operating conditions. Although the optical measurement systems (including both the hardware and image processing software), calibration methods and utilized operational modal analysis techniques were not specifically designed and optimized to be used for monitoring large wind turbines, the accuracy reached in this feasibility study is very promising. It is believed that this accuracy can easily be increased further by utilizing more specialized hardware and data processing methods.

References

1. M. Ozbek, D.J. Rixen, O. Erne, G. Sanow, Feasibility of monitoring large wind turbines using photogrammetry. Energy **35**, 4802–4811 (2010)
2. M. Ozbek, F. Meng, D.J. Rixen, MJL. van Tooren, in *Identification of the Dynamics of Large Wind Turbines by Using Photogrammetry*. Proceedings of the IMAC International Modal Analysis Conference, Jacksonville, 2010
3. M. Ozbek, D.J. Rixen, in *Wind Farm Monitoring Based on Computer Vision and Laser Optical Measurement Systems*. Proceedings of SDEWES 2012, the 7th Conference on Sustainable Development of Energy, Water and Environment Systems, Ohrid, 1–7 Jul 2012
4. K. Schroeder, W. Ecke, J. Apitz, E. Lembke, G. Lenschow, A fiber Bragg grating sensor system monitors operational load in a wind turbine rotor blade. Meas. Sci. Technol. **17**, 1167–1172 (2006)

Chapter 36
Aero-elastic Parameter Estimation of a 2.5 MW Wind Turbine Through Dynamic Analysis of In-Operation Vibration Data

Muammer Ozbek and Daniel J. Rixen

Abstract Aero-elastic parameters of a 2.5 MW—80 m diameter—wind turbine were extracted by using the in-operation vibration data recorded for various wind speeds and operating conditions. The data acquired by 8 strain gauges (2 sensors on each blade and 2 sensors on the tower) installed on the turbine was analyzed by using OMA (Operational Modal Analysis) methods while several turbine parameters (eigenfrequencies and damping ratios) were extracted. The obtained system parameters were then qualitatively compared with the results presented in a study from literature, which includes both aeroelastic simulations and in-field measurements performed on a similar size and capacity wind turbine.

36.1 Introduction

Growing energy demands require wind turbine manufacturers to design more efficient and higher capacity wind turbines which inevitably results in larger and larger new models to be put into service. However, an important consequence of this increase in size and flexibility of the structure is the complicated dynamic interaction between different parts of the turbine. Motion of the blades interacts with aerodynamic forces, electro-magnetic forces in the generator and the structural dynamics of several turbine components (drive train, nacelle and tower). Understanding these dynamic interactions, the corresponding structural behavior

M. Ozbek (✉)
Faculty of Mechanical, Maritime and Materials Engineering, Engineering Dynamics Section, Delft University of Technology, Mekelweg 2, 2628CD Delft, The Netherlands
e-mail: ozbek.muammer@yahoo.com; m.ozbek@tudelft.nl

D. J. Rixen
Institute of Applied Mechanics, Technische Universität München, Boltzmannstr 15, D-85748 Garching, Germany
e-mail: rixen@tum.de

A. Y. Oral et al. (eds.), *International Congress on Energy Efficiency and Energy Related Materials (ENEFM2013)*, Springer Proceedings in Physics 155,
DOI: 10.1007/978-3-319-05521-3_36,

and response characteristics is essential for optimizing the energy produced, ensuring safe and reliable operation and increasing the life-time of the system. Therefore, more attention is paid to developing theoretical models for estimating the behavior of new wind turbines.

Contemporary aeroelastic simulation tools coupled with structural dynamics models enable designers to detect, understand and solve most of the possible problems at very early stages and optimize their designs [1, 2].

Considering the fact that only the models based on real response measurements are able to represent the complicated interactions among different parts of the structure, several tests have been applied on both parked and rotating turbines. Although there are numerous studies conducted on wind turbines at parked condition, the information related to dynamic testing and modal analysis of MW scale large wind turbines during operation is quite limited. This work aims at making a contribution to this challenging field of experimental and operational modal analysis by presenting the results of the in-field vibration tests performed on a 2.5 MW—80 m diameter—wind turbine and the corresponding data analyses.

For this purpose, the dynamic response of the test turbine was monitored by using 3 different measurement systems namely, conventional strain gauges, photogrammetry and laser interferometry, while the turbine was both at parked condition and rotating. The recorded data was analyzed by using OMA methods and eigenfrequencies and damping ratios were extracted. The obtained system parameters were then qualitatively compared with the results presented in a study from literature [3], which includes both aeroelastic simulations and in-field measurements performed on a similar size and capacity wind turbine.

36.2 Analysis Results and Identified System Parameters

Researchers [2–5] agree on the fact that performing modal analysis on a rotating turbine is much more challenging than performing the same analysis on a parked turbine due to the facts that;

- For a rotating wind turbine some of the important turbine modes have very high aeroelastic damping ratios ranging between 10 and 30 % (in terms of critical damping ratio) which makes them very difficult to be detected by most of the identification algorithms that are currently in use.
- For a rotating turbine, integer multiples of rotational frequency (also called P harmonics where P denotes the rotational frequency) always dominate the response of the structure. These frequencies can be effective up to 24P and sometimes coincide with the real eigenfrequencies of the system [1, 2].
- Besides, some important requirements of OMA algorithms, time invariant system, and steady state random excitation assumptions are difficult to accomplish for rotating wind turbines because of the rotation of the blades and yawing, pitching motion of the turbine.

36.2.1 Tests on Parked Turbine: Strain and LDV

Within the scope of the research, twelve different turbine modes were successfully extracted from the measurements taken on the parked turbine by using strain gauges and LDV (laser Doppler vibrometer). Table 36.1 summarizes the modal parameters (frequencies and damping ratios) calculated by using strain gauge and LDV measurements. As can be seen in the table, frequency values are relatively stable and do not change depending on the measurement block analyzed. However, damping values may differ slightly. Since the turbine is kept at a fixed orientation during the tests, the relative angle between the effective wind direction and the normal of the rotation plane continuously changes depending on the instantaneous wind direction resulting in a different aerodynamic coupling for each measurement.

These modal parameters are very important for tuning and validation of numerical models and for the verification of prototype designs. They can also be used for health monitoring applications.

36.2.2 Tests on Rotating Turbine: Strain and Photogrammetry

This section summarizes the results of the analyses of strain gauge and photogrammetry measurements taken on the rotating turbine. During the test period, the response of the turbine was continuously measured by strain gauges. Therefore, modal parameters could be extracted for various operating conditions and wind speeds. Calculated modal parameters were then compared with the results presented by Hansen et al. [3] The work mentioned includes the results of both aeroelastic simulations performed by the stability tool HAWCSstab [6] and the real measurements taken on a wind turbine which has a similar size and capacity as the test turbine in our work.

Figure 36.1 shows the change in aeroelastic damping ratio calculated for side to side tower mode as a function of wind speed. Identified values are in a very good agreement with the HAWCStab simulation results both in terms of trend and magnitude. The values found are less than 1 % through different operating conditions and wind speeds.

The same comparison is made for the fore aft tower mode and the results are shown in Fig. 36.2. Although the two tower modes have almost the same frequencies, aero-elastic damping calculated for the fore aft mode is greater due to the motion of the tower in the direction perpendicular to rotor plane.

Comparison of aeroelastic damping ratios found for the first BW edgewise mode is shown in Fig. 36.3. The extracted damping ratios are slightly higher than the HAWCStab results, but are very close to the estimations [3] made by using in-field vibration data.

Table 36.1 Modal parameters calculated for the parked turbine

Mode	Frequency (Hz)	Damping
1st longitudinal tower	0.345	0.003
1st lateral tower	0.347	0.003—0.009
1st yaw (BW flapwise)	0.902	0.010—0.020
1st tilt (FW flapwise)	0.974	0.011—0.020
1st symmetric flapwise	1.077	0.010—0.020
1st vertical edgewise (BW)	1.834	0.004
1st horizontal edgewise (FW)	1.855	0.004
2nd tilt (FW flapwise)	2.311	0.005
2nd yaw (BW flapwise)	2.430	0.004
2nd symmetric flapwise	3.00	0.005
2nd edgewise	6.36	0.005
Tower torsion mode (needs further verification)	6.154	0.005

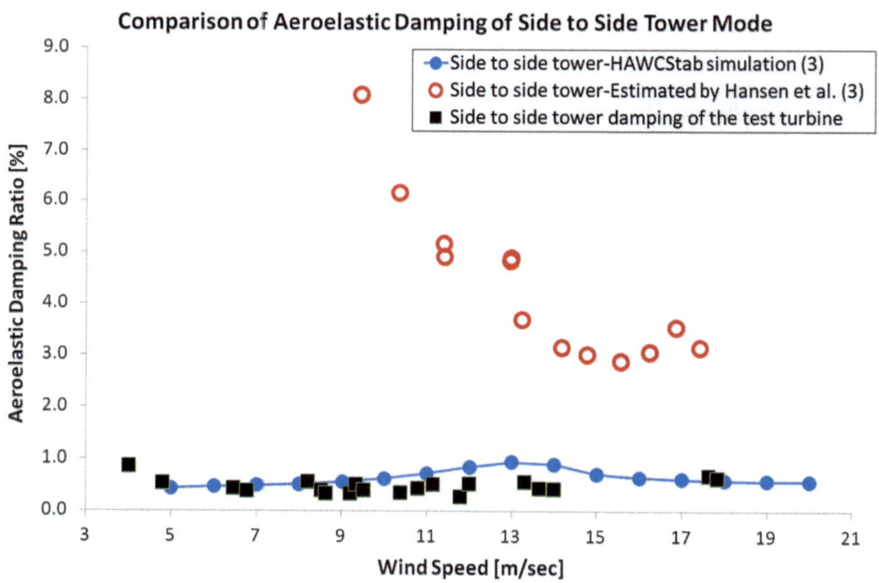

Fig. 36.1 Side to side tower mode damping comparison

Similarly, Fig. 36.4 displays the same damping comparison for the 1st edgewise FW mode. Acquired damping ratios are again very close to both simulations and estimations given by Hansen et al. [3].

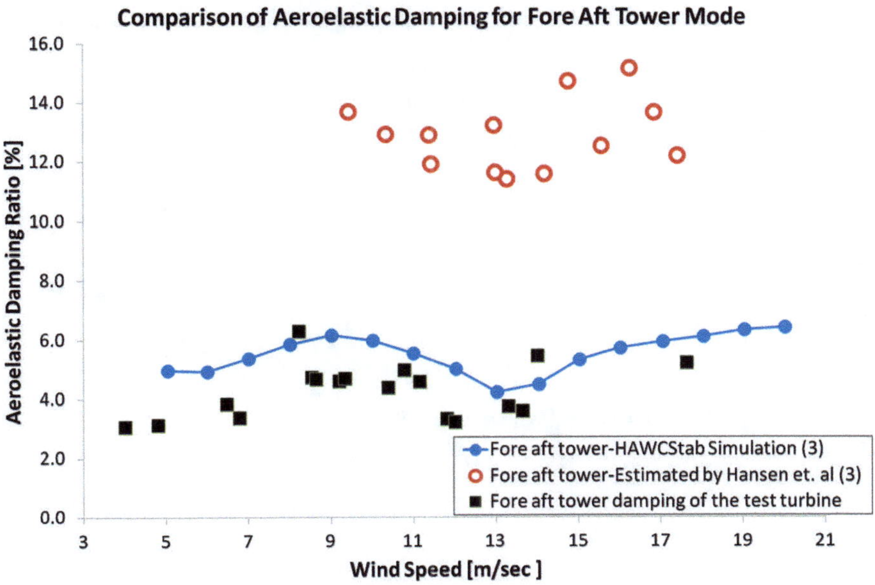

Fig. 36.2 Fore aft tower mode damping comparison

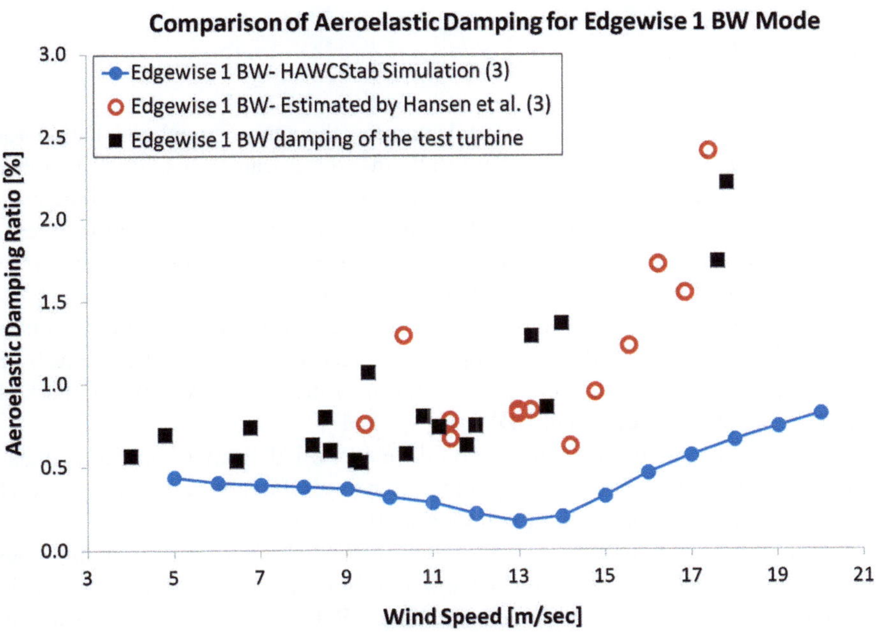

Fig. 36.3 The 1st edgewise BW mode damping comparison

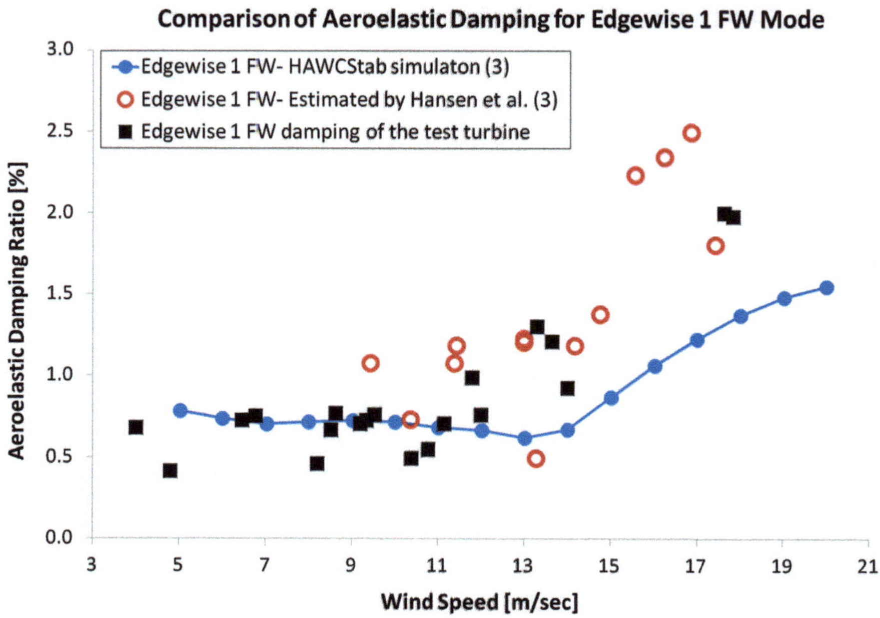

Fig. 36.4 The 1st edgewise FW mode damping comparison

36.3 Conclusions

OMA (Operational Modal Analysis) tools namely, the analysis methods that do not require the forces acting on the system to be measured, can be a solution to some important problems encountered in dynamic testing of wind turbines. Since estimation of the modal parameters is solely based on the use of measured response, these methods can easily be used to extract the dynamic properties of these large structures excited by natural environmental inputs (winds).

Analyses performed by using OMA methods seem very promising in extracting the modal parameters. Within the scope of the research, twelve different turbine modes were successfully calculated from the measurements taken on the parked turbine using strain gauges and LDV.

Similarly, several turbine modes could be identified from in-operation measurements using strain gauges and photogrammetry. Obtained results are in good coherence with those presented in similar studies in literature.

Performing modal analysis on a rotating turbine is much more challenging than performing the same analysis on a parked turbine due to the high aeroelastic damping of some important modes, rotational P harmonics that dominate the dynamic response and the difficulties in fulfilling some important system identification assumptions such as time invariant system and steady state random excitation.

During the analyses, it was observed that frequency values are more easily identified and the calculated values are mostly stable and reliable. However, significant scatter can be encountered in estimated damping ratios. This scatter can be caused by physical factors such as the change in operating conditions or mathematical uncertainty related to the applied algorithms.

References

1. M. Ozbek, D.J. Rixen, Operational modal analysis of a 2.5 MW wind turbine using optical measurement techniques and strain gauges. Wind Energy **16**, 367–381 (2013)
2. M. Ozbek, F. Meng, D.J. Rixen, Challenges in testing and monitoring the In-Operation vibration characteristics of wind turbines. MSSP Mech. Syst. Signal Process. (2013). doi: 10.1016/j.ymssp.2013.07.023
3. M.H. Hansen, K. Thomsen, P. Fuglsang, Two methods for estimating aeroelastic damping of operational wind turbine modes from experiments. Wind Energy **9**, 179–191 (2006)
4. S. Chauhan, M.H. Hansen, D. Tcherniak, in Application of operational modal analysis and blind source separation/Independent component analysis techniques to wind turbines. *Proceedings of IMAC International Modal Analysis Conference*, Orlando, FL, (2009)
5. M. Ozbek, D.J. Rixen, T.W. Verbruggen, in Remote monitoring of wind turbine dynamics by laser interferometry:Phase1. *Proceedings of IMAC International Modal Analysis Conference*, Orlando, FL, (2009)
6. M.H. Hansen, Aeroelastic stability analysis of wind turbines using an eigenvalue approach. Wind Energy **7**, 133–143 (2004)

Chapter 37
A Study of Energy Conversion Efficiency of a Savonius Type Wave Energy Converter System

Mustafa Tutar and Ceyhan Erdem

Abstract In the present study, two-dimensional, two-phase and turbulent flow around a horizontal axis 3-bladed Savonius rotor is considered. Numerical wave tank (NWT) simulations based on FVM/FDM technique in association with volume of fluid (VOF) element method are performed for specified values of wave heights for no-rotor flow case. Once validated against the theoretical data, the numerical simulations are extended to investigate the overall performance of the turbine over a very large range of wave height conditions for the rotor-flow case.

37.1 Introduction

Savonius type hydraulic turbines, in which the fluid energy is captured through a hydraulic mechanism rather than an aerodynamic mechanism, are considered to be simple, efficient with good starting capabilities and to operate at relatively low rotational speeds. There have been both experimental and computational studies [1–3] of energy efficiency and power performance analysis of Savonius turbines, which spin due to differential drag on the curved surface [2]. The net driving force, which can be attributed to the differential drag reproduced between the advancing blade(s) and the returning blade(s) can be increased by either reducing the reverse force on the returning blade(s) or increasing the positive force on the advancing blade(s).

M. Tutar (✉)
Department of Mechanical and Manufacture, Mondragon Goi Eskola Politeknikoa, Loramendi 4, 20500 Arrasate, Spain
e-mail: mtutar@mondragon.edu

M. Tutar
IKERBASQUE, Basque Foundation for Science, 48011 Bilbao, Spain

C. Erdem
Department of Aerospace Engineering, Middle East Technical University, 06531 Ankara, Turkey
e-mail: ceyhanerdem@gmail.com

A. Y. Oral et al. (eds.), *International Congress on Energy Efficiency and Energy Related Materials (ENEFM2013)*, Springer Proceedings in Physics 155,
DOI: 10.1007/978-3-319-05521-3_37,

The motion of the waves usually sets the water particles in orbital motion which can be considered to a combination of both longitudinal and traverse motions of water waves. In longitudinal motion, the water particles oscillate back and forth parallel to the wave propagation direction while in transverse motion, these particles oscillate up and down in their position. These two motions are later combined together to reproduce the overall orbital motion. The kinetic energy of the water particle's orbital motion can be used to rotate the blades [2]. The relative power performance of the Savonius rotor is usually determined with the shape and size of the orbital motion with respect to the rotor diameter and their shapes are subject to change depending on the wave length to water depth aspect ratio. Therefore, the study of orbital motion of the mechanic waves and their characteristics with respect to size and positioning of the rotor is essential for a better optimization of design of such devices.

The present work aims at studying the planar regular wave propagation and its interaction with a horizontal Savonius rotor using numerical methods in a numerical wave tank (NWT) at different wave heights. These studies construct a basis for further investigation of the effect of different governing parameters on the performance of such ocean wave energy conversion devices for further application.

37.2 Equation of Fluid Motion

The governing flow equations for the present 2-D turbulent flow behaviour are continuity, momentum i.e. Reynolds averaged Navier–Stokes (RANS) equations and turbulence transport equations. These conservation equations in non-linear differential form of vector notation for incompressible, viscous fluid flow conditions can be summarized below.

$$\frac{\partial \overline{u}_i}{\partial x_i} = 0 \tag{37.1}$$

$$\rho \frac{\partial \overline{u}_i}{\partial t_i} + \rho \bar{u}_j \frac{\partial \overline{u}_i}{\partial x_j} = -\frac{\partial \overline{P}}{\partial x_i} + \frac{\partial}{\partial x_j}\left(\mu \frac{\partial \overline{u}_i}{\partial x_j} - \rho \overline{u_i' u_j'}\right) + \rho g_i \tag{37.2}$$

In the above equations, ρ is the fluid density, $\bar{u}_i$ is the time averaged velocity, x_i is the coordinate direction, u_i' is the deviation from the time averaged velocity, $\bar{P}$ is the time averaged pressure, g_i is gravity acceleration, μ is the dynamic viscosity of the fluid, $-\rho\overline{u_i' u_j'}$ is the Reynold's stress tensor which is required to be modelled using a turbulence model for closure of RANS equations. The temporal and spatial co-ordinates correspond to t and x_i, respectively. In 2-D Cartesian coordinates the continuity and RANS equations can be re-defined by simply dropping the over bar for brevity for the time averaged quantity. In eddy viscosity based k-ε turbulence models, Reynolds stress tensor is related to the mean flow straining field for incompressible flow as below:

$$-\rho\overline{u_i'u_j'} = \frac{2}{3}\rho K\delta_{ij} + 2\mu_t\bar{S}_{ij}. \tag{37.3}$$

where k denotes the turbulence kinetic energy, μ_t is the eddy viscosity related to turbulence kinetic energy, k and its dissipation rate, ε, and $\bar{S}_{ij}$ is the time averaged strain rate tensor related to mean velocity gradient in the flow. The turbulent kinetic energy, k and its dissipation rate, ε for isothermal are then defined by the turbulence transport equations to determine the eddy viscosity term, which is used to calculate the Reynolds stress term to closure the RANS equations for the present RNG k-ε model.

For the present two-phase flow, the volume of fluid (VOF) element method initially proposed by Hirt and Nichols [4] is used for free surface tracking a surface in a fixed Eulerian mesh. In the VOF method, a single set of momentum equations is shared by the fluids and the volume fractions of each of the fluids in each computational cell are tracked through domain. Interface tracking where variables and properties in any given cell are purely represented by either one of phases or mixture of phases depending upon the volume fraction values:

$$\begin{cases} f_i = 0 & \text{The cell is empty (of } i\text{th cell)} \\ f_i = 1 & \text{The cell is full (of } i\text{th cell)} \\ 0 < f_i < 1 & \text{The cell contains the interface} \end{cases} \tag{37.4}$$

The fractional volume function is governed by a transport equation and this equation determines the movement of interface position such that:

$$\frac{\partial f}{\partial t} + v \cdot \nabla f = 0 \tag{37.5}$$

37.3 Computational Details

A 2-D schematic diagram of a NWT which is constructed as a representation of the experimental wave tank (EWT) study of Hindasageri et al. [5] is shown in Fig. 37.1. The Savonius rotor, which is placed at water sub-merged level of z = 0 m in the NWT, is generated with the same geometric dimensions in accordance with that used in the experiment. The boundaries are also illustrated in Fig. 37.1. At the left boundary, inflow boundary conditions are imposed to generate a wave train coming from the x = 0 m. A sinusoidal wave boundary condition with a Second-order Stokes wave formulation is implemented here.

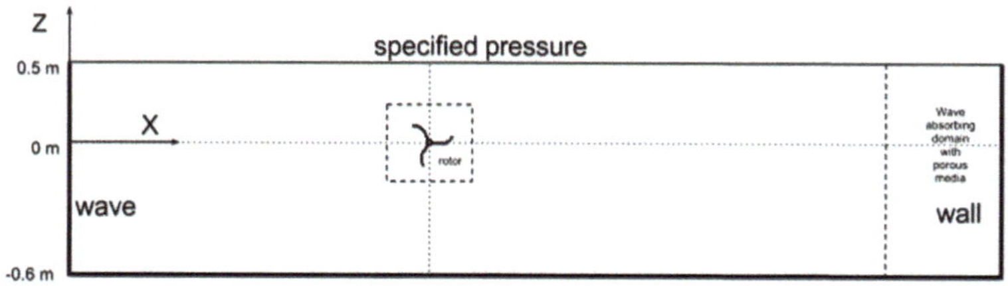

Fig. 37.1 A schematic of NWT generated in the present numerical study

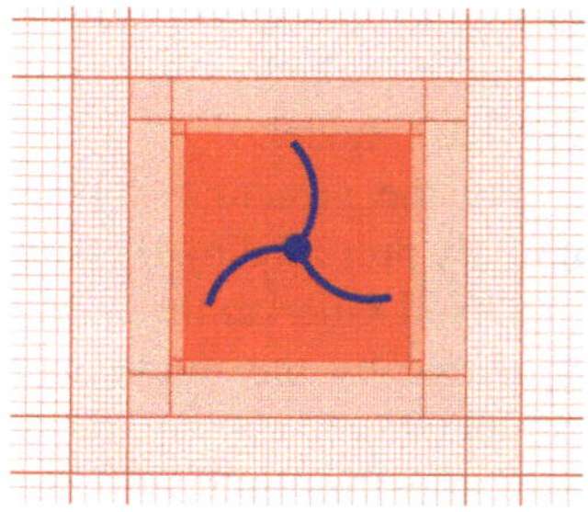

Fig. 37.2 The multi block meshes employed in the computational domain

Based on this formulation, at the inflow boundary, the following below velocity components are employed as inflow velocity boundary conditions [6]:

$$u = \frac{H}{2}\frac{gk}{\omega}\frac{\cosh k(h+z)}{\cosh kh}\cos(kx-\omega t) + \frac{3H^2\omega k}{16}\frac{\cosh 2k(h+z)}{\sinh^4(kh)}\cos 2(kx-\omega t) \quad (37.6)$$

$$w = \frac{H}{2}\frac{gk}{\omega}\frac{\sinh k(h+z)}{\cosh kh}\sin(kx-\omega t) + \frac{3H^2\omega k}{16}\frac{\sinh 2k(h+z)}{\sinh^4(kh)}\sin 2(kx-\omega t) \quad (37.7)$$

where H is the wave height, ω is the wave frequency, k is the wave number, and h is the mean wave depth. The wave absorbing domain-porous media on the other hand is defined in the flow exit zone to prevent the wave reflection into computational domain. No-slip boundary conditions at the bottom surface and the flow exit domain i.e. outflow boundary are also imposed. At the free surface kinematic and dynamic boundary conditions with the specified atmospheric pressure condition is imposed to make sure that no transport equations are resolved in the air region (i.e. air is not treated as a fluid but rather as avoid).

The multi-block meshing modeling is utilized for more efficient use of computational resources and fast flow solution. The computational domain is constructed with non-uniformly spaced 2-D quadrilateral Cartesian mesh elements (Fig. 37.2) with fine resolution near wall surfaces and interface between air and

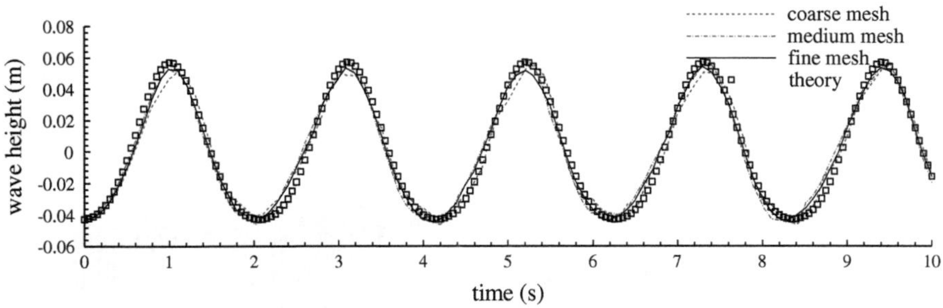

Fig. 37.3 Wave elevation history at a probe position of x = 30 m for a wave height of $H = 100$ mm

water to successfully resolve the air–water surface movement due to wave propagation and to improve the numerical accuracy for measuring velocity and pressure gradients.

A FDM/FVM based numerical flow modelling approach of FLOW-3D [7] is used to solve the partial differential equations (continuity, momentum and energy equations) governing rotor movement and surface tracking here. To increase the convergence rate, momentum equations and the pressure based continuity equation are also coupled with a pressure–velocity coupling scheme of Generalized Minimal Residual Solver (GMRES) scheme [8] and the first order upwind scheme for discretization of the momentum equation. The one fluid VOF model is chosen for free surface, the Fractional Areas/Volumes Obstacle Representation (FAVOR) is chosen for efficient geometry definition.

37.4 The Results and Discussion

Initial computations are performed for no-rotor flow problem with three different mesh resolutions (maximum number of cells is around 115,000) at wave height of H = 100 mm for wave period of 2.1 s, water height of 0.6 m, and a wave length of 4.62785 m. As seen in Fig. 37.3, the numerical data corresponds well with the theory for all mesh resolution with a better agreement with the final mesh configuration of around 115,000 cells, which is found to be in very good correspondence with the theory. The shape of the surface waves is found to be almost sinusoidal throughout the domain.

Distinguished phase contours and corresponding velocity vectors obtained for each case as in Fig. 37.4 can be attributed to differing wave propagation conditions and energy conversion rate with varying wave height. The maximum rotational speed obtained in the clock wise direction and slightly higher rotational speed is obtained as the wave height increases as an indication of positive effect of wave height increase on the rotational torque as shown in Fig. 37.5.

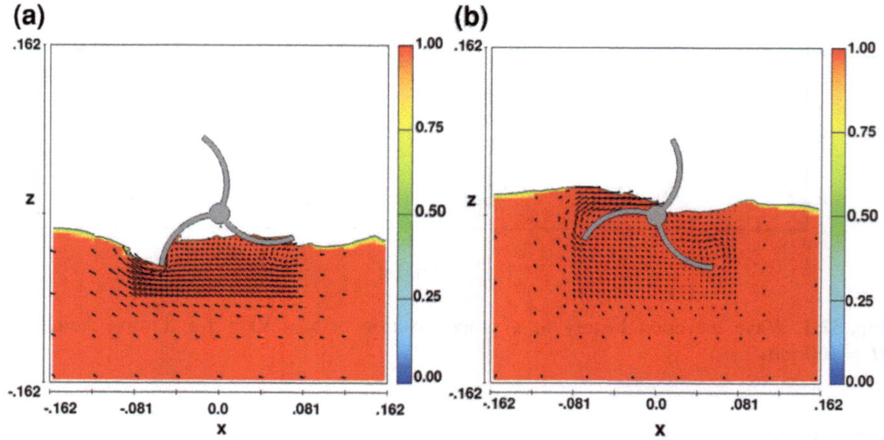

Fig. 37.4 Phase contours and local velocity vectors at $t = 12.1$ s; **a** $H = 100$ mm; **b** $H = 160$ mm

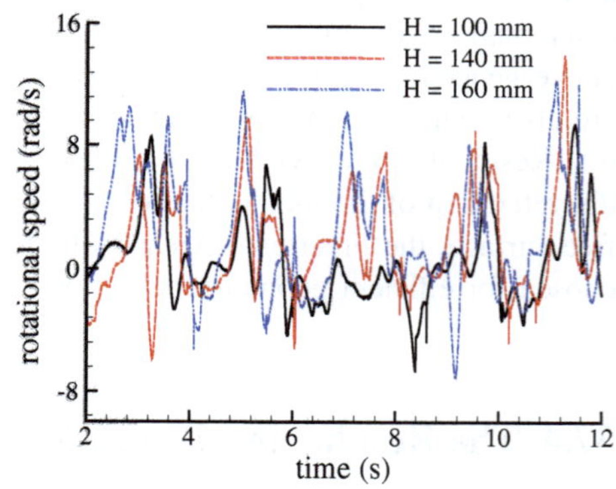

Fig. 37.5 Time evolution of rotational speed at different wave height values

37.5 Conclusions

The present numerical model successfully reproduces planar mechanic wave propagation. The higher wave heights the higher rotational speed of the Savonius rotor. Non-continuous flow through the rotor causes fluctuating rotational motion.

Acknowledgments This work has been funded under the project of the Gobierno Vasco-Basica y Aplicada-PI 2011-8.

References

1. M. Nakajima, S. Lio, T. Ikeda, Performance of double-step Savonius rotor for environmentally friendly hydraulic turbine. J. Fluid Sci. Tech. **3**(3), 410–419 (2008)
2. M. Faizal, M. Ahmed, Y.-H. Lee, On utilizing the orbital motion in water waves to drive a Savonius rotor. Renew. Energy **35**, 164–169 (2010)
3. M.A. Zullah, Y.-H. Lee (2012) Performance evaluation of a direct drive wave energy converter using CFD. Renew. Energy, 1–5
4. C.W. Hirt, B.D. Nichols, Volume of fluid (VOF) method for the dynamics of free boundaries. J. Comput. Phys. **39**, 201 (1981)
5. V. Hindasageri, H. Ramesh, S.C. Kattimani, Performance of Savonius rotors for utilizing the orbital motion of ocean waves in shallow waters. J. Sustain. Energy Environ. **2**, 117–119 (2012)
6. R.G. Dean, R.A. Dalrymple, *Water Wave Mechanics for Engineers and Scientists*, vol. 2 (World Scientific, Singapore, 1991)
7. FLOW 3D, User Manual Version 10.0.1, Flow Science Inc., Santa Fe (2012)
8. Y. Saad, M.H. Schultz, GMRES: a generalized minimal residual algorithm for solving nonsymmetric linear systems. SIAM J. Sci. Stat. Comput. **7**, 856–869 (1986)

Chapter 38
Impact of Phosphorus Diffusion Gettering on HEM Multicrystalline Silicon Wafers Taken from Different Ingot Regions

Nabil Khelifati, Djoudi Bouhafs, Seddik-El-Hak Abaidia, Abd El-Ghani Boucheham and Baya Palahouane

Abstract In this present contribution, we investigate the effect of extended phosphorus diffusion (PD) gettering on the electrical quality of p-type HEM mc-Si wafers. The study was made for three sets of wafers taken from top, center and bottom of the same ingot. The gettering process was performed through two annealing plateaus; the first is standard and the second is extended at different temperatures. Wafers characterization was mainly made by quasi-steady state photoconductance (QSSPC) technique. The obtained results indicated that the gettering effectiveness mainly depends on the position of wafers in the ingot and also the temperature of gettering process. Furthermore, appropriate modeling of QSSPC lifetime curves using Hornbeck–Haynes model reveals that the origin of the electrical property improvement is due to the reduction of the density of specific metallic impurities in the bulk.

38.1 Introduction

The multicrystalline silicon (mc-Si) wafers are widely used as precursor elements in solar cells manufacturing, and constitute more than half of the overall industrial market. This is due to their low cost of manufacturing as compared to that of monocrystalline silicon (c-Si) [1].

However, the high contamination level of mc-Si wafers by metallic impurities (Fe, Cr, Mn, Cu, Ni, Co, etc.) during the elaboration process is considered as their major disadvantage. In the case of cast mc-Si grown by Heat Exchanger Method

N. Khelifati (✉) · D. Bouhafs · A. E.-G. Boucheham · B. Palahouane
Research Center in Semi-conductors Technology for the Energetic, Bd. 2 Frantz Fanon, les sept merveilles, B.P. 140, Algiers, Algeria
e-mail: n.khelifati@gmail.com

N. Khelifati · S.-E.-H. Abaidia
University of M'hamed Bougara, Faculté des Sciences, Boumerdès 35000, Algeria

A. Y. Oral et al. (eds.), *International Congress on Energy Efficiency and Energy Related Materials (ENEFM2013)*, Springer Proceedings in Physics 155,
DOI: 10.1007/978-3-319-05521-3_38,

(HEM), the quality of wafer depends on the location in the ingot from where it is sourced. Indeed, the wafers cut out of top, bottom and sides regions of an ingot show an inferior quality compared to those from the center region. The low quality of the bottom region is mainly due to oxygen-related defects [2, 3], high density of dislocations [4, 5], and higher concentration of metals due to the solid-state diffusion from the crucible during and after crystallization [6]. However, the low lifetime in the top region of the ingot, which freezes at the end, is attributed mainly to the segregation of the metallic impurities. It has been shown that the concentrations of transition metals are determined by the segregation from the liquid-to-solid phase in the central regions of the ingots, which produces high concentrations of these metals in the top region of the ingot [6].

The presence of such impurities in different ingot region can provoke a high carrier recombination activity and greatly limiting the efficiency potential of mc-Si based solar cells.

Nevertheless, most metallic impurities dissolved or precipitated can be partially removed from the whole bulk of a Si wafer, using extrinsic gettering method. It refers to a thermal process step that activates the diffusion of interstitial impurities from the bulk of the wafer to less important superficial regions of high solubility. These regions are generally created by phosphorus diffusion (PD) and/or by aluminum-silicon (Al-Si) alloying [7–12].

In this study, we report on the investigation of the response of wafers taken from top, middle and bottom regions of HEM mc-Si ingot to the PD gettering. The process was carried out through two annealing plateaus. Wafers characterization was mainly made by QSSPC techniques and the obtained results were inspected using Hornbeck–Haynes model [13].

38.2 Experimental Details

The wafers investigated in this study were 1.5 Ωcm, p-type mc-Si grown by HEM [14]. Three sets of sister wafers, with thickness of ~300 μm were chosen from top, middle and bottom of the same ingot. The wafers from the top and bottom regions were adjacent to the discarded region (usually few centimeters between top and bottom) of the ingot.

Before any experimental process, the as-cut wafers were slightly etched in a bath of NaOH:H_2O (30 %) to remove the saw damage. After this step, the wafers have undergone PD on both sides, at 900 °C for 20 min (standard gettering), in a tube furnace using a phosphorus oxychloride ($POCL_3$) liquid source. Subsequently, wafers were subjected to an extended annealing under a nitrogen flow (extended gettering) during 120 min and at temperatures T_L varied between 600 and 900 °C. Some wafers were processed without any PD or annealing, they are considered references. Following to PD and annealing steps, a stripping of ~10 μm from each side of wafers was made, using NaOH:H_2O solution, to remove the phosphorus-silicate-glass (PSG) layer.

The main characterization technique used in this study is QSSPC measurements. It was performed by *WCT-120* tester. Before any lifetime measurement, the wafers surfaces have been passivated by Iodine-Ethanol (I-E) solution 0.08 mol. These measurements allow determining the injection level dependent minority carrier lifetime. Lifetime values were taken at 1×10^{15} cm^{-3} excess carrier density to exclude the influence of trapping artefacts.

38.3 Results and Discussion

Figure 38.1 shows the results obtained by QSSPC measurements and the performed fits employing Hornbeck–Haynes model (solid lines).

We observe clearly an improvement of carrier lifetime for all wafers of three sets; top, centre and bottom, but with different orders of magnitude. Qualitatively, the more important improvement is observed for the wafers taken from the top of ingot, where the low lifetime is attributed mainly to the segregation of the metallic impurities, indicating an effective PD gettering.

Indeed, for the top region the carrier lifetime measured at $\Delta n = 1 \times 10^{15}$ cm^{-3} (see Fig. 38.2) varies significantly by 3,500 %; from 0.7 μs for the reference to 24.5 μs for the wafers gettered at 900 °C. The improvement is less important in the case of wafers taken from centre and bottom region, where the increment was estimated by 350 and 690 %. However, we note that the higher carrier lifetime observed (47 μs) is referred to wafers of bottom region gettered at 900 °C. This effect is certainly due to an effective PD gettering through the diffusion of undetermined metals impurities towards phosphorus doped region.

Using Hornbeck–Haynes model, a complete modeling of injection-level dependent lifetime curves (see Fig. 38.1) allows to estimate the recombination center density N_r and the electron to hole capture cross-sections ratio so called symmetry parameter ($k = \sigma_n / \sigma_p$), and then to conclude the origin of dominant recombination activities after gettering process. The basic formula used for modeling measured QSSPC data can be found in [15]. More details about the calculation formula and the fitting procedure are reported elsewhere in [16–18].

The variation of estimated N_r (see Fig. 38.3) is in agreement with the evolution of effective lifetime presented in Fig. 38.2, and then confirm the explanation above based on PDG phenomenon. This can be especially observed between the as-cut wafers and those gettered at 900 °C, where the increment of τ_{eff} is accompanied by a significant decrease of N_r from 1×10^{13} cm^{-3} to 9×10^{10} cm^{-3}, from 1.2×10^{12} cm^{-3} to 1.1×10^{11} cm^{-3} and from 2×10^{12} cm^{-3} to 1.2×10^{11} cm^{-3} for top, center and bottom region, respectively.

Figure 38.4 illustrates the variation of symmetry parameter k ratio estimated from modeling as a function of the extended gettering temperature T_L for top, center and bottom wafers. For wafers of top region, it is obviously shown that the k values vary independently between those associated to Fe_i on the one hand, and to FeB, CrB pairs, Co_s (Acceptor) and W_i on the other hand. So, this result indicates

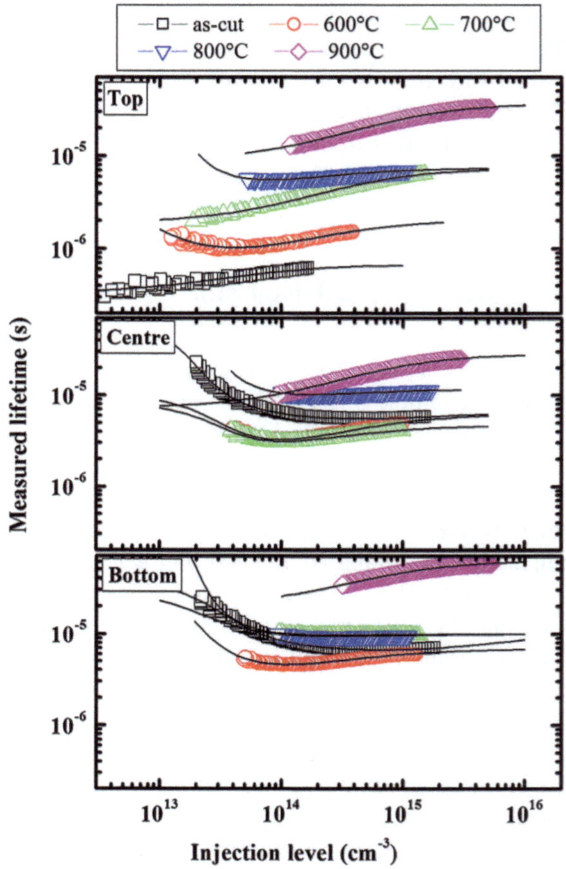

Fig. 38.1 Evolution of apparent minority carrier lifetime versus minority carrier density curves as a function of extended gettering temperature T_L for investigated wafers taken from different ingot regions. The *solid lines* represent the fits obtained by Hornbeck–Haynes model

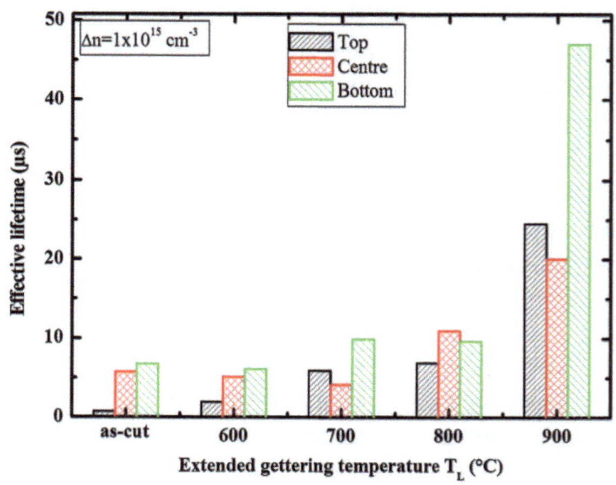

Fig. 38.2 Variation of the effective lifetime measured at $\Delta n = 1 \times 10^{15}$ cm^{-3} as a function of gettering temperature T_L, for wafers taken from *top*, *centre* and *bottom* of the ingot

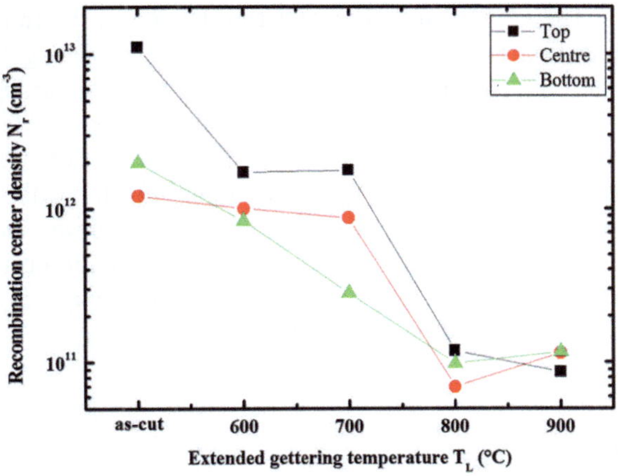

Fig. 38.3 Variation of the recombination center density N_r estimated by fitting, as a function of gettering temperature T_L

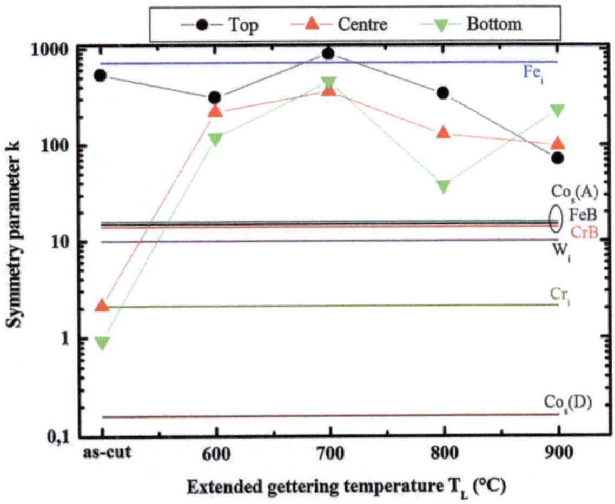

Fig. 38.4 Variation of the electron to hole capture cross-sections ratio as called symmetry parameter ($k = \sigma_n / \sigma_p$) versus extended gettering temperature T_L

that in spite of the reduction of its density, the type the dominate recombination center remain practically unchanged.

For the wafers of middle and bottom, the same result can be observed, with the exception of the important shift of k from the value associated to Cr_i towards those of Fe_i and this is just after the gettering processed at 600 °C. These results indicate the quasi-dominance of Fe_i, FeB, CrB pairs, Co_s (Acceptor) and W_i recombination activities and probably an effective PD gettering of Cr_i atoms.

38.4 Conclusion

In summary, we have studied the effect of extended PDG on recombination activity of minority charge carrier in p-type HEM mc-Si wafers taken from top, middle and bottom of the ingot.

The QSSPC measurement results showed an improvement of lifetime in high injection level indicating an effective extended gettering. This effect was especially noticed for the wafers of the top region strongly contaminated by metallic impurities, where the effective minority carrier lifetime was improved by more than 3,500 %. But the higher value of carrier lifetime (47 μs for $\Delta n = 1 \times 10^{15}$ cm^{-3}) has been found for the wafers of bottom treated at 900 °C. Hornbeck–Haynes model fitting of QSSPC curves showed that this improvement is accompanied by a decrease of the recombination center density and that probably FeB, CrB pairs, Co_s (Acceptor) and W_i impurities are the dominant sources of these centers after each gettering temperature.

References

1. Cuevas, in *Materials 2003 Conference*, The Institute of Materials Engineering Australia, Sydney (2003)
2. D. Karg, G. Pensl, M. Schulz, C. Hassler, W. Koch, Phys. Stat. Sol. (b). **222**, 379–87 (2000)
3. M. Ghosh, D. Yang, A. Lawerenz, S. Riedel, H.J. Moller, in *Proceedings of the 14th European Photovoltaic Solar Energy Conference*, Barcelona, Spain, 1997, pp. 724–72
4. B.L. Sopori, L. Jastrzebski, T. Tan, *Proceedings of the 25th IEEE photovoltaic Specialists Conference*, (Washington, DC, 1996), pp. 625–628
5. A. Bentzen, H. Tathgar, R. Kopecek, R. Sinton, A. Holt, in *Proceedings of the 31st IEEE Photovoltaic Specialists Conference*, Orlando, 2005, pp. 1074–1077
6. D. Macdonald, A. Cuevas, A. Kinomura, Y. Nakano, L.J. Geerligs, J. Appl. Phys. **97**, 033523 (2005)
7. E. Spiecker, M. Seibt, W. Schröter, Phys. Rev. B **55**(15), 9577 (1997)
8. P. Manshanden, L.J. Geerligs, Sol. Energy Mater. Sol. Cells **90**, 998–1012 (2006)
9. A. Bentzen, A. Holtb, Mater. Sci. Eng. B **159–160**, 228–234 (2009)
10. S.M. Joshi, U.M. Gosele, T.Y. Tan, Appl. Phys. Lett. **77**(8), 3858 (1995)
11. S. Alcantara, C. Canizo, A. Luque, Sol. Energy Mater. Sol. Cells **87**, 411 (2005)
12. V. Kveder, W. Schroter, A. Sattler, M. Seibt, Mater. Sci. Eng. B **71**, 175–181 (2000)
13. J.A. Hornbeck, J.R. Haynes, Phys. Rev. **97**, 311–321 (1955)
14. D. Ouadjaout, Y. Gritli, L. Zair, M. Boumaour, Rev. Energ. Ren. **8**, 49–54 (2005)
15. A. Bentzen, in Phosphorus diffusion and gettering in silicon solar cells. Ph.D. thesis, University of Oslo 2006
16. D. Macdonald, A. Cuevas, Appl. Phys. Lett. **74**, 1710 (1999)
17. D. Macdonald, A. Cuevas, Appl. Phys. Lett. **75**, 1571 (1999)
18. K.R. McIntosh, B.B. Paudyal, D. Macdonald, J. Appl. Phys. **104**, 084503 (2008)

Chapter 39
Photochemical Degradation of Polybrominated Diphenylether BDE209 Under Ultraviolet Irradiation

Yassine Agguine, Nadjia Laouedj, Ahmed Bekka, Zohra Bouberka, Abdelouahab Nadim, Said Eddarir and Ulrich Maschke

Abstract The photodegradation of decabromodiphenylether (BDE209) in Tetrahydrofuran was investigated under UV light, employing a xenon source with 150 W. The degradation reactions obey a first-order rate law. The photolytic chemical reduction of BDE209 in THF under UV irradiation can rapidly degrade BDE209 to form lower bromine substituted diphenylethers. The reductive debromination mechanism of the photolytic degradation of BDE209 can facilitate the design of remediation processes and help to predict their fate in the environment.

39.1 Introduction

Polybrominated diphenylethers (PBDEs) represent a group of persistent organic pollutants that are of international concern because of global distribution, persistence, and toxicity. PBDEs have been extensively used as flame retardants (FRs) in many areas of human activities such as electric and electronic equipment industry, polymers, plastics and textile materials [1]. Brominated flame retardants (BFRs) are

N. Laouedj · A. Bekka
Laboratoire de synthèse et caractérisation des oxydes, Université des Sciences et de la Technologie d'Oran «USTO», BP 1505El-Mnaouer 31000 Oran, Algeria

Z. Bouberka
Laboratoire physico-chimie des matériaux-catalyse et environnement, Université des Sciences et de la Technologie d'Oran «USTO», BP 1505El M'naouer 31000 Oran, Algeria

Y. Agguine · A. Nadim · U. Maschke (✉)
Unité Matériaux et Transformations—UMET (UMR CNRS N°8207), Bâtiment C6, Université Lille 1—Sciences et Technologies, 59655 Villeneuve d'Ascq Cedex, France
e-mail: ulrich.maschke@univ-lille1.fr

S. Eddarir
Laboratoire de Chimie Bioorganique et Macromoléculaire (LCBM), Faculté des Sciences et Techniques, Université Cadi Ayyad, Guéliz, Marrakech, Morocco

A. Y. Oral et al. (eds.), *International Congress on Energy Efficiency and Energy Related Materials (ENEFM2013)*, Springer Proceedings in Physics 155,
DOI: 10.1007/978-3-319-05521-3_39,

currently the largest market group of FRs due to their low cost and high performance efficiency. Commercial PBDEs are manufactured by bromination of diphenylethers resulting in a mixture of diphenylethers containing tetra-, penta-, hepta-, octa-, and deca-congeners in various percentages [2–4]. The five major marketed FRs are tetrabromobisphenol A (TBBPA) with the largest production volume, followed by three technical mixtures of polybrominated diphenylethers (PBDEs), which are known as decabromodiphenylether (decaBDE), octabromodiphenylether (octa-BDE), and pentabromodiphenylether (penta-BDE) [4, 5]. Unlike the other two PBDE mixtures, decaBDE is almost exclusively BDE209. DecaBDE is primarily used as an additive in electrical and electronic appliances, and as a polymer backcoast in textiles [6].

BDE209 remains one of the most important chemical and also environmental pollutant. Therefore, it is important to understand its degradation processes and fate in the environment, including both photodegradation kinetics and photoproducts. Research investigations showed that BDE209 can break down to produce many lighter-weight PBDE congeners, including three highly accumulative PBDEs associated with Penta BDE [7]. Photolysis is an important degradation pathway for some persistent organic pollutants including PBDEs in the environment, and has been reported to occur under a variety of conditions [8–12]. Watanabe et al. presented a study on the photolysis of BDE209 in a mixture of hexane, benzene, and acetone (8:1:1), employing both sunlight and artificial UV light. Their results indicated that BDE209 was debrominated down to lower brominated PBDEs and polybrominated dibenzofurans (PBDFs) [13]. Recently, another investigation on the photolysis of decaBDE in toluene and adsorbed on silica gel, sand, soil, and on sediments, also showed similar results of photolytic debromination [14–18]. Chen et al. [19] found that the photochemical reaction rate decreases with decreasing bromination degree and that it might be also affected by the positions of substituted Br-atoms on the aromatic rings.

Only few studies were reported on the photolysis of strongly brominated PBDEs in Tetrahydrofuran (THF). THF has a very good solvent activity compared to other solvents and is known to be a good hydrogen donor, which facilitates debromination of PBDEs [1]. The primary objective of this study was to investigate the photolytic degradation experiments of BDE209 to understand the photodegradation kinetics and mechanisms in THF under UV light.

39.2 Experimental Part

Decabromodiphenyl ether (BDE209, purity 98 %), and Octabromodiphenylether ether (OctaBDE, purity 98 %) were purchased from Albermarle corporation. Tetrahydrofuran (THF) was obtained from Merck in HPLC quality. The molecular structures of Decabromodiphenylether (BDE209) and octabromodiphenylether (OctaBDE) are shown in Fig. 39.1.

The irradiation process was carried out using standard 10 mm thick Hellma quartz cells, and the irradiation source was fixed through the central axis of the

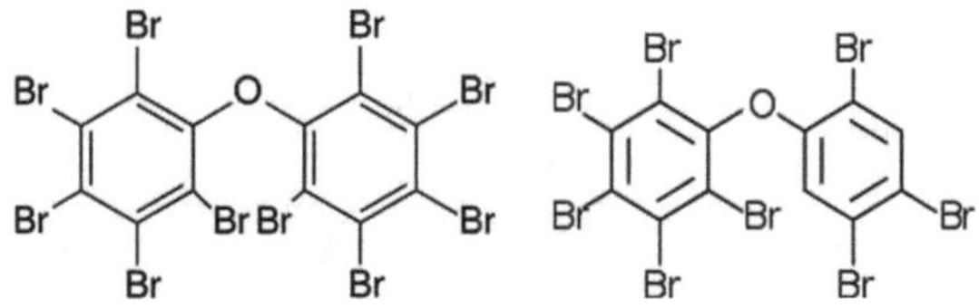

Fig. 39.1 Molecular structures of decabromodiphenylether (BDE209) (*on the left hand side*), and octabromodiphenylether (OctaBDE) (*on the right hand side*)

cell. The distance between UV source and quartz cell was kept at 3 cm. A 150 W Xenon light source (Hamamatsu model LC8) was used for the irradiation experiments.

During photodegradation experiments, aliquots of solutions containing BDE209 were placed in the Hellma cells and submitted to UV irradiation. The initial concentration of BDE was kept at 0.13 g/L in THF unless otherwise stated. At appropriate sampling times, the photodegradation was finished by removing the vials from the light exposure and immediately initiating analysis.

Absorption spectra were recorded using a Varian Cary 100 UV-visible double-beam spectrophotometer, equipped with a Peltier accessory for precise temperature control (±0.1 °C). All UV absorption spectra were recorded with THF for the determination of quantum yields and computed from the calibration curves in the range 200–400 nm, using the above mentioned Hellma cells at the wavelength of maximum absorption of BDE209 and OctaBDE.

39.3 Results and Discussion

Figure 39.2 shows the changes of the absorption spectra of BDE209 solutions in THF under UV irradiation. It can be seen that the main absorbance band of BDE209 is located at 228 nm, assigned to the π–π^* electron transition of the C=C band in the benzene ring [20]. The peak intensity attributed to 228 nm gradually decreases within irradiation time, indicating the progress of the BDE209 decomposition. It should be pointed out that the rate of decomposition of BDE209 is quite fast, in the range of seconds, in contrast to results from current research [11, 21]. Previous studies have suggested that the hydrogen-donating ability of the solvent significantly affects the rate of photodegradation of PBDE [22]. These findings are in good agreement with our results since THF represents a high hydrogen donating ability. After 4 min of irradiation, the main absorption band blue shifted to 209 nm and this effect was more pronounced with time. A loss of BDE209 was observed with concomitant formation of photodegradation products. The initial BDE209 concentration was reduced to 33 % after 4 min irradiation time. This decrease was considered as the first evidence of photochemical reaction of BDE209 in THF solution. The oxidants produced from the interactions of BDE with light will

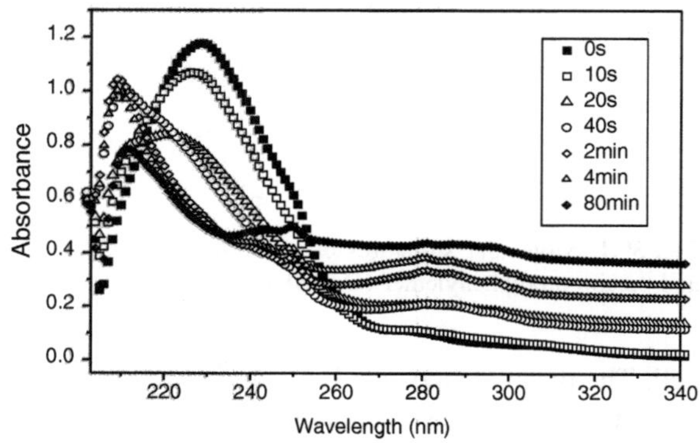

Fig. 39.2 Photodegradation of BDE209 in THF using UV-visible irradiation (initial concentration of BDE: 0,13 mg/L)

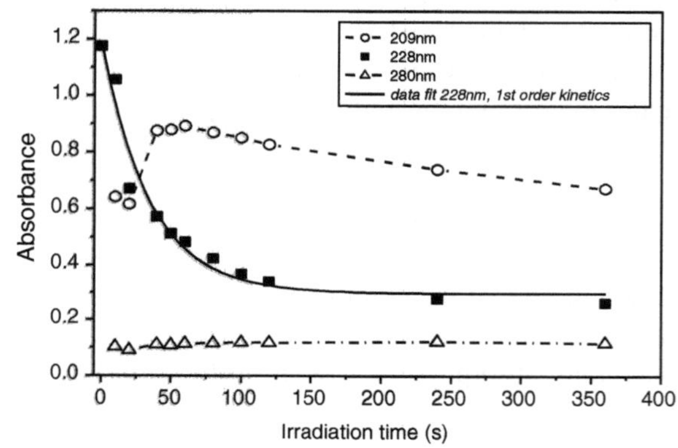

Fig. 39.3 The absorbance of BDE209 and its photoproducts at different irradiation times (initial concentration of BDE209: 0,13 mg/L)

kinetically have the highest probability to react with THF. It was shown that the reductive dehalogenation, in order to form other congeners, represents the main photolytic pathway for these compounds in a variety of media and conditions [8–13, 15]. Several publications have reported on the formation of polybrominated dibenzofurans (PBDFs) as an alternative degradation pathway for photolysis of PBDEs [11–15, 21]. Erikson et al. [11] found relatively high concentrations of di- and tri-BDFs during photodecomposition studies on decaBDE in a mixture of methanol and water.

In all cases cited here, consecutive bromine atom losses were mentioned as an important photodegradation pathway. This photolysis reaction is called reductive debromination. The first step of this reaction is the cleavage of the carbon–bromine bond. The strength of the aryl–Br bond is lower than that of the C–O bond [11]. Figure 39.3 shows the concentration profiles of initial BDE209 and its photoproducts at different irradiation times. In particular Fig. 39.3 represents the disappearance of the 228 nm band characterizing BDE209, as well as the appearance of two new bands at 209 and 280 nm. Qualitative and quantitative data clearly indicate that the 228 nm band strongly decreases, yielding 80 % decomposition at 240 s. An attempt was made to rationalize the disappearance of the 228 nm band as a function of irradiation time, which corresponds to a decrease of the concentration of BDE209. Since the concentration of BDE209 varied exponentially with time, these data were best fitted using 39.1, equal to a first order reaction. The rate constants of the process were obtained by non-linear least-square fitting of the concentration-time data to 39.1:

$$\frac{A_0}{A} = e^{-kt} \tag{39.1}$$

where A_0 represents the initial absorbance of the reactant, A is the absorbance of the reactant at time t, t is the irradiation time, and k is the reaction rate constant (s^{-1}). From the fitting procedure, the reaction rate constant was evaluated to 1.23×10^{-2} s^{-1}. The reaction half time of degradation was only 56.3 s as compared to 6.3 min in hexane [23], and 3.0 min in acetonitrile [24].

39.4 Conclusion

An efficient new route for the degradation of the polybrominated diphenylether BDE209 in THF was found and monitored by UV-visible spectroscopy studies. Indeed, irradiation of BDE209 leads to fast photolytic reactions yielding mainly lower brominated compounds. A certain number of experimental parameters, like the irradiation time, were optimized in order to obtain high yields between 80 and 90 % of photoproducts. After photochemical reaction times of some tenth of seconds, lower brominated molecules were formed from BDE209 which could undergo further debromination reactions.

References

1. EHC-122, Brominated diphenyl ethers. *International Program on Chemical Safety*. (World Health Organization, Switzerland, 1994)
2. C. Lassen, S. Løkke, L.I. Andersen, Brominated flame retardants: substance flow analysis and assessment of alternatives. *Environmental Project Nr. 494*. Danish Environmental Protection Agency (1999)
3. P.O. Darnerud, G.S. Eriksen, T. Jóhannesson, P.B. Larsen, M. Viluksela, Polybrominated diphenyl ethers: occurrence, dietary exposure, and toxicology. Environ. Health Perspect. **109**, 49–68 (2001)
4. L.S. Birnbaum, D.F. Staskal, Brominated flame retardants: cause for concern? Environ. Health Perspect. **112**, 9–17 (2004)
5. EHC-192, Flame retardants: a general introduction. *International Program on Chemical Safety*. (World Health Organization, Switzerland, 1997)
6. A.F. Lagalante, T.D. Oswald, Analysis of polybrominated diphenyl ethers (PBDEs) by liquid chromatography with negative-ion atmospheric pressure photoionization tandem mass spectrometry (LC/NI-APPI/MS/MS): application to house dust. Anal. Bioanal. Chem. **391**, 2249–2256 (2008)

7. K. Betts, Does a key PBDE break down in the environment? Environ. Sci. Technol. **42**, 6781 (2008)
8. J. Bezares-Cruz, C.T. Jafvert, I. Hua, Solar photodecomposition of decabromodiphenyl ether: products and quantum yield. Environ. Sci. Technol. **38**, 4149–4156 (2004)
9. P.H. Peterman, C.E. Orazio, K.P. Feltz, Sunlight photolysis of 39 mono-hepta PBDE congeners in lipid. Organohalogen Compd. **63**, 357–360 (2003)
10. M. Barcellos da Rosa, H.-U. Krüger, S. Thomas, C. Zetzsch, Photolytic debromination and degradation of decabromodiphenyl ether, an exploratory kinetic study in toluene. Fresenius Environ. Bull. **12**, 940–945 (2003)
11. J. Eriksson, N. Green, G. Marsh, A. Bergman, Photochemical decomposition of 15 polybrominated diphenyl ether congeners in methanol/water. Environ. Sci. Technol. **38**, 3119–3125 (2004)
12. W.-U. Palm, R. Kopetzky, W. Sossinka, W. Ruck, C. Zetzsch, Photochemical reactions of brominated diphenylethers in organic solvents and adsorbed on silicon dioxide in aqueous suspension. Organohalogen Compd. **66**, 4101–4105 (2004)
13. I. Watanabe, R. Tatsukawa, Formation of brominated dibenzofurans from the photolysis of flame retardant decabromobiphenyl ether in hexane solution by UV and sun light. Bull. Environ. Contam. Toxicol. **39**, 953–959 (1987)
14. S. Ohta, H. Nishimura, T. Nakao, O. Aozasa, H. Miyata, Characterization of the photolysis of decabromodiphenyl ether and the levels of PBDEs as its photoproducts in atmospheric air of Japan. Organohalogen Compd. **52**, 321–324 (2001)
15. G. Söderström, U. Sellström, C.A. de Wit, M. Tysklind, Photolytic debromination of decabromodiphenyl ether (BDE 209). Environ. Sci. Technol. **38**, 127–132 (2004)
16. I. Hua, N. Kang, C.T. Jafvert, J.R. Fábrega-Duque, Heterogeneous photo-chemical reactions of decabromodiphenyl ether. Environ. Toxicol. Chem. **22**, 798–804 (2003)
17. J.A. Tokarz 3rd, M.Y. Ahn, J. Leng, T.R. Filley, L. Nies, Reductive debromination of polybrominated diphenyl ethers in anaerobic sedment and a biomimetic system. Environ. Sci. Technol. **42**, 1157–1164 (2008)
18. Y.-S. Keum, Q.X. Li, Reductive debromination of polybrominated diphenyl ethers by zero-valent iron. Environ. Sci. Technol. **39**, 2280–2286 (2005)
19. D. Chen, B. Mai, J. Song, Q. Sun, Y. Luo, X. Luo, E.Y. Zeng, R.C. Hale, Polybrominated diphenyl ethers in birds of prey from Northern China. Environ. Sci. Technol. **41**, 1828–1833 (2007)
20. A.M. Geller, Ph.D. thesis, University of Bayreuth, Germany, 2008
21. L. Sanchez-Prado, M. Llompart, M. Lores, C. Garcia-Jares, R. Cela, Investigation of photodegradation products generated after UV-irradiation of five polybrominated diphenyl ethers using photo solid-phase microextraction. J. Chromatogr. A **1071**, 85–92 (2005)
22. S. Rayne, M.G. Ikonomou, M.D. Whale, Anaerobic microbial and photochemical degradation of 4, 4′-dibromodiphenyl ether. Water Res. **37**, 551–560 (2003)
23. Y.H. Shih, C.K. Wang, Photolytic degradation of polybromodiphenyl ethers under UV-lamp and solar irradiations. J. Hazard. Mat. **165**, 34–38 (2009)
24. C. Sun, D. Zhao, C. Chen, W. Ma, J. Zhao, TiO_2-mediated photocatalytic debromination of decabromodiphenyl ether: kinetics and intermediates. Environ. Sci. Technol. **43**, 157–162 (2009)

Chapter 40
Influence of Heat Treatment on Structure and Charge Capacity of Sol–Gel Produced TiO_2 Films

Orhan Özdemir, Fatma Pınar Gökdemir, Bahadır Keskin and Kubilay Kutlu

Abstract Titanium dioxide thin films were synthesized by sol–gel route from titanium isopropoxide (TIP) with acetic acid. Prior to the heat treatment, the films were amorphous phase. Above 400 °C, phase transition took place from amorphous into anatase phase. Electrochromic properties of each phase indicated reversible coloration upon Li^+ ion intercalation in cyclic voltammetric measurements. Nevertheless, both charge capacity and energy band gap of films begun to decrease with increase in annealing temperature due to the crystallization.

40.1 Introduction

TiO_2 is gained a great attention due to its interesting chemical, electrical and optical properties, which directly depends on the crystalline phase. TiO_2 have three phases such as anatase, rutile and brookite where rutile phase is optically active and the most stable one whereas anatase phase has attracted much attention because of its higher photocatalytic activity for photocatalysis [1] and solar energy conversion [2, 3]. Sol gel is one of the best way to fabricate TiO_2 films onto different substrates at low temperature with lower cost. Mostly, as deposited films exhibit amorphous phase and by thermal heat treatment, phase transition occurs from amorphous to anatase phase around 500°C. However other deposition techniques may cause anatase phase accompany by rutile and brookite traces. Amorphous to anatase phase transformation not only depends on annealing but also species that get involved in sol synthesis. TiO_2 sol preparation usually has three approaches; Ti alkoxide precursors in aqueous [4–6] or organic media [7], hydrolysis of $TiCl_4$ at low temperatures [8] and non-hydrolytic combination of

O. Özdemir (✉) · F. P. Gökdemir · B. Keskin · K. Kutlu
Department of Physics, Yıldız Technical University, 34210 Esenler, Istanbul, Turkey
e-mail: ozdemir@yildiz.edu.tr

A. Y. Oral et al. (eds.), *International Congress on Energy Efficiency and Energy Related Materials (ENEFM2013)*, Springer Proceedings in Physics 155,
DOI: 10.1007/978-3-319-05521-3_40, © Springer International Publishing Switzerland 2014

Ti alkoxide and $TiCl_4$ [9]. In this work, sol was prepared by a modified route from titanium isopropoxide with acetic acid, which controls hydrolysis and condensation reactions by reducing the availability of groups. In addition to that the synthesized sol was investigated by thermogravimetric/differential thermal analysis (TG/DTA) to ascertain phase transition temperatures. Furthermore, predetermined temperatures were selected as annealing temperature and influence of the temperature on both structural and electrochemical properties were traced by relevant investigating tools such as UV–Vis. Transmission measurement, Attenuated Total Reflection Fourier transform (ATR-FTIR) specroscopy and cyclic voltammetry (CV).

40.2 Experimental

Coating sol prepared by adding 4.15 ml of isopropyl alcohol into 1.60 ml TIP solution and followed by mixing at 70 ± 10 °C on a magnetic stirrer. 5.15 ml acetic acid and 12 ml methanol were introduced and then solution was transformed immediately to a white precipitate. Finally it was dissolved in 12.5 ml methanol and synthesized on Corning, SLG and ITO coated glass substrates with a constant speed of 100 mm/min by dip coater. Before deposition, substrates were cleaned in an ultrasonic bath using acetone, isopropyl alcohol and DI water at 30 °C, respectively. As deposited samples were dried at room temperature before initialization of heat treatment (at 100 °C for 20 min). Resuming the same processes for double layer, heat treatment was applied under air at 200 and 500 °C for 1 h. Coating sol was left to dry for TG DTA measurements that were performed with a Seiko SII Exstar 6,300 model thermal analysis system using an alumina crucible in static air ambient with a heating rate of 10 °C/min. For ATR-FTIR, a Bruker Tensor27 model IR spectrophotometer was used. Transmission experiments were made by using a PG Instruments T80 model UV Vis spectrophotometer and optical parameters of the films were determined using OptiChar software. Surface morphology and particle size of the films were investigated through atomic force microscopy (AFM) that operated at non-contact mode. Electrochemical analyses were carried out using cyclic voltammetry experiments with an Autolab PGSTAT30 potentiostat galvanostat.

40.3 Results and Discussion

Figure 40.1a showed TG DTA curves where two mass losses associated with endothermic and exothermic events. The first endothermic event took place around 90 °C, corresponding to the elimination of water while following exothermic events were due to the volatilization and combustion of other species such as CH_3OH $(CH_3)_2CHOH$ and CH_3COOH. Nearly localized two peaks in the DTA

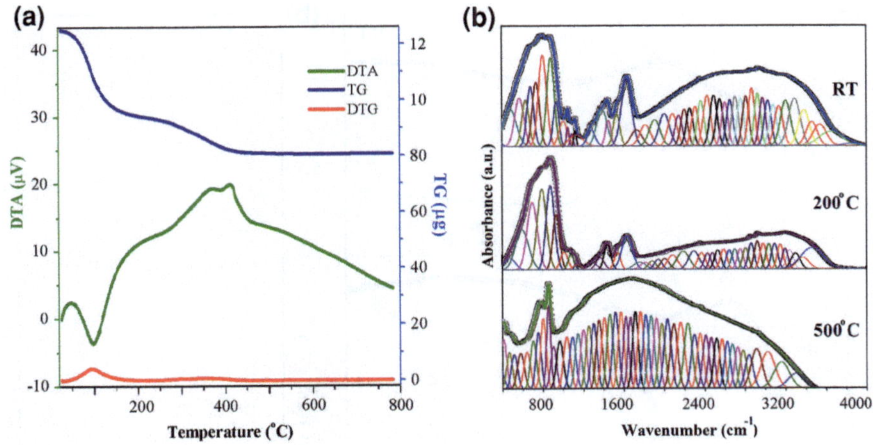

Fig. 40.1 **a** TG-DTA curves of the sol. **b** FTIR spectra of the as deposited and annealed films on ITO coated glass substrates

curve at 367 and 410 °C, were ascribed to the transition of the amorphous phase into anatase phase. There were no associated signals in TG curve, confirming the anatase rutile phase transition between 400 and 800 °C. FTIR absorption spectra with the results of deconvolution processes (solid lines) for the films were given in Fig. 40.1b. The presence of Ti–O–Ti and Ti–O polymeric chains were clearly shown from the bands at 471 and 789 cm^{-1}. Also, vibration of the Ti–O–O band was identified from the band at 693 cm^{-1} [10]. Moreover, the bands at 1,009, 1,122, and 1,138 cm^{-1} were asscociated to stretching of Ti–O–C [11] while LO mode of amorphous TiO_2 [12] appeared at 874 cm^{-1}. The broadband from 3,000 to 3,600 cm^{-1} denoted the stretching vibration modes of hydroxyl groups [13], and begun to broaden after the addition of water at 1,635 cm^{-1}.

The bands appeared at 1,288 and 1,368 cm^{-1} were the vibration mode of C–O–O group [14]. The doublet in 1,441 and 1,538 cm^{-1} designated the symmetric-asymmetric stretching vibration of the carboxylic group coordinated to Ti as a bidentate ligand [15]. Moreover, the separation between these signals proposed that acetate behaved preferentially as a bidentate rather than as a bridging ligand between two Ti atoms [11, 16] since the latter possibility required higher seperation value greater than 150 cm^{-1}. Optical transmittance spectra of the films deposited on SLG substrates were given in Fig. 40.2.a. The optical absorption coefficient (α) of the film was calculated through a relation "$\alpha = 4\pi k/\lambda$" where k was the extinction coefficient which was obtained through a fitting to the experimental data by OptiChar software by varying film refractive index (n), k and thickness of the film. In the analysis, dispersive glass and surface as well as bulk inhomogeneity were taken into account. The generated transmittance (solid lines) was also shown in Fig. 40.2a. Hence, variation of n and k as function of λ was also given as insets. On the other hand, optical band gap were determined with Tauc plots using absorption coefficients [17]. According to Tauc's law $\alpha h\nu = A(h\nu\text{-}E_g)^n$ where A is a constant, $h\nu$ is

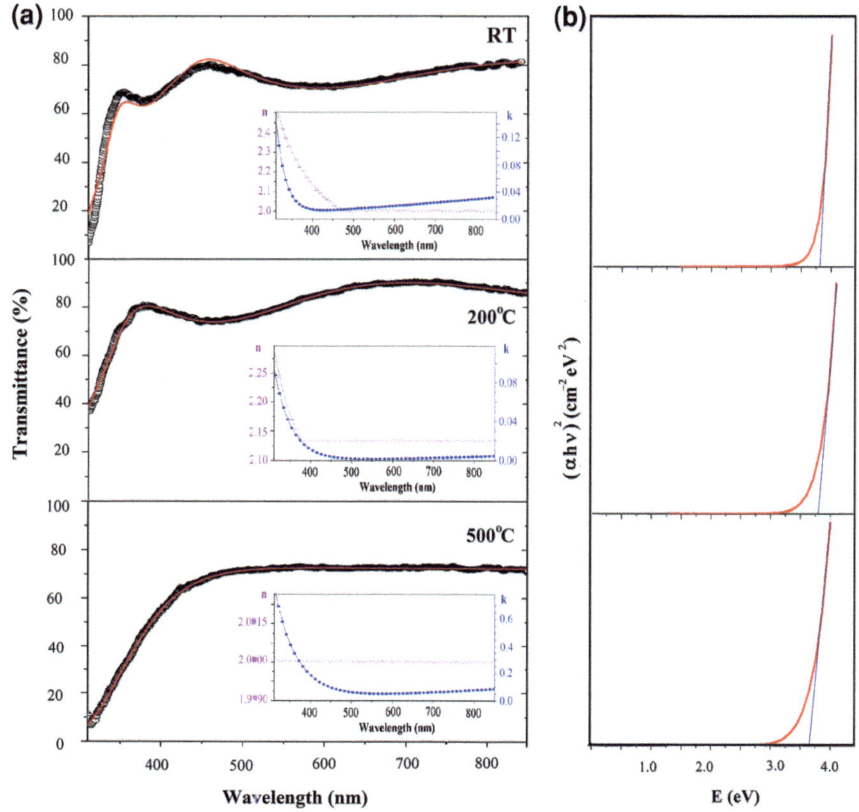

Fig. 40.2 **a** Transmittance and n, k spectra of the films deposited on Corning glass substrates. **b** Tauc plots of the films for direct allowed transition

photon energy, E_g is optical energy band gap and n is the fingerprint of the transition (n = 1/2, 3/2, 2 and 3 for; direct allowed, direct forbidden, indirect allowed and indirect forbidden, respectively). Figure 40.2.b indicated that best fit for the films was given by a direct allowed transition so gap values were determined as 3.85, 3.80 and 3.65 eV for the as deposited, 200 °C and 500 °C annealed films, respectively. However, in the case of indirect allowed transition, band gap of the films were 3.38, 3.11 and 2.84 eV. In either case, it was certain that band gap decreases with the increase in annealing temperature. Such a situation might be due to quantum confinement. Therefore, AFM analysis was performed to resolve the proposition by examining granule size of the films.

Figure 40.3a displayed AFM analysis of as deposited and annealed films. Increasing in annealing temperature resulted to grow of the granule sizes. In that case, the increase of the granule size should lead to increase in charge capacity. Hence, electrochemical experiments were carried out in a conventional three-electrode electrochemical cell where a platinum wire and a silver wire were used

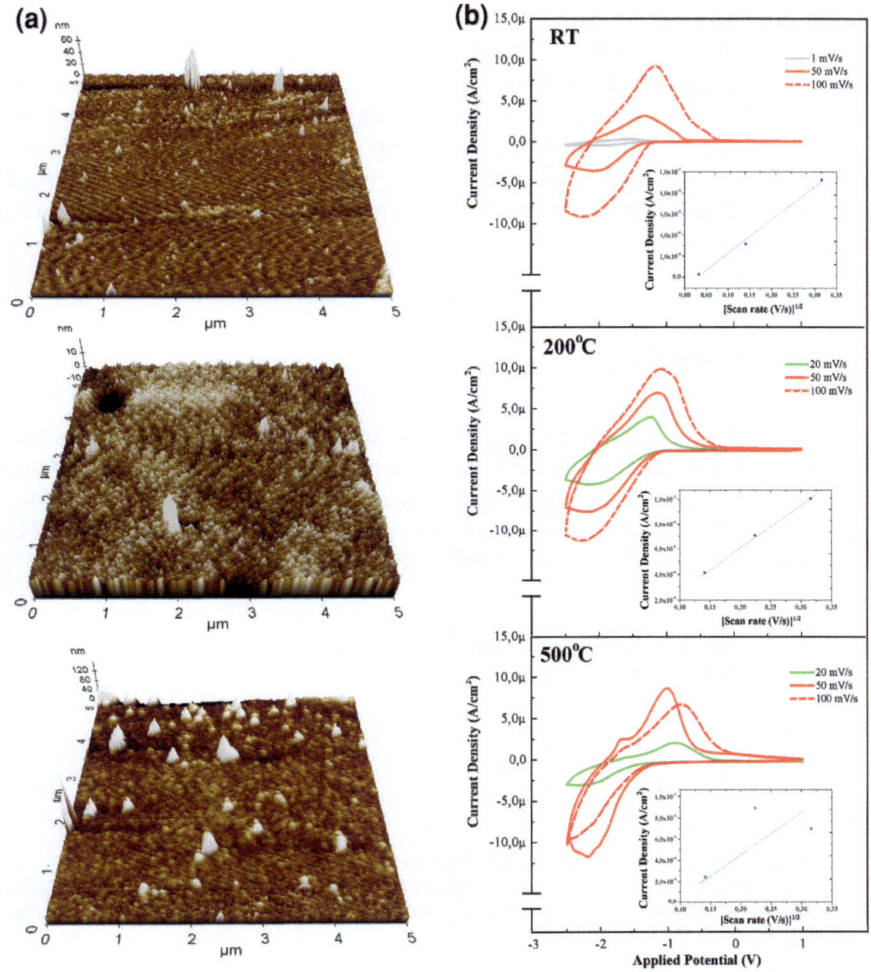

Fig. 40.3 **a** AFM pictures of the films deposited on SLG substrates. **b** Cyclic voltammograms of the films deposited on ITO coated glass substrates (d seklini ekle)

as a counter and reference electrode, respectively in a 1 M $LiClO_4$/PC carbonate electrolyte. TiO_2 films showed cathodic electrochromism and changed color transparent to blue with the injection of electrons and Li^+ ions into the oxide matrix by following reaction [18–20], $TiO_2 + x(Li^+ + e^-) \Leftrightarrow Li_xTiO_2$. Cyclic voltammograms at different scan rates between + 1 V and −2.5 V and Randles-Sevcik plots (for anodic peaks) of TiO_2 films were given in Fig. 40.3b.

However inserted/extracted charge values (mC/cm^2) during coloration/bleaching ($Ti^{4+} \Leftrightarrow Ti^{3+}$) calculated from cyclic voltammograms (for 20 mV/s) were found as 17.18:13.15, 24.20:16:20 and 13.63:10.40, for the as deposited, and annealed films at 200 °C and 500 °C, respectively. It seemed that by increasing annealing temperature up to 500 °C, TiO_2 films became more compact and charge capacities differed with a small increment. That might be originated from easier Li^+ insertion without undergoing two phase transition from the tetragonal to orthorhombic phase, which caused internal stress in the crystalline anatase structure [21] for annealing temperatures lower than phase transition temperature. According to cyclic voltammograms, anodic peak position was much more negative for an as deposited films compared to heat treated films. This was interpreted as opener insertion of Li ions [22] was easier.

40.4 Conclusions

Ex-situ heat treatment on as grown films had drastically impacted on both structural and electrochemical properties: due to the quantum confinement, energy band gap of the films began to decrease while granule size of the films started to grow. Upon a transition of amorphous to anatase phase in TiO_2 film around 500 °C, charge capacity of the film dropped gradually.

Acknowledgments This study is financially supported by Yildiz Technical University projects under a contract of 2012-01-01-DOP02 and 2011-01-01-KAP03. The authors thank to Assoc. Prof. Dr. Nevim SAM for FTIR measurements.

References

1. J. Zhang, Q. Xu, Z. Feng, M. Li, C. Li, Angew. Chem. Int. **47**, 1766 (2008)
2. G.K. Mor, K. Shankar, M. Paulose, O.K. Varghese, C.A. Grimes, Nano Lett. **6**(2), 215 (2006)
3. M. Adachi, Y. Murata, J. Takao, J. Jiu, M. Sakamoto, F. Wang, J. Am. Chem. Soc. **126**, 14943 (2004)
4. P. Kumar, S. Badrinarayanan, M. Sastry, Thin Solid Films **358**, 122 (2000)
5. B.E. Yoldas, J. Mater. Sci. **21**, 1986 (1087)
6. N. Kotov, F. Meldrum, J.H. Fendler, J. Phys. Chem. **98**, 8827 (1994)
7. D.E. Skinner, D.P. Colombo Jr, J.J. Cavaleri, R.M. Bowman, J. Phys. Chem. **99**, 7853 (1995)
8. N. Serpone, D. Lawless, R. Khairutdinov, J. Phys. Chem. **99**, 16646 (1995)
9. P. Arnal, R.J.P. Corriu, D. Leclercq, P.H. Mutin, A. Vioux, Chem. Mater. **9**, 694 (1997)
10. Y. Gao, Y. Masuda, Z. Peng, T. Yonezawa, K. Koumoto, J. Mater. Chem. **1**, 608 (2003)
11. R. Parra, M.S. Góes, M.S. Castro, E. Longo, P.R. Longo, J.A. Varela, Chem. Mater. **20**(1), 143 (2008)
12. R.J. Gonzalez, R. Zallen, Physical Review B **55**(11), 7014 (1997)
13. A.J. Maira, J.M. Coronado, V. Augugliaro, K.L. Yeungz, J.C. Conesa, J. Soria, J. Catalysis **202**, 413 (2001)
14. M. Burgos, M. Langlet, Thin Solid Films **349**, 19 (1999)
15. N.T. Nolan, M.K. Seery, S.C. Pillai, J. Phys. Chem. C **113**(36), 16151 (2009)
16. R. Urlaub, U. Posset, R. Thull, J. Non-Cryst. Solids **265**(3), 276 (2000)
17. J. Tauc, F. Abeles, "Opt. prop. solids", Amsterdam, 372 (1972)
18. C.G. Granqvist, *Handbook of inorganic electrochromic materials* (Elsevier, Amsterdam, 1995)

19. C.K. Dyer, J.S.L. Leach, J. Electrochem. Soc. **125**(1), 23 (1978)
20. T. Ohtsuka, M. Masuda, N. Sato, J. Electrochem. Soc. **134**(10), 2406 (1987)
21. W.H. Ryu, D.H. Nam, Y.S. Ko, R.H. Kim, H.S. Kwon, Electrochim. Acta **61**(1), 19 (2012)
22. Z. Wang, X. Hu, Thin Solid Films **352**(1–2), 62 (1999)
23. Ü.Ö. Akkaya Arıer, F.Z. Tepehan, Surf. Coat. Technol. **206**, 37–42 (2011)
24. K. Bange, T. Gambke, Adv. Mater. **2**(1), 10 (1990)
25. C.M. Wang, S.Y. Lin, J. Solid State Electrochem. **10**, 255 (2006)

Chapter 41
Investigation of Natural Draft Cooling Tower Performance Using Neural Network

Qasim S. Mahdi, Saad M. Saleh and Basima S. Khalaf

Abstract In the present work Artificial Neural Network (ANN) technique is used to investigate the performance of Natural Draft Wet Cooling Tower (NDWCT). Many factors are affected the rang, approach, pressure drop, and effectiveness of the cooling tower which are; fill type, water flow rate, air flow rate, inlet water temperature, wet bulb temperature of air, and nozzle hole diameter. Experimental data included the effects of these factors are used to train the network using Back Propagation (BP) algorithm. The network included seven input variables (T_{wi}, h_{fill}, m_w, T_{aiwb}, T_{aidb}, v_{low}, v_{up}) and five output variables (ma, T_{aowb}, T_{wo}, Δp, ε) while hidden layer is different for each case. Network results compared with experimental results and good agreement was observed between the experimental and theoretical results.

41.1 Introduction

A cooling tower cools water by a combination of heat and mass transfer. The hot water to be cooled is distributed in the tower by spray nozzles, splash bars, or film-type fill, which exposes a very large water surface area to atmospheric air. A portion of the water absorbs heat and it is changed to a vapour at constant pressure.

Q. S. Mahdi (✉) · B. S. Khalaf
Mechanical Engineering Department, College of Engineering,
Al-Mustansiriyah University, Baghdad, Iraq
e-mail: qasim602006@yahoo.com

B. S. Khalaf
e-mail: smir_eng@yahoo.com

S. M. Saleh
Mechanical Engineering Department, College of Engineering,
Baghdad University, Baghdad, Iraq
e-mail: almashatsaad@yahoo.com

A. Y. Oral et al. (eds.), *International Congress on Energy Efficiency and Energy Related Materials (ENEFM2013)*, Springer Proceedings in Physics 155,
DOI: 10.1007/978-3-319-05521-3_41,

This latent heat of vaporization has long been used to transfer heat from the water to the atmosphere. Additional heat is taken away by the air by virtue of its temperature increase but this sensible heat exchange is minor compared to the latent component provided by the water's phase change [1]. Cooling towers of interest play an important role in the cool-end system of power plant, and its cooling capacity can affect the total power generation capacity directly. The cooling efficient is highly sensitive to environmental conditions, particularly for most cases under the cross-wind conditions that may reduce dry-cooling towers up to (40 %) of the total power generation capacity. However, for the conventional design of cooling towers, the impact of cross-wind, which actually exists in most cases, has not been paid more attention [2].

41.2 Natural Draft Cooling Tower

The natural draft wet cooling tower (NDWCT) or hyperbolic cooling tower, Fig. 41.1 makes use of the difference in temperature between the ambient air and the hotter air inside the tower. As hot air moves up wards through the tower (because hot air rises), fresh cool air is drawn into the tower through an air inlet at the bottom. Due to the layout of the tower, no fan is required and there is almost no circulation of hot air that could affect the performance. Concrete is used for the tower shell with a height of up to (200 m). These cooling towers are mostly only for large heat duties because large concrete structures are expensive. Wet cooling towers have a water basin with a cold water outlet at the base. These are both large engineered structures, able to handle counter-flow, cross-flow, or parallel-flow modes of heat transfer, as indicated in Fig. 41.1. The fill construction inside the tower is a conventional frame structure, always prefabricated as illustrated in Fig. 41.2. It carries the water distribution, a large piping system, the spray nozzles, and the fill-package. Often dripping traps are applied on the upper surfaces of the fill to keep water losses through the uplift stream under 1 %. Finally, noise protection elements around the inlet decrease the noise caused by the continuously dripping water [3].

41.3 Model of ANN

41.3.1 Three-Layer BP Network

Three-layer BP network used as shown in Fig. 41.3. ($V_{j,i}$) represents the weights between the input layer vectors and hidden layer vectors, and ($W_{k,j}$) represents the weights between the hidden layer vectors and output layer vectors. The input layer has seven neurons, including dry bulb temperature of inlet air (T_{aidb}), wet-bulb temperature (T_{aiwb}), circulating water inlet temperature (T_{wi}), circulating water

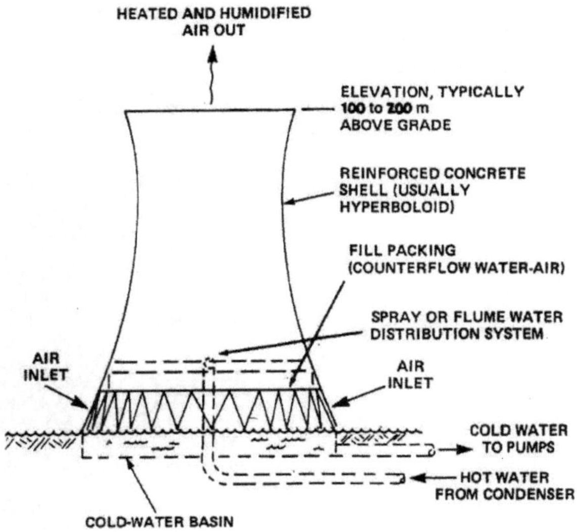

Fig. 41.1 Natural draft cooling towers [4]

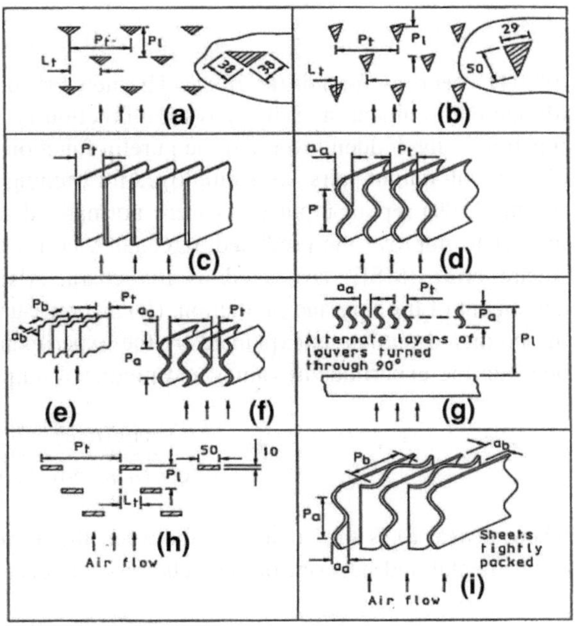

Fig. 41.2 Types of fill in counter cooling tower [3]

inlet mass flow rate (m_w), film fill at different height (h_{fill}) and inlet air velocity at bottom and top of tower (v_{low}, v_{up}) respectively. Output layer has five neurons, including circulating water outlet temperature (T_{out}), dry bulb temperature of outlet air (T_{aowb}), air mass flow rate (m_w), pressure drop (Δp) and cooling effectiveness (ε). The hidden layer has two layers at (1,047) neurons for film fill and

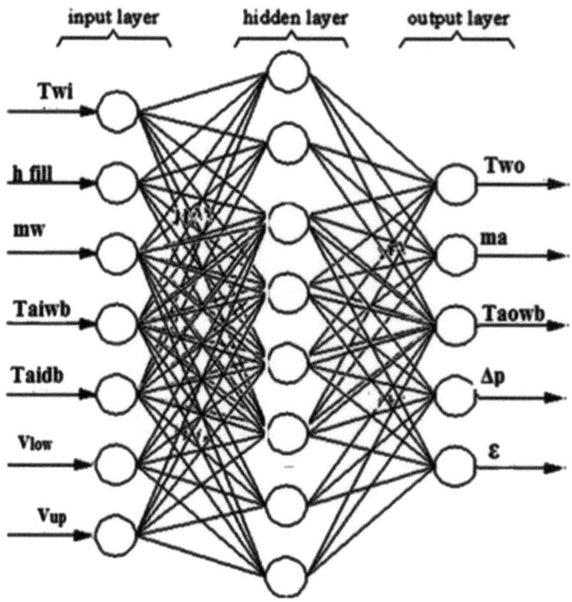

Fig. 41.3 The structure of ANN for modeling the experimental model tower

(93,041) neurons for nozzle effect. The network part is implemented under the Matlab environment, and the activation function is chosen as the tangent sigmoid function in the hidden layer and the purelin function in the output layer. There are (200) input–output pairs, were employed for training set as the testing the network. All the (200) input output pairs were normalized to fall in the interval [−1, 1] inorder to improve the predicted agreement, correlation coefficient (R) and mean square error (MSE) are acted as the characteristic parameters to assess the agreement of training and prediction. (R) is a measure of how well the variation in the predicted outputs is explained by the experimental values, and the (R) value between the experimental values and predicted outputs is defined by [5].

$$R = \frac{\mathrm{cov}(a, b)}{\sqrt{\mathrm{cov}(a, a).\mathrm{cov}(b, b)}} \tag{41.1}$$

where cov(a, b) is the covariance between the a and b sets which represent the experimental and network output sets, respectively, and is given by

$$\mathrm{cov}(\mathrm{a},\ \mathrm{b}) = \mathrm{E}[(\mathrm{a} - \mu\mathrm{a})(\mathrm{b} - \mu\mathrm{b})] \tag{41.2}$$

where E is the expected value, μa and μb are the mean value of a set and b set, respectively. In addition, cov(a, a) and cov(b, b) are the auto covariances of (a) and (b) sets, respectively, and are expressed by

$$\mathrm{cov}(\mathrm{a},\ \mathrm{a}) = \mathrm{E}\left[(\mathrm{a} - \mu\mathrm{a})^2\right] \tag{41.3}$$

$$\mathrm{cov}(\mathrm{b},\ \mathrm{b}) = \mathrm{E}\left[(\mathrm{b} - \mu\mathrm{b})^2\right] \tag{41.4}$$

The (R) values closer to (+1) indicate a stronger agreement of training and predicted values, while the values closer to (−1) indicate a stronger negative relationship between training and prediction. The mean square error is calculated from:

$$MSE = \frac{1}{N}\sum_{i=1}^{N}(a_i - b_i)^2 \tag{41.5}$$

41.3.2 Determination of the Number of Nodes in the Hidden Layer

The performance of an ANN is affected by the characteristic of the network, such as the number of hidden layer and the number of nodes in hidden layer. But up to now, there is not a definite method to choose the optimal number of hidden layer and the number of nodes in hidden layer. General speaking, the (ANN) model which has one hidden layer can meet with simulative requirements [6]. So the model of one hidden layer is used in his study. Three formulas which are used to determine the node number in hidden layer are given by (41.6) and (41.7) [7].

$$\sum_{i=1}^{n} C_{n1}^{i} \geq K \tag{41.6}$$

where K is the sample number, if ($\mathrm{i} > \mathrm{n1}$, $\mathrm{C_{in}} = 0$)

$$n1 = \sqrt{n+m} + c \tag{41.7}$$

where c is a constant which belongs to [8].

$$\mathrm{n1} = \log_2 \mathrm{n} \tag{41.8}$$

When there is only one hidden layer in this (ANN) model, according to [9], the calculated formula of node number in hidden layer is defined by,

$$n1 = \sqrt{mn} \tag{41.9}$$

Another empirical formula is put forward in [10]. It can be used to determine the node number in hidden layer, and the expression is given by

$$nl = \sqrt{0.43\text{mn} + 0.12\text{m}^2 + 2.54\text{n} + 0.77\text{m} + 0.86} \quad (41.10)$$

In the above five formulas (41.6)–(41.10), (n_1) is the node number in hidden layer, m is the node number in output layer, (n) is the node number in input layer.

41.4 Results and Discussion

41.4.1 Artificial Neural Network Results

Three programs have been written which are feed forward Neural Network Back Propagation (FNNBP) using three dataset; water flow rate effect, fill effect and nozzle effect. The results for each dataset were compared and analyzed based on the convergence rate and the accuracy of results. The training process continues until the slop conditions are met the minimum error (10^{-7}) or maximum number of iteration. Neural networks have ability to learn highly nonlinear relationship. The training of the network (BP) using input and output data, The size of search space in X–Y dimension for (BP) represented by the range of initial weight, biases and maximum dimension.

41.4.2 Cooling Tower Range

Figure 41.4 shows a comparing between experimental [11], and neural results for film fill at height (60 mm). For each value of water inlet temperature, as the water flow rate is increased, the cooling tower range is decreased .The theoretical results show the rate of heat transfer is decreased gradually along the tower, good agreement between experimental and neural predicted results was found.

Figure 41.5 shows a comparing for range between experimental and neural results for film fill at different nozzle hole with heating air. The purpose of heating air to cover wet bulb temperature for air at any condition of weather.

The effect of air flow rate on cooling tower range, for different values of the water flow rate is illustrated in Fig. 41.6. For each value of water flow rate, as the air flow rate is increased, the cooling tower range is increased.

41.4.3 Cooling Tower Effectiveness

Figure 41.7 shows the comparing for effectiveness between experimental and neural results for film fill at height (60 mm) for different water inlet temperature. Good agreement between experimental and neural predicted results was found.

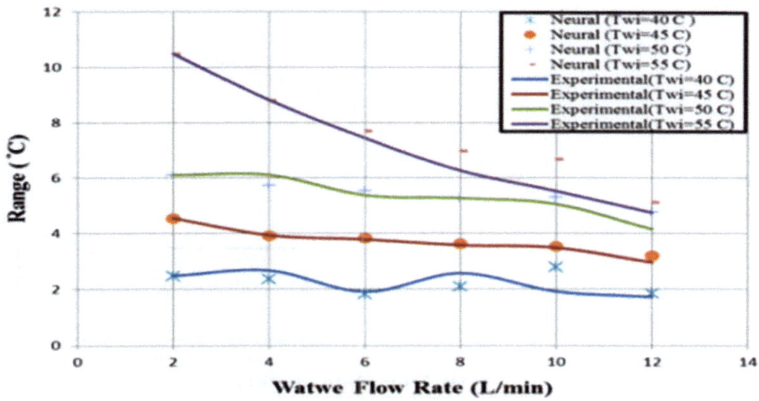

Fig. 41.4 Comparing range between experimental and neural results for film fill at height (60 mm)

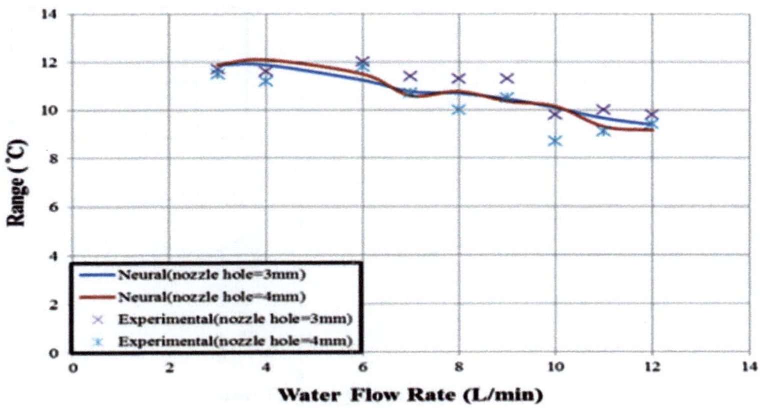

Fig. 41.5 Comparing range between experimental and neural results for film fill at different nozzle hole

Figure 41.8 shows the comparing for effectiveness between experimental and neural results for different nozzle hole. This figure show that effectiveness increase with increasing water flow rate.

The main function of the tower fill is to increase the contact area between the water and air and to maximize the resident time for the water in the tower for this reason when film fill at height (120 mm) effectiveness is more than the other height of fill as shown in Fig. 41.9.

The cooling tower effectiveness increases with the increase in air mass flow rate and decreases with the increase in water flow rate. Therefore, the best cooling tower effectiveness is achieved at the highest flow rate of air and at the lowest water flow rate which is agreement with neural prediction results.

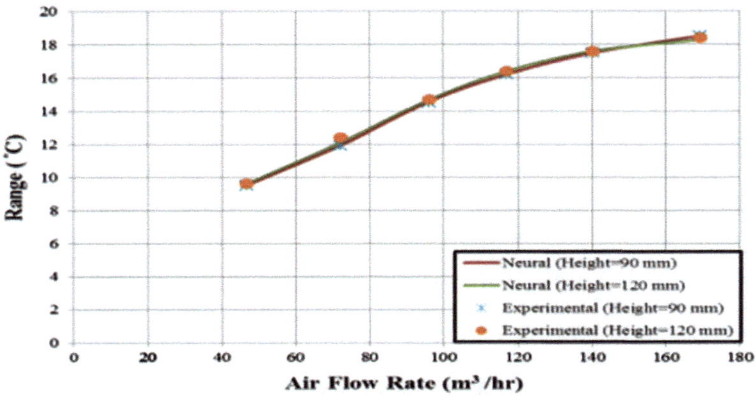

Fig. 41.6 Comparing range between experimental and neural results for film fill at water flow rate (5 L/min)

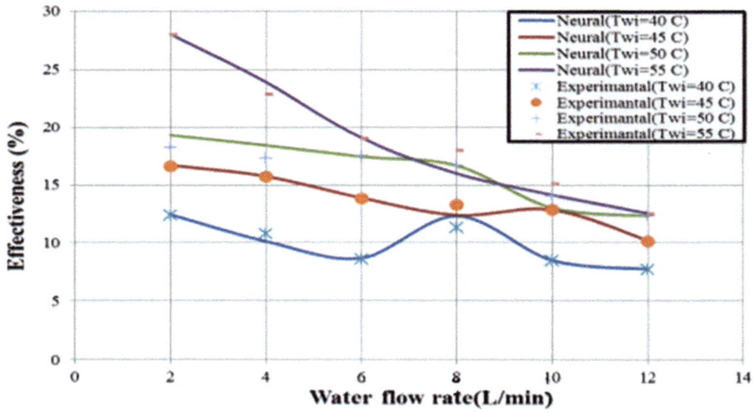

Fig. 41.7 Comparing effectiveness between experimental and neural results for film fill at height (60 mm)

41.4.4 Cooling Tower Approach

Figure 41.10, shows that a comparing approach between experimental and neural results for film fill at height (60 mm). As shown in this figure, the outlet water temperature approach to wet bulb temperature increases with increasing water flow rates. This increase due to the increase of water flow rate. These results are obtained because of the heat and mass transfer coefficients decreasing which gives a good agreement between experimental and neural results.

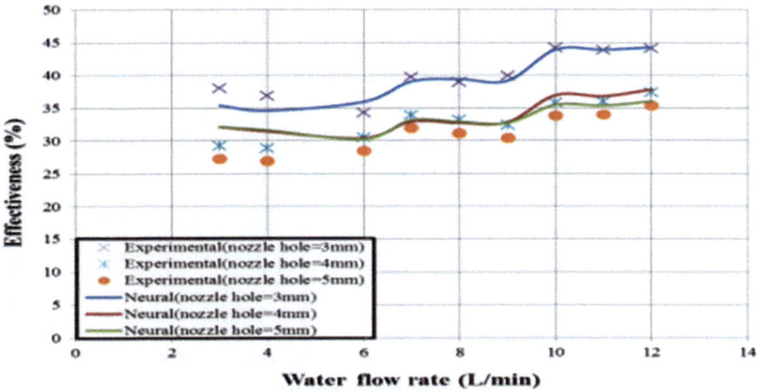

Fig. 41.8 Comparing effectiveness between experimental and neural results for film fill at different nozzle hole

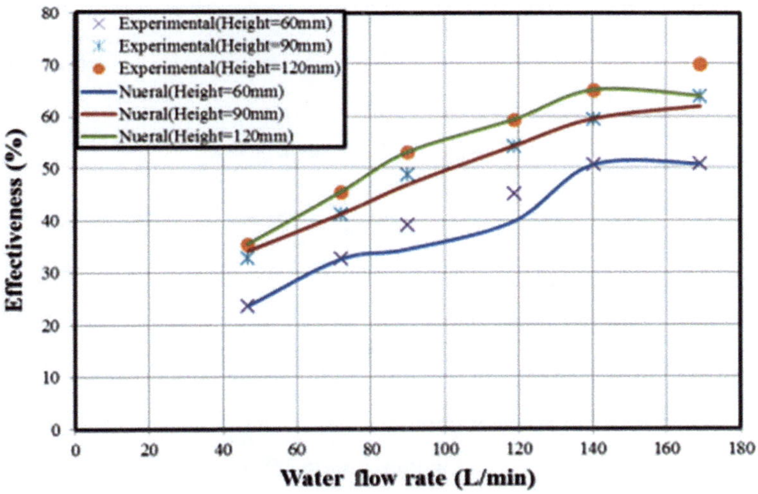

Fig. 41.9 Comparing effectiveness between experimental and neural results for film fill at water flow rate (5 L/min)

Figure 41.11 shows the comparing between experimental and neural results for film fill at water flow rate (5 L/min).approach decrease with increasing air water flow rate at constant water flow rate. The best fill performance is occur at lowest approach where the film fill (height = 120 mm), are achieved that result.

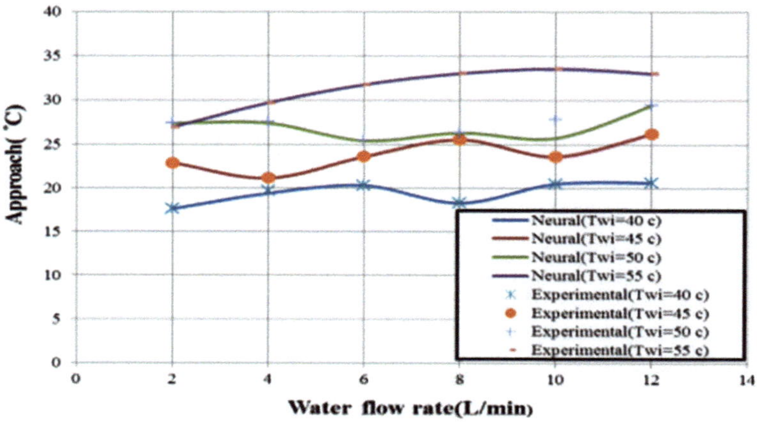

Fig. 41.10 Comparing approach between experimental and neural results for film fill at height (60 mm)

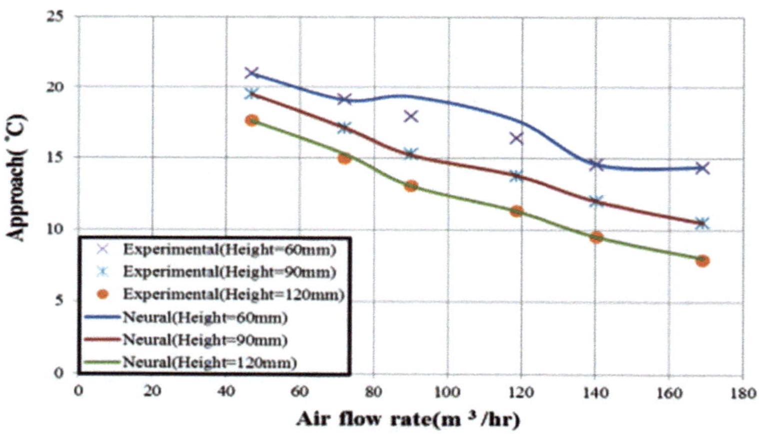

Fig. 41.11 Comparing approach between experimental and neural results for film fill at water flow rate (5 L/min)

41.4.5 Cooling Tower Pressure Drop

Pressure drop is one of the prime concerns while dealing with internal flow in pipe or duct because of the power required by the pump or fan is directly proportional to the pressure drop. As a result, the pressure drop increases almost with the square of the air velocity. This is seen as shown in Figs. 41.12 and 41.13. These graphs are exponentially increasing with the air velocity for both experimentally and theoretically. The higher pressure drop means the higher power requirement by the cooling tower and this raises the operational cost.

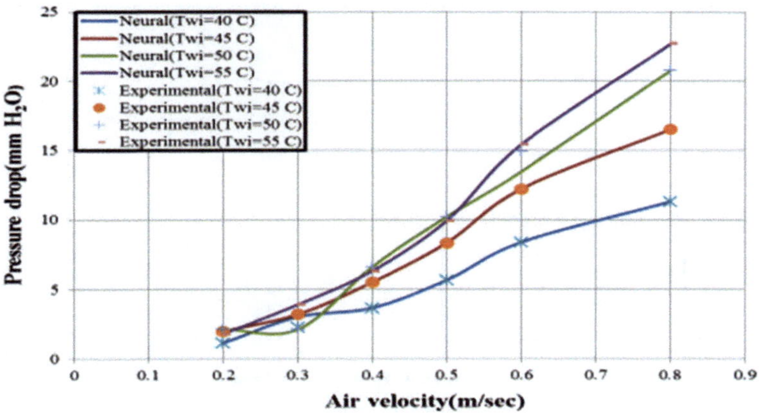

Fig. 41.12 Comparing pressure drop between experimental and neural results for film fill at height (60 mm)

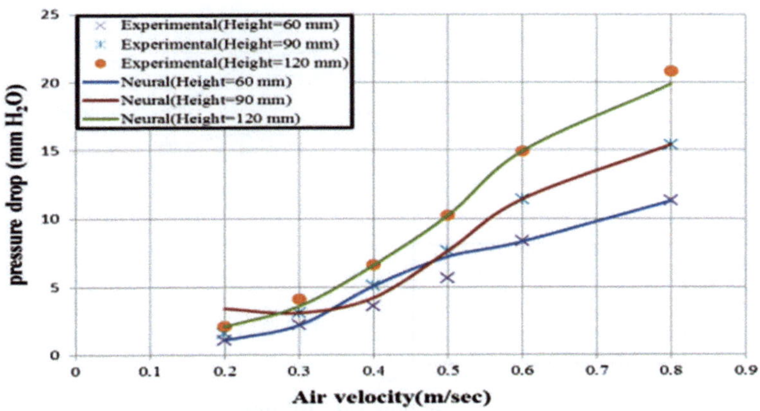

Fig. 41.13 Comparing pressure drop between experimental and neural results for film fill at water flow rate (5 L/min)

Where the agreement with neural prediction results for film fill more agree at constant water flow rate. Pressure drop across the packing increases with increasing air velocity at constant height of fill.

41.4.6 Cooling Tower Outlet Water Temperature

The draft through the natural draft cooling tower is strongly coupled to the water outlet temperature. Figure 41.14 shows the comparing for outlet water temperature between experimental and neural results for film fill at height (60 mm), where the air

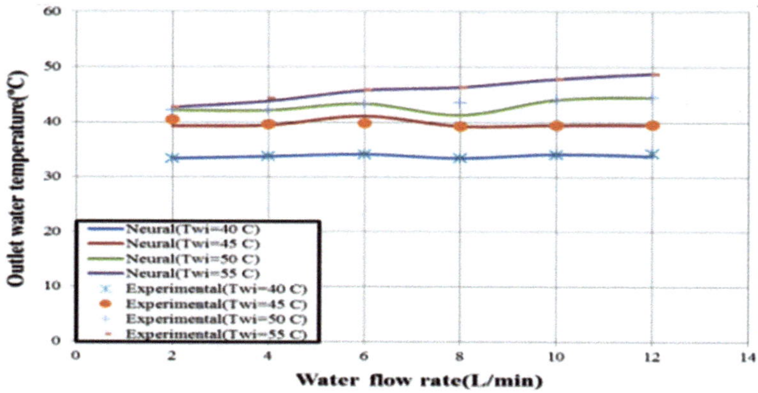

Fig. 41.14 Comparing exit water temperature between experimental and neural results for film fill at height (60 mm)

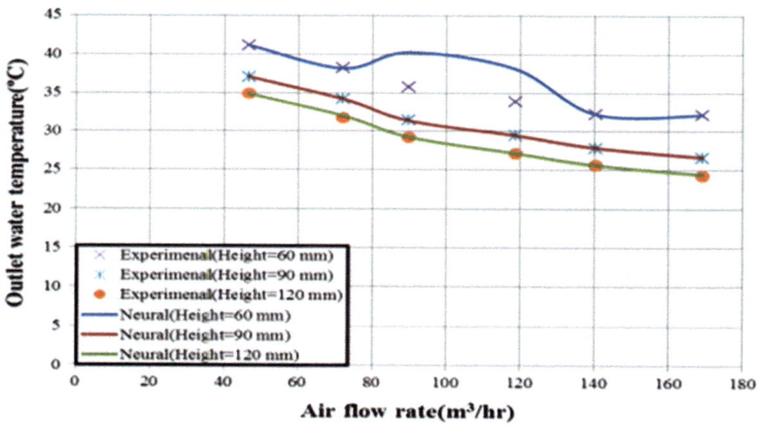

Fig. 41.15 Comparing exit water temperature between experimental and neural results for film fill at water flow rate (5 L/min)

flow rate is related to water outlet temperature. This is because the density of the air inside the cooling tower is a function of the air temperature that agreement with neural prediction results. Figure 41.15 shows a comparing of outlet water temperature between experimental and neural results, where outlet water temperature increase with increasing water flow rate at constant air flow rate (46.8 m^3/h).

Because increasing water flow rate need more cold air flow rate due to exchange heat transfer from one to another. A good the agreement with neural prediction results was found.

41.5 Conclusions

The conclusions drawn from the experimental and theoretical analysis are given as follows:

1. The tower characteristic is improved when the cooling range is decreased and the tower approach is increased at constant tower volume.
2. Increasing the air mass flow rate ($\dot{m}_a$) causes decrease in the temperature of the outlet water, while increasing the water mass flow rate ($\dot{m}_w$) causes increase in the temperature of the outlet water.
3. Increasing the air mass flow rate ($\dot{m}_a$) causes decrease in the wet bulb temperature of the outlet air, while increasing the water mass flow rate ($\dot{m}_w$) causes increase in the wet bulb temperature of the outlet air.
4. The maximum rate of mass and heat transfer occurs at the top stage of the tower. The rate of mass and heat transfer between the water and the bulk air is increased when the inlet water temperature is increased, while the rate is decreased when the inlet air wet bulb temperature is increased.

References

1. M. M. El-Wakil, *Power Plant Technology* (McGraw-Hill, New York, 1988)
2. M. Gao, F. Sun, K. Wang, Y. Shi, Y. Zhao, Performance Prediction of wet cooling tower using artificial neural network under cross-wind conditions. Int. J. Therm. Sci. **48**, 583–589 (2009)
3. K. Hooman, School of Mechanical and Mining Engineering, The University of Queensland, Qld4072, Australia (2010)
4. American Society of Heating, in *Refrigerating and air-conditioning engineers*, ASHRAE System and Equipment Handbook (SI) (2008)
5. M. Hosoz, H.M. Ertunc, H. Bulgurcu, Performance prediction of a cooling tower using artificial neural network. Energy Convers. Manag. **48**, 1349–1359 (2007)
6. R. Hecht-Nielson, in Theory of back propagation neural networks, *Proceedings of the International Joint Conference on Neural Networks*, Washington, vol 1, pp. 593–605 (1989)
7. F. Xin, *Basic Theory and Method of Neural Net Intelligence* (Southwest Jiaotong University Press, Chengdu, 2000)
8. E. Ding, *Air Cooling Techniques in Power Plants* (Water and Electric Power Press, Beijing, 1992)
9. Q. S. Xie, *Neural Net Method in Mechanical Engineering* (China Machine Press, Beijing, 2003)
10. Y. B. Yao, J.L. Wang, Research on raising BP network training speed. Inf. Technol. **1**, 4–6 (2002)
11. B. S. Khalaf, Experimental and theoretical study of natural draft cooling tower performance, Ph.D. thesis, Baghdad University (2012)

Chapter 42
Exergo-Economic Analysis of an Experimental Aircraft Turboprop Engine Under Low Torque Condition

Ramazan Atilgan, Onder Turan and Hakan Aydin

Abstract Exergo-economic analysis is an unique combination of exergy analysis and cost analysis conducted at the component level. In exergo-economic analysis, cost of each exergy stream is determined. Inlet and outlet exergy streams of the each component are associated to a monetary cost. This is essential to detect cost-ineffective processes and identify technical options which could improve the cost effectiveness of the overall energy system. In this study, exergo-economic analysis is applied to an aircraft turboprop engine. Analysis is based on experimental values at low torque condition (240 N m). Main components of investigated turboprop engine are the compressor, the combustor, the gas generator turbine, the free power turbine and the exhaust. Cost balance equations have been formed for all components individually and exergo-economic parameters including cost rates and unit exergy costs have been calculated for each component.

42.1 Introduction

Energy analysis (also known as first law analysis) states conservation of energy, and that is not enough for identifying the real inefficiencies and imperfections. It is a measure of quantity of energy only. First law based analysis can be misleading alone. Besides, exergy analysis (also known as second law analysis) complements

R. Atilgan (✉) · O. Turan
Faculty of Aeronautics and Astronautics, Anadolu University, 26470 Eskisehir, Turkey
e-mail: ratilgan@anadolu.edu.tr

O. Turan
e-mail: onderturan@anadolu.edu.tr

H. Aydin
TUSAS Engine Industries, Eskisehir, Turkey
e-mail: tei.hakan@gmail.com

A. Y. Oral et al. (eds.), *International Congress on Energy Efficiency and Energy Related Materials (ENEFM2013)*, Springer Proceedings in Physics 155,
DOI: 10.1007/978-3-319-05521-3_42,

energy analysis and, identifies and quantifies locations, magnitudes and sources of irreversibilities through exergy destruction rates of each engine component. This combination gives a chance to designers and operators to use non-renewable resources, such as kerosene ($C_{12}H_{23}$) which is used in aircraft engines as fuel, effectively and also to reduce operating costs besides environmental impacts [1, 2]. Exergy analysis is a measure of both quantity and quality of energy. Exergo-economic analysis is an unique combination of exergy analysis and cost analysis conducted at the component level. In exergo-economic analysis, cost of each exergy stream is determined. Inlet and outlet exergy streams of the each component are associated to a monetary cost.

Exergo-economic analysis is widely used to evaluate cost efficiency of energy conversion systems, but application of these analyses to the aircraft engines are only performed on some studies [3–6]. In this study, exergo-economic analysis is applied to an experimental aircraft turboprop engine that operates under 240 N m torque condition.

42.2 System Description

Turboprop engines are well-known with their low fuel consumption especially in short-haul flights (under 1,000 miles) [7] and have the highest propulsive efficiency at flight speeds of 400 mph ($\cong$650 kph) compared to turbojet and turbofan engines [8]. A study showed that an advanced turboprop aircraft concept cruising at between 640 and 700 kph will result 25 % cut of CO_2 emissions by reducing fuel consumption [9].

Turboprop engines use a gas turbine core to drive an external propeller which produces higher bypass ratios (BPR) nearly the same as turbofan engines. This higher BPR can be an explanation of how turboprop engines have high fuel efficiency. Because, higher BPR means smaller amount of air flow burned inside the engine core and that makes possible for turboprops to generate a lot of thrust at low fuel consumption [10].

The external propeller of turboprop engine should rotate at very low speed in comparison with its driving turbine; therefore a gear box is used in turboprop engines to provide a speed reduction of almost 1:15. This reduction is necessary, because no propeller can resist stresses and forces occur as a consequence of centrifugal forces when it rotates at the same speed of turbine. Moreover, at high subsonic flights (over 0.7 Mach) the tips of blades can approach supersonic speeds, flow separates and shock waves form. This causes to decreased propeller efficiency and poor overall engine performance, hence turboprops give best performance at subsonic flights (under 650 kph) [8].

In this study, a turboprop aircraft engine (CT7-9C) is investigated. This engine is commercial turboprop version of GE T700 turboshaft engine. CT7-9 has been operated in several aircrafts since 1984, and more than 1,500 engine have been produced. Over 30 million flight hours are logged for this engine all around the World [11].

Main components of CT7-9C are compressor (consist of five axial stages and one centrifugal impeller stage), combustion chamber (annular type), gas turbine (two axial stages), power turbine (two axial stages) and an exhaust nozzle. Control volume of this engine is shown in Fig. 42.1 [1]. Components of the engine and two shafts can be seen in this schematic. CT7-9C turboprop engine has free turbine which means the propeller is driven by a designated turbine named power turbine. Another turbine named gas turbine drives compressor, and thus engine start mechanism does not requires rotating the propeller and accessory gearbox. It only drives compressor and gas turbine and this provides to reduce torque loads [12].

42.3 Methodology and Analysis

Exergy analysis provides to assess sources, locations and true magnitudes of exergy destructions. As stated before, exergy analysis complements first law based energy analysis and combines first law analysis and second law analysis together. Exergy can be divided into four parts: physical exergy, chemical exergy, potential exergy and kinetic exergy. Potential and kinetic exergies are neglected in this study. Physical exergy and chemical exergy (only for combustion chamber) are taken into account in calculations [13].

42.3.1 Assumptions

- Exergy is defined as energy that is out of equilibrium with the reference environment [14]. Hence, determining properties of the term environment accurately is one of the key issues such exergy based analysis. In this study, temperature and pressure of the environment (T_0 and p_0) are taken as 279 K and 93 kPa, respectively.
- CT7-9C turboprop engine lifetime of 30 years and 700 total annual numbers of hours of engine at full load per year is assumed,
- $\dot{C}_{W_{comp}} = \dot{C}_{W_{ggt}}$,
- $\dot{C}_4 = \dot{C}_{45} = \dot{C}_5$,
- All system units are assumed to be adiabatic,
- Air and exhaust gases in the engine are ideal gases,
- The combustion reaction is complete,
- 5 % of the compressor discharge air is assumed to be used as cooling and seal pressurising purposes,

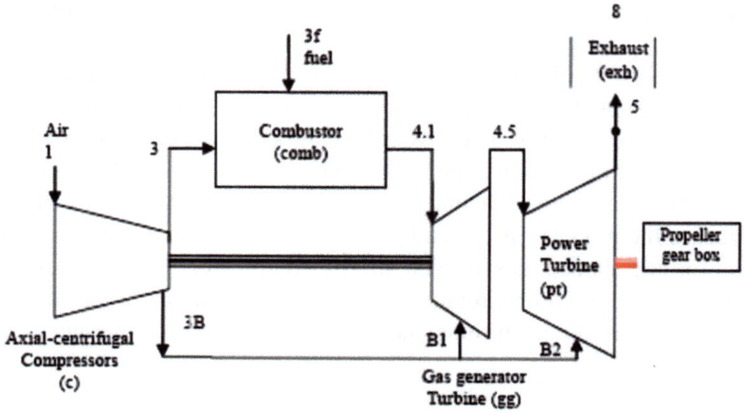

Fig. 42.1 Control volume of turboprop engine [1]

- Pumps (fuel, oil and hydraulic) and heat exchangers (fuel = air or fuel = oil) are not included in the analysis,
- The kinetic and potential exergies are neglected,
- Chemical exergy is neglected other than combustion chamber [3].

42.3.2 Exergo-Economic Equations

Main equations for exergo-economic analysis can be written as follows [3]:

$$\sum_{in} \dot{C}_k + \dot{Z}_k^{TOT} = \sum_{in} \dot{C}_k + \dot{C}_w \tag{42.1}$$

$$\dot{C}_k = c\dot{E}x_k \tag{42.2}$$

$$\dot{C}_w = c\dot{W} \tag{42.3}$$

$$\dot{Z}_k^{TOT} = \dot{Z}_k^{CIC} + \dot{Z}_k^{OM} \tag{42.4}$$

where $\dot{C}_k$ and $\dot{C}_w$ are the exergy costs of the flow of *k*th component and power; c is the unit exergy costs of stream and power; $E\dot{x}_k$ and $\dot{W}$are the exergy of stream and power entering and leaving related the control volume; $\dot{Z}_k^{CIC}$, $\dot{Z}_k^{OM}$ and $\dot{Z}_k^{TOT}$ are hourly levelised costs of capital investment cost, operating and maintenance and the total levelised cost of equipment inside the control volume, respectively.

Table 42.1 Matrix system for linear equation system

c_3	c_4	$c_{W_{gt}}$	$c_{W_{pt}}$	*const.*
−4.075	0	4.068	0	−96.25
4.075	−10.692	0	0	−459.81
0	0.637	−4.835	0	−102.46
0	4.975	0	−2.030	−51.23

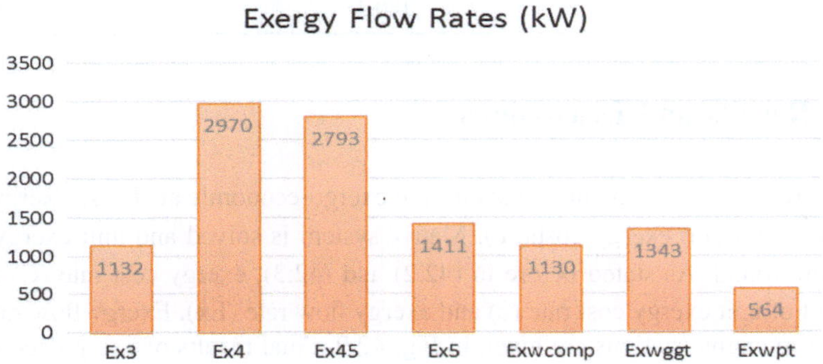

Fig. 42.2 Exergy flow rates

42.3.3 Cost Balance Equations

Cost balance equations are formed by using main equations and control volume of the engine, and are given below in (42.5) (42.6) (42.7) (42.8) and (42.9) for the compressor, combustion chamber, gas generator turbine, power turbine and exhaust nozzle, respectively:

$$\dot{C}_1 + \dot{C}w_{comp} + \dot{Z}_{comp}^{TOT} = \dot{C}_3 + \dot{C}_B \tag{42.5}$$

$$\dot{C}_3 + \dot{C}_f + \dot{Z}_{cc}^{TOT} = \dot{C}_4 \tag{42.6}$$

$$\dot{C}_4 + \dot{Z}_{ggt}^{TOT} = \dot{C}_{W_{ggt}} + \dot{C}_{45} \tag{42.7}$$

$$\dot{C}_{45} + \dot{C}_B + \dot{Z}_{ppt}^{TOT} = \dot{C}_{W_{pt}} + \dot{C}_5 \tag{42.8}$$

$$\dot{C}_5 + \dot{Z}_{ex}^{TOT} = +\dot{C}_8 \tag{42.9}$$

After cost balance equations above are formed, a linear equation system is obtained with 5 equations and 8 variables. This equation system can be solved with the aid of a matrix system which is given in Table 42.1.

Table 42.2 Exergy cost rates and unit exergy cost rates

Station number	$\dot{c}$ ($/Gj)	$\dot{C}$ ($/h)
3	53.1	216.4
4	63.2	676.2
45	63.2	635.9
5	63.2	321.2
w_{comp}	29.5	120.1
w_{ggt}	29.5	142.7
w_{pt}	180.2	365.9

42.4 Results and Discussions

There are roughly two main outputs for the exergo-economic analysis: Exergy cost rates ($\dot{C}$) and unit exergy costs ($\dot{c}$). Matrix system is solved and unit exergy cost rates are found. As stated before in (42.2) and (42.3), exergy cost rate ($\dot{C}$) is the product of unit exergy cost rate ($\dot{c}$) and exergy flow rate ($\dot{E}x$). Exergy flow rates of turboprop engine stations are given in Fig. 42.2. Final results of exergo-economic analysis are shown in Table 42.2.

42.5 Conclusion

In this paper, exergo-economic analysis of an aircraft turboprop engine is performed. Cost performance of engine components are investigated individually. Power turbine produces shaft horse power to drive propeller, and exergy cost rates and unit exergy cost rates for power turbine of the engine is calculated as 365,9$/h and 180,2$/Gj respectively. The highest value for exergy cost rate is found for inlet stream of gas generator turbine (station number 4) with the value of 676.2$/h, while the highest unit exergy cost value is found for power turbine with the value of 180.2$/Gj. These results can help to operators and designers to identify cost-ineffective processes and identify technical options which could improve the cost effectiveness of the overall turboprop engine and engine components.

References

1. R. Atılgan, O. Turan, O. Altuntas, H. Aydın, K. Synylo, Environmental impact assessment of a turboprop engine with the aid of exergy. Energy (2013)
2. M.J. Moran, H.N. Shapiro, *Fundamentals of engineering thermodynamics* (Wiley, USA, 2008)
3. H. Aydin, O. Turan, A. Midilli, T.H. Karakoc, Exergetic and exergo-economic analysis of a turboprop engine: a case study for CT7-9C. Int. J. Exergy **11**(1), 69–88 (2012)

4. E.T. Turgut, T.H. Karakoc, A. Hepbasli, Exergoeconomic analysis of an aircraft turbofan engine. Int. J. Exergy **6**(3), 277–294 (2009)
5. O. Balli, H. Aras, N. Aras, A. Hepbasli, Exergetic and exergoeconomic analysis of an aircraft jet engine (AJE). Int. J. Exergy **5**(5–6), 567–581 (2008)
6. O. Altuntas, T.H. Karakoc, A. Hepbasli, Exergetic, exergoeconomic and sustainability assessment of piston-prop aircraft engine. Int. J. Therm. Sci. Technol. **32**(2), 133–143 (2012)
7. M.S. Ryerson, M. Hansen, The potential of turboprops for reducing aviation fuel consumption. Transp. Res. Part D **15**, 305–314 (2010)
8. A.F. El-Sayed, *Aircraft propulsion and gas turbine engines* (CRC Press, USA, 2008)
9. J. Akermann, Sustainable air transport-on track in 2050. Transp. Res. Part D **10**, 111–126 (2005)
10. M.J. Kroes, T.W. Wild, *Aircraft Powerplants* (Glencoe/McGraw-Hill, USA, 1995)
11. H. Aydin, Doctoral thesis, Natural and Applied Science, Anadolu University, Eskisehir, Turkey (2010)
12. H. Aydin, O. Turan, T.H. Karakoc, A. Midilli, Component-based exergetic measures of an experimental turbopop/turboshaft engine for propeller aircrafts and helicopters. Exergy **11**(3), 322–347 (2012)
13. P.M. Sforza, *Theory of aerospace propulsion* (Elseiver, Amsterdam, 2011)
14. D.S. Scott, Exergy. Hydrogen Energy **28**, 369–375 (2003)

Chapter 43
The Potential of Solar as Alternative Energy Source for Socio-Economic Wellbeing in Rural Areas, Malaysia

Rashidah Zainal Alam, Chamhuri Siwar and Norasikin Ahmad Ludin

Abstract Malaysia's energy sector is highly dependent on fossil fuels as a primary energy source. Economic growth and socio-economic wellbeing also rely on the utilization of energy in daily life routine. Nevertheless, the increasing cost for electricity and declining fossil fuels resources causes various negative impacts to the people and environment especially in rural areas. This prompted Malaysia to shift towards alternative energy sources such as solar energy to ensure social, economic and environmental benefits. The solar energy is one of the potential renewable energy sources in tropical countries particularly in Malaysia. The paper attempts to analyze the benefits and advantages related to energy efficiency of solar for sustainable energy use and socio economic wellbeing in rural areas, Malaysia. The paper uses secondary sources of data such as policies, regulations and research reports from relevant ministries and agencies to attain the objectives. As a signatory country to the UN Convention on Climate Change and the Kyoto Protocol, Malaysia has taken initiatives for decreasing energy dependence on oil to reduce greenhouse gas emissions (GHG) for sustainable development. The paper shows solar energy becomes one of the promising alternative energy sources to alleviate energy poverty in Malaysia for rural areas. Finally, solar energy has increased socio-economic wellbeing and develops green potential and toward achieving energy efficiency in energy sector of Malaysia by preserving environment as well as reducing carbon emission.

R. Z. Alam (✉) · C. Siwar
Institute for Environment and Development (LESTARI), Universiti Kebangsaan Malaysia (UKM), Bangi, Selangor, Malaysia
e-mail: shida_rza@yahoo.com

C. Siwar
e-mail: csiwar@ukm.my

N. A. Ludin
Solar Energy Research Institute (SERI), Universiti Kebangsaan Malaysia (UKM), Bangi, Selangor, Malaysia
e-mail: shikin_46@yahoo.com

A. Y. Oral et al. (eds.), *International Congress on Energy Efficiency and Energy Related Materials (ENEFM2013)*, Springer Proceedings in Physics 155,
DOI: 10.1007/978-3-319-05521-3_43,

43.1 Introduction

Population of this world is increasing rapidly years by years. This number is expected to dramatically boost with the demand of energy in worldwide. Energy is a major factor for the economic growth and benefits to socio-economic wellbeing in global market scenario [1]. In Malaysia, energy become main contributor for rapid growth by increase about 20 % from 13,000 MW in year 2000 to 15,000 MW in year 2009 [2]. People utilize the energy to facilitate the daily life routine [3]. Most of the countries are depending on fossil fuels for meeting their energy demand [4]. Nevertheless, the increasing cost for electricity and declining fossil fuels resources causes various negative impacts to the people and environment especially in rural areas.

Therefore, renewable energy provide a better solution accommodate the demand of energy to beneficial sustainable development. It including people in rural areas which always being related with lack of basic needs [5] or usually suffered from energy poverty [6]. The source which is infinite and available also abundant source become the major factors of choice to supply the energy in rural areas [7].

Solar energy is one of the potential energy that could be harvested for energy uses especially in rural areas. The location of Malaysia which at latitude of 3.164-N and 101.7-E is considered totally equatorial [6] and Malaysia's condition also which in the tropical region [8, 9]. In order to make solar energy progressively provide benefits to rural areas, elements of energy efficiency needed to put into consideration. This is because energy efficiency has helped in unnecessary waste of energy as well as cost and prevents environmental degradation [10].

People in rural areas are the community that always being neglected their needs for the development. Using traditional fossil fuels energy like cerosin are obviously give negative impacts to the environment and health. Lower income people and poor are the major communities living in rural areas [11]. Solar energy could provide a better option to rely on in order to increase socio-economic wellbeing and reduce the poverty of energy in rural areas. The present study explores the potential of solar energy becomes one of the promising alternative energy sources for Malaysia in rural areas and how solar energy has increased socio-economic wellbeing and develops green potential in energy sector of Malaysia by preserving environment as well as reducing carbon emission.

43.2 Methods

The study used secondary sources such as policies, acts and regulations from relevant ministries and agencies of the Malaysian Government. Relevant published materials such as research reports, articles, books, annual reports as well as websites were also reviewed in order to accumulate secondary data.

43.3 Results and Discussion

43.3.1 Solar Energy as a Potential Alternative Source

Solar has become as one of the potential energy source after Government of Malaysia (GoM) embarked the National Policy on RE also as fastest growing on renewable energy [12]. Furthermore, solar has present to be a sufficient and convenient energy resource especially in rural areas [13]. Solar also become popular and act as perfect choice because it only need minimal maintenances, quick times to install [12] and placed in the strategic geographical location, which has abundant sunshine (Fig. 43.1) [14] with average irradiance per year of 1,643 kWh/m^2 become promising condition for the development of solar energy [15] in rural areas, Malaysia. Solar is projected to has highest cumulative capacity among other RE up to 9 GW in Malaysia (Fig. 43.2) [6] and good for the place where electricity grid extension is impractical [16, 17] and to overcome the expensive cost to install for the small population which always being the obstacle to supply electricity. Besides, solar energy also reduce carbon emission that make it relevant as sustainable energy [13]. Table 43.1 shows that there are increasing demand of solar energy in rural areas, Malaysia [18]. Besides that, growing number of technologies for solar energy reducing energy efficiency gap as an opportunity for meeting many social and environmental objectives [19].

43.3.2 Solar Energy: Potential for Socio-Economic Wellbeing

Recent studies put the center of discussion mostly about potential of solar as alternative energy [20] but not clearly defined or debate on how solar actually benefits people in term of socio-economic wellbeing especially in rural areas. Research conducted by [17, 25, 26, 28] proven that solar being as aid to socio-economic wellbeing in rural area in Malaysia. As a signatory country to the UN Convention on Climate Change and the Kyoto Protocol, Malaysia has taken initiatives for decreasing energy dependence on oil to reduce greenhouse gas emissions (GHG) for sustainable development [21]. Malaysia has a population of 28.3 million people and about 71.3 % of Malaysia's population lived in urban areas while 28.7 % live in rural areas [22]. Based on the statistics in 2009, 3.8 % of them live below the poverty line [23]. So, how solar could change and improve socio-economic wellbeing in rural areas, Malaysia?

People in rural areas normally come from lower income people and poor. These groups relying on traditional sources such as charcoal, firewood and farm residue to meet their energy demand and basic daily life routine [24]. The grid electricity connection is out of coverage and very expensive [3]. Using traditional source from fossil fuel emitted a lot of hazardous chemical that not just effect the

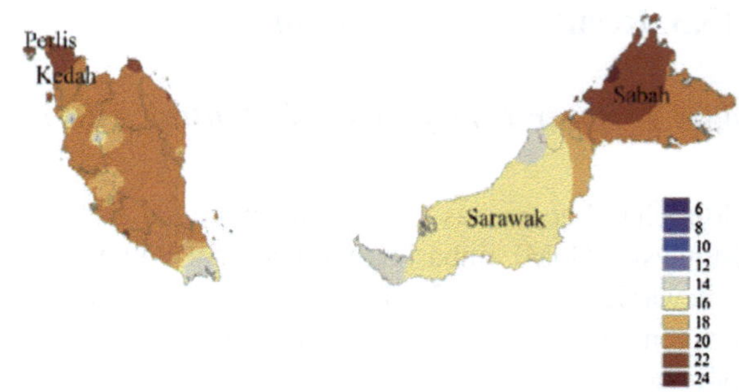

Fig. 43.1 Annual average solar radiation in Malaysia (MJ/m^2/day) [14]

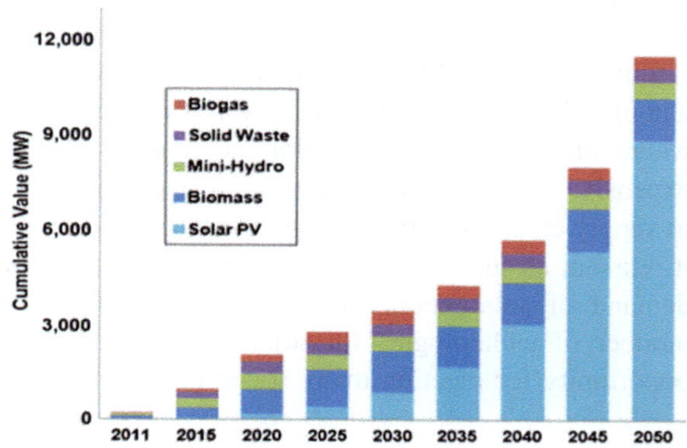

Fig. 43.2 Cumulative value of renewable energy in Malaysia [6]

Table 43.1 Solar in rural areas, Malaysia [18]

Year	State	Project name	No. of houses	Capacity installed (kW)
2005	Pahang	Kg Org Asli Ganuh, Rompin	56	100
	Kelantan	Kg Org Asli Blau, Gua Musang	54	100
		Kg Org Asli Aring 5, Gua Musang	70	100
	Sabah	Kg Pegalungan, Nabawan	76	100
		Kg Monsok Ulu, Tambunan	19	45
		Kg Sinulihan, Tuaran	44	50
2006	Sabah	Kg Meligan, Sipitang	140	200
		Kg Lubukan, Semporna	40	75
	Johor	Kg Orang Asli Sg Peroh, Kluang	18	45
		Kg Org Asli Tg Tuan, Mersing	24	45
		Kg Org Asli Tanah Abang, Mersing	81	150
		Kg Peta, Mersing	58	100
		Kg. Punan, Mersing	60	100
2007	Sabah	Kg Karakit dan Pekan Karakit	63	850
2009–2010	Sabah	Kg Kuamut	270	1,000

environment but also bad for human's health [24]. The solar energy is a good option to overcome the problem. A good health is important to continue life routine to make it efficient. Studies conducted by [16] shows that solar energy could increase health and to improve the life routine.

Women are usually associated with the slavery at their house mostly in rural areas [25]. With bunch of works at home, women lost their purpose of living and do not have extra time for themselves. Solar bring ray of hope for women in rural areas. Solar energy could increase women empowerment [26]. Women traditionally have to work long hours at home for cooking, wash clothing and cleaning, now could have extra time to communicate and have the social interaction with the neighbours and the community [25, 16]. Women also could increase their incomes by build cottage industries with the assistance of solar energy in rural areas.

Education is a critical element to increase socio-economic wellbeing in rural areas. Unfortunately, the short supply of electricity limited the study hours for children especially during night times. Therefore, solar energy is relevantly educated people by providing electricity in rural areas [27, 28].

Solar energy shows its potential for socio-economic wellbeing in rural areas. This potential is covered from improve human health, increase women empowerment by vanished gender inequality, enhance children's education also developing strong self-esteem and confident levels. These potential's elements are drivers for poverty eradication especially in rural areas, Malaysia.

43.4 Conclusions

Solar energy is renewable source of energy and has a great potential to be developed. Efforts to improve solar energy are needed such as integration, participation and collaboration between the government, private sector, non-governmental organizations (NGOs) and communities, particularly for rural areas. It is necessary to develop solar potential in rural areas throughout Malaysia. Support and assistance especially in terms of financial assistance, skills transfer to operate and maintenance the solar system to the local population are necessary to ensure not only the sustainability of the use of solar power for a long period but also to increase the delivery of efficiency of solar system to the community in rural areas. Education and awareness about the importance of solar energy should also be instilled among Malaysians, especially those in rural areas so that the potential of this energy source can be developed further and could preserving environment as well as reducing carbon emission.

Acknowledgments Financial assistance provided by the Fundamental Research Grant Scheme (FRGS), on "Greening the Economy", Institute for Environment and Development (LESTARI), Universiti Kebangsaan Malaysia (Ref. No. FRGS/1/2012/SSO7/UKM/01/3) headed by Prof. Emeritus Chamhuri Siwar is gratefully acknowledged.

References

1. E. Kentel, E. Alp, Hydropower in Turkey: Economical, social and environmental aspects and legal challenges. Environ. Sci. Policy **31**, 34–43 (2013)
2. M. Siti Indati, Y.P. Leong, H.H. Amir, Issues and challenges of renewable energy development: A Malaysian experience, PEA-AIT international conference on energy and sustainable development: Issues and strategies, ESD, 2010
3. Rashidah, Z. Aa, Siwar, C. Ahmad Ludin, N.: Potensi Mikro Hidro Sebagai Sumber Tenaga Alternatif ke arah Mencapai Kelestarian Ekologi di Kawasan Pedalaman, Malaysia. Prosiding Seminar Antarabangsa Ke-6 Ekologi, Habitat manusia and erubahan Persekitaran di Alam Melayu (2013)
4. M.S. Firdaus, R.I. Roberto, A.B. Siti Hawa, G.M. Scott, An evaluation of the installation of solar photovoltaic in residential houses in Malaysia: Past, present and future. Energy Policy, 7975–7987 (2011)
5. R.K. Akikur, R. Saidur, H.W. Ping, K.R. Ullah, Comparative study of stand-alone and hybrid solar energy systems suitable for off-grid rural electrification: A review. Renew. Sustain. Energy Rev. **27**, 738–752 (2013)
6. H. Borhanazad, S. Mekhilef, R. Saidur, G. Boroumandjazi, Potential application of renewable energy for rural electrification in Malaysia. Renew. Energy **59**, 210–219 (2013)
7. Z.A. Rashidah, M.A.H. Bhuiyan, C. Siwar, N.A. Ludin, *Potential of biomass as alternative energy source in Malaysia*. 4th Social, Development and Environmental Studies International Conference, Universiti Kebangsaan Malaysia (UKM), Bangi, Malaysia (2013)
8. S. Ahmad, M.Z.A.A. Kadir, S. Shafie, Current perspective of the renewable energy development in Malaysia. Renew. Sustain. Energy Rev. **15**(2), 897–904 (2011)
9. S.C. Chua, T.H. Oh, W.W. Goh, Feed-in tariff outlook in Malaysia. Renew. Sustain. Energy Rev. **15**(1), 705–712 (2011)
10. C.C. Shing, H.O. Tick, Solar energy outlook in Malaysia. Renew. Sustain. Energy Rev. **16**, 564–574 (2012)
11. B. Mainali, S. Silveira, Financing off-grid rural electrification: Country case Nepal. Energy (2010). doi:10.1016/j.energy.2010.07.004
12. Malaysian Building Integrated Photovoltaic Project (MBIPV). Final evaluation report (2011), Available at http://www.mbipv.net.my/content.asp?zoneid=9&categoryid=49
13. A.A. Lahimer, M.A. Alghoul, F. Yousif, T.M. Razykov, N. Amin, K. Sopian, Research and development aspects on decentralized electrifications options for rural household. Renew. Sustain. Energy Rev. **24**, 314–324 (2013)
14. B. Abdul Malek, *Renewable energy development in Malaysia*. 34th Expert Group on New and Renewable Energy Technologies (EGNRET). Kuala Lumpur, Malaysia, 1–46 (2010)
15. A.H. Haris, *MBIPV Project: Catalyzing Local PV Market* (Finance and Investment Forum on PV Technology, Malaysia, 2008)
16. N. Wamukonya, Socio-economic impacts of rural electrification in Namibia: Comparisons between grid, solar and unelectrified households. Energy Susutain. Dev. (5),3 (2001)
17. United Nations Development Programme (UNDP), Energy and poverty in Malaysia: Challenges and the way forward. UNDP Regional Centre in Bangkok (2007)
18. Ministry of Rural and Regional Development (2012), Available at htto://www.rurallink.gov.my/electricity (2012)
19. G. Calum, Energy efficiency vs economic efficiency. Energy Policy **2**, 241–254 (1997)
20. N.L. Panwar, S.C. Kahushik, S. Kothari, Role of renewable energy source in environmental protection: A review. Renew. Sustain. Energy Rev. **15**, 1513–1524 (2011)
21. S. Selamat, C.Z. Azner Abidin, Renewable energy and kyoto protocol: Adoption in Malaysia (2010), Available at http://publicweb.unimap.edu.my
22. Department of Statistics 2010, Report population distribution and basic demographic characteristics (2010)

23. World Bank (2009), Countries and economic, Malaysia, Available at http://data.worldbank.org/country/malaysia
24. J.A. Cherni, Y. Hill, Energy and policy providing for sustainable rural livelihoods in remote locations-the case of Cuba. Geoforum **40**, 645–654 (2009)
25. I. Mahat, Implementation of alternative energy technologies in Nepal: Towards the achievement of sustainable livelihoods. Energy Sustain. Dev. **8**, 2 (2004)
26. S. Twomlow, D. O'Neill, J.T. Ellis-Jones, Rural development and engineering perspective on sustainable smallholder farming in developing countries. Biosyst. Eng. **81**(3), 355–362 (2002)
27. M. Gustavsson, Educational benefits from solar technology—Access to solar electric services and changes in children's study routines, experiences from eastern province Zambia. Energy Policy. **35**, 1292–1299 (2007)
28. S. Mekhilef, A. Safari, W.E.S. Mustaffa, R. Saidur, R. Omar, M.A.A. Younis, Solar energy in Malaysia: Current state and prospects. Renew. Sustain. Energy Rev. **16**, 386–396 (2012)

Chapter 44
Economic and Environmental Assessment of a 1 MW Grid Connected Rooftop Solar PV System for Energy Efficient Building in Bangladesh

Sanjib Chakraborty, Rubayet Hosain, Toufiqur Rahman and Ahmead Fazle Rabbi

Abstract This paper evaluates the potentiality of a 1 MW grid connected rooftop solar PV system for an Energy Efficient Building in Bangladesh, which was estimated by utilizing NASA SSE solar radiation data, PVsyst simulation software and RETScreen simulation software. Economic and environmental viability for a ten-storied building with roof area of 6,500 m^2 in the Capital City of Bangladesh, Dhaka was assessed by using the RETScreen simulation software. The yearly electricity production of the proposed system was 1,581 MWh estimated by PVsyst where the technical prospective of gird-connected solar PV in Bangladesh was calculated as about 50,174 MW. The economic assessments were determined the simple payback in such a way that the generated electricity first fulfills the demand of the building, and then the rest of the energy is supplied to the grid. The result indicates that the roof top solar PV system for an Energy efficient building in Dhaka city has a favorable condition for development both in economic and environmental point of view.

S. Chakraborty (✉) · R. Hosain · T. Rahman · A. F. Rabbi
Department of Management and Engineering, Linköping University, Linköping, Sweden
e-mail: sanch062@student.liu.se

R. Hosain
e-mail: rubho892@student.liu.se

T. Rahman
e-mail: mdra347@student.liu.se

A. F. Rabbi
e-mail: ahmra344@student.liu.se

A. Y. Oral et al. (eds.), *International Congress on Energy Efficiency and Energy Related Materials (ENEFM2013)*, Springer Proceedings in Physics 155,
DOI: 10.1007/978-3-319-05521-3_44,

44.1 Introduction

Our local energy supply and energy use have direct influence on our global energy system of transitions and or global climate changes. These changes could be measured and identified through an intensive study on the energy use of the buildings we live in. On the basis of energy efficient solar buildings concepts [1] the source of electricity is an important part to supply of primary energy. Solar energy systems therefore have an important role to play.

In the urban areas of Bangladesh thousands of building are being constructed every year, by utilizing the buildings properly it is possible to produce electricity from solar PV system. If we look through the solar PV technology diffusion in Bangladesh, it is worth to say that Bangladesh is celebrating installation of 2 million solar PV home systems recently [2]. The established renewable energy (solar PV) network through Infrastructure Development Company Limited (IDCOL) with various Partner organizations (PO) has signified in the case of Bangladesh. In the line of the RE policy the government of Bangladesh has planned to develop 800 MW power from renewable energy by 2015 [3]. In Bangladesh government has recently provided directive to include certain percentage of solar power in commercial and residential buildings as a pre-condition to connect to the grid. Hence, it's an appropriate time to develop an energy and economically optimized Solar PV system for the buildings of the urban area in Bangladesh.

44.2 Solar Energy Potential in Bangladesh

Bangladesh is the country with in area of 147,500 km^2, which is situated between 20.300 and 26.380 north latitude and 88.040 and 92.440 east longitude, which is considered as an ideal location for the utilization of solar energy. GIS-based Geospatial Toolkit (GsT) and NASA Surface Meteorology and Solar Energy (SSE) data are used for the estimation of the potential of solar energy in Bangladesh. It was found that the range of solar radiation about 4–5 $kWh/m^2/day$ which is on about 94 % area of Bangladesh (Fig. 44.1). Theoretically Bangladesh receives near about 70 PWh of solar energy every year, which is approximately 3,000 times higher than the current electricity generation in Bangladesh [4].

The range of average annual solar radiation is typically from 100 to 300 W/m^2, hence with a solar PV efficiency of 10 % with 3–10 km^2 area is required for the establishment of an average electricity output of 100 MW, that is considered as 10 % of a large coal or nuclear power plant [5]. Near about 10,000 km^2 land of Bangladesh is necessary for the power generation from solar PV system to fulfill the electricity demand of 3,000 kWh/capita/year [6]. In Dhaka city, 7.86 % of total land is suitable for the generation of solar PV electricity [7]. Taking into account the grid availability, only 1.7 % of the land in Bangladesh is assumed technically suitable for solar photovoltaic power generation [8].

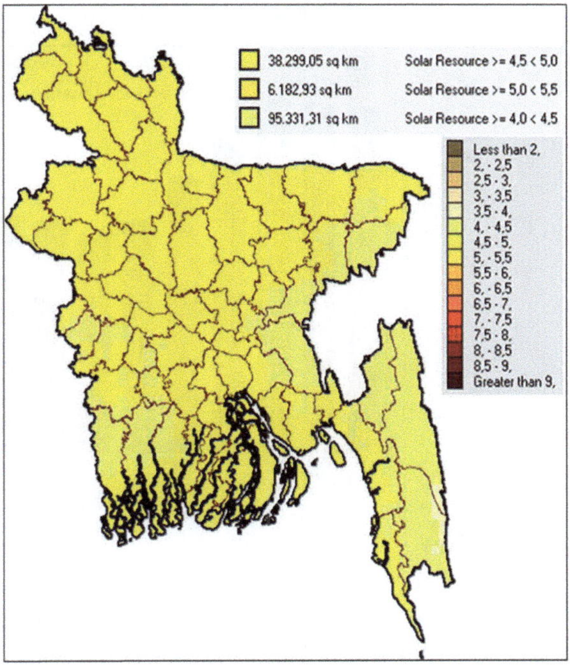

Fig. 44.1 Solar radiation (kWh/m^2/day) and area of Bangladesh with highest potential for solar energy utilization [4]

Table 44.1 Daily load data for the building

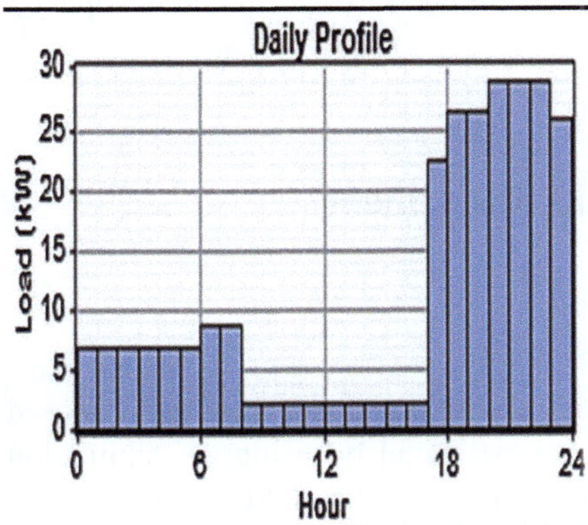

	Baseline	Scaled
Average [kWh/d]	265	265
Average (kW)	11.0	11.0
Peak (kW)	28.8	28.8
Load factor	0.384	0.384
Per year load requirement	100 MW	

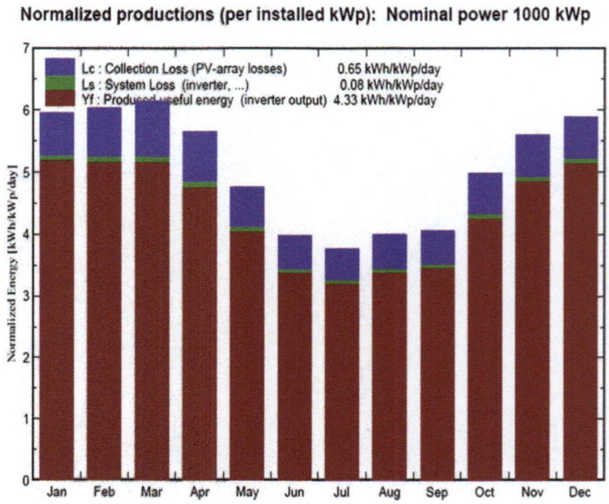

Fig. 44.2 Energy generation per day and losses

44.3 System Overview

Area of the roof will be more than 6,000 m^2 which is quite normal for the Dhaka city and the height of the building will be approximately ten stories. Total number of apartment in the building will be forty. Total load for the building has been collected from a field data with a similar size building in the Dhaka city and daily load data (estimated by HOMAR optimization software) is presented (Table 44.1).

The proposed system consists of 3075 fixed Solar PV panels with efficiency of 15.6 %.The panels are faced south and at an angle equal to the sight latitude. Azimuth angle was taken zero. To convert the direct current into alternating current two unit of inverter were used in the proposed system with a total capacity of 500 kW each and efficiency of 98.5 %.

44.4 Economic and Environmental Assessment

44.4.1 Electricity Generation

The amount of usable energy after considering the system losses from the proposed 1 MW grid connected solar PV system were estimated by using the simulation package PVSyst [9]. The proposed system will produce 1,581 MWh/year energy after considering 12 % approximate energy loss. It seems from the Fig. (44.2), that the energy production is maximum in the month of February and March approximately 5.3 kWh/kWp/day and minimum in the month of July approximately 3.4 kWh/kWp/day after considering all kind of losses. Throughout the year the performance ratio of the plant is almost constant with the average value 85 %.

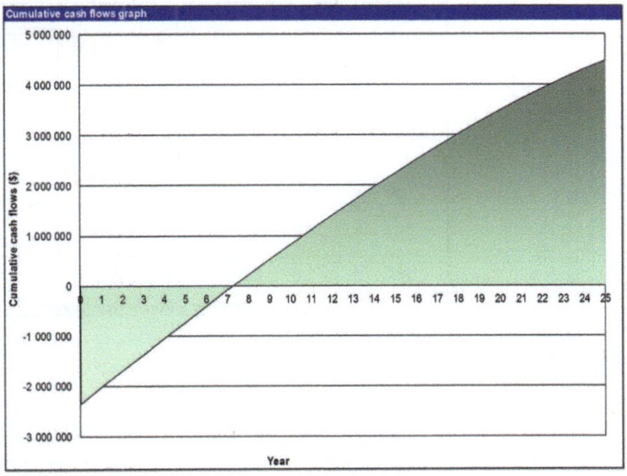

Fig. 44.3 Break-even analysis for proposed system

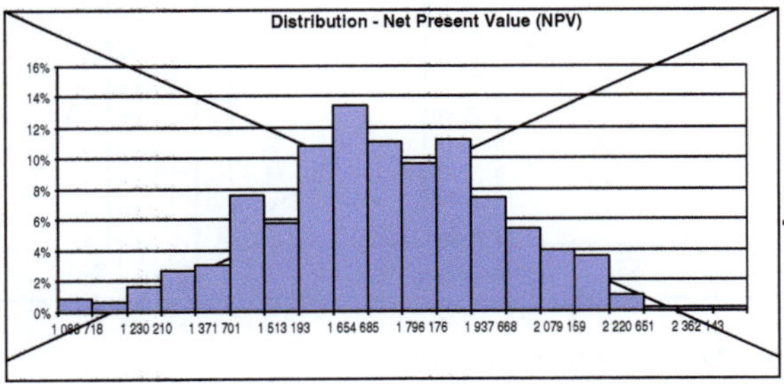

Fig. 44.4 Distribution of net present value

44.4.2 Economic Assessment

To evaluate the potentiality of proposed solar PV system from economic point of view several analyses has been done over the lifetime of the system by using RETScreen [10].

Break even analysis: The number of years for a project to equalize the cash flow to the total investment cost can be determined by a Break-even analysis [11]. Initially the cash flow of the project is always negative and after break-even point the cash flow becomes positive. In the case of proposed Solar PV, the initial cost can be recovered within 7.3 running year (Fig. 44.3).

Net Present Value (NPV): NPV represents the difference between the cash inflows and cash outflows in a present value of a project, considering all kind of discount rate. Positive NPV indicates the potentiality of a feasible project [12]. For the proposed Solar PV system NPV was calculated by using the RETScreen simulation tool and presented in Fig. (44.4) and the net present value of the proposed Solar PV system is 1.7 Million USD.

Cost of Energy (COE) Production: It is the avoided cost of energy that brings the NVP to zero [4]. The energy production cost per MWh is obtained from

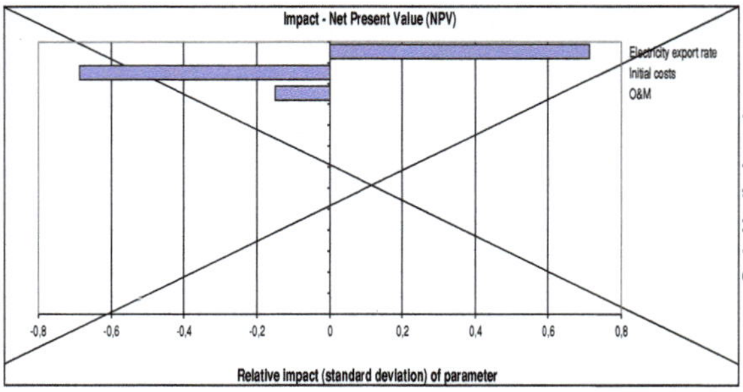

Fig. 44.5 Risk analysis

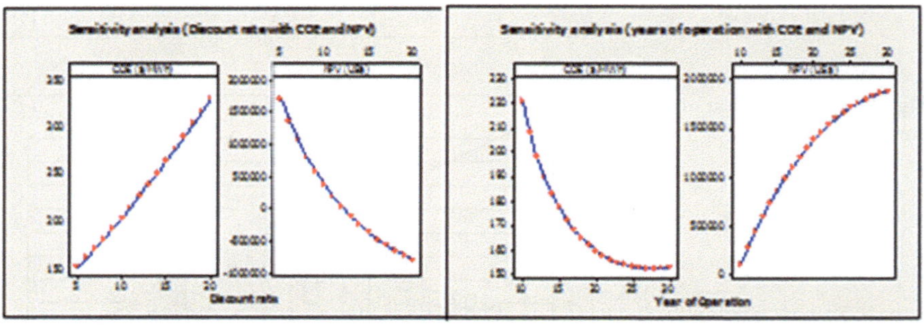

Fig. 44.6 Sensitivity analysis

RETScreen assuming all financial parameter are kept constant. In the proposed Solar PV system the cost of energy per MWh equal to 153.21 US$ in Dhaka.

Risk Analysis: Risk analysis has been performed based on the Net present value to get a deeper insight of the risk associated with the proposed Solar PV system. From Fig. (44.5) it can be illustrated that electricity export rate has positive impact on the NPV and on the other hand Initial cost, O&M (Operation and maintenance) has negative impact on the net present value (NPV)

Sensitivity Analysis: Sensitivity analysis was performed for some important parameters such as Initial cost, Electricity export rate, discount rate etc. In every economic action with a vibrant feature discount rate is an important issue as well for the proposed Solar PV case [11]. Discount rate was varied from 5 to 20 % to examine effect on the cost of energy production (COE) and Net present value (NPV), while the year of operation was kept constant at 25 year. Also by varying year of operation from 10 to 30 year by keeping the Discount rate constant the response of the cost of energy production (COE) and Net present value (NPV) was analyzed. From Fig. (44.6) it can be illustrate that at initially COE and NPV are more sensitive to discount rate than Year of operation.

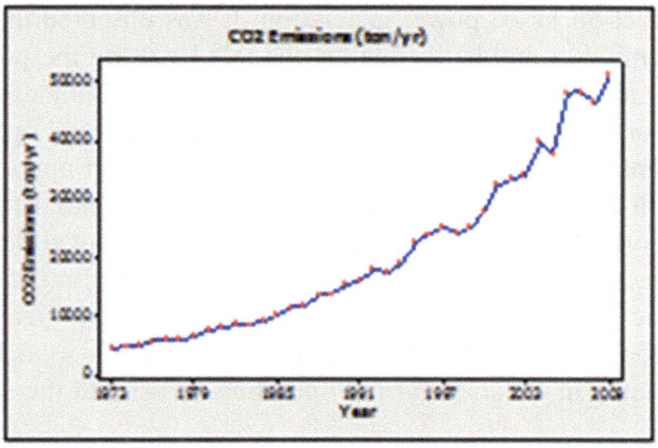

Fig. 44.7 CO_2 emission data Bangladesh

44.5 Environmental Assessment

The contribution of GHGs from Bangladesh per capita in past is one of the lowest in the world, but in recent times the emission of GHGs is increasing in an extreme rate due to the radical increase of power production, vehicle use and industrialization [13]. CO_2, CH_4 and N_2O are considered as GHGs and one of the major findings of this paper is to estimate how much GHG gas can be reduced by the proposed solar PV system. The analysis has been done by using RETScreen and only CO_2 reduction given as a Priority. The status of the CO_2 emission for Bangladesh from 1973 to 2009 is illustrated in the following Fig. 44.7.

The calculation of the annual reduction of CO_2 emissions due to the implementation of the proposed 1-MW solar PV system was performed based on the emission factor (0.584 tCO_2/MWh) of Bangladesh. Around 928 tons/year CO_2 emission mitigation was observed in the proposed PV system. It seems that, the use of solar PV system for the power production would be a great prospect for the reduction of the CO_2 emission in Bangladesh.

44.6 Conclusion

The paper examines the technical potential of solar PV electricity generation and feasibility of rooftop solar PV system for an Energy Efficient Building in Bangladesh. It is estimated that about 1,581 MWh/year power can be produced from the proposed system. Among the whole power near about 100 MWh/year power will consumed by the ten-storied building with roof area of 6,500 m^2 and rest of the energy about 1,481 MWh/year will be injected to the grid. It was also found that it will take about 7.3 years to recover all the initial costs of the proposed Solar PV system. The cost of energy for the proposed system has been estimated about 0.15 US$ per unit, which is cost competitive with conventional grid-connected fuel-oil based power generation. It was observed that, around 928 tons/year CO_2 emission mitigation can be possible by using the proposed rooftop solar PV system. The use of rooftop solar PV system in Bangladesh would be a great prospect for the reduction of the CO_2 emission and it will open a great opportunity for the green development and will create green job opportunities for Bangladesh. It is clear for this study that there is a great potential for the grid connected rooftop solar PV system in Bangladesh. In practical not all the building in Bangladesh will be suitable for rooftop solar PV system and also not all the building owner would be willing to lease their roofs for twenty-five years period, while with a little forethought and careful attention the rooftop solar leasing could poetically be a win-win situation for both building owner and the solar project developers.

References

1. A.G. Hestnes, *Building integration of solar energy systems*. (University of Science and Technology, Trondheim, 1999) Received 2 Sept 1999; revised version accepted 16 Mar 2000
2. http://www.idcol.org/cited on 5 April, 2013, 10.00 am
3. Renewable energy policy Bangladesh, 2008
4. M.A.H. Mondal, A.K.M.S. Islam, Potential and viability of grid-connected solar PV system in Bangladesh. Renew. Energy **36**, 1869–1874 (2011). Elsevier
5. R. Gottschalg, in *Future Prospects for PV Technologies*, ed. by A.K.M.S. Islam, D.G. Infield. Photovoltaic Technology for Bangladesh UK, (Bangladesh University of Engineering and Technology, Dhaka and Loughboroogh University, 2001), pp. 251–264
6. S. Islam, A.U. Huda, Proper utilization of solar energy in Bangladesh: effect on the environment, food supply and the standard of living. Renew. Energy **17**, 255–263 (1999). Elsevier
7. M.H. Kabir, W. Endlicher, J. Jagermeyr, Calculation of bright roof-tops for solar PV applications in Dhaka mega city, Bangladesh. Renew. Energy **35**, 1760–1764 (2010). Elsevier
8. B. Sorensen, GIS management of solar resource data. Sol. Energy Mater. Sol. Cells **67**, 503–509 (2001). Elsevier
9. http://www.pvsyst.com/en/ cited on 25 Feb, 2013, 11.00 am
10. RETScreen. RETScreen International, natural resources Canada (1997–2008), http://www.retscreen.net/ang/home.php. Cited on 10 March 2013, 9 am
11. E.S. Hrayshat, Viability of solar photovoltaic as an electricity generation source for Jordan. Sol. Energy **74**, 19–31 (2003). Elsevier
12. D.S. Remer, A.P. Nieto, A compendium and comparison of 25 project evaluation techniques. Part 1: Net present value and rate of return methods. Int. J. Prod. Econo. **42**, 79–96 (1995). Elsevier
13. A.U. Ahmed, K. Islam, M.R. Uddin, in *An Inventory of Greenhouse Gas Emissions in Bangladesh: Initial Results*, vol 25 (Springer, Heidelberg, 1996), pp. 300–303

Chapter 45
An Experimental Study on the Effect of Using Fresnel Lenses on the Performance of Solar Stills

Tarek I. Abdelsalam and Bahy Abdel-Mesih

Abstract The global water concern is mainly about the scarcity of fresh water resources despite the abundance of saline and brackish water in oceans, seas, and underground. Solar desalination offers a worthy solution to produce fresh water by using solar radiation, which also lessens the energy concern by offering a renewable source of energy to alter the consumption of fossil fuels and other non-renewable resources. One of the solar desalination technologies is the solar still system, which is a portable unit capable of producing distilled water by evaporating brackish or saline water by using solar thermal energy. The steam is then condensed on the inside of the glass cover and collected as fresh water. Solar stills are easy to manufacture and install using local materials and workmanship, which suits underprivileged remote communities that face difficulties in finding clean potable water, while locating near a source of saline water. However, efficiency and productivity of solar stills are still feeble when compared to other traditional desalination techniques. As an attempt to overcome these issues, an upgraded system is proposed and tested experimentally to augment the incoming solar radiation falling on the top glass surface of the still by concentrating extra solar radiation to preheat the flowing feedwater to the solar still system. The results of the experimental study showed that the integration of linear Fresnel lenses has approximately tripled the productivity of distilled water and improved efficiency of a solar still, by about 68.76 %, when compared to a conventional non-concentrating solar still.

T. I. Abdelsalam (✉) · B. Abdel-Mesih
Centre for Renewable Energy (CRE), Mechanical Engineering, The British University in Egypt (BUE), Sherouk, Cairo, Egypt
e-mail: tarek.ihab@bue.edu.eg

B. Abdel-Mesih
e-mail: bahi.Saad@bue.edu.eg

A. Y. Oral et al. (eds.), *International Congress on Energy Efficiency and Energy Related Materials (ENEFM2013)*, Springer Proceedings in Physics 155,
DOI: 10.1007/978-3-319-05521-3_45,

45.1 Introduction

Egypt enjoys one of the highest solar irradiation intensities worldwide. Simultaneously it suffers from scarcity of fresh water resources due to high population growth rates. Nevertheless, almost all countries of MENA regions have access to seas or oceans which raise the issue of seawater desalination to tackle the problem of fresh water demand. This brings up the idea of harnessing the sun's energy to desalinate seawater. Trieb and Müller-Steinhagen [1] have described the potential of concentrated solar power desalination in MENA countries, and predicted that in Egypt it will be around 2.87 billion m^3/year in 2020 and reach 64.57 billion m^3/year by 2050. These figures were the highest among other MENA regions.

One of the main advantages of solar stills is their ease of manufacturing and installation using local materials. Thus, they are suitable for underprivileged communities with potable water shortages. Yet, this comes at the expense of efficiency and productivity, which are low compared to traditional techniques.

Among other studies on concentrated solar desalination, Tanaka and Nakatake [2] have introduced a vertical multiple-effect diffusion-type still, which was proposed with the addition of a flat-plate mirror. The still was made of 10 partitions with 10 mm diffusion gaps between partitions. The system was predicted to produce up to 34.5 kg/m^2 day of distilled water on winter solstice days on the equator. The factors affecting the productivity of multiple-effect diffusion solar stills with flat reflector have been theoretically investigated [3] and experimentally validated [4]. It was found that the reflectivity of reflectors plays an important role as the distilled water production decreases 10 % with the loss of surface reflectivity from 0.95 to 0.8. The deviation of the solar still from the optimum azimuth angle affects production by up to 12 %. The outdoor testing showed that daily productivity of experiments matched with the theoretical predictions with a margin of error ± 7 % except for the results on a cloudy day. As a variation to the solar still with flat-plate mirror, Tanaka and Nakatake [5] have introduced a basin type still with both internal and external reflectors. This system showed an increase of the daily water production of 48 % for the entire year when both reflectors were used, and only 22 % enhancement when the internal reflector was added. It is worth mentioning that during summer the benefit of these reflectors was negligible due to shadow cast by the external reflector on the basin liner. Abdel-Rehim and Lasheen [6] have modified a solar desalination system to enhance its performance. The modification unit included a solar parabolic trough with focal pipe and serpentine heat exchanger with oil inside as the working fluid. This resulted in an increase in fresh water productivity by an average of 18 %. Also, Nassar et al. [7] have achieved more than 303 % improvement in the fresh water productivity over. This was done by incorporating a concave mirror where a metallic elliptical solar still under vacuum was located at its focus. Vacuum conditions assured that raw water in the solar still will have a lower boiling point. As far as this research is concerned, the authors of this paper have not come across a previous effort in incorporating Fresnel lenses to improve the performance of traditional single flat-plate solar stills.

45.2 Setups Design

Two identical solar stills were manufactured for simultaneous testing; one with and another without the proposed improvement. Each solar still was made of a glass basin with the dimensions 468 mm × 1,260 mm, with their bases padded with black EPDM rubber for enriched thermal energy storage. The basin was inserted inside a wooden insulation box, as shown in Fig. 45.1, while allowing for a 5 mm gap between each corresponding sides, which was filled with rubber rags for glass protection and to store heat escaping from the basin. The box was mounted on four caster wheels with brakes for better portability. The glass cover was inclined at Cairo's average solar altitude angle of 30°. The fresh water collecting chamber had an inclined base towards the outlet tap to enhance water collectivity.

For the improved setup, a saline water tank was manufactured to store and preheat the feed water. The tank was positioned at a higher altitude than the solar still to instigate water flow through the pipes under gravitational force. Moreover, four identical linear Fresnel lenses were purchased; each lens was 400 mm × 320 mm, with a thickness of 2 mm, and a focal length of about 65 cm. The lenses were made from PMMA, which is a light, transparent, and shatter-resistant thermoplastic material usually used as an alternative to glass. The lenses concentrate solar beams on the main feed pipe and held in position by a wooden support structure, as shown in Figs. 45.2 and 45.3, to adjust the focal line on the main feed pipe. Finally, there is a tap at the end of the piping system, for flow control, and a flexible hose connecting to tap with the solar still basin. The pre-heated components, tank and pipes, were painted in black to enhance solar beams absorptivity.

45.3 Experimental Procedure

All experiments were performed during February and March atop of building A at the British University in Egypt. The experiment starts after pouring the specified saline water quantity, 12.5 L, in each of the traditional solar still basin, and the improved setup's water tank. The two setups were oriented to face south, and then closed with the cover glasses at 9:30 am to start the experiment that ends after six hours at 3:30 pm. At 10:00 am the feed water tap, of the modified system, was opened slightly to allow for droplets of preheated water to flow into the basin water. Manual sun tracking was implemented for the Fresnel lenses structure every 30 min, in addition to a regulation of the feed water flow rate by opening slightly the feed water tap of the improved setup. All experimental runs were performed under similar design parameters, but with two levels of water input salinity: saline water from the Suez Canal that represented the high level of salinity, and brackish water from a swamp that represented the low level of salinity. According to the

Fig. 45.1 CAD model for traditional setup

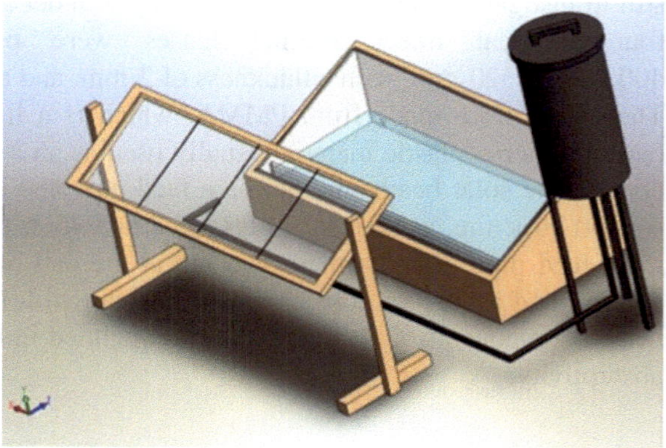

Fig. 45.2 CAD model for improved setup

Fig. 45.3 The two setups working concurrently

Fig. 45.4 Collection of output fresh water

experimental scheme, each run was performed three times under same conditions, which means that each setup experienced six runs for the two salinity levels, with a total of 12 runs for both setups. The output fresh water was eventually collected in a calibrated plastic pot to be measured as shown in Fig. 45.4. The solar intensities and ambient temperatures were measured by the campus's Weather Station, while the glass and basin water temperatures were measured by thermocouples. All environmental data were averaged and belong only to the experiment's 6-h period.

In order to assure reliable and consistent comparative results throughout each experimental day a procedure of preparatory actions were implemented. For instance, glass covers were cleaned with sheets of a newspaper in order to enhance the condensation process, ensure high clarity and transmissivity for solar beams, and to pave the way for condensate droplets to trickle along the glass without dirt obstacles that can drips them back to the basin water. Additionally, saline water tank and pipes were cleaned by water injection in the opposite direction to the usual flow direction. Moreover, it was ensured that the saline water was poured into the basin of the solar still via a calibrated plastic jug, while maintaining a slow rate of discharge to avoid splashing of saline water into the fresh water collection chamber. The linear Fresnel lenses were positioned as was shown in Fig. 45.2, while keeping the flat lenses surfaces toward the pipes and the grooved surfaces towards the sun for a better-quality focal line along the main pipe.

45.4 Results

The results of the measurements were tabulated, as shown in Table 45.1, which exhibits sample results for two experimental days. The efficiency of each run was calculated by using the following formulae:

Table 45.1 Sample results obtained for two experimental days with different salinity levels

Day	Run	Setup	Average peak hour solar intensity G (W/m^2)	Input salinity level	Average ambient temperature, T_a (°C)	Measured glass temperature, T_{ig} (°C)	Measured basin water temperature T_{bw} (°C)	Desalinated water output, Q_{fresh} (ml/m^2/day)	Efficiency η (%)	Efficiency increase (%)
1	7	Improved	697.5	Low	23	36.3	57.6	2,350	30.60	57.57
	8	Traditional				36.1	46.2	750	19.42	
2	11	Improved	670	High	28	36.4	54.7	1,950	26.43	63.44
	12	Traditional				36.2	45.8	600	16.17	

$$\eta_{traditional} = \frac{Q_{\text{fresh}} \times \text{LH}_w \times 1,000,000\, J/MJ}{\text{G} \times t \times A_{\text{b}} \times 1,000\, ml/l} \times 100 \tag{45.1}$$

$$\eta_{improved} = \frac{Q_{\text{fresh}} \times \text{LH}_w \times 1,000,000\, J/MJ}{\text{G} \times t \times (A_{\text{b}} + A_{preheat}) \times 1,000\, ml/l} \times 100 \tag{45.2}$$

where Q_{fresh} is the run's measured generated fresh water volume in ml, LH_w is the latent heat of water vaporization ($\approx$ 2.3 MJ/kg), G is the experimental day's average measured peak solar intensity during the run's period in W/m^2, t is the run's period (=21,600 s), A_b is the basin surface area of the solar still ($\approx$0.5897 m^2), $A_{preheat}$ is the preheat system surface area exposed to the sun ($\approx$0.5829 m^2). While the percentages increase in efficiency were calculated using this equation:

$$\text{Efficiency increase} = \frac{\eta_{improved} - \eta_{traditional}}{\eta_{traditional}} \times 100 \tag{45.3}$$

By averaging all runs results, it was calculated that the proposed improvement has increased the average efficiency by about 68.76 %; with 76.67 % improvement for the high salinity water input runs, and 60.85 % for low salinity water input runs. Correspondingly, the improved setup has increased the average productivity by 231.17 %, from 641.7 to 2,125 ml per 6 hours. Also, it increased the average measured basin temperature from 44.7 to 55.2 °C. Nevertheless, such achievement is less prominent when the sky is cloudy or unclear due to the attenuation of beam radiation, which significantly affects the Fresnel lenses effectiveness. This observation was concluded from a separate experiment set during a cloudy day with results shown in Table 45.2. Another concern was the increase in capital costs, by about 127.53 %, resulted predominantly because of importing the Fresnel lenses.

Figures 45.5 and 45.6 demonstrate the advantageous effects on productivity, final output and accelerated fresh water generation, of the improved setup when compared with traditional solar still. Additionally, the effect of input water salinity can be observed from the differences between the two plots.

Table 45.2 Results of two experimental runs during a cloudy day

Setup	Average peak hour solar intensity (W/m^2)	Input salinity level	Average ambient temperature, T_a (°C)	Measured glass temperature, T_{ig} (°C)	Measured water temperature T_{bw} (°C)	Desalinated water output, Q_{fresh} (ml)	Efficiency η (%)	Efficiency increase (%)
Improved	560	High	33	39.2	47.2	650	6.89	8.96
Traditional				38.8	43.8	300	6.32	

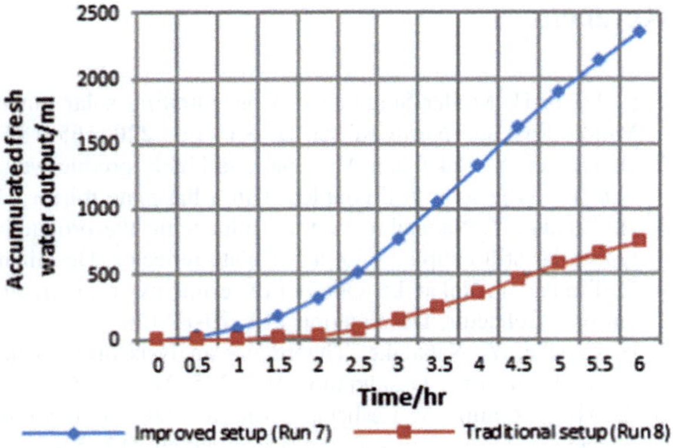

Fig. 45.5 Measured accumulated fresh water output for a low input salinity run

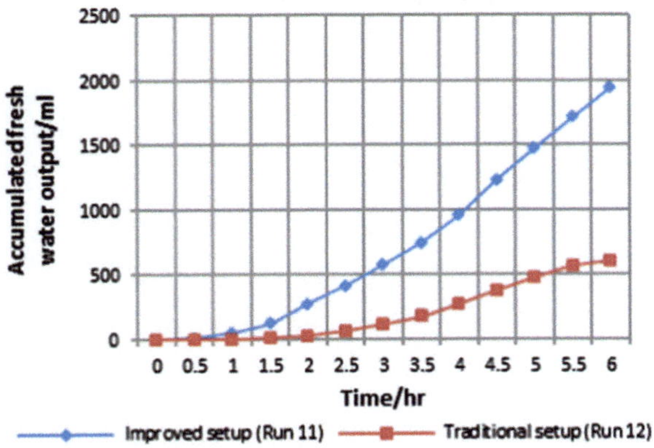

Fig. 45.6 Measured accumulated fresh water output for a high input salinity run

45.5 Conclusion

The experimental results indicated that the proposed idea has succeeded to enhance the desalination performance by averagely improving the productivity by 231.17 % and efficiency by 68.76 %. Additionally, this technique has also succeeded to accelerate the start of fresh water generation. However, for the proposed improved setup design, further experimentally-based investigations and development are required to overcome the sun tracking and high capital costs drawbacks. Firstly, it is recommended to improvise an appropriate and economical sun tracking mechanism for the solar concentrators and study its effectiveness. Secondly, it is suggested to assess the economic feasibility of the proposed improvement on large-scaled units, while persuading local optical factories to start manufacturing solar concentrators to eliminate much of the added costs allocated to the lenses.

References

1. F. Trieb, H. Müller-Steinhagen, Concentrating solar power for seawater desalination in the Middle East and North Africa. Desalination **220**, 165–183 (2008)
2. H. Tanaka, Y. Nakatake, A simple and highly productive solar still: a vertical multiple-effect diffusion-type solar still coupled with a flat-plate mirror. Desalination **173**, 287–300 (2005)
3. H. Tanaka, Y. Nakatake, Factors influencing the productivity of a multiple-effect diffusion-type solar still coupled with a flat plate reflector. Desalination **186**, 299–310 (2005)
4. H. Tanaka, Y. Nakatake, Outdoor experiments of a vertical diffusion solar still coupled with a flat plate reflector. Desalination **214**, 70–82 (2007)
5. H. Tanaka, Y. Nakatake, Theoretical analysis of a basin type solar still with internal and external reflectors. Desalination **197**, 205–216 (2006)
6. Z. Abdel-Rehim, A. Lasheen, Experimental and theoretical study of a solar desalination system located in Cairo. Egypt, Desalination **217**, 52–64 (2007)
7. Y. Nassar, S. Yousif, A. Salem, The second generation of the solar desalination systems. Desalination **209**, 177–181 (2007)

Chapter 46
Solar Energy for Rural Egypt

Tarek I. Abdelsalam, Ziad Darwish and Tarek M. Hatem

Abstract Egypt is currently experiencing the symptoms of an energy crisis, such as electricity outage and high deficit, due to increasing rates of fossil fuels consumption. Conversely, Egypt has a high solar availability of more than 18.5 MJ daily. Additionally, Egypt has large uninhabited deserts on both sides of the Nile valley and Sinai Peninsula, which both represent more than 96.5 % of the nation's total land area. Therefore, solar energy is one of the promising solutions for the energy shortage in Egypt. Furthermore, these vast lands are advantageous for commissioning large-scaled solar power projects, not only in terms of space availability, but also of availability of high quality silicon (sand) required for manufacturing silicon wafers used in photovoltaic (PV) modules. Also, rural Egypt is considered market a gap for investors, due to low local competition, and numerous remote areas that are not connected to the national electricity grid. Nevertheless, there are some obstacles that hinder the progress of solar energy in Egypt; for instance, the lack of local manufacturing capabilities, security, and turbulent market in addition to other challenges. This paper exhibits an experience of the authors designing and installing decentralized PV solar systems, with a total rated power of about 11 kW, installed at two rural villages in at the suburbs of Fayoum city, in addition to a conceptual design of a utility scale, 2 MW, PV power plant to be installed in Kuraymat. The outcomes of this experience asserted that solar PV systems can be a more technically and economically feasible solution for the energy problem in rural villages.

T. I. Abdelsalam (✉)
Centre for Renewable Energy (CRE), Mechanical Engineering, The British University in Egypt (BUE), 43, Sherouk, Cairo 11837, Egypt
e-mail: tarek.ihab@bue.edu.eg

Z. Darwish
Enactus BUE, 43, Sherouk, Cairo 11837, Egypt
e-mail: ziad3037@bue.edu.eg

T. M. Hatem
Mechanical Engineering, The British University in Egypt (BUE), 43, Sherouk, Cairo 11837, Egypt
e-mail: tarek.hatem@bue.edu.eg

A. Y. Oral et al. (eds.), *International Congress on Energy Efficiency and Energy Related Materials (ENEFM2013)*, Springer Proceedings in Physics 155,
DOI: 10.1007/978-3-319-05521-3_46,

46.1 Introduction

The electricity demand is in a continuous rise, while the supply is limited due to the predominant dependence on non-renewable energy resources. Therefore, a breakthrough in the electricity production through new and renewable energy technologies has become the key to overcome this pattern. Egypt is qualified to tackle its current electricity shortage by exploiting its high solar energy availability and vast and vacant deserts. According to ROBAA's measurements, Egypt has very few cloudy days and high global solar radiation, attributable to its presence at the solar belt, which can reach up to 30 MJ/m^2 during summer [1]. Moreover, El-Sadek has appraised around 99 % of the total Egyptian population is concentrated in an area representing about 5.5 % of the total nation's land area, which is mainly located around the Nile valley and Delta, leaving behind massive unexploited deserts with few or no inhabitants [2].

46.2 Decentralized System

A project commenced on an agreement between the British University in Egypt and Misr El-Kheir Charity foundation. The purpose of this project was mainly to supply electrical to selected critical service and residential buildings in two secluded villages—Sydna Al-Khidr & Saydna Mousa—located at Wadi El Rayan, Fayoum, where approximately 2,000 residents have no access to the national electricity grid. The two villages were located in a natural protectorate more than 30 km away from the national grid, where pollution is strictly prohibited to preserve its natural beauty.

The decentralized systems were off-grid systems designed to supply 11.04 kW to five buildings in the two villages: a 3.36 kW system for the main medical healthcare center, shown in Fig. 46.1, 2.4 kW system for the central school, two 1.2 kW systems for each village's elementary school, and a 2.88 kW system for three residential households, which were aimed to encourage teachers and physicians to reside in the villages. Furthermore, an additional prototype system of 1.2 kW was constructed and tested in Cairo for analysis and training purposes before commencing the other systems in Fayoum. The power loads in different buildings varied from lighting and fans to computers and refrigerators, thus a separate system sizing for each building was prepared to determine the suitable components properties and sizes based on the loads' power ratings and space availability at the roof as shown in Fig. 46.2.

46.3 Centralized System

Another centralized PV system was designed to supply 2 MW to a rural community near Kuraymat, 90 km south Cairo. This system was planned to feed a local grid of a nearby solar thermal power plant in order to evade the necessity of incorporating additional electricity storage system.

Fig. 46.1 Medical center

Fig. 46.2 Polycrystalline PV a top of school

The project is to be implemented by the New and Renewable Energy Authority (NREA) in consultancy with the Electrical Power Systems (EPS) and BUE team. At the moment, the BUE team has finalized the design for civil work, components sizing and selection, distribution of panels, and electrical circuits for the project. Finally, a financial study was implemented to assess the economic feasibility of the project and estimate its breakeven price per kWh.

46.4 Results

The two systems were analyzed technically and economically to be used as references for future similar projects in the two directions.

46.4.1 Decentralized System

The prototype costs and technical data were not taken into this analysis, since it was constructed and tested in Cairo, which does not represent rural Egypt.

46.4.1.1 Technical Results

The following figures exhibit the data concerning one of the elementary school systems, with a rated power and voltage of 1.2 kW and 48 V respectively. The batteries average temperatures were measured using BTS-01 sensors connected to Vario Track charge controller and Xcom-232i data logger, shown in Fig. 46.3, to record the data on a SD memory card. Figure 46.4 presents the batteries' average temperatures throughout July 2013, which was recorded to be the hottest month in Egypt. However, Fig. 46.5 presents the temperature of a sample day, on 2nd October 2013, when the measured ambient temperature was ranging between 23 and 37 °C during that day.

Similarly, the Vario Track charge controller measured the input voltage from the PV system and output voltage to the batteries during the sample day, as shown in Figs. 46.6 and 46.7, and delivered the data to the Xcom-232i data logger.

The critical points for the four measurements were probed and tabulated as shown in Table 46.1.

The Xtender inverter was also connected to the Xcom-232i data logger through a Communication Bus connection, while measuring and delivering data concerning important electrical parameters of the system. Figure 46.8 displays the power generated from the PV panels and power consumed by the loads. One can notice from this figure that there was only one AC input to the system, from the solar system, since there was no auxiliary AC input from any other source.

46.4.1.2 Economical Results

The costs of the four main components of the five systems were tabulated as shown in Table 46.2. The total capital costs budget for the systems was 482,500 EGP, whereas connecting the villages to the national grid was estimated to cost more than 14 million EGP due its remote location.

46.4.2 Centralized System

The results of the centralized system included measured primary data concerning the site as well as an economic study.

46.4.2.1 Technical Results

Primary data were measured from the site to study the expediency of commencing a mega-sized PV power plant in Kuraymat. The monthly direct normal irradiances (DNI) and temperatures throughout the year were averaged and plotted as shown

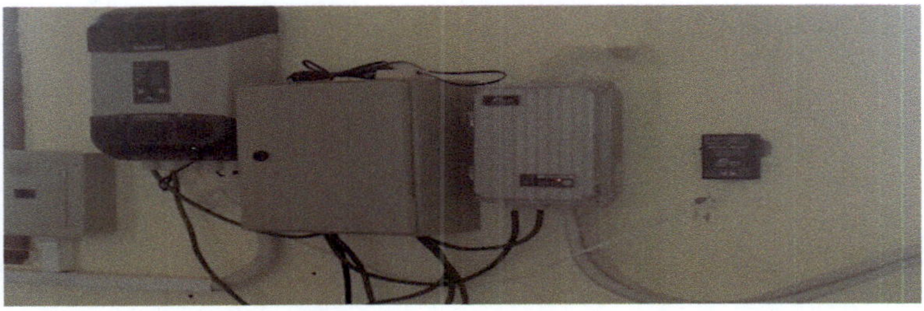

Fig. 46.3 From *right* to *left* (Xcom-232i data logger, Vario Track charge controller, combination box, Xtender inverter, switches box)

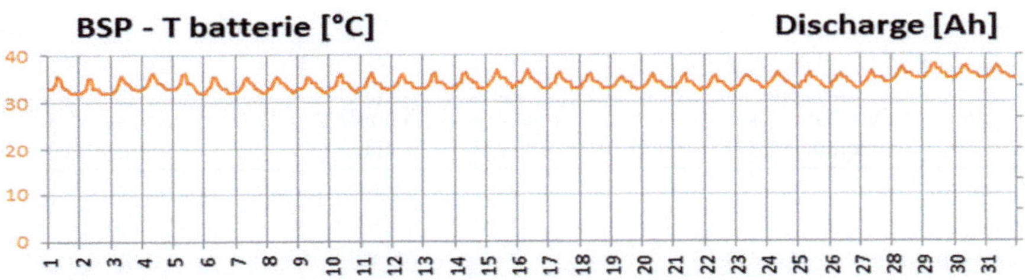

Fig. 46.4 Batteries average temperature throughout July

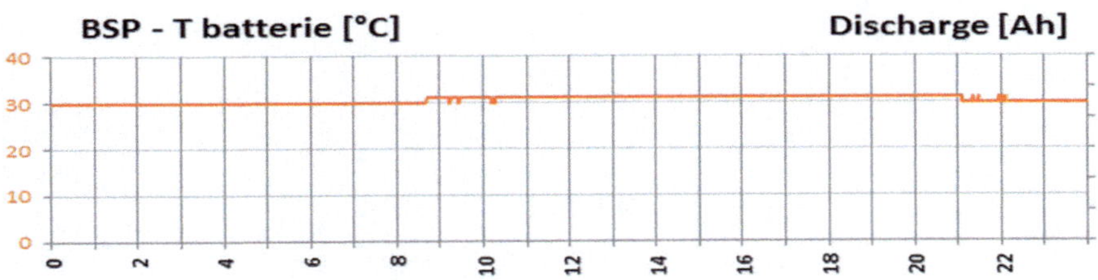

Fig. 46.5 Batteries average temperature throughout the sample day

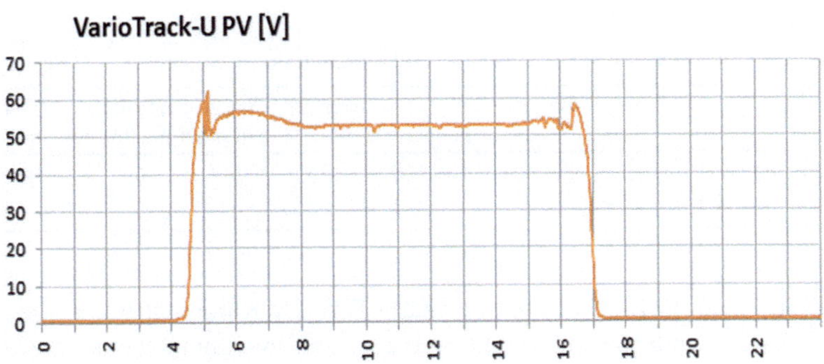

Fig. 46.6 PV average voltage

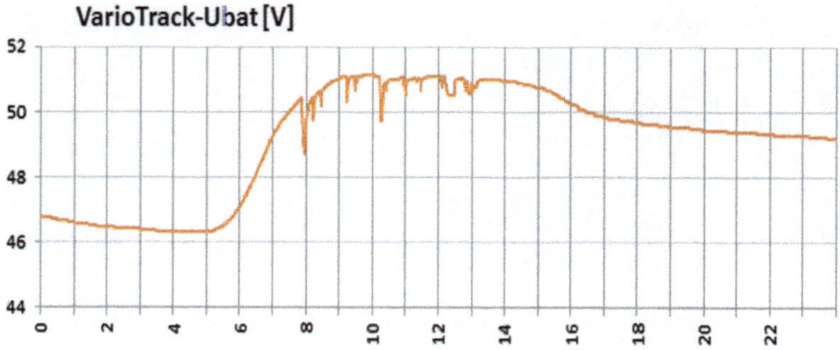

Fig. 46.7 Batteries average voltage

Table 46.1 Critical points for the graphs

Critical Point	July's batteries temperature (°C)	Sample day batteries temperature (°C)	Sample day batteries voltage (V)	Sample day PV voltage (V)
Maximum	39	31	51.1	62.16
Average	34.1	30.5	49.1	27.94
Minimum	31	30	46.3	0.88

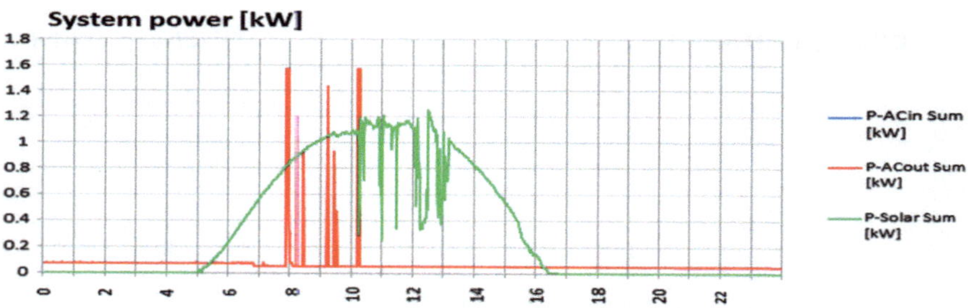

Fig. 46.8 Elementary school system power input and output

Table 46.2 Capital cost distribution for the 11 kW decentralized systems

Component	Batteries	PV panels	Inverters	Charge controllers and other electronics	Civil work and others
Allocated cost (EGP)	221,950	106,150	72,375	57,900	24,125
Percentage of capital costs (%)	46	22	15	12	5

in Figs. 46.9 and 46.10. The average yearly DNI for Kuraymat was estimated as 2,431 kWh/m^2, while the minimum, average, and maximum temperatures were recorded as 2, 22, and 42 °C respectively.

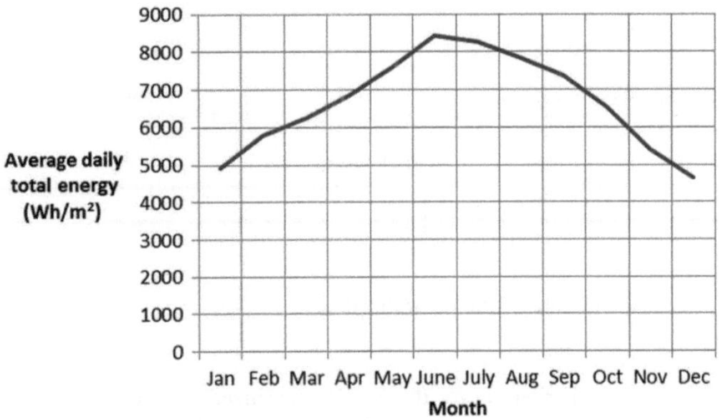

Fig. 46.9 Annual DNI distribution

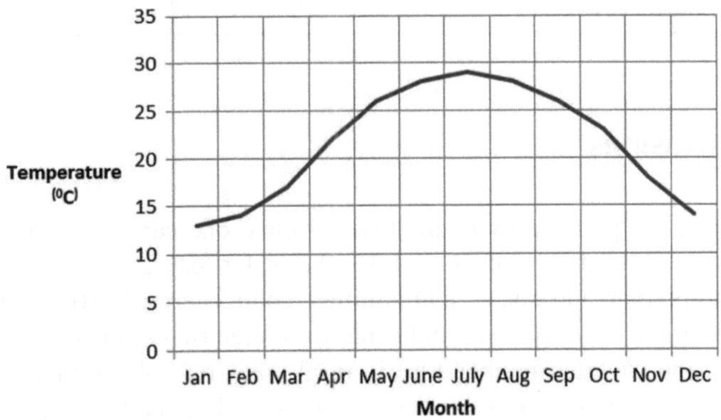

Fig. 46.10 Annual temperatures distribution

46.4.2.2 Economical Results

Figure 46.11 shows the cash inflows and outflows throughout the typical 25 years lifetime of the centralized PV system. The capital costs were estimated to be around 36 million EGP, distributed as shown in Table 46.3, while the annual running costs were estimated to be around 540,000 EGP. The annual interest rate was assumed to be 10 %, while an annual increase in electricity price due to rising demand over limited supply, excluding the inflation effect, was set as 4 %. The breakeven price per kWh was computed to start with 63.7 piasters, at the first year, and then rise annually to finally reach 169.8 piasters after 25 years, while the current electricity price in Egypt during peak hours reaches around 45 piasters.

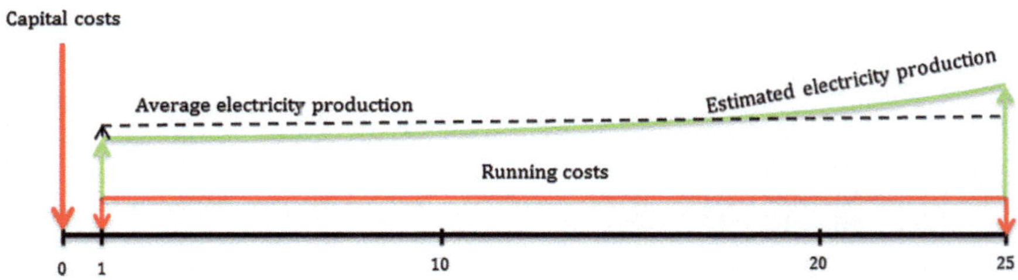

Fig. 46.11 Cash flow timeline of the centralized system

Table 46.3 Capital cost distribution for the 2 GW centralized system

Component	PV panels	Civil work	Inverters	Transformer and other electronics	Others
Allocated cost (EGP)	15,280,000	8,000,000	6,000,000	6,000,000	720,000
Percentage of capital costs (%)	42	22	17	17	2

46.5 Conclusions

After studying the practicality of the two systems, one can conclude that despite the promising advantages attributed to the solar energy solution in Egypt, including high solar and space availabilities, solar energy is still a significantly expensive alternative to conventional centralized electricity generation techniques. Technically, the elevated temperature in rural areas resembled another weakness point to the performance of system components such as batteries and PV panels. Nevertheless, despite the necessity of a costly electricity storage system, decentralized systems expressed a versatile and quick solution for small rural villages scattered in the desert and far away from the national electricity grid. Besides, as time goes, the feasibility of solar energy solutions will improve as a result of the continuously rising demand and limited supply of conventional energy resources, in addition to the foreknown evolution in storage and system performance that anticipated to be achieved through extensive and tailored research and development efforts R&D.

References

1. S.M. Robaa, *A Study of Solar Radiation Climate at Cairo Urban Area* (Royal Meteorological Society, Giza, 2006)
2. A. El-Sadek, *Water Desalination: An Imperative Measure for Water Security in Egypt* (Elsevier B.V, Cairo, 2009)

Chapter 47
Experimental Research of Pyrolysis Gases Cracking on Surface of Charcoal

Valentin Kosov, Vladimir Kosov and Victor Zaichenko

Abstract For several years, in the Joint Institute for High Temperatures of Russian Academy of Sciences, two-stage technology of biomass processing has been developing [1]. The technology is based on pyrolysis of biomass as the first stage. The second stage is high-temperature conversion of liquid fraction of the pyrolysis on the surface of porous charcoal matrix. Synthesis gas consisted of carbon monoxide and hydrogen is the main products of the technology. This gas is proposed to be used as fuel for gas-engine power plant. For practical implementation of the technology it is important to know the size of hot char filter for full cracking of the pyrolysis gases on the surface of charcoal. Theoretical determination of the cracking parameters of the pyrolysis gases on the surface of coal is extremely difficult because the pyrolysis gases include tars, whose composition and structure is complicated and depends on the type of initial biomass. It is also necessary to know the surface area of the char used in the filter, which is also a difficult task. Experimental determination of the hot char filter parameters is presented. It is shown that proposed experimental method can be used for different types of biomass.

47.1 Introduction

Biomass is the largest and the most important renewable energy source in the world. World production of biomass is estimated at 150–200 billion metric tons a year [2]. Biomass includes all variety of natural organics, agricultural wastes, industrial wastes, especially forestry and wood industry, paper mills, fast-growing plants, organic part of municipal waste.

V. Kosov (✉) · V. Kosov · V. Zaichenko
Joint Institute for High Temperatures of the Russian Academy of Sciences (JIHT RAS), Izhorskaya, str. 13, 125412 Moscow, Russia
e-mail: kosov@ihed.ras.ru

A. Y. Oral et al. (eds.), *International Congress on Energy Efficiency and Energy Related Materials (ENEFM2013)*, Springer Proceedings in Physics 155,
DOI: 10.1007/978-3-319-05521-3_47,

Biomass can be converted to energy in three ways: thermal conversion, chemical conversion, and biochemical conversion. In order to use the most effective of dry biomass thermochemical technologies such as burning, gasification and pyrolysis, which in spite of its effectiveness have some disadvantages.

Gasification is the most effective way to convert biomass is its conversion into combustible gas. However, the existing technologies of biomass gasification have several disadvantages. Air gasification is the easiest method to convert biomass into the gas. However, the resulting gas contains up to 60 % nitrogen and 40 % carbon dioxide. The calorific value of the gas is generally around 4–5 MJ/m^3, which is too low for efficient use. Overall efficiency of gasification gas power plant is limited to 20 % [3].

Oxygen and steam gasification allow increasing the calorific value of the gas which contains no nitrogen and a small amount of carbon dioxide. The maximum gas yield reaches 1.3 nm^3 per kg of raw material and its calorific value is about 11 MJ/m^3 [4]. Steam gasification is the widespread process because of its simplicity. The main disadvantage of the process is concerned with necessity of steam generation, which reduces overall effectiveness of power plant. Use of oxygen for the purpose of gasification demands an air separation unit in technological chain that leads to rise in price of the end product. It should also be noted that purification of the gas from tar and ash is an urgent problem for all methods of gasification.

Along with gasification, pyrolysis is an effective method of thermochemical processing of biomass, industrial and domestic waste and at the same time one of the least-developed technologies of bioenergy. Pyrolysis is the of thermal decomposition process organics without oxygen. It runs at relatively low temperatures (500–800 °C) as compared with the processes of gasification (800–1,300 °C) and combustion (900–2,000 °C). Modern technologies of pyrolysis are focused to obtain a liquid product, so-called pyrolysis oil. Pyrolysis oil is a thick black tarry liquid close in composition to the biomass, and consists of a complex mixture of hydrocarbons with substantial water content (up to 30 %) and having a calorific value of about 20 MJ/kg [5]. Pyrolysis oil theoretically can be used directly as a fuel, but its wide use as a furnace fuel or motor requires more complex processing technology.

The technology of biomass thermal processing with synthesis gas production developed in the JIHT RAS is based on two stage thermal conversion of biomass with pyrolysis as first stage and high-temperature processing of the pyrolysis gases and pyrolysis oil on surface of char as second stage [1]. The method is similar to the one suggested in [6] and used in [7] for processing of wood waste and peat. Carbon, hydrogen and oxygen content of condensable and non-condensable volatiles allows them to be converted into synthesis gas consists of hydrogen and carbon monoxide in practically equal parts.

For practical implementation of the technology it is important to know the size of hot char filter for full cracking of the pyrolysis products on the surface of char. The most difficult is to determine the cracking parameters of tars, since their structure is complicated and depends on the type of initial biomass. To solve this problem, there were a series of experiments showed the possibility of determining the minimum size of the hot char filter for the complete decomposition of volatile pyrolysis products.

47.2 The Experimental

The experimental set-up (Fig. 47.1) consisted of a high-temperature two-chamber fixed-bed reactor and a system of extraction and analysis of gas and vapor forming as a result of heating an initial raw material.

The reactor was a stainless steel tube with an inside diameter of about 37 mm, which was placed within two-section furnace with independent heaters for each section. The chambers were 300 mm length each. Raw material was placed into the chamber 1. Char obtained by pyrolysis of the same raw material was placed in the chamber 2. There were series of experiments with different amount of char placed in the camber 2. Pyrolysis process was explored when chamber 2 was empty. Before experiments the top chamber was heated up to temperature 1,000 °C that was held further at the constant level. After that the temperature of the bottom chamber was raised at the rate 10 °C/min.

Pyrolysis gases formed during pyrolysis of initial raw material passed through the porous carbon bed with the fixed temperature. As a result of homogeneous and heterogeneous chemical reactions in the high-temperature zone, a decomposition of torrefaction gases took place. Conversion degree depended both on the temperature in the top chamber and on the residence time in a high-temperature zone. Non-condensable gas came into the volume meter (eudiometer). The samples of the gas were chromatographed.

Softwood pellets were used as a raw material for pyrolysis. Carbonized softwood pellets were used as the hot char filter. The properties of the pellets are shown in Table 47.1.

47.3 Results and Discussion

The results of measurements of the volume gas produced during pyrolysis per 1 kg of softwood pellets and subsequent cracking of tars shown in Fig. 47.2. From these data it follows that the cracking of tar on the surface of coal increases the volume of gas obtained by seven times.

The composition of non-condensable pyrolysis gas is shown in Table 47.2.

In the process of heterogeneous cracking on the surface of coal, methane and carbon dioxide decompose into hydrogen and carbon monoxide, forming the synthesis gas. Similar processes occur in the cracking of tars, acids, alcohols and water. The gas obtained in the cracking of tars consists almost entirely of hydrogen and carbon monoxide in roughly equal proportions.

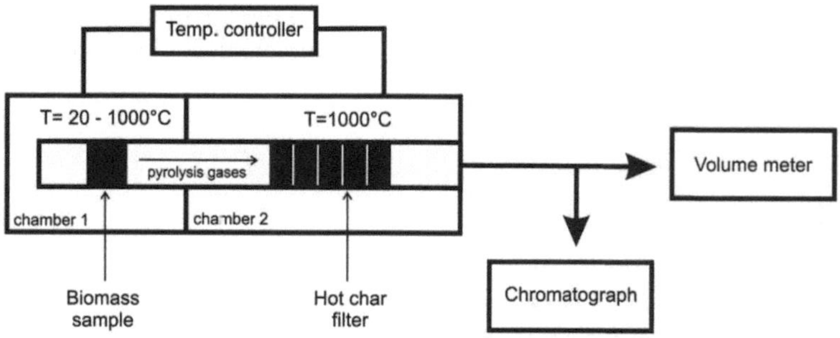

Fig. 47.1 Scheme of the experimental reactor

Table 47.1 Properties of softwood pellets (dry material, wt%)

Moisture	8.3
Ash	1.4
Volatiles	84.6
C	47
H	6
N	0.2
O	44

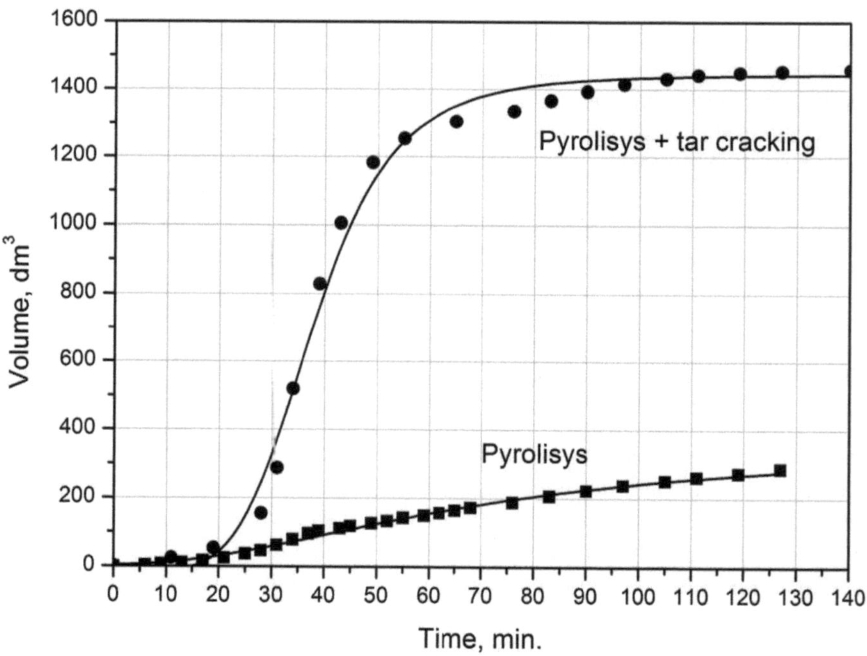

Fig. 47.2 Gas yield per 1 kg of raw material during pyrolysis of softwood pellets and subsequent cracking of tars

Table 47.2 Non-condensable pyrolysis gas composition

H_2	CO	CO_2	CH_4	N_2
32.6 %	22.4 %	24.1 %	16.3 %	4.6 %

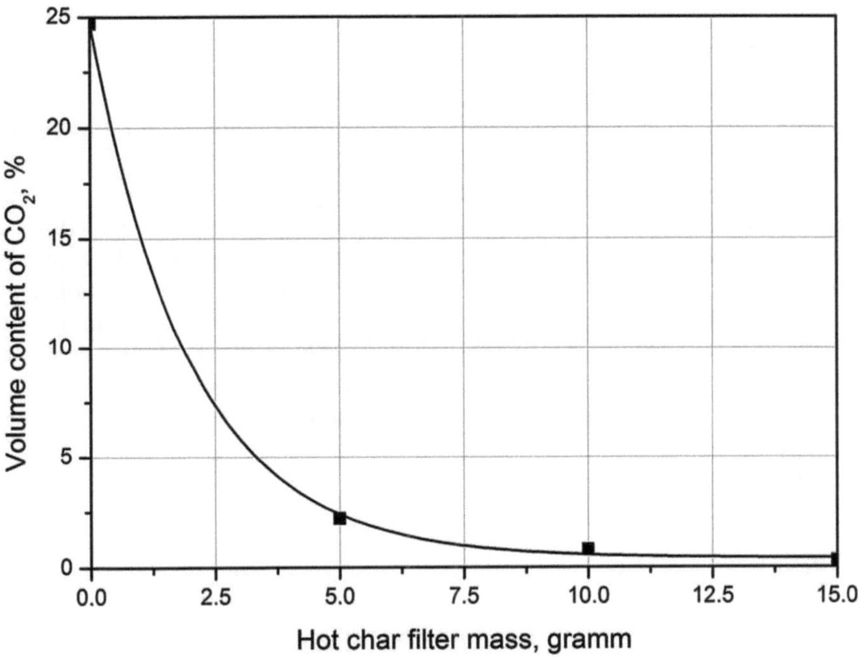

Fig. 47.3 The amount of carbon dioxide in the gas leaving the reactor at a different quantity of char in the cracking chamber

The cracking process of the pyrolysis products can be estimated by the amount of carbon dioxide in the gas leaving the reactor at a different quantity of char in the cracking chamber 2. This assumption is based on the fact that the activation energy for the secondary cracking of pyrolysis tars [8] in two times less than the activation energy for the Boudoir reaction [9].

For the estimation of the carbon dioxide amount, serial experiments were conducted, in which the chamber 2 was placed charcoal with mass of 5, 10 and 15 g. The mass of initial biomass in the chamber 1 was 5 g. The amount of carbon dioxide in the gas leaving the reactor at a different quantity of char in the cracking chamber 2 is shown in Fig. 47.3.

From these data it follows that charcoal in the chamber 2 works as a catalyst accelerating the decomposition process of carbon dioxide. Using charcoal filter that equal to the amount by weight of the initial biomass reduces the amount of carbon dioxide in the exhaust gas by more than order of magnitude.

Thus, using the charcoal filter is equal to the initial weight of the biomass is possible to achieve a high degree of decomposition of the pyrolysis products. Further increase in coal mass filter does not significantly increase the degree of decomposition.

47.4 Conclusion and Future Work

The data obtained show the possibility of experimental determination of parameters of carbon filter to achieve the desired characteristics of the gas produced two-step conversion of biomass. The method of determination the degree of decomposition condensed pyrolysis products on the content of carbon dioxide allows to estimate the required amount of carbon in the cracking chamber 2. However, the experiments were not considered processes of tars volumetric decomposition in the chamber 2. Additional experiments with increased sample weight of initial biomass for the same reactor size are necessary to be carried out. Moreover, for the application of this technology as a method of gas cleaning it is necessary to define the characteristic residence time of tars in the char filter that is sufficient for their complete decomposition.

References

1. V.F. Kosov, V.A. Sinelshchikov, V.M. Zaichenko, New technology for integrated processing of biomass and natural gas with production of hydrogen and pure carbon materials, in *The Proceedings of 16th European Biomass Conference and Exhibition* (Valencia, Spain, 2008), pp. 1171–1175
2. M. Balat, A. Gunhan, Biomass energy in the world, use of biomass and potential trends. Energy Sources **27**, 931–940 (2005)
3. M. Lisý, P. Kohout, Z. Skála, M. Balás, J. Moskalík, Biomass gasification and cogeneration, in *16th European Biomass Conference* (2008), pp. 843–849
4. L. Pengmei, Y. Zhenhong, M. Longlong, W. Chuangzhi, C. Yong, Z. Jingxu, Hydrogen-rich gas production from biomass air and oxygen/steam gasification in a downdraft gasifier. Renewable Energy **32**(13), 2173–2185 (2007)
5. T. Bridgwater, Pyrolysis of biomass, in *Proceedings of 17th European Biomass Conference and Exhibition* (Hamburg, Germany, 2009)
6. S.K. Chembukulam, A.S. Dandge, N.L. Kovilur, R.K. Seshagiri, R. Vaidyeswaran, Smokeless fuel from carbonized sawdust. Ind. Eng. Chem. Prod. Res. Dev. **20**, 714–719 (1981)
7. V.V. Kosov, V.F. Kosov, I.L. Maikov, V.A. Sinelshchikov, V.M. Zaichenko, High calorific gas mixture produced by pyrolysis of wood and peat, in *The Proceedings of 17th European Biomass Conference and Exhibition*, 29 June–3 July 2009 (Hamburg, Germany, 2009), pp. 1085–1088
8. A.G. Liden, F. Berruti, D.S. Scott, A kinetic model for the production of liquids from the flash pyrolysis of biomass. Chem. Eng. Comm. **65**, 207–221 (1988)
9. M. Barrio, J.E. Hustad, CO2 gasification of birch char and the effect of CO inhibition on the calculation of chemical kinetics, in *Progress in Thermochemical Biomass Conversion*, ed. by A.V. Bridgwater (Blackwell Science Ltd, Oxford, 2008), pp. 47–60

Chapter 48
High-Calorific Gas Mixtures Produced from Biomass

Valentin Kosov, Vladimir Kosov, Vladimir Sinelschikov and Victor Zaichenko

Abstract Investigations in the field of processing of different types of biomass into high calorific gas fuel are carried out in the Joint Institute for High Temperatures of RAS. Design of technology for effective processing of low-grade solid fuel to gas mixtures which can be used as fuel for power plants is rather actual problem both in the view of natural resources conservation and in the view of the development of autonomous energy unit for thermal and electric power supply. The experimental data on quantity and composition of the gaseous products formed in the process of peat pyrolysis are presented at various operating parameters of the process. It is shown that as a result of peat pyrolysis and the subsequent cracking of emanating products at temperature 1,000 °C it is possible to receive 1.4 m^3 of gas with specific caloricity 11.7 MJ/m^3 per 1 kg of original raw material.

48.1 Introduction

Developing the technologies for conversion of different kinds of biomass into gas fuel and pure carbon materials is rather an actual problem both from the point of view of natural resources conservation, and from the point of view of creating autonomous plants to produce both electricity and heat operating on local kinds of fuel.

Gasification and pyrolysis are the most popular methods that developed for producing gas fuel from biomass. Both of them have some advantages and disadvantages. Air gasification is the easiest method to convert biomass into gas. However the resulting gas contains up to 60 % carbon dioxide and nitrogen and

V. Kosov · V. Kosov (✉) · V. Sinelschikov · V. Zaichenko
Joint Institute for High Temperatures of the Russian Academy of Sciences (JIHT RAS), Izhorskaya str. 13, 125412 Moscow, Russia
e-mail: vladimir.f.kosov@mail.ru

A. Y. Oral et al. (eds.), *International Congress on Energy Efficiency and Energy Related Materials (ENEFM2013)*, Springer Proceedings in Physics 155,
DOI: 10.1007/978-3-319-05521-3_48,

its calorific value is generally around 4–5 MJ/m^3 which is too low. Overall efficiency of gasification gas power plant is limited to 20 %. The gasification products also contain undesirable substances such as tars and dust which are need to be removed [1]. To increase a calorific value of the gasification products oxygen or water steam gasification can be used. The gas produced by oxygen and steam gasification contains no nitrogen and small amount of carbon dioxide. The maximum gas yield reaches 1.3 nm^3 per kg of raw material and its calorific value is about 11 MJ/m^3 [2]. Steam gasification is the widespread process because of its simplicity. The main disadvantage of the process is concerned with necessity of steam generation which reduces overall effectiveness of power plant. Use of oxygen for the purpose of gasification demands an air separation unit in technological chain that leads to rise in price of end product.

Calorific value of gas produced in the process of biomass pyrolysis is considerably higher than calorific value of gas produced in the air gasification process. Insignificant degree of processing of initial raw materials into pyrolysis gas is main disadvantage of this technology. Significant increase the gas yield can be achieved by high-temperature conversion of liquid fraction. For this purpose both the catalytic [3] and non-catalytic [4, 5] methods can be used. In the present paper the scheme similar to the scheme offered in [6] for processing of wood sawdust is considered. Experimental feasibility of the technology based on such scheme was partially presented in [6]. Synthesis gas consisted from carbon oxide and hydrogen and char are the end products of this technology.

48.2 Experimental Conditions

All the experiments were carried out at the setup similar to used in [6, 7] and schematically presented in Fig. 48.1. It consisted of a high-temperature reactor A, a gas supply system and a gas expense controller G at the entrance of reactor, and a system of extraction and analysis of gas at the exit of reactor including condenser B, eudiometer C and chromatograph. The samples of the gas for chromatography were taken out in the point D. The reactor was a stainless steel tube with an inside diameter of about 37 mm, which was placed within two-section furnace with independent heaters for each section. Each section was 300 mm length. The heaters allowed varying the temperature inside the reactor from 100 to 1,200 °C. Such construction of the furnace made possible to use the reactor in one and two chamber regimes.

In the first case the top chamber of reactor was not heated up and was empty. During experiments the behavior of different raw materials situated in bottom chamber at the stage of pyrolysis, activation and also the process of pyrocarbon deposition from gaseous hydrocarbons on the surface of char were investigated. The gas supply system made it possible to carry out experiments in different gas media and monitor the volumetric gas flow through the reactor. The pressure in reactor was equal to 105 Pa.

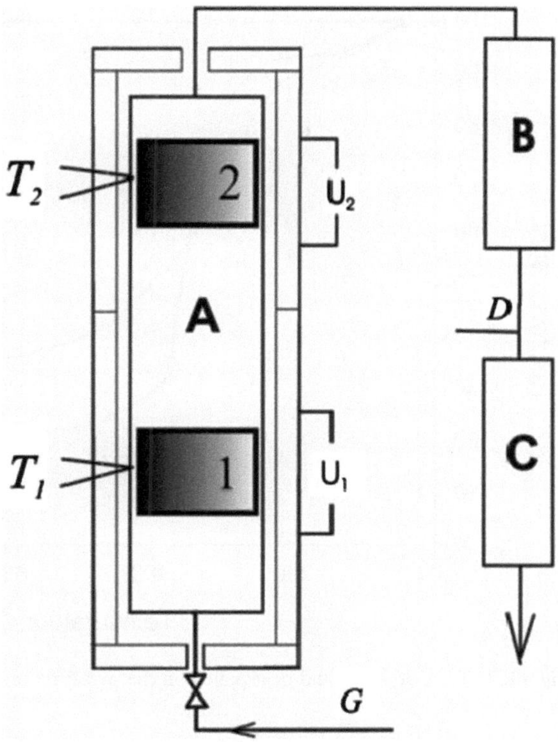

Fig. 48.1 Experimental installation scheme

The two chamber regime was used for investigation of pyrolysis of different raw materials, composition and yield of gas produced in process of its thermal conversion. Raw material 1 (see Fig. 48.1) was placed into the bottom chamber. Char 2 obtained by pyrolysis of the same raw material was placed in the top chamber. The depth of char bed was equal to 50 mm. Before experiments the top chamber was heated up to temperature T_2 that was held further at the constant level.

After that the temperature of the bottom chamber was raised at the rate 10 °C/min. Thus, gases and vapour formed during pyrolysis of initial raw material passed through the porous carbon bed with the fixed temperature T_2 (further this mode is designated as «pyrolysis with cracking»). As a result of homogeneous and heterogeneous chemical reactions in the high-temperature zone a decomposition of pyrolysis gases and vapour took place. Conversion degree depended both on the temperature T_2 and on the residence time in a high-temperature zone. The output gas mixture passed through the condenser B. Non-condensable gas came into the eudiometer C. Wood and peat pellets were used as initial raw material Humidity of the samples was 6 and 10 % for wood and peat respectively.

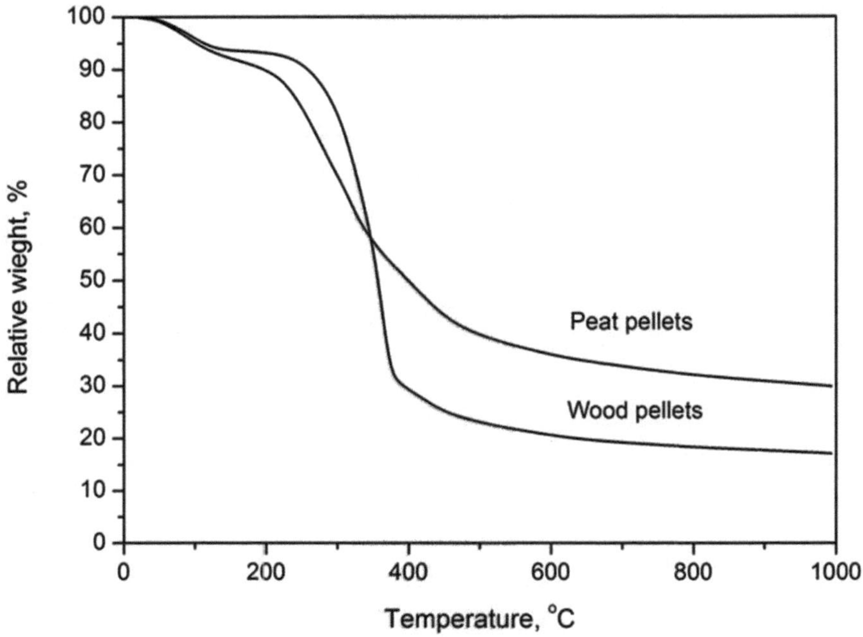

Fig. 48.2 TGA of wood and peat pellets in the pyrolysis process

48.3 Pyrolysis of Wood and Peat Pellets

The different raw materials pyrolysis experiments were carried out in one chamber regime of reactor. During the experiments raw material was heated in oxygen-free medium from room temperature to 1,000 °C. The fulfilled experimental investigations showed similar behavior of peat and wood pellets during their thermal processing. The weight loss curves for wood and peat pellets during pyrolysis process are presented in Fig. 48.2. The data shown in Fig. 48.2 correspond to a heating rate of approximately 10 °C/min. It can be seen that the loss of mass occurs mainly in the temperature interval from 200 to 600 °C and ranges from 65 (for peat pellets) to 85 % (for wood pellets) of the initial mass. The change in the mass with a further increase in the temperature is insignificant and is equal to around 5 % of the initial mass in the temperature interval from 600 to 1,000 °C. The obtained char is a brittle porous material containing around 80 and 90 % of carbon for peat and wood pellets, respectively. This difference mainly depends on the difference of ash level of the initial raw materials. Changing the heating rate from 2 to 35 °C/min resulted in the mass of char decreasing by no more than 10 % and the specific volume of open pores increasing by around 20 %. The specific volume of open pores for char from wood and peat pellets is equal to 0.5 and 0.28 cm^3/g, respectively.

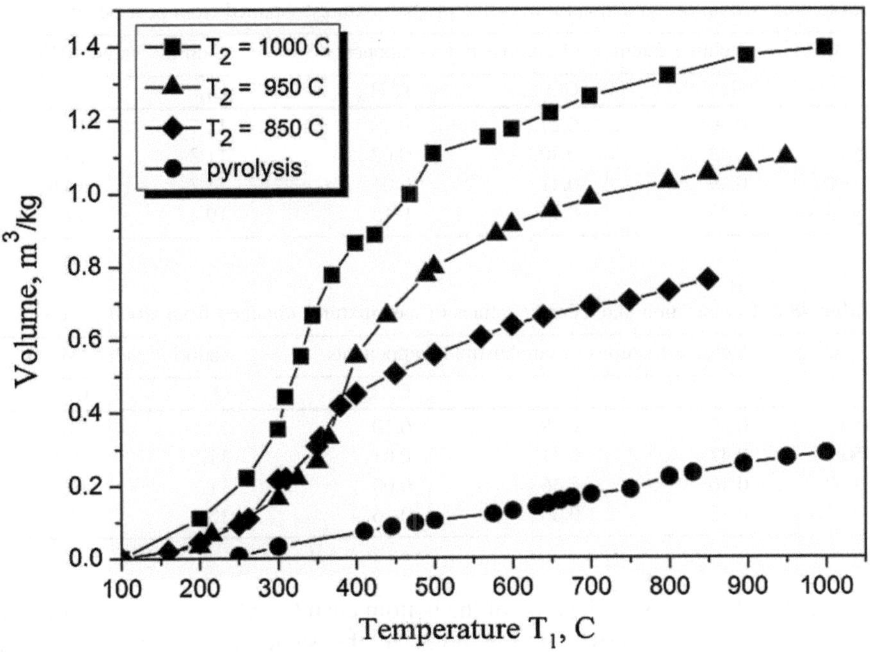

Fig. 48.3 Gas yield per 1 kg of raw material during thermal processing of peat pellets versus temperature of the bottom chamber for different regimes

The chromatographic analysis of composition of the gases evolved during pyrolysis of wood and peat pellets was fulfilled in the temperature range 250–1,000 °C at the heating rate 10 °C/min. Gas yield and its composition for wood and peat pellets are very similar. Major gas components are H_2, CO, CO_2 and C_nH_m. Note that methane constituted the main part in the hydrocarbon mixture C_nH_m. At temperature above 500 °C the carbon dioxide content substantially decreased whereas the content of combustible gases (methane, hydrogen and carbon monoxide) increased that led to increase of the calorific value of the product gas mixture. The overall gas yield was equal to 0.29 m^3 per kg of initial raw material. The lower calorific value of product gas was equal to $Q_L = 9.6$ MJ/m^3. Extraction of the gas evolved in the temperature range 500–1,000 °C leads to the appreciable increase of the calorific value of product gas up to 13 MJ/m^3 but the gas yield decreases to 0.18 m^3 per kg of initial raw material.

48.4 Pyrolysis and Tars Cracking

Increasing the conversion degree of raw material into a combustible gas can be achieved as a result of high-temperature thermal processing of gases and vapors emanating in process of pyrolysis. The data on gas volume (per kg of peat pellets)

Table 48.1 Composition and calorific value of gas mixtures obtained from peat pellets

T_2, C	Volume fractions of combustible components			Calorific value, MJ/m^3	
	H_2	CO	C_nH_m	Q_H	Q_L
850	0.40	0.27	0.08	11.7	10.6
950	0.43	0.40	0.02	11.3	10.4
1,000	0.49	0.41	0.01	11.7	10.6
Pyrolysis	0.23	0.19	0.13	10.4	9.6

Table 48.2 Composition and calorific values of gas mixtures obtained from wood pellets

T_2, C	Volume fractions of combustible components			Calorific value, MJ/m^3	
	H_2	CO	C_nH_m	Q_H	Q_L
850	0.39	0.28	0.10	12.5	11.5
950	0.47	0.41	0.01	11.5	10.6
1,000	0.46	0.46	0.00	11.7	10.9
Pyrolysis	0.28	0.26	0.16	13.4	12.2

obtained in the process of heating of the bottom chamber of the reactor at different temperatures T_2 are shown in Fig. 48.3. In the same figure the similar data received in «pyrolysis» mode are also shown.

As follows from the presented data the volume of the gas produced in mode «pyrolysis with cracking» was much more than the volume of the gas produced in «pyrolysis» mode. This difference increased essentially with increasing the top chamber temperature T_2. Simultaneously reduction of quantity of the liquid fraction collected in the condenser B was observed. For the temperature $T_2 = 1{,}000$ °C there was no liquid fraction in the condenser that was evidence of full tar and pyroligneous liquor conversion into gas.

In Tables 48.1 and 48.2 data on content of combustible components in the gas mixtures obtained by thermal processing of peat and wood pellets for different temperatures of the top chamber («pyrolysis with cracking» mode) and corresponding high Q_H and low Q_L calorific values are presented. Similar data for gas mixture obtained in «pyrolysis» mode are shown at the same tables.

48.5 Conclusion

As a result of the pyrolysis and the subsequent cracking of emanating volatile products over a char at the temperature 1,000 °C it is possible to increase the conversion degree of initial raw materials and to receive about 1.4 m^3 of gas with calorific value approximately 11.7 MJ/m^3 from 1 kg of wood or peat pellets.

References

1. M. Lisý, P. Kohout, Z. Skála, M. Balás, J. Moskalík, Biomass Gasification and Cogeneration, in *The Proceedings of 16th European Biomass Conference*, Valencia, Spain, June 2008, pp. 843–849
2. P. Ly, Z. Yuan, L. Ma, C. Wu, Y. Chen, J. Zhu, Hydrogen-rich gas production from biomass air and oxygen/steam gasification in a downdraft gasifier. Renewable Energy **32**(13), 2173–2185 (2007)
3. D. Dayton, A review of the literature on catalytic biomass tar destruction-milestone completion report, 2002, NREL/TP-510-32815, National Renewable Energy Laboratory, web address: www.osti.gov/bridge
4. M. Boroson, J. Howard, J. Longwell, W. Peters, Heterogenous cracking of wood pyrolysis tars over fresh wood char surfaces. Energy Fuels **3**, 735–740 (1989)
5. S.K. Chembukulam, A.S. Dandge, N.L. Kovilur, R.K. Seshagiri, R. Vaidyeswaran, Smokeless fuel from carbonized sawdust. Ind. Eng, Chem. Prod. Res. Dev. **20**, 714–719 (1981)
6. V.V. Kosov, V.F. Kosov, I.L. Maikov, V.A. Sinelshchikov, V.M. Zaichenko, High calorific gas mixture produced by pyrolysis of wood and peat, in *The Proceedings of 17th European Biomass Conference and Exhibition*, Hamburg, Germany, 29 June–3 July 2009, pp. 1085–1088
7. Kosov V.F., Sinelshchikov V.A., Zaichenko V.M., New technology for integrated processing of biomass and natural gas with production of hydrogen and carbon materials, in *The Proceedings of 16th European Biomass Conference and Exhibition*, Valencia, Spain, 2–6 June, 2008, pp. 1171–1175

Chapter 49
Aspects Regarding Design of Wind Power Plants Foundation System

Vasile Farcas and Nicoleta Ilies

Abstract During the past years wind power plants projects have become very important all over the world. Considering the height of the superstructure and the technological conditions which impose limit displacements and settlements, one of the most important verification for those foundations is at serviceability limit state. The value of the displacements must be calculated for dynamic loads, especially wind charge. The article present the particularities of the wind power plants foundations design, but also considerations about the reinforcement and the concrete used for the foundation.

49.1 Introduction

Wind energy is becoming more and more important in Romania, large projects were contracted, the most part in Dobrogea (a region in south eastern part of Romania), but also important projects were expecting authority's approval. According to European Wind Energy Association (EWEA) [8], wind power installed in Europe is increasing every year, as seen in Table 49.1.

Wind power plants foundations are a challenge for engineers due to the special design requirements. Generally, wind power plants foundations are strip foundations, with circular or ring shape. The design of the foundations is made according to European norm—Eurocode 7—Geotechnical Design [3] according to Romanian norm.

V. Farcas (✉) · N. Ilies
Department of Structures, Faculty of Civil Engineering, Technical University of Cluj Napoca, 400027 Cluj Napoca, Romania
e-mail: vasile.farcas@yahoo.com

N. Ilies
e-mail: nicoleta.ilies@dst.utcluj.ro

A. Y. Oral et al. (eds.), *International Congress on Energy Efficiency and Energy Related Materials (ENEFM2013)*, Springer Proceedings in Physics 155,
DOI: 10.1007/978-3-319-05521-3_49,

Table 49.1 Wind power installed in Europe by the end of 2012

Country	Austria	Germany	Netherlands	Romania	Spain
Capacity (MW) installed in 2012	296	2,415	119	923	1,122
Capacity (MW) installed at the end of 2012	1,378	31,308	2,391	1,905	22,796

Ring shape surface is used mostly in the situation when is proposed that the electrical cables to be droved by the central area of the foundation and they are sectioned under the foundation bottom. On the foundation socket is executed the wind power plant body, having metallic tube cross section. Foundation bottom diameter is approximately (0.25–0.30) H, where H—is the total height of the wind power plant. Foundation bottom cross section is approximately 0.15 H and the socket cross section is $\sim 1{,}20$ m larger than the tower cross section.

49.2 Design Considerations

Elastic circular foundation is made by reinforced concrete, having minimum C25/30 strength class, in order to assure the minimum exposure class XC4 and XF1, according to Romanian norm and [2]. In case of the presence of chemical aggressive waters or soil it is necessary to consider exposure class XA (according to [2]). In order to determine plane surface of the foundation bottom it is applied European norm [3]. Actions value on the foundation bottom and safety factors is determined for the design approaches according to EN 1990—Basis of structural design norm [1]. Safety factors are determined for Ultimate Limit State (ULS) and Serviceability Limit State (SLS).

Considering V_z, Q_{xy} and M_{xy} for the axial load, horizontal load and the bending moment acting on the foundation bottom, there are two important design cases:

- For the design case where $V_{z,max}$ and $M_{xy,af}$ it is required that the entire foundation bottom to be active. In this case is possible to check a pressures condition [3] (49.1), where p_{acc} is the value of accepted pressures on the foundation ground, and σ_{max} is the maximum pressure on the foundation bottom:

$$\sigma_{\max} \leq p_{acc} \tag{49.1}$$

There is also required to check the value of the loads eccentricity which imposes to maintain axial load inside the cross section core limit (Fig. 49.1), $e \leq r_{e1} = 0.25 \cdot r_a \cdot (1 + r'^2)$, where $r' = \frac{r_e}{r_a}$.

The value of the effective pressures acting on the foundation bottom is calculated with the relation:

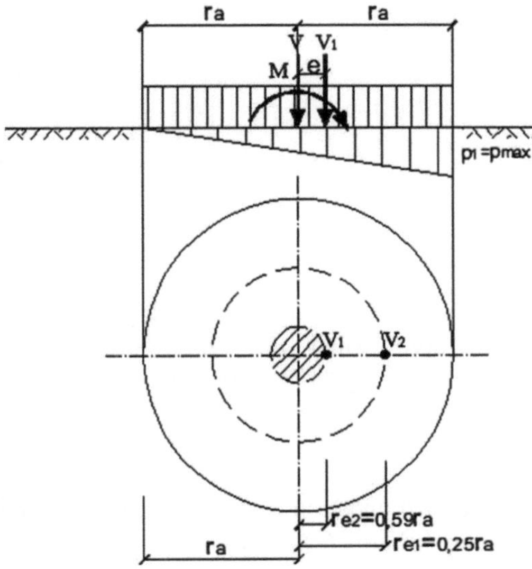

Fig. 49.1 Loads distribution on a circular foundation

$$\sigma_{\min}^{\max} = \frac{V}{r_a^2} \cdot \frac{1}{J_1(1 - r'^2)} \cdot \left(1 \pm \frac{e}{r_{e1}}\right) \tag{49.2}$$

- For the design case where: M_{xymax}, $V_{z,af}$ it is considered the loading of a certain surface from the foundation bottom.

In this case it is imposed that resultant force to act inside the core of the cross section. Maximum pressure value acting on the ground is given by the relation:

$$\sigma_{\max} = \frac{V}{r_a^2} \cdot \frac{2}{\pi(1 - r'^2)} \cdot \frac{e}{r_{e1}} \left[1 - 0.7\left(\frac{e}{r_{e1}} - 1\right) \cdot \left(1 - \frac{e}{r_{e2}}\right) \cdot (1 + r')\right] \tag{49.3}$$

For the design phase is mandatory to verify bearing capacity limit state (ULS) and deformation limit state (SLS), according to SR EN 1997 [3]. For the limit state in the foundation ground the verification is given by the relation (49.4), where: V_d—design vertical load on the foundation ground (on the foundation bottom) calculated for an ultimate limit state, R_d—design bearing capacity of the foundation ground, for drained or undrained conditions. It is imposed that $\frac{e}{r_a} \leq 0.59$. Also, it is necessary a sliding check (ULS).

$$V_d \leq R_d \tag{49.4}$$

The check is performed for design combination $Q_{xz,max}$ and $F_{z,af}$.

The most important check is on deformation limit state (SLS). Taking into consideration the large height of the structure and the technological conditions it is required that the tower inclination to have a lower value than the maximum

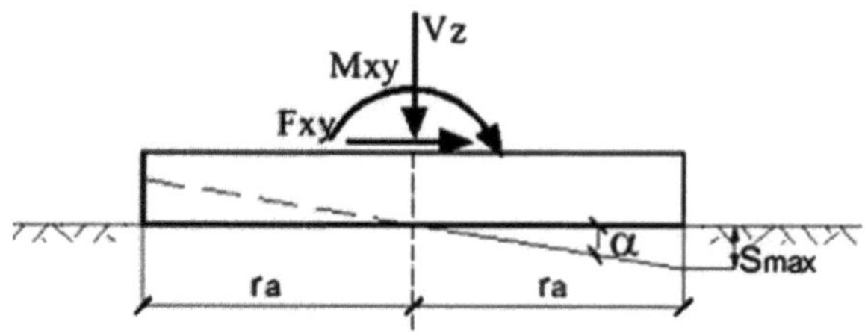

Fig. 49.2 Foundation rotation due to applied loads

allowable value. To guarantee the stability of the structure it is required to respect the condition (49.5), where the value of $tg\alpha_{adm}$ is considered to have a maximum value of 3 mm/m ($\alpha \leq 0.17°$) (Fig. 49.2):

$$tg\alpha < tg\alpha_{adm} \tag{49.5}$$

The value of foundation inclination on dynamic loads is calculated using the relation (49.6), where: $K_{\phi,din}$—dynamic coefficient of subgrade reaction on rotation for the foundation soil, I_f—inertia moment for the foundation bottom ($= \frac{\pi D^4}{64}$).

$$tg\alpha = \frac{M_{xy}}{K_{\phi,din} \cdot I_f} \tag{49.6}$$

$$K_{\phi,din} \geq \frac{M_{xy}}{tg\alpha_{adm} \cdot I_f} \tag{49.7}$$

From the relation (49.5) and (49.6) it is calculated the required minimum value of dynamic coefficient of subgrade reaction on rotation (49.7) [4–7], which represents the minimum value necessary for the foundation soil, so that to be impossible to have a larger inclination than the allowable value. The value of static coefficient of subgrade reaction on rotation is calculated using the relation (49.8) [7], where $\varpi_\phi = 1.98$.

$$K_\phi = \frac{\varpi_\phi \cdot E_0}{\sqrt{A} \cdot (1 - \nu^2)} \tag{49.8}$$

By knowing the value of required $K_{\phi,din}$ it is necessary to determine the minimum required value of the elastic deformation modulus for the foundation ground, using the relation (49.9) [5] for circular foundation. The same way it is calculated the required value of static deformation modulus (49.10) [5], where ν is Poisson's coefficient, determined in dynamic conditions.

$$E_{s,din} = K_{\phi,din} \frac{3}{4} \cdot \frac{1}{r_a^3} \cdot \frac{(1 + \upsilon) \cdot (1 - \upsilon)^2}{1 - \upsilon - 2\upsilon^2} \tag{49.9}$$

$$E_s = K_{\phi,static} \cdot \frac{3}{4} \cdot \frac{1}{\left(r_a^3 - r_i^3\right)} \cdot \frac{(1+\upsilon)\cdot(1-\upsilon)^2}{1-\upsilon-2\upsilon^2} \tag{49.10}$$

When the value of dynamic deformation modulus required $E_{S,din}$ is smaller than the value of dynamic deformation modulus of the foundation soil, it is allowed to utilize direct (shallow) foundations. When $E_{s,din} > E_{s,soil}$ is necessary to implement consolidation solutions for the foundation soil or to design deep foundation systems.

49.3 Foundations or Consolidation Solutions

Some possible solutions to consolidate foundation soil are: Deep consolidation of the foundation soil using gravel columns or stabilized gravel plots; Foundation soil consolidation by reinforced gravel mattresses (with geo grids). A major solution is the one of deep foundation systems, using raft foundations bearing on a row of piles (inclined piles, having 8:1 inclination) disposed along the foundation perimeter. In reinforcement design it is important to tag on: foundation reinforcement design, punching shear design, design on shearing forces, cracks opening design, design of bolts anchorage in the foundation body. To calculate foundation reinforcement, it is considered a circular or ring shape surface of the foundation, divided into 10 ring shape areas and it is calculated the radial and tangential bending moments for the entire span. It can be used numerical and FE methods.

49.4 Conclusions

Wind energy is a renewable energy source of the power, generated by the wind power. Although now is a relatively minor source of electricity for most countries, wind energy production has increased almost five times between 1999 and 2012, leading, in some countries the share of wind energy in total energy consumption to be significant: in Denmark (23 %), Spain (8 %). At the same time, studies show that this energy production is continuously growing, therefore foundation design will be a challenge and a necessity for civil or geotechnical engineers.

References

1. EN 1990—Eurocode. Basis of structural design
2. EN 1992—Eurocode 2. Design of concrete structures
3. EN 1997-1—Eurocode 7. Geotechnical Design. Part 1. General rules

4. V. Farcas, N. Ilies, A. Popa, Ecological systems to produce electrical power. Design of the foundations system for wind power plants. In *Proceedings of the International Conference "Mineral Resources and Environment"*, Romania, pp. 69–70. ISBN 978-606-536-116-4 (2010)
5. A. Popa, N.Ilies, *Fundatii*. (Casa Cartii de Stiinta, 2013)
6. A. Popa, V. Farcas, *Geotehnica*. (UTPress, Cluj Napoca , 2004)
7. A. Popa, F. Roman, *Calculul structurilor de rezistenta pe mediu elastic*. (UTPress, Cluj Napoca, 1998).
8. http://www.ewea.org/fileadmin/files/library/publications/statistics/Wind_in_power_annual_statistics_2012.pdf

Chapter 50
Aspects Regarding Soil Investigation and Foundation Design for Photovoltaic Power Plants

Vasile Farcas and Nicoleta Ilies

Abstract Between all sources of green energy, the photovoltaic power plants are among the best solutions encountered nowadays. Despite all the advantages given by this solution, the major problem remains the large surface of terrain required to build the entire project. As a result, instead of consuming good agricultural soils for the use of photovoltaic power plants, other categories of soils can be exploited. In order to protect good agricultural terrains the photovoltaic power plants are mostly displaced in areas with difficult soil conditions such as soft soils or height slopes. The paper presents the particularities of photovoltaic panels power plants, designed on difficult soil condition. Moreover, the paper describes special aspects regarding solar power plants foundations and geotechnical investigations on slopes and soft terrain.

50.1 Introduction

The first step necessary in the process of making a photovoltaic power plant is to find a good site, from geographical point of view, slope inclination, cardinal orientation, altitude, number of sunny days per year etc. Choosing the site is relatively simple because the requirements for obtaining maximum power outputs are defined by panels producer. It remain the problem of the site characterization from geotechnical point of view and some other limitations like constructions

V. Farcas (✉) · N. Ilies
Department of Structures, Faculty of Civil Engineering, Technical University of Cluj Napoca, 400027 Cluj Napoca, Romania
e-mail: vasile.farcas@yahoo.com

N. Ilies
e-mail: nicoleta.ilies@dst.utcluj.ro

A. Y. Oral et al. (eds.), *International Congress on Energy Efficiency and Energy Related Materials (ENEFM2013)*, Springer Proceedings in Physics 155,
DOI: 10.1007/978-3-319-05521-3_50,

interdiction on agricultural terrain, natural reservations etc. The main technical problem which can define the success of a photovoltaic plant is the soil. The document who have to answer to this problem is the Geotechnical Report.

50.2 General Requirements for a Geotechnical Study: Aims

The geotechnical report must provide all necessary geotechnical, geological and hydrological information about the site, in order to perform the design, the construction and the exploitation of photovoltaic farm. Also the report must provide enough data about the local weather, frost depth, seismic risk etc.

The soil characterization from laboratory and in situ testing and the elaboration of the Geotechnical Report must be performed after the requirements of Eurocode 7—Geotechnical Design [8, 9].

To establish the volume and type of investigation required it is necessary to have the elements which will be constructed. In general all photovoltaic plants have common elements how follows:

- Foundations for panels;
- Roads—for access inside and trough the plant;
- Small buildings—usually only one level height, with relatively small charge, max $1.0 \div 1.5$ to/m^2;
- Trenches for electrical cables;
- Slope; usually the plant is erected on a hill and slope stability must be certified.
- Other elements: fences, gates etc.

The team which carries out the geotechnical report must be a legal society formed by qualified and competent personal, certified by their country authorities for this kind of work. It is mandatory to have equipment for site investigations, laboratory equipment, software's and of course a minimum experience.

50.2.1 The Geotechnical Study Content

All tests must be conducted in order to establish the foundation condition for buildings and slopes. The volume of works shall provide enough, but not unnecessary information.

- In situ tests

All works are focused in establishing the next information:

- Stratigraphy; determinations of layers boundary, thickness, layers type and geotechnical characteristics;

Table 50.1 Required number of drillings

Surface	<1 ha	2 ha	5 ha	10 ha	20 ha	50 ha
No. of drill	4	6	8	15	24	40

- Water levels (piezometric and hydrostatic), maximal and minimal levels, chemical aggressiveness;
- Geophysics characteristics (resistivity, seismic) etc.

The geotechnical study will include all kind of in situ test needed:

- drillings (rotary drilling, percution drillingetc.) fitted for the soil type;
- CPT—cone penetration tests, CPT-U—cone penetration test with pore pressure measurement, SPT standard penetration test, DP—dynamic penetration test, PLT—plate load test, DMT—dilatometer tests (Marchetti), PMT—pressure-meter test (Menard) etc.
- Geophysical research—geoelectrical methods, geoseismic methods, georadar methods.
- CBR test—Californian Bearing Resistance Ratio, for road construction etc.

The number of tests is different from site to site, but a minimum number of drillings can be found in Table 50.1 (of course the number depends on every country regulation) [10]:

- Laboratory test

From the drillings, samples for laboratory testing will be taken. The laboratory will perform physical and mechanical tests against the samples as follow:

- Particle size analysis by sedimentation and sieving or combined methods,
- Density—specific gravity, apparent and relative density,
- Friction angle, cohesion for undrained and drained conditions,
- Compressibility test, oedometer tests,
- Determination of natural moisture,
- Atterberg limits,
- Testing of organic matter in soils,
- Proctor Test: Moisture content /dry density value,
- Test of free swelling- for swelling clays,
- Soluble salt content (Mg, Cl etc.) in soil and water if phreatic level exits (water table),
- CBR test on natural or stabilized soil sample.

The tests number may be increased or decreased depending on the geological homogeneity of the site. To perform the various laboratory tests should pay attention to sampling and sampler devices, using a sampler to keep unchanged the following soil properties: structure, density, moisture, grain size, plasticity and chemical components stable. Some soils like gravels, present real difficulties to sampling; in these cases in situ test are preferred, for example DP test.

Table 50.2 Soil parameters for mail soil layers

Nr.	Depth	Soil type	γ (kN/m^3)	φ (°)	c (kN/m^2)
1	0.00…−1.00	Silty sand	18.0	29.0	5
2	−1.00…−1.50	Sand with traces of fines, dense	17.5	31.5	0
3	−1.50…−8.00	High plasticity clay, very stiff	20.0	15.0	70

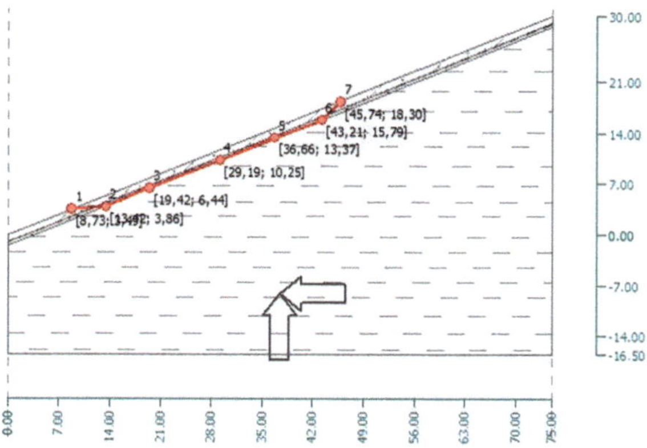

Fig. 50.1 Slope stability analysis for a polygonal sliding surface

50.3 Slope Stability Case Study

The purpose of the slope stability analysis is to determine if the solution proposed is able to solve the stability problems found on the slope. Also this study's aim is to prove that the foundations for the photovoltaic power plants may stabilize the slope. In order to perform the stability analysis it is considered a slope having 75 m length, 22° inclinations and a succession of soil layers as in Table 50.2. This considered stratigraphy it is characteristic for Transylvanian slopes [1, 3, 4–7], therefore with small adjustments this study may be used for regular design. The ground water table is found in the layer of sand, at −1.10 m depth from the natural ground level.

Based on the soil stratigraphy and ground water table, the conclusions arising is that the most probable sliding surface will occur in the sandy layer of soil, influenced also by the water table, which modifies the soil parameters. The analysis to find a stabile profile was performed using Slope Stability Module of Geo Fine software [2], considering different probable situations. In this case the most suitable analysis uses Sarma method, for slopes with probable polygonal sliding surfaces and Fellenius-Petterson method, for slopes with probable circular sliding surface. According to the Romanian and European norms, for a slope, in order to be stable, the minimum factor of safety required is $F_S = 1.50$. This case study considers therefore the safety factors approach.

The analysis using Geo5 software proved that the sliding surface with the smallest safety factor occurs in the sandy layer of soil, as in Fig. 50.1, considering

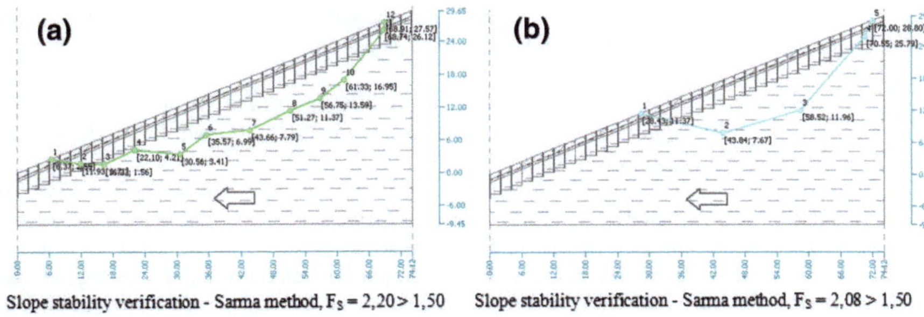

Fig. 50.2 Slope stability analysis with the foundations: **a** general stability, **b** local stability—polygonal sliding surface

also a seismic action for Transylvanian region, with the vertical seismic coefficient $k_v = 0.08$.

By realizing the foundations for the photovoltaic power plant, a row of stiff metallic piles, having 110 mm diameter, embedded into the stiff clay layer, placed at every 2 m, these piles acting like a retaining system, the entire slope is stabilized, as seen in Fig. 50.2.

This soil consolidation measures have to be completed with an appropriate drainage system located around the foundations. Ground water table may influence significantly soil parameters and respectively soil active pressure and stability conditions.

50.4 Conclusions

One of the most important stages for this type of construction works is soil investigation. Hydro-geological aspects represent an important issue in predicting the hazards in the constructed area; therefore all the precaution measures need to be accomplished. Underground water has a negative effect on mechanical characteristic of the soil, by reducing them, which is increasing soil instability. Also seismic action may lead to important decrease of safety factors.

References

1. N. Chira, V. Farcas, G. Chiorean, Using ecological solutions for slope stability at Cojocna salty, in *Advances in Environmental and Geological Science and Engineering 3rd International Conference on Environmental and Geological Science and Engineering*, Constantza Maritime University, ISSN 1792-4685, ISBN 978-960-474-221-9 (2010), pp. 39–44
2. Geotechnical software suite Geo 5. User's guide (2010)

3. I. Molnar, N.M. Ilies, Constitutive models used in geotechnical engineering design, in *Proceedings of the 11th International Scientific Conference VSU' 2011*, vol II. 2–3 June 2011, (Sofia, Bulgaria, 2011), pp. VI- 38–42
4. O.C. Muresan, N.M. Ilies, A. Popa, Consolidation of an active surface landslide, in *Proceedings of the 11th International Scientific Conference VSU' 2011*, vol II. 2–3 June 2011 (Sofia, Bulgaria, 2011), pp. VI- 49–53
5. A. Popa, N. Ilies, C. Chiorean, Utilization of analytical methods in slope stability and consolidation, in *Proceedings of 14th Danube-European Conference on Geotechnical Engineering from Research to Design in European Practice* (Bratislava, Slovakia, 2010)
6. A. Popa, F. Lacatusu, N. Ilies, Consolidarea amplasamentului Muzeului Etnografic al Transilvaniei. Simpozionul Infrastructuri eficiente pentru transporturi terestre, Zilele Academice Timisene, Editia a X-a, (Timisoara, Romania, 2007) (in Romanian)
7. A. Popa, N.M. Ilieş, Fundarea unei construcţii pe un versant instabil, Lucrările conferintei: A XI-a Conferinţă Naţională de Geotehnică şi Fundaţii, Timişoara, pp. 49–59, ed. Politehnica, (2008) (in Romanian)
8. SR EN 1997-1. Eurocode 7: Geotechnical design—Part 1: General rules (2006)
9. SR EN 1997-1/NB. Eurocode 7: Geotechnical design—Part 1: General rules. National Annex (2007)
10. A. Stanciu, I. Lungu, Foundations, ed. Tehnica, Bucharest (2006)

Chapter 51
Development of New Technologies of Solid and Gaseous Biofuel Production

Victor Zaichenko

Abstract Perspective direction of complex usage of biomass is connected with technologies of combined processing of organic fossil fuels and biomass with production of energy and carbon materials of high purity which can be used as high-calorific fuel and raw material for industrial technologies. Various directions of combined processing of a biomass are considered. The technology of pyrolysis of wood waste and peat and natural gas with productions of pure carbon materials and power gas with high content of hydrogen is presented. It is shown, that the combined technology of processing of biomass and natural gas is allowed to solve the problems connected with hydrogen production for power use.

51.1 Introduction

For electric power industry basic change of the existing power generation schemes aimed to reduce adverse environmental sequences can be realized only if they are economically justified. Absence of practical steps in the field of hydrogen penetration into the power industry is connected with those circumstances. To solve this problem it is necessary to develop new approaches to the production of hydrogen from natural gas, namely simultaneous production of carbonaceous materials, which can be regarded as valuable by-products.

Simultaneously with developing hydrogen power engineering the processes and technologies of carbonaceous materials industrial implementations should be developed. Only in the case of positive economic effect both of hydrogen using in power units and carbonaceous materials, received from natural gas in different industrial technologies it is possible to count for wide using of hydrogen energy

V. Zaichenko (✉)
Joint Institute for High Temperatures of the Russian Academy of Sciences (JIHT RAS),
Izorskaya str. 13, 125412 Moscow, Russia
e-mail: zaitch@oivtran.ru

A. Y. Oral et al. (eds.), *International Congress on Energy Efficiency and Energy Related Materials (ENEFM2013)*, Springer Proceedings in Physics 155,
DOI: 10.1007/978-3-319-05521-3_51,

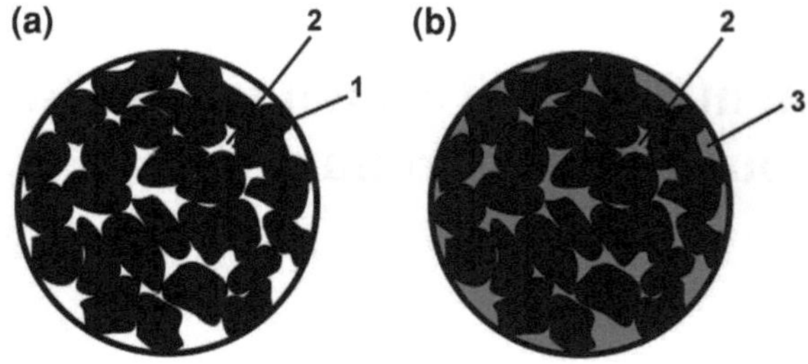

Fig. 51.1 Principal scheme of stuffing of the porous carbon structure by pyrocarbon using the natural gas pyrolysis process: **a**—the initial structure of granulated dispersed carbon sample; **b**—the same structure after stuffing with pyrocarbon. *1*—particles of dispersed carbon, *2*—inner pores free and stuffed, *3*—the exterior pores stuffed with pyrocarbon

methods. When thermal decomposition of natural gas is carried out for hydrogen and black carbon production it is necessary to minimize the pyrocarbon yield that in this case is a by-product and lead to considerable environmental pollution. This problem was considered in detail early [1, 2]. The dispersible carbon produced from natural gas has restricted applications in industries. If the carbonaceous materials mechanical characteristics would meet the requirements of various industries, the system of complex using hydrogen and carbon may be realized.

51.2 New Process of Carbonaceous Materials Production from Natural Gas

The considerations mentioned above were the background for development in the Joint Institute for High Temperatures of the Russian Academy of Sciences a new technology for producing from natural gas carbon material, named «granulated pyrocarbon». This technology is formed on the basis of a two-stage pyrolysis of natural gas. At the first stage thermal decomposition of natural gas is carried out following the homogeneous mechanism with forming hydrogen and dispersible carbon. At the second stage the obtained previously dispersible carbon are granulated and thermal decomposition of natural gas are carried out in this porous matrix bed.

By thermal decomposition (pyrolysis) of natural gas in the porous bed matrix of granulose dispersible carbon the process is carried out according to the heterogeneous mechanism with pyrocarbon formation, which fills up the pores and convert the dispersible carbon into a solid monolith. Figure 51.1 illustrates the essence of the stuffing process. During the pyrolysis process the interior and exterior pores are stuffed with pyrocarbon forming a monolithic carbonaceous material.

The developed process producing from natural gas a new type of carbonaceous materials, which were named «granulated pyrocarbon» is realized as follows. At

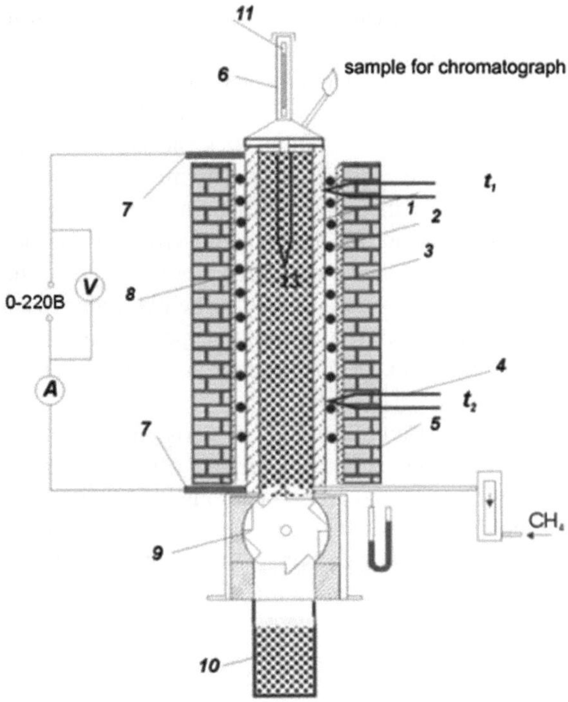

Fig. 51.2 Experimental installation for studying granulated pyrocarbon production from natural gas

the first stage natural gas pyrolysis is carried out using the homogeneous mechanism to produce disperse carbon and hydrogen. For this purpose, for example, the natural gas thermal decomposition technology, which was described earlier, may be used. Further on the obtained dispersible carbon is granulated in granules of 2–4 mm diameter. The granulated disperse carbon is loaded in a vertical furnace, where temperature about 1,300 K (at the bottom and in the middle cross section) is maintained. In the second stage methane flows through the furnace from the bottom upward in a countercurrent with granulated porous carbon. Methane is decomposed on the hot porous carbon particles following the heterogeneous root. The evolving pyrocarbon is deposited at the inner and outer surfaces of disperse carbon.

51.3 Experimental Installation for Studying the New Method of Carbonaceous Materials Production from Natural Gas

The process was studied in a laboratory scale installation (Fig. 51.2). The working section was manufactured from an alumina tube (1). First, the working section was loaded by granulose of granulated disperse carbon obtained by natural gas homogeneous thermal decomposition process. During experiments in accordance with the outlet of stuffed material from the reactor volume a fresh material (8) from above was filled in. The tube was heated by an electrical resistance heater (2) with current leads (7). The heater was isolated from the furnace body by a glass cloth (3). The temperature distribution was measured by thermocouples (4). The thermal isolation of the alumina tube was manufactured from refractory bricks (5). For loading of fresh material and checking the bed movement the feeding bin was made from a silica tube (6) equipped with a measuring rule (11). The unloading of stuffed material at the bottom was carried out by a gear wheel (9) to a receiving bin (10).

The natural gas was fed from below under the bed, and the gaseous reaction products were removed from the upper section of the installation and were burned in an open flame. In this section gas samples for chromatographic analysis could be taken.

The developed technology of two-stage natural gas pyrolysis allows producing ultra-pure carbonaceous materials, which can be utilized in different industries. The deficit in clean carbon materials in various production spheres is significant in particular in metallurgy [3]. Our technical economical calculations shows that production of pure carbon granulose from natural gas at presents pure carbon materials prices are economically justified without taking into account usage of hydrogen, which is received simultaneously with carbon for energy purposes.

51.4 New Technology of Integrated Thermal Processing of Carbon-Containing Waste and Natural Gas (Example of Wood Wastes)

The described above process of production of carbonaceous material from natural gas can be also used for production of solid carbon materials by integrated processing of carbon containing waste including woods.

The wood is a low-calorific fuel. The specific expenditures for using woods as energetic fuel are much higher than for coal, oil and natural gas. The developed technology of stuffing of a porous matrix with pyrocarbon allows receiving high quality energy fuel composed of carbon from the wood and carbon from natural gas. The main goal of the developed process is to modify the wood feedstock by adding carbon from natural gas.

Along with energy application the process may be used to yield materials with high carbon content in various industries for example in metallurgy.

Dry wood of different sorts has practically identical composition: carbon, on the average—49 %, hydrogen—6 %, oxygen—44 %, nitrogen—0.1–0.3 % and 0.2–0.8 % of ash.

The process of carbonaceous materials production from wood, or in general, from carbon containing wastes, can be realized in an installation described above and shown on Fig. 51.2. The wood feedstock is loaded into the furnace from

Table 51.1 The characteristics of carbon materials produced by using developed technology

Characteristics	Carbon materials from wood	Carbon materials from black carbon
Moisture content,%	0.24	0.01
Volatile, %	1.04	0.43
Ash, %	1.23	0.04
Sulphur, %	0.02	0.02
C,%	97.84	99.56
H, %	0.22	0.25
Calorific value, MJ/kg	32.95	33.18
Density, g/cm^3	0.76	1.62

above. The gaseous products of natural gas thermal decomposition pass from the bottom upwards through the wood particles. In the upper sections of the furnace the hot pyrolysis reaction products moving in a countercurrent with the wood particles are gasifying them with formation of porous structure of charcoal. The charcoal particles move down to the high temperature section. Here the upward moving natural gas is decomposed and the appearing pyrocarbon fills the charcoal pores.

As experiments have shown, depending on the residence time of charcoal particles in contact with the decomposing natural gas it is possible to receive various degree of porous charcoal matrix stuffing with pyrocarbon namely 20, 50 and 70 % and more. In some experiments we obtained a stuffed material, which consists by 96–99 % of carbon, with ash content (depending on sorts of wood)—0.2–1.5 %.

Thus the developed process allows to receive from wood wastes high-quality energy fuel with high carbon content. Along with power production those carbonaceous particles can be used in industries where high purity carbon materials are indispensable.

As a source of hydrocarbon gases instead of methane various hydrocarbons including oil gas may be used. The developed process can be considered as a technology for utilization of oil gases on the deposits where those gases are simply flared. If there are also sources of wood wastes they can be processed together to produce valuable products.

A separate direction of possible usage of the developed technology is the hydrogen energy. Solid carbonaceous materials and combustible gases with hydrogen content up to 80 % are produced in the developed process. The schema of complex usage of these products namely when solid carbon will be used in some industry and for energetic purposes, and gaseous products with high content of hydrogen—for power engineering, will allow reducing pollution of the environment (Table 51.1).

The photo of particles of carbon materials produced from natural gas black carbon and wood waste presented on Fig. 51.3. In the table the properties of the produced carbon materials are presented.

Fig. 51.3 New carbon materials produced by developed technology: **a**—solid carbon materials from natural gas (granulated black carbon from natural gas stuffed by pyrocarbon), **b**—solid carbon materials produced as a result of combined thermal processing of natural gas and wood waste products

The proposed technology gives the opportunity to realize a process for production of hydrogen for energy usage in conditions when all received components can be practically used and do not need disposal.

51.5 Conclusion

Economical justifying is a main factor, defining further development of new technologies.

Technology of using each of type fossil fuels passes usually through two stages. First, a fuel used in the raw non-processed form. Afterwards technologies are developed, aimed to raise efficiency of its use in energy installation. Development of these technologies in the times of primary using coal and oil was dictated by need to produce from fossil fuels raw materials required for other industries. Products of natural gas procession (hydrogen) is considered now not only as a raw material for chemical technologies but as a perspective fuel with a significant role to reduce environmental hazards of fuel-energy complex.

Use of carbonaceous materials produced from natural gas by the proposed technology in metallurgy will allow reducing ecological pollutions connected with production and use of coke because it will be possible to substitute it with carbon from the natural gas. Technical economical estimates show, that assuming existing prices of natural gas and pure carbon material, it's production with the help of proposed technology will be economically justified.

Hydrogen energy (based on production hydrogen from natural gas) should be considered as part of the total problem connected with development of new methods and technologies of complex natural gas processing. Further work in this direction certainly must be continued.

Acknowledgments This work was supported by grant 14.515.11.0096 from the Ministry of Science and Education of the Russian Federation.

References

1. E.E. Schpilrain, V.Y. Shtrenberg., V.M. Zaichenko, Possible ways of hydrogen energy development-2, in *The Proceedings of the 4th International Symposium on Hydrogen Power "HYPOTHESIS-IV"*, (Stralsun, Germany, 2001), pp. 304–309
2. R.G. Popov, E.E. Shpilrain, V.M. Zaichenko, Natural gas pyrolysis in regenerative gas heater, Part I: Natural gas thermal decomposition on heat saving matrix of regenerative gas heater. Int. J. Hydrogen Energy **24**, 327–334 (1999)
3. B. Gauderback, S. Lynum, Hydrogen from natural gas without release of CO_2 to the atmosphere, in *Proceedings of the 11th World Hydrogen Energy Conference*, (Stutgart, Germany, 1996), pp. 511–523

Chapter 52
Data File of the Building Site's Renewable Energy Characteristics

Vygantas Žėkas, Vytautas Martinaitis and Giedrius Šiupšinskas

Abstract After choosing the engineering solution of Integrated whole Building Design Process (IWBDP) and Global Optimized Local Designed (G.O.L.D) one of the most important elements of the creative process becomes the input data. There is a lack of comprehensive approach to the locally disposed building site's RE and higher overall exploitation of this potential. That is why there is a need to have a prepared data file presenting building site's energy potential. The aim of the described work is to create the input data file, to present the file creation proces and the key characteristics. Building site data must be handy not only for the initial conception creation stage, but also for the subsequent operating and maintenance of the building on going through the BIM systems. Data must be based on the long-term observations and target research of the building site. Data file is required to provide information about the standard nature processes occurring in the defined geometric boundaries of the site. The results of the work are the proposed forms of input data file, structure and main energy characteristics of the building site. When planning the development of the building site, creating the design concept, and selecting the combination of RE integration technologies, data array proposed in the article will allow a more objective selection of the optimal solution of RE integration.

V. Žėkas (✉) · V. Martinaitis · G. Šiupšinskas
Department of Building Energetics, Vilnius Gediminas Technical University, Vilnius, Lithuania
e-mail: vygantas.zekas@vgtu.lt

V. Martinaitis
e-mail: vytautas.martinaitis@vgtu.lt

G. Šiupšinskas
e-mail: giedrius.siupsinskas@vgtu.lt

A. Y. Oral et al. (eds.), *International Congress on Energy Efficiency and Energy Related Materials (ENEFM2013)*, Springer Proceedings in Physics 155,
DOI: 10.1007/978-3-319-05521-3_52,

52.1 Introduction

One of the main challenges of technological development is the assurance of growing energy needs. Energy consumption increase is accompanied by two major problems. The first problem is the threat of fossil fuel shortages. The second is an alternative energy absorption capacity increase. At the same time, it is a constant challenge—how to reduce the risk of energy shortages by increasing alternative energy, normally renewable energy, amount. Renewable energy, generally perceived as wind, solar, biomass, water and soil sources of energy. Their defining feature is that they are anywhere in the world just in different composition and amount. In order to use these resources in separate or combined way, it is necessary to look for engineering responses, which are generated by complex optimization way. In the published strategic plan of International Energy Agency "Towards Near-Zero Primary Energy Use and Carbon Emissions in Buildings and Communities" [1] the 2030 energy user future vision "GOLD (Global Optimized Local Designed) is presented. This idea reflects a clear need in the planning stage to take into account the global impact of solutions and scale, but at the same time meet the needs by using local resources taking into account local surround influence factors. In the path of realization of idea it is possible to identify a broader approach to the construction process, as an example an integrated whole building design process (IWBD, Integrated Whole Building Design) [2] or the creation of an integrated project (IPD, integrated project delivery) [3], along with the performed analysis of building and its engineering systems. In addition we cannot ignore that the need may also be controlled [4]. There are plenty of methods and instruments allow analysing the cases of the RE use: RE use planning tools [5, 6], the initial evaluation tools of the RE [7], RE development and hybrid RE transformer modelling computer tools and algorithms [8–10]. Banos et al. [11] provides RE questions resolved by optimization methods and performed optimization work examples. In this context, power engineering should be prepared to provide quality care for the building conception stage. Primary engineering contribution is the collection and presentation of starting data. Having in mind the renewable energy sources, the input data becomes the basis of meteorological data and the technical characteristics characterizing the uniqueness of the RE. In the age of information technology is increasingly being used the satellite information processing to generate any location's meteorological data. Since the satellite information includes the total land surface area, so with the same level of detail and reliability it is possible to generate the necessary data and fairly reasonably compare alternative RE plant development sites use or solutions. This approach provides a uniform and reliability, but inaccurate data in a single building site scale. Natural conditions and RE place potential depends on the individual local geographical conditions. The main problem is that the satellite information level of detail is insufficient. More often meteorological stations weather data is used, which then later has to be recounted into RE streams at the same time has to be

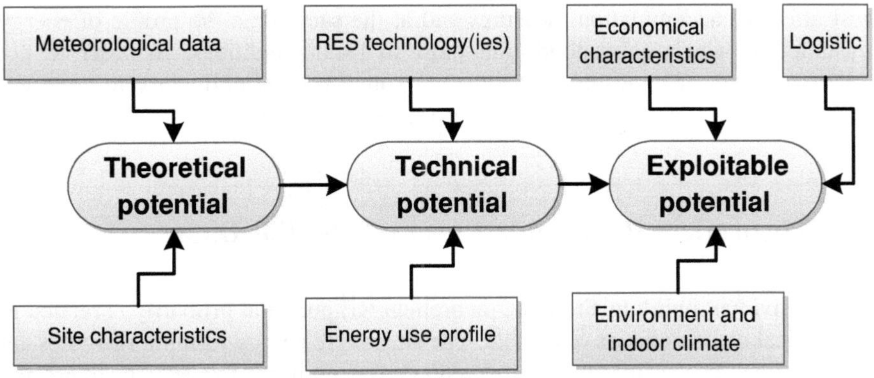

Fig. 52.1 Evaluation of renewable energy potential levels

assessed the influence of local conditions and formed the data source file. It has to be useful, not only at the planning stage, but also for solving tasks of hybrid RE solutions modelling.

52.2 Renewable Energy Potential

For determination of renewable energy potential different tools and methods are applied [12]. The main evaluation principle is that the potential (accessibility) can be set at different levels:

- theoretical (gross energy) potential level;
- technical (fraction of gross energy that can be harvested by the energy conversion system);
- economic (fraction of energy that can be useful to use).

The theoretical potential is understood as existing in the nature some type of energy level that is determined by the physical formulas, technical potential—human activity, plant efficiency, availability, and primary energy consumption quantity changed in influenced by other factors; economic level—the amount of energy which is cost effective to master taking into account logistics, environmental impact and economic indicators. Arrange the levels of potential at the same time are a cycle of steps required to identify opportunities for energy self-sufficiency within the site. The first level, the theoretical potential is connected only to assessment of local conditions and having in mind RE, first of all it is related to meteorological data. In the second stage technological solutions, energy user, energy and values ensuring standard microclimate are presented (Fig. 52.1).

Figure (52.1) scheme shows that meteorological data for energy engineer becomes as a base when offering the solutions combined technological combinations. In later phases it has to be selected active, passive RE technology and the most affective accumulation measures and at the same time the profile of energy consumption which corresponds the most to local conditions. In order to not restrict the way of choosing technological solution, it is obligatory to have RE potential data which has to be in uniform detail and affected by local conditions.

52.3 Theoretical Renewable Energy Potential Data

In the individual building site case, theoretical RE potential primarily is related to the physical characteristics of the site. The site where energy forming activities are expected first of all primarily is perceived as a space limited to a specific area—the site area. Usually it is not a case to evaluate the space above the site or on the surface and the following conclusions about this area as in the total volume of the site existing RE potential. When imaginary site volume limits are set without passing through volume RE flow (solar, wind, air heat, heat into the ground soil layer, heat from ground to the air, precipitation reflected solar energy end other), we can clearly identify the other important characteristics of the site: the volume of selected depth soil layer or air flow rate.

That way, the site reaching, described by the meteorological data RE flow is seen as a building site RE potential. Potential and his character describing data and potential quantity and quality influenced encirclement factors becomes the main theoretical RE potential detection task. As a result in providing the input data file.

The site volume is recommended to set up according planned underground and ground engineering communications, potential zones of influence. Presenting shadow involutes of neighbouring objects, it has to be presented 360° on the horizon visible range of obstacles dimensions.

52.3.1 *Meteorological Data*

Meteorological data as primary RE engineering expression is captured by measuring equipment in a selected range and then processed. In computer hybrid RE systems modelling programs in RE use tools and methods have been used already processed meteorological data, i.e. determinate the day, month and year values. It can be seen that the planning tasks area are used for a longer period values, i.e. monthly or annual average values. There are used both historical climate data as well as probability estimates to determine the future of the period data. The meteorological data must be included in the data input file.

52.3.2 Derivative Indicators of RE Resources

When meteorological data is specified with the physical the additional data processing must be made. There is performed a quantitative (energy) [13] and qualitatively (exergetic) [14], resource evaluation as well as statistical data processing. Statistical data processing can be performed in the primary meteorological data generating stage, as well as in the stage of recalculating them into the quantitative/qualitative indicators. As shown by some of the most recent examples, the statistics as a method is widely applied. First of all it is applied because of chaotic nature of value variation of RE the flow. The statistical analysis provides additional characteristics that enable to evaluate RE inconsistent flow variation in time.

Using derivative RE resource indicators, power supply reliability, can be seen without a computer simulation. For power supply reliability evaluation as statistical analysis base is determinate the most frequently recurrent flow values; the range of values, as well as the median, which eliminates the influence of extreme values.

52.4 Conclusion

Increasing integrity of technical solutions and alternative energy sources use degree promotes a comprehensive look at the disposed available RE potential in the site. Getting popular integrated design and planning methods, improving as an advantage computer simulation system. There are created storage of information data assessable to energy engineers. In this context, the energy engineer must be prepared at the first stage of development of the concept of building to offer for energy user all the necessary information. Initial information on the building site and the existing energy potential had to be submitted to a uniform level of detail and without limitations in available technical solutions. This paper presents a broader approach to the initial RE use planning stage. It is proposed to prepare the input data file, which could include:

1. Technical land information influencing RE potential;
2. Meteorological site information as the primary numeric information about RE potential;
3. Renewable energy potential—qualitative and quantitative evaluation of individual flows, each dynamics in selected time interval; the total RE flow and its dynamics in the selected time interval;
4. Relative size—comparable characteristics describing the size of the RE potential having for site square meter;
5. Statistical evaluation of the RE potential data—presentation of statistical characteristics and evaluation of RE flow prevailing values energy supply reliability.

The data file must provide a uniform engineering level of detail evaluated and described the site existing RE forms. Data time step should be sufficient detail to address the complex challenges of energy supply, at the same time cumulative indicators are presented reflecting long term RE potential and its characteristics.

The proposed input data file would be user-friendly prime source of information for both, RE use planning, as well as in the design stages. Use of input data file which reflects potential of the place would enable to reach greater reliability of RE energy use in energy supply.

Acknowledgements This research was funded by a grant (No. ATE-03/2012) from the Research Council of Lithuania.

References

1. International Energy Agency, Energy Conservation in Buildings and Community Systems Programme. 2007–2012 Strategic Plan. Towards Near-Zero Primary Energy Use and Carbon Emissions in Buildings and Communities
2. Integrated Whole Building Design, Guidelines ISBN: 978-0-478-33130-1 (electronic). Published in October 2008 by the Ministry for the Environment Wellington, New Zealand http://www.mfe.govt.nz/publications/sus-dev/integrated-whole-building-design-guidelines/integrated-building-guidelines.pdf. Downloaded 05 May 2013
3. AIA, *Integrated Project Delivery: A Guide (Version 1)*. (AIA, California, 2007)
4. Pedro S. M., Anibal T. de Almeida. (2010) Multi-objective optimization of a mixed Renewable system with demand-side management. Renewable and Sustainable Energy Reviews 14: 1461-1468
5. H. Omar, B. Kankar, Optimal planning and design of a renewable energy based supply system for microgrids. Renewable Energy **45**, 7–15 (2012)
6. M. Ranjevaa, A.K. Kulkarnia, Design optimization of a hybrid, small, decentralized power plant for remote/rural areas. Energy Proceedia **20**, 258–270 (2012)
7. I. Nygaard et al., Using modelling, satellite images and existing global datasets for rapid preliminary assessments of renewable energy resources: the case of Mali. Renew. Sustain. Energy Rev. **14**(8), 2359–2371 (2010)
8. D. Connolly et al., A review of computer tools for analysing the integration of renewable energy into various energy systems. Appl. Energy **87**(4), 1059–1082 (2010)
9. P. Pezzini, O. Gomis-Bellmunt, A. Sudria-Andreu, Optimization techniques to improve energy efficiency in power systems. Renew. Sustain. Energy Rev. **15**(4), 2028–2041 (2011)
10. O. Erdinc, M. Uzunoglu, Optimum design of hybrid renewable energy systems: overview of different approaches. Renew. Sustain. Energy Rev. **16**(3), 1412–1425 (2012)
11. R. Banos et al., Optimization methods applied to renewable and sustainable energy: a review. Renew. Sustain. Energy Rev. **15**(4), 1753–1766 (2011)
12. A.A. Dimakis et al., Methods and tools to evaluate the availability of renewable energy sources. Renew. Sustain. Energy Rev. **15**, 1182–1200 (2011)
13. W. Liao et al., Thermodynamic analysis of human-environment systems: a review focused on industrial ecology. Ecol. Model. **228**, 76–88 (2012)
14. G. Wall, G. Gong, On exergy and sustainable development—part 1: conditions and concepts. Exergy Int. J. **1**(3), 128–145 (2001)

Chapter 53
A Multiagent Energy Management System for a Small Microgrid Equipped with Power Sources and Energy Storage Units

Weronika Radziszewska and Zbigniew Nahorski

Abstract An Energy Management System (EMS) for a small microgrid is presented, with both demand and production side management. The microgrid is equipped with renewable and controllable power sources (like a micro gas turbine), energy storage units (batteries and flywheels). Energy load is partially scheduled to avoid extreme peaks of power demand and to possibly match forecasted energy supply from the renewable power sources. To balance the energy in the network on line, a multiagent system is used. Intelligent agents of each device are proactively acting towards balancing the energy in the network, and at the same time optimizing the cost of operation of the whole system. A semi-market mechanism is used to match a demand and a production of the energy. Simulations show that the time of reaching a balanced state does not exceed 1 s, which is fast enough to let execute proper balancing actions, e.g. change an operating point of a controllable energy source. Simulators of sources and consumption devices were implemented in order to carry out exhaustive tests.

53.1 Introduction

The way of producing and using electricity is changing. In new approach energy may flow in both directions, traditionally from large producers to end consumers, but also in the opposite way. New types of power grids are under construction, e.g. microgrids, and smart grids.

A *smart grid* is a concept in which exchange of information between different elements of electrical grid (consumers, producers, storage units and prosumers) is

W. Radziszewska (✉) · Z. Nahorski
Systems Research Institute, Polish Academy of Sciences, Warsaw, Poland
e-mail: Weronika.Radziszewska@ibspan.waw.pl

Z. Nahorski
e-mail: Zbigniew.Nahorski@ibspan.waw.pl

A. Y. Oral et al. (eds.), *International Congress on Energy Efficiency and Energy Related Materials (ENEFM2013)*, Springer Proceedings in Physics 155,
DOI: 10.1007/978-3-319-05521-3_53,

introduced to control and coordinate supply and demand of energy, in order to: (1) ensure the quality of electric energy supply; (2) reduce the cost; (3) give priority to renewable energy sources.

A *microgrid* is a part of a grid that has the ability to connect or disconnect to/from the external power grid and balance the energy within itself. Microgrid is usually equipped with power sources that might be producing renewable energy, e.g. wind farms, photovoltaic panels, micro *hydroelectric power plant*; or might use fuel like gas or biomass, e.g. micro gas turbines, engines, cogeneration units. In this paper we consider microgrid that is connected to the external power supplier, which is able to supply additional power in case of deficit or buy the overproduction of the microgrid.

Energy management systems (EMS) are made to manage the sources and consumers of energy, to optimize the production schedule and control energy usage. Centralized or decentralized EMS has been considered. The system presented in this paper is a decentralized one and uses multiagent approach to energy management in a microgrid. The system implements both demand and production controls, by planning energy consumption and optimising energy production. The considered microgrid consists of 5 buildings, equipped with internal energy sources (wind turbines, photovoltaic panels, gas turbines, a reciprocating engine, a hydropower plant), storage units (flywheels and batteries) and consuming nodes (e.g. energy research laboratories, offices, conference halls, hotel rooms, etc.) [1]. The microgrid is simulated on a computer, including all installations of the produced power by renewable sources (with simulating the weather conditions) and levels of consumption on each node of the simulated electrical grid.

In the next sections short description of chosen elements of the system, followed by test results and conclusions are presented.

53.2 System of Complex Energy Management

Due to small sizes of microgrids, advanced mechanisms have been implemented to ensure balancing of the power at all times of their operation and maintaining the power quality. We focused on creating a system which implements this idea and validating its usefulness by simulation. For this, a virtual environment has been constructed. It includes: (1) models of power grid and devices; (2) simulating the behaviour of producers (implemented by simulating weather conditions for renewable sources) and consumers; (3) scheduling of planned events that define some profiles of energy usage; and (4) short-time energy balancing system. The outline of elements of the system is presented in Fig. 53.1.

Simulation of production: Creation of realistic test cases requires certain amount of test data. As the required amount of measured data was not sufficient for data-demanding experiments (data were needed for the specific geographical location), they had to be generated in a way that preserves the statistical qualities of the real life phenomenon. Such generators were constructed for the solar irradiance, wind

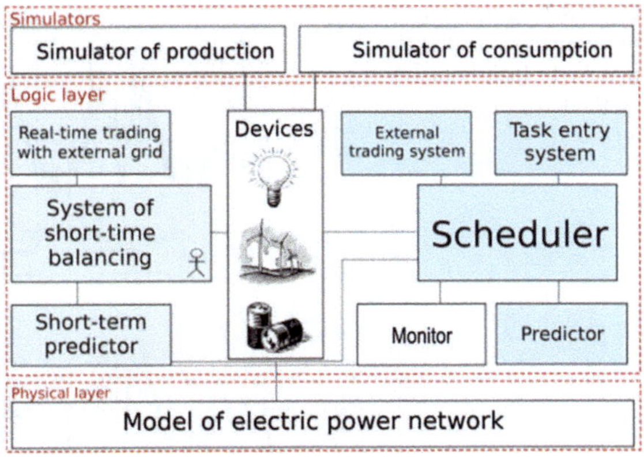

Fig. 53.1 Schema of the developed EMS system

speed and river water flow. Detailed description of the method can be found in [2]. A matched-block bootstrap method with fitness proportionate selection was used to generate appropriate time series [3, 4]. They were found statistically indistinguishable from the original data. The method allows for creating test sequences of an arbitrary length. The real data are divided into blocks and a new sequence is constructed from them, in such a way that the transition between the blocks is as smooth as possible. The idea of this method is illustrated in Fig. 53.2.

Simulation of consumption: Consumed energy simulators in most projects are simple, and basically use macro profiles for power usage. For testing purposes in our system, a fast changeable power levels have to be simulated, and that is why detailed power usage of each particular device is considered. It is much more accurate, makes the simulation less abstract, and gives a possibility to use existing devices' parameters, which may be measured or found in the literature. It is possible to measure the way the devices use the energy, e.g. such data for refrigerators, washing machines or projectors are known. There is, however, insufficient knowledge about how frequent and in what way users operate the devices. Modeling users' behavior of electric equipment usage is the most difficult part of this simulation, which currently can be only approximated based on research in different countries, as e.g. in [5].

The created simulator is using the devices' behavior described in various probability-dependent ways: switch-on and -off rules, working time rules, short work profiles and daily profiles. The schema of this is presented in Fig. 53.3. As a result, the simulator derives unrepeatable consumption profiles that have a random effect included, thus imitating real life situations. More details can be found in [6].

Scheduler: The EMS consists of two modules. A scheduler makes longer plans of energy usage and production, based on their long-time profiles and submitted tasks of planned events. A task is an event that is happening in a microgrid, e.g. reservation of a room in a hotel, a seminar, a scientific experiment, or a meeting in a room. The scheduler analyzes the list of submitted tasks and suggests the best times for realizing them; taking into account limitations defined by the user, and

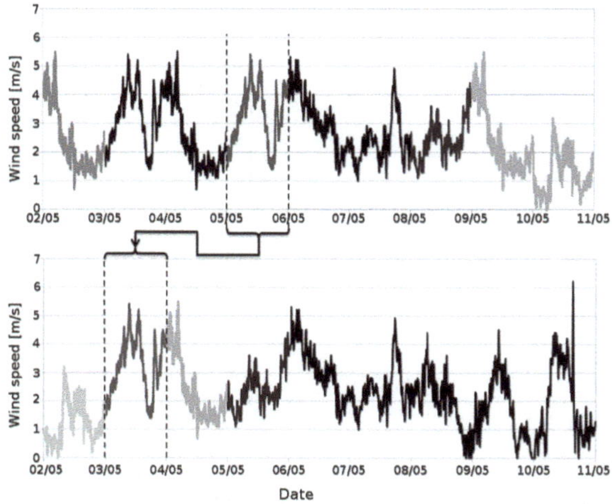

Fig. 53.2 Graphical representation of the matched-block bootstrap method

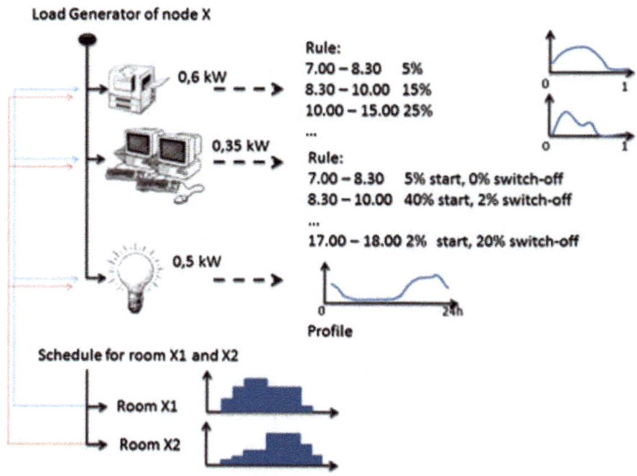

Fig. 53.3 Schema of different power consumption representations, which consider the character of the device's usage

constructs a schedule using heuristic algorithms. The schedule is suboptimal, but it guarantees obtaining a result within a specified time. Thanks to that the system knows the optimal operating point of all controllable sources and can use it as a baseline for further balancing. What is more, such knowledge allows the system to make more beneficial, long-term deals with external power provider when microgrid cannot produce enough power to operate. Our solution is different from that described in [7], where only cost of the energy purchased from the external operator and power capacity are minimized, and then deviations from the plan by the users are penalized. In our approach scheduling is used to minimize exchange of energy with the external operator, and the deviations from the plan are balanced internally by the system of short-time balancing.

System of short-time balancing: System of short-time balancing deals with unpredictable activity of some sources, non-exact realization of planned events and with unregistered energy consumption by using short-time forecast, and negotiating distribution of energy to minimize the use of polluting or expensive sources and to maximize use of the renewable energy. It is implemented as a multiagent system [8]. Such systems have proved their utility in simulating energy systems [9]. In the considered EMS each device is represented by its agent.

Agent is an autonomous program that has an ability of proactively acting to fulfill a defined task. In this case the task is to actively balance the changes of production/consumption of the energy by the device that agent is connected to. Agent is realizing it by interrogating all agents that have regulatory capacity if they can balance its device: cover the deficit or take excess power from the device. Having received the offers, agent chooses the source that is the most preferable. Preferences are defined as a "price of usage", which depends on the cost of producing energy by the device and its ecological impact on environment. The most preferable are wind turbines, photovoltaic panels, and hydropower. When the choice is done, the power is virtually send and the controllable units are requested to change their operating points accordingly. The operation of the agents is not synchronized in time. It means that agents start acting as soon as they receive information of state change of their devices.

Importance of storage capacity in microgrids is underlined in [10, 11]. Two types of energy storage units are considered: a flywheel and a battery. The character of these sources is different: a flywheel can take or give a large amount of power, but for very limited time, battery charges and discharges much slower, but can keep the power for long time. The preference of choosing storage systems for balancing power changes depends also on the state of the storage. When battery is fully charged, all requests for discharging (sending power to the system) are preferred, even before renewable sources. When battery is discharged, the charging mode becomes the most preferable choice. The agent of the storage units is also preventing batteries from deep discharge by requesting energy when the charge level becomes too low.

The operation of the system is based on market models and allows for smooth operation in the synchronized mode of the microgrid (connected to the external power supplier). The detailes of solutions used are presented in [12,13]. The same mechanism works also in the island mode, but in this situation, coping with global deficit of energy has to be managed separately. It can be done by e.g. switching off all less important devices to reach the maximum potential power level that can be supplied from available sources or dynamically deciding what should be switched on and off.

53.3 Balancing Tests and Results

A preliminary test run on 20 devices, each located on a separate computer, showed that the system balanced the energy and all agents behaved properly. The states of imbalance were very short, below 1 s, which nevertheless gives enough time for the physical device to reach a new operating point. Data from sample simulations are presented in Fig. 53.4.

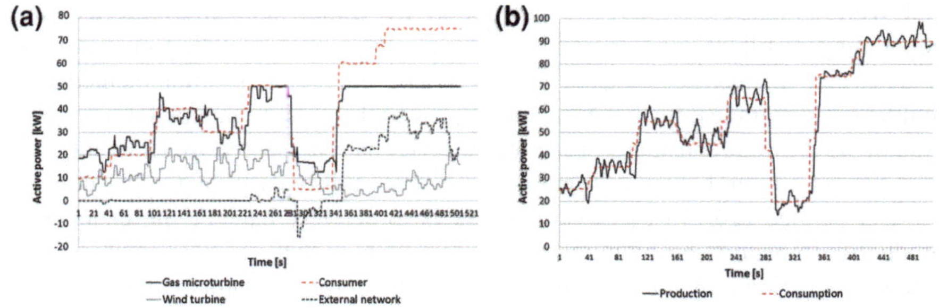

Fig. 53.4 Diagrams representing the balancing process. **a** The operation curves of all considered devices. **b** The curves aggregated to production and consumption categories

53.4 Conclusions

When creating this system many different side problems aroused. Simulation of production is very dependent on local conditions. The same is with consumption, and moreover, there is not enough research done to recognize patterns of human behavior. One of reasons to that are the privacy issues and high cost of installation of smart meters. An interesting problem is the island operation mode of the microgrid. Optimization of power usage and power balancing in this case is in general a complex problem, as under deficit of power the order of switching off devices has to be considered. Long-term testing of the proposed EMS will be performed and statistical analysis of the system operation will be made. Positive results presumably allow us to check performance of this system on a real installation.

Acknowledgments The research was supported by the Polish Ministry of Science and Higher Education under the grant N N519 580238, and by the Foundation for Polish Science under International PhD Projects in Intelligent Computing. The project is financed from The European Union within the Innovative Economy Operational Programme 2007-2013 and European Regional Development Fund.

References

1. J. Wasilewski, M. Parol, T. Wojtowicz, Z. Nahorski, A microgrid structure supplying a research and education centre—Polish case. In: *3rd IEEE PES International Conference and Exhibition on Innovative Smart Grid Technologies* (ISGT Europe), pp. 1–8 (2012)
2. W. Radziszewska, Z. Nahorski, Generation of inputs to renewable energy sources using matched-block bootstrap approach with fitness proportionate selection. EnviroInfo **2013**, 727–735 (2013)
3. E. Carlstein, K.-A. Do, P. Hall, H. Künsch, Matched-block bootstrap for dependent data. Bernoulli J. **4**(3), 305–328 (1998)

4. B. Efron, R. Tibshirani, *An Introduction to the Bootstrap* (Chapman & Hall, New York, 1993)
5. Atlas de la demanda eléctrica española. RED Eléctrica de España. Technical Report (1999)
6. W. Radziszewska, Z. Nahorski, Simulation of energy consumption in a microgrid for demand side management by scheduling. In *Proceedings of the 2013 Federated Conference on Computer Science and Information Systems,* pp. 679–682 (2013)
7. M. Vasirani, S. Ossowski, A collaborative model for participatory load management in the smart grid. In *Proceedings of 1st International Conference on Agreement Technologies*. CEUR, pp. 57–70 (2012)
8. M.J. Wooldridge, *Introduction to Multiagent Systems* (Wiley, New York, 2001) pp.21–38
9. S. McArthur, E. Davidson, V. Catterson, A. Dimeas, N. Hatziargyriou, F. Ponci, T. Funabashi, Multi-agent systems for power engineering applications. IEEE Trans. Power Syst. **22**(4), 1743–1759 (2007)
10. P. Vytelingum, T.D. Voice, S.D. Ramchurn, A. Rogers, N.R. Jennings, Agent-based micro-storage management for the smart grid. In *Proceedings of the 9th International Conference on Autonomous Agents and Multiagent Systems*, AAMAS'10, vol. 1. International Foundation for Autonomous Agents and Multiagent Systems, Richland, SC, pp. 39–46 (2010)
11. R. Palma-Behnke, C. Benavides, E. Aranda, J. Llanos, D. Saez, Energy management system for a renewable based microgrid with a demand side management mechanism. In *2011 IEEE Symposium on Computer Intelligence Applications in Smart Grid (CIASG),* Paris, pp. 11–15 (2011)
12. W. Radziszewska, Z. Nahorski, M. Parol, P. Pałka, Intelligent computations in an agent-based prosumer-type electric microgrid control system. In *Issues and Challenges of Intelligent System and Computational Intelligence*, vol. 530 (Springer, Berlin, 2014), pp. 293–312
13. P. Pałka, W. Radziszewska, Z. Nahorski, Balancing electric power in a microgrid via programmable agents auctions. Control Cybern. **4**(41), 777–797 (2012)

Chapter 54
Comparative Study Between Wind and Photovoltaic (PV) Systems

Wesam Taha

Abstract This paper reviews two renewable energy systems; wind and photovoltaic (PV) systems. The common debate between the two of them is to conclude which one is better, in terms of cost and efficiency. Therefore, comparative study, in terms of cost and efficiency, is attempted. Regarding total cost of both, wind and PV systems, many parameters must be taken into consideration such as availability of energy (either wind or solar), operation and maintenance, availability of costumers, political influence, and the components used in building the system. The main components and parameters that play major role in determining the overall efficiency of wind systems are the wind turbine generator (WTG), gearbox and control technologies such as power, and speed control. On the other hand, in grid-connected PV systems (GCPVS), converter architecture along with maximum power point tracking (MPPT) algorithm and inverter topologies are the issues that affects the efficiency significantly. Cost and efficiency analyses of both systems have been carried out based on the statistics available till today and would be useful in the progress of renewable energy penetration throughout the world.

54.1 Introduction

Due to concerns regarding fossil fuels depletion, renewable energy sources have gained a lot of attention in the current era. Additionally, seeking cleaner energy sources in the sake of carbon footprint reduction has made this topic significantly interesting. Furthermore, from a third prospective, renewable energy is assumed to be, economically, less expensive than fossil fuels, which is a crucial issue these days since the price of natural oil and gas are in a continuous increase [1]. Hence, scientists recommend future engineers to be well educated and highly trained in

W. Taha (✉)
Department of Electrical Engineering, The Petroleum Institute, Abu Dhabi, UAE
e-mail: weataha@pi.ac.ae

A. Y. Oral et al. (eds.), *International Congress on Energy Efficiency and Energy Related Materials (ENEFM2013)*, Springer Proceedings in Physics 155,
DOI: 10.1007/978-3-319-05521-3_54,

establishing projects based on different types of renewable energy sources such as solar, wind, ocean and biomass, as they believe that renewable energy will be the dominant energy source in the coming decade.

This paper focuses, only, on two types of renewable energy sources: wind and solar. The reason behind choosing those two sources is their importance among the other sources [2]. Figure 54.1 demonstrates this by showing investments spent of variety of renewable energy sources in 2010.

After narrowing the scope of the study (i.e. wind and PV systems only), the question is which one is better? And better in terms of what? Certainly, what concerns industries, investors and capitals the most is the cost and then the cost with respect to the efficiency obtained. Such comparison has to be taken carefully as one comes at the expense of the other. In other words, the demand on highest efficiency requires more advanced and sophisticated equipment which shall in turn raise the total cost of the system being tested. Therefore, this paper attempts to draw a conclusion that shall answer such questions.

The paper is organized as follows. Sections 54.2 and 54.3 describe wind and PV systems, respectively, arguing the most important parts and parameters in both systems. Then, Sect. 54.4 discusses a methodology to compare between wind and PV systems, in terms of cost and efficiency. Lastly, Sect. 54.5 concludes the paper.

54.2 Wind Turbine Energy

As demonstrated in Fig. 54.1, wind energy is the most growing renewable energy source. Many parts, parameters and topologies must be considered in evaluating the cost and efficiency of wind turbine systems. The power produced by a wind turbine is influenced by wind velocity, blade design, blade pitch, cross-sectional area, etc. [3]. Wind machines type can be categorized into horizontal shaft and vertical shaft [3].

This paper focuses on the horizontal shaft as it has been demonstrated in the covered literature that it has more rotating speed, efficiency and stability [3–5]. The horizontal shaft machine can be classified as facing upwind or downwind (see Fig. 54.2). In general, horizontal shaft wind machines are the upwind type because they have longer operating life [3].

The following subsections describe the main parts and parameters, with different topologies that contribute the most to the overall system's efficiency and cost.

54.2.1 Wind Generator Technology

Wind turbine generators (WTGs) can be categorized into fixed speed generators and variable speed generators [4]. A squirrel-cage induction generator is, generally, used in fixed speed WTG. However, variable speed generators are now the

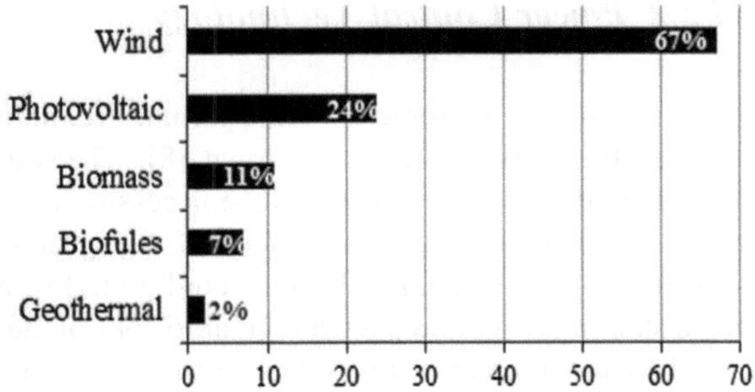

Fig. 54.1 Financial investments (US$/billion) on renewable energy sources in 2008–2009 [2, p. 13]

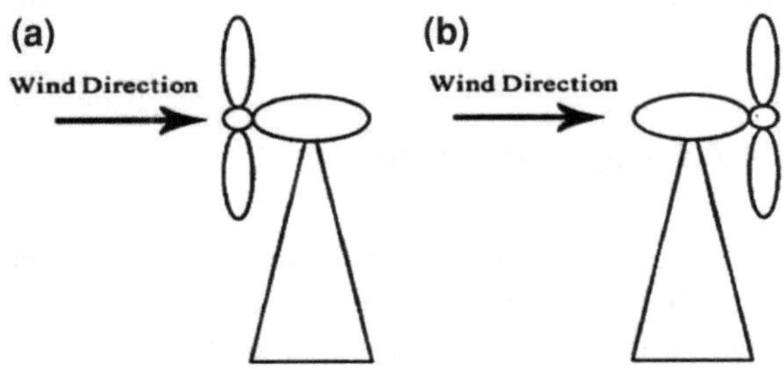

Fig. 54.2 Horizontal shaft wind machine type. **a** Upwind. **b** Downwind [3, p. 256]

dominant generators because of their higher efficiency, along with better power quality [6–8]. Thus, this paper focuses on analyzing variable speed WTGs. The electricity produced by WTGs depends on several factors such as cut-in wind speed, rated wind speed, cut-out wind speed, hub height and tower height [9, 10].

The commonly used WTGs in small wind turbines are the permanent magnets generators [7]. Other types of WTGs are synchronous generator, induction generator [6–8]. Reference [6] proves that synchronous generators have an incremental efficiency of 2 % over induction generators.

Currently, the typical generators used in the variable speed generator technology, at the mega-watt scale, are the double-fed induction generator (DFIG) and permanent magnet synchronous generator (PMSG) [11, 12].

DFIG is the dominant generator nowadays as it possesses significance advantages over the self-exciting induction generator and synchronous generator. DFIG can operate at over-synchronous and sub-synchronous speed. Additionally, output power can exceed the rated power [3]. Hence, DFIG has a higher efficiency over the aforementioned WTGs. However, PMSG is a more developed technology. Without the gearbox installed in the DFIG system, the PMSG system has less mechanical losses and better stability [11]. Thus, it is deemed to be promising, however, not widely adopted by wind power industry yet [13].

54.2.2 Power Control Technology

Utilizing power electronics in wind turbine technology is becoming more significant as their prices are going down [8]. The significance of enhancing wind turbines with power electronics is the necessity of controlling the output power of wind machine to ensure the safety of the gearbox and the WTG and avoid occurrence of any damage [3, 7]. According to [3], control procedure is classified into a stall regulator and a pitch regular to control the amount of wind energy being absorbed and transferred to produce power. Under those two control categories, [8] demonstrated that the best topologies of converters used in wind turbine are the matrix and multilevel converters, in addition to back-to-back converter.

54.2.3 Capacity Factor

Rated power is one of the most important characteristics of any electrical equipment. However, power systems never operates at this value, they deliver much less power instead. Thus, employing the mean output electric power provides a more useful and reliable data rather than the rated power [10]. Mathematically, the mean output electric power is defined as the product of the rated power and capacity factor (CF). In wind systems, CF is a dimensionless value that connects between the rated power and the actual power delivered by the WTG [14]. The value of CF mainly depends on the average wind speed under certain values for cut-in and cut-out wind speed. Reference [14] developed a linear model for estimating the CF of wind turbine as a function of average wind speed between 7 and 12 m/s. It demonstrated that the value of CF is directly proportional with the wind speed. However, this is viable to a certain extent. If the value of the average wind speed is too high from the rated wind speed the value of CF dwindles. Additionally, CF is also directly proportional to the tower height. As the tower's height increases, the value of CF increases [9, 10].

54.2.4 Efficiency Analysis

Seeking highest efficiency is always a crucial aim in electrical systems. However, electric power generated from wind systems is considered unstable [15]. The output power and losses are not linear elements and they are dependent on wind speed. Therefore, predicting the power levels based on average wind speed are not that viable and hence the estimated efficiency [15]. Commonly, the proposed methods used to calculate the total efficiency calculate the output power as a function of wind speed. Wind speed has a Weibull probability distribution [7]. A proposed methodology for calculating the total efficiency of wind generators, using induction generators, rated at 5 MW, is presented in [15]. The results show that the total efficiency is around 96.7 %. However, using the conclusion drawn from [6] which states that using synchronous generator instead of induction generator could result in incremental efficiency of 2 %, therefore, the total efficiency is expected to rise to 98.7 %. Additionally, since DFIGs have better adjustment, in terms of voltage and frequency, they are nominated to be the optimal selection for wind generators [3].

54.2.5 Cost Analysis

There is no certain cost of energy for wind farms [2]. The economic viability of wind power depends on the following factors [2, 16]:

- Availability of wind
- Installed capital costs and cost of capital
- Operation and maintenance costs
- Wind integration costs
- Transmission and grid upgrade costs
- Availability of costumers
- Political influence and governmental subsidies
- Capacity factor.

Parameters that fall under wind integration costs are the costs of blades, hub, nacelle, gearbox, generator, control systems, tower, electric instruments and infrastructure costs [9, 10]. Using analysis of cost of energy method by [9], the cost of wind energy is, approximately, US$236.2/kW.

Referring to [17], Eq. (54.1) estimates the total annual average cost (F_c) of a stand-alone wind system supplying a typical house on a ranch located at South-Central Montana is presented. The stand-alone wind system consists of two WTG rated at 10 kW. An interest rate of 60 % was used, in addition to life expectancy of 20 years for the wind turbines and 4 years for the batteries. Additionally, balance of system cost was estimated to be 25 %.

$$\begin{aligned} Fc_{\text{wind}} = {} & [\text{cost of wind turbine} \times 1.25 + \text{cost of batteries} \times (5\ \text{installments})] \\ & (0.08718\ \text{compound interest factor}) + \text{US\$0.2}\ (2) \\ & (25.63\ \text{kWh/turbine/day})\ (356\ \text{days}) \end{aligned} \tag{54.1}$$

Using Eq. (54.1), the total cost of the described wind system in [17] is US$5,574. In Canada, a gearless direct-drive synchronous generator wind system costs US$1,312/kW, which is a quite difference from the aforementioned costs [16]. However, as mentioned earlier in Sect. 2.1, the commonly used WTGs are DFIGs rather than synchronous, which are of a higher cost. Thus, a raise in the total cost is expected.

54.3 Photovoltaic System

Grid-connected PV system (GCPVS) configuration is the commonly used and researched configuration rather than the stand-alone. In terms of efficiency, PV systems are highly exposed to shadows, dust and mismatch of their electrical parameters, which makes seeking high efficiency significantly challenging [18]. Mainly, GCPVSs compose of dc–dc inverter with a maximum power point tracking system (MPPT) and an inverter [19]. The dc–dc converter is controlled by MPPT system to operate the PV array near the MPP and the inverter is controlled by synchronous reference frame in order to obtain low harmonic distribution from the grid current. Some advanced systems, takes it a step further by considering the system communication in order to increase the feasibility of the system and to reach to a predictive maintenance. This topic is covered in fair details in [20]. The following subsections describe the main parts and parameters, with different topologies, which play a major role in the system's efficiency and cost.

54.3.1 Converter Topologies

The considered architecture of the dc–dc converter affects the overall efficiency of the system. Researchers diverge in classifying the different topologies used. Figure 54.3 shows different possible converter architecture.

In the past decade, the centralized architecture (C1) was widely used and deemed to be the best architecture (Fig. 54.3a) [19]. However, when this converter is exposed to partial shadow, the overall efficiency drops severely as the shaded PV modules act as passive loads [19, 20]. This issue is solved in the modular architecture (C2) as the shaded modules act as short circuit since each module is integrated with its own MPPT and boost converter which allow the other modules to remain unaffected (Fig. 54.3b). Such feature provide 20 % efficiency raise from the centralized architecture when exposed to partial shadow [20]. The AC module technology (C3), also known as the decentralized inverter, shown in Fig. 54.3c, PV modules are integrated with dc–ac converter and all are connected directly to the grid. Besides the privilege of having each module operating independently just as the modular architecture, capacity enlargement is easily achieved in the AC module technology unlike the modular architecture [18]. However, this module requires more complex circuit topologies to achieve high-voltage amplification.

54.3.2 Maximum Power Point Tracking

It is very important that the PV system operates near the maximum power point to ensure a high efficiency for the PV array [19]. One of the algorithms used to obtain MPP is the perturbation and observation (P&O) [21]. The P&O method sets a

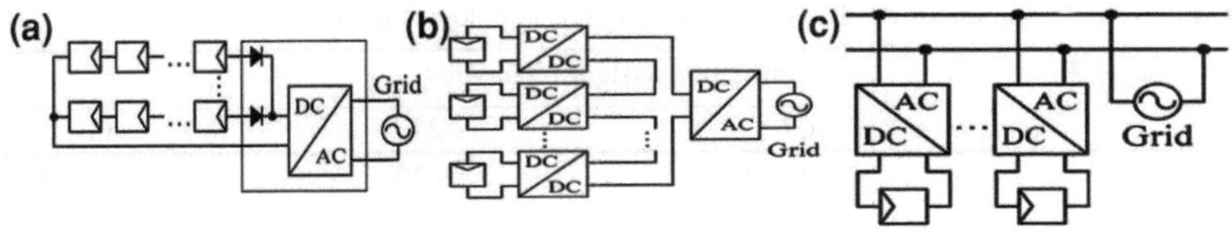

Fig. 54.3 Converter topologies for PV systems. **a** Centralized. **b** Modular. **c** AC module

reference voltage and then compares the power of the previous step with the current step and, consequently, responds by increasing or decreasing the voltage. Hence, the reference value is changed by a constant factor; however, it results in continuous oscillations while operating to reach the MPP, which makes it unreliable method [20].

An incremental conductance method was developed to overcome the drawbacks of the P&O method [20]. Incremental conductance method is based on the fact that the derivative of the output power is zero at the MPP, positive on the left and negative on the right of the MPP [22]. Therefore, MPP is reached faster. Literature [20] proved that incremental conductance method can reach up to 98 % efficiency.

54.3.3 Capacity Factor

As mentioned earlier in Sect. 2.5, CF refers to the ratio of average power to rated power. Similar to wind systems, CF in PV systems helps selecting the optimum PV module for a certain place. The significance of CF is that it can make the power extracted from a PV system at a particular time of the day measureable, which is quite important to companies and investors in PV systems. In reference [23], five different PV modules were compared in terms of CF to determine the most suitable module for Logan Airport in US for the four seasons. The selected PV modules in [23] have, approximately, the same cost of maximum power watt. Therefore, cost analysis is not an issue. One of the most important findings is that CF is independent of the rated power of the PV module. Instead, it depends on the manufacturer's specifications such as the series resistance, temperature coefficients and reference temperature [23].

54.3.4 Efficiency Analysis

Total efficiency of a PV system depends on the converter topology, MPPT system and inverter architecture [19, 20]. Table 54.1 shows an efficiency comparison between PV systems with different dc–dc converter topologies operating at 50 kHz (refer to Fig. 54.3). PV system with decentralized inverter (C3) presents the best efficiency among the converter topologies. The modular topology presents the

Table 54.1 Efficiency of the PV system for different topologies at 50 kHz

Efficiency (%)	Conversion system topologies		
	C1	*C2*	*C3*
PV array	99.30	99.39	99.90
Converter	95.16	85.06	94.94
Overall	**94.50**	**84.54**	**94.84**

worst efficiency as there are two conversion steps. However, if the modules are exposed to partial shadow, the efficiency of C1 drops severely and becomes less than the modular topology. The overall efficiency shows that the inverter operating at 50 kHz has better results. It has to be noted that the figures in Table 54.1 were recorded using P&O for MPPT which has a lower efficiency than the incremental conductance method. Thus, substituting the MPPT system with the incremental conductance method could enhance the total efficiency of the PV system.

54.3.5 Cost Analysis

The most two influential elements in determining the cost of the PV system are the price per peak Watt (Wp) of PV modules and the conversion efficiency of modules (W/m^2) [24, 25]. However, the price per kW is used in the next section for comparison purposes between wind and PV systems. Currently, PV modules can be purchased by around US$2.50/Wp. PV modules represent about 25–50 % of the PV system total cost.

Other costs are represented in balancing the system components such as inverters, mounting hardware, wiring and installation labor cost. A detailed methodology for calculating the cost of a GCPVS on a life cycle of 5, 10 and 15 years is introduced in [24]. Equation (54.2) estimates the initial cost for the system proposed in this literature by assuming the cost for a PV module is US$4.5/Wp, US$1/A-h, the cost of flooded lead acid battery and US$1.0/W. Furthermore, the writer assumed additional 20 % cost for balancing the system. Additionally, taxes imposed on customers are also taken into consideration.

$$\begin{aligned}\text{Initial cost} &= (\text{cost of PV module} + \text{cost of battery} + \text{cost of inverter}) \\ &\quad \times 1.2 - \text{Rebate/Tax Credits} \\ &= [\$4.5/\text{Wp} \times \text{S} + \$1/\text{A h} \times (3 \times \text{E}/12\ \text{V}) + \$1/\text{W} \times \text{S}\ (1 + 25\ \%)] \\ &\quad \times 1.2 - \text{Rebate/Tax Credits}\end{aligned} \tag{54.2}$$

where S is the size of the PV system in Watts and E is the average daily household electricity in Watt-hour. Referring to (54.2), the total cost of a GCPVS composed of 5.46 Wp PV array size, 9,559 A-h and 6,014 W inverter is, approximately, US$8,200.

54.4 Comparative Analysis

From the covered literature in this paper, comparison between wind and PV systems, in terms of efficiency and cost, has to be conducted carefully due to the variety of parameters that influence both systems. Additionally, efficiency and cost may deviate, significantly, in locations other than the location where the studies were conducted. As instance, availability of solar radiation differs from place to place and even at the same place depending on the weather conditions. Furthermore, demand on electricity is different from one country to another which influences the price of electricity per kilowatt-hour. However, a rough comparison is sought.

In terms of total efficiency, according to [15], a total energy efficiency of 96.8 % was reached using induction generator, rated at 5 MW, in Erimo, Japan. If this system is compared to the PV system covered in [19], 94.84 % total efficiency using decentralized converter and O&P method for MPPT, wind system would be the prevailing system. However, Erimo is famous of its high wind speeds, which means that the efficiency stated, cannot be generalized to other locations. Furthermore, as mentioned earlier in Sect. 3.5, if MPPT is changed from O&P to incremental conductance method in [19], an overall efficiency increase of the system is supposed to occur. Generally, PV systems' efficiency is higher than 90 % with a maximum of 95 %, whereas wind systems has an efficiency between 82 and 92 % and a maximum of 96 % when wind speed is very high (e.g. 13 m/s).

In terms of total cost, both systems have to be of the same rating to conduct a viable comparison. In [17], a methodology to compare annual cost of standalone wind and PV systems, both rated at 10 kW, is pursued. The paper concluded that wind system is cheaper with a cost of US$0.23/kW, whereas the cost of the PV system is US$0.38/kW. Utilizing the cost per kilo-watt from [17], cost of both systems at selected ratings is shown in Fig. 54.4. This shows that PV systems are much more expensive than wind systems at mega-watt rating. However, at kilo-watt ratings, it cannot be concluded than wind systems are cheaper than PV since the difference is not significant and the cost (US$/kW) is not linear for both systems. Figure 54.5 shows the cost (US$/kW) for WTG's rated at different watt-level obtained from [9] which proves the nonlinearity of the trend. Similarly, the nonlinearity applies to PV systems too. The total cost of a wind and PV systems rated at 400 kW are obtained from [9, 25], respectively. The wind system costs US$54,568 while the PV system costs US$51,429 which contradicts with [17]'s conclusion. This difference in cost of PV system can be explained by the dramatic continuous decrease of power electronics' cost that are used in PV systems ([17] was conducted in 1997, while [9] was conducted in 2007).

It should be noted that PV systems can only generate power during daytime and hence less energy production, regardless of the system's efficiency. This explains the vast difference of investments spent on both systems as wind systems are better in terms of investments in addition to the bulk power extracted from wind systems (refer to Fig. 54.1).

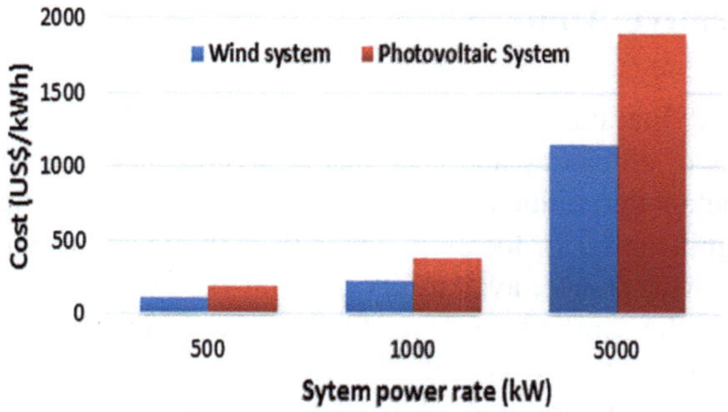

Fig. 54.4 Cost of wind system versus PV system at selected ratings

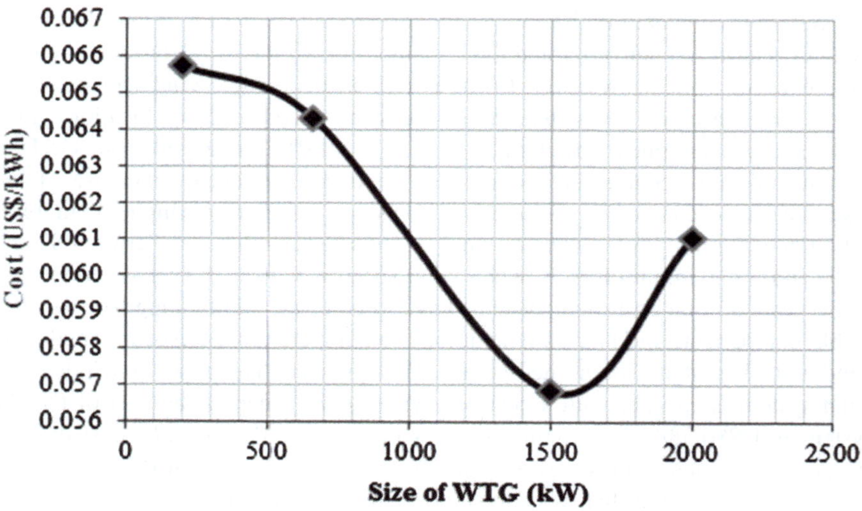

Fig. 54.5 WTG's cost at different ratings

54.5 Conclusions

This paper has reviewed two renewable energy sources, wind and solar, considering the major components and parameters that influence the overall efficiency and cost of both systems. Specifically, the paper focused on horizontal wind machines. The commercially available generators used in wind energy extraction are induction, synchronous and DFIG. Recent study shows that DFIG is the best among other generators though it is expensive compared to induction machine. Control schemes such as power and speed can enhance the system's stability and efficiency. On the other hand, grid connected PVs' are mainly composed of solar panel, converter with MPPT system, and inverter. Converter topologies along with MPPT algorithms play the biggest role in determining the overall efficiency.

The paper revealed that the decentralized inverter topology along with incremental conductance method is the best combination for the PV system.

To be able to compare overall efficiency between wind and PV system, wide research including all different parameters for the two systems has been taken into consideration. Generally, the overall efficiency of PV systems is higher than wind systems. However, it is not always applicable as efficiency relies on the technological revolution of individual components use in the PV or wind system.

Generally, wind systems are more expensive than PV system at low levels (up to 500 kW), however, cheaper at high levels (above 500 kW) since WTG are designed to operate at high power level rather than the electronics used in PV systems, which are more effective at low power levels. Thus, wind systems are more convenient than PV systems when the aim of the system is to produce energy in the level of mega-watt, whereas PV systems, normally, aim for kilo-watt level power production. However, PV systems are much more convenient in urban and suburban environments due to space restriction.

Consideration of both the cost and efficiency usually impose the user to compromise which is more important than the other. Increasing efficiency means applying developed techniques and installing more expensive equipment which result in raising the total cost of the system.

References

1. J.E.J. Cottrill, Economic assessment of the renewable energy sources. IEE Proc. Phys. Sci. Meas. Instrum. **127**(5), 279–288 (1980)
2. W.S. Oliveira, A.J. Fernandes, Optimization model for economic evaluation of wind farms—how to optimize a wind energy project economically and technically. Int. J. Energy Econ. Policy **2**(1), 10–20 (2012)
3. D.-C. Sue, High efficiency double-fed induction generator applied to wind power generator technical analyses. Energy Power Eng. **3**(3), 253–261 (2011)
4. G. Lator, A. Mullane, M. O'Malley, Frequency control and wind turbine technologies. IEEE Trans. Power Syst. **20**(4), 1905–1912 (2005)
5. D. Carli, D. Brunelli, D. Bertozzi, L. Benini, *A* high-efficiency wind-flow energy harvester using micro turbine. in *International Symposium Power Electronics, Electrical Drives, Automation and Motion*, Pisa, 2010, pp. 778–783
6. L. Piegari, R. Rizzo, P. Tricoli, in *High Efficiency Wind Generators with Variable Speed Dual-Excited Synchronous Machines*. International Conference on Clean Electrical Power, Capri, 2007, pp. 795–800
7. F.M. Rodrigo, L.C. de Lucas, S. de Pablo Gomez, J.M. de la Fuente, Analysis of the efficiency improvement in small wind turbines when speed is controlled. in *IEEE International Symposium Industrial Electronics*, Vigo, 2007, pp. 437–442
8. L.H. Hansen, P.H. Madsen, F. Blaabjerg, H.C. Christensen, U. Lindhard, K. Eskildsen, *Generators and Power Electronics Technology for Wind Turbines*. The 27th Annual Conference of the IEEE. (Industrial Electronics Society, Denver, CO, 2001), pp. 2000–2005
9. T.-H. Yeh, L. Wang, in *Benefit Analysis of Wind Turbine Generators Using Different Economic-Cost Methods*. International Conference on Intelligent Systems Applications to Power Systems, Toki Messe, 2007, pp. 1–6

10. L. Wang, T.-H. Yeh, W.-J. Lee, Z. Chen, Benefit evaluation of wind turbine generators in wind farms using capacity-factor analysis and economic-cost methods. IEEE Trans. Power Syst. **24**(2), 692–704 (2009)
11. J. Faiz, B.M. Ebrahimi, M. Rajabi-Sebdani, A. Khan, in *Optimal Design of Permanent Magnet Synchronous Generator for Wind Energy Conversion Considering Annual Energy Input and Magnet Volume*. International Conference on Sustainable Power Generation and Supply, Nanjing, 2009, pp. 1–6
12. J. Chen, A. Xue, C. Tian, T. Bi, C. Gao, in *Unified Model for Coordinated Control of Permanent Magnet Synchronous Wind Generator System*. in Proceedigs of 30th Chinese Control Conference, Yantai, 2011, pp. 5114–5119
13. S. Li, T.A. Haskew, R.P. Swatloski, W. Gathings, Optimal and direct-current vector control of direct-driven PMSG wind turbines. IEEE Trans. Power Electron. **27**(5), 2325–2337 (2012)
14. A. Zahedi, in *Developing a Linear Model for Estimating the Capacity Factor of Wind Turbines*. 21st Australasian Universities Power Engineering Conference, Brisbane, QLD, 2011
15. A. Inoue, R. Takahashi, T. Murata, J. Tamura, M. Kimura, M.-O. Futami, K. Ide, A calculation method of the total efficiency of wind generators. Electr. Eng. Jpn. **157**(3), 946–954 (2006)
16. D.L. Hoffman, T.S. Molinski, How new technology developments will lower wind energy costs. in *CIGRE/IEEE PES Joint Symposium Integration of Wide-Scale Renewable Resources into the Power Delivery System*, Calgary, AB, 2009
17. W.D. Kellogg, M.H. Nehrir, G. Venkataramanan, V. Geraz, Generation unit sizing and cost analysis for stand-alone wind, photovoltaic, and hybrid wind/PV systems. IEEE Trans. Energy Convers. **13**(1), 70–75 (1998)
18. B. Liu, S. Duan, T. Cai, Photovoltaic DC-building-module-based BIPV system—concept and design considerations. IEEE Trans. Power Electron. **26**(5), 1418–1429 (2011)
19. M.C. Cavalcanti, G.M.S. Azevedo, B.A. Amaral, K.C. de Oliveira, F.A.S. Neves, Z.D. Lins, in *Efficiency Evaluation in Grid Connected Photovoltaic Energy Conversion Systems*. IEEE 36th Power Electronics Specialists Conference, Recife (IEEE, New York, 2005), pp. 269–275
20. E. Roman, R. Alonso, P. Ibanez, S. Elorduizapatarietxe, D. Goitia, Intelligent PV module for grid-connected PV systems. IEEE Trans. Ind. Electron. **53**(4), 1066–1072 (2006)
21. C. Hua, J. Lin, C. Shen, Implementation of a DSP-controlled photovoltaic system with peak power tracking. IEEE Trans. Ind. Electron. **45**(1), 99–107 (1998)
22. S. Balathandayuthapani, C.S. Edrington, S.D. Henry, J. Cao, Analysis and control of a photovoltaic system: application to a high-penetration case study. IEEE Syst. J. **6**(2), 213–219 (2012)
23. A. Jiang, Y. Zhu, Life cycle cost analysis of residential grid-connected solar photovoltaic systems in Florida. Int. J. Constr. Educ. Res. **7**(1), 71–81 (2011)
24. Z.M. Salameh, B.S. Borowy, A.R.A. Amin, Photovoltaic module-site matching based on the capacity factors. IEEE Trans. Energy Convers. **10**(2), 326–332 (1995)
25. W.-F. Su, S.-J. Huang, C.-E. Lin, Economic analysis for demand-side hybrid photovoltaic and battery energy storage system. IEEE Trans. Ind. Appl. **37**(1), 171–176 (2001)

Chapter 55
The Graphene Oxide Polymer Composites with High Breakdown Field Strength and Energy Storage Ability

Yang Li, Jun Hu, Jinliang He and Lei Gao

Abstract The crystalline structure of poly(vinylidene fluoride-co-hexafluoropropylene) (PVDF-HFP) is strongly related with its breakdown characteristic and energy storage capability. A graphene oxide (GO) polymer composite, making use of the specific interaction between GO and PVDF-HFP that can induce the formation of β-phase when crystallizing from solution, was developed. The results indicate that the breakdown field strength of the composites with GO nanosheets can reach more than 350 MV/m with small variance. And the composites exhibit a moderate dielectric constant (>13) which can obtain a maximum energy storage density of near 10 J/cm^3 and discharged 5 J/cm^3. Besides GO can alter the crystalline structure of PVDF-HFP from the non-polar phase to the polar phase. As a result, the composites with GO can achieve the similar results by mechanical stretching and avoid the necking effect in stretching.

55.1 Introduction

High energy density dielectric materials are of crucial important for wind and solar power storage system. Because the energy density is closely related with the electrical breakdown field, a higher electrical breakdown field can lead to higher electrical energy density.

The phase transition and physical properties of polyvinylidene fluoride and its copolymers have been studied since the 1960s [1, 2]. The β-phase of PVDF is usually obtained via uniaxial or biaxial stretching, melt crystallization under high pressure or high electric field. A GO nanosheets/PVDF nanocomposite films results a purely β-phase with 0.1 wt.% GO content and enhances the electrical, thermal and mechanical properties of such films [3]. GO nanosheets including

---

Y. Li (✉) · J. Hu · J. He · L. Gao
State Key Laboratory of Power Systems, Department of Electrical Engineering,
Tsinghua University, Beijing 100084, China
e-mail: yangli12@mails.tsinghua.edu.cn

A. Y. Oral et al. (eds.), *International Congress on Energy Efficiency and Energy Related Materials (ENEFM2013)*, Springer Proceedings in Physics 155,
DOI: 10.1007/978-3-319-05521-3_55,

carboxyl, hydroxyl and epoxy groups [4, 5], which can be used as nucleating agent for polymer to produce high performance nanocomposites materials. In this paper, PVDF-HFP/GO prepared through solution casting method were characterized in term of structure, breakdown and energy storage properties.

55.2 Experimental

The dried graphene oxide sheet was dispersed in DMF solvent (0.5 mg/mL) and ultrasonicated for 4 h. PVDF-HFP/GO nanocomposite films were obtained by dissolving PVDF-HFP in DMF under magnetic stirring for 6 h and then adding the desired amount of suspended graphene oxide nanosheets (GOn) in DMF. The films were prepared by solvent casting on glass substrates (Fig. 55.1). Pure PVDF-HFP sample was prepared using the same procedure as the nanocomposite. Films thickness was around 20 μm.

55.3 Results and Discussion

55.3.1 Structural Properties

An instable polar structure is essential for a crystal to be ferroelectric [1]. The most common structure is the α-phase which consists of trans-gauche (TGTG') molecule conformation and is the only nonpolar. The other three, β, γ and δ-phase, are polar.

The FTIR spectra of pure PVDF-HFP and nanocomposite film containing 0.1 wt.% GO are presented in Fig. 55.2. For samples with GO, the FTIR characteristics of α-phase (i.e., 490, 531, 615, 763, 796 and 976 cm^{-1}) tend to decrease. At the same time, the intensities of γ-phase bands at 510 and 812 cm^{-1} appear and β-phase bands at 839 and 1,072 cm^{-1} continue to increase. In order to quantify the polar phase content present in each sample, infrared absorption bands at 510 (γ-phase), 763 (α-phase) and 839 (β-phase) cm^{-1} and a similar calculation procedure presented in [6–8] was used. The relative fraction of β-phase, $F(\beta)$ and γ phase, $F(\gamma)$ can be calculated according to:

$$
\begin{aligned}
f(\beta) &= \frac{X_\beta}{X_\alpha + X_\beta} = \frac{A_\beta}{(K_\beta/K_\alpha)A_\alpha + A_\beta} \\
f(\gamma) &= \frac{X_\gamma}{X_\alpha + X_\gamma} = \frac{A_\gamma}{(K_\gamma/K_\alpha)A_\alpha + A_\gamma} \\
F(\beta) &= \frac{f(\beta) - f(\beta)f(\gamma)}{1 - f(\beta)f(\gamma)} \\
F(\gamma) &= \frac{f(\gamma) - f(\beta)f(\gamma)}{1 - f(\beta)f(\gamma)}
\end{aligned}
\qquad (55.1)
$$

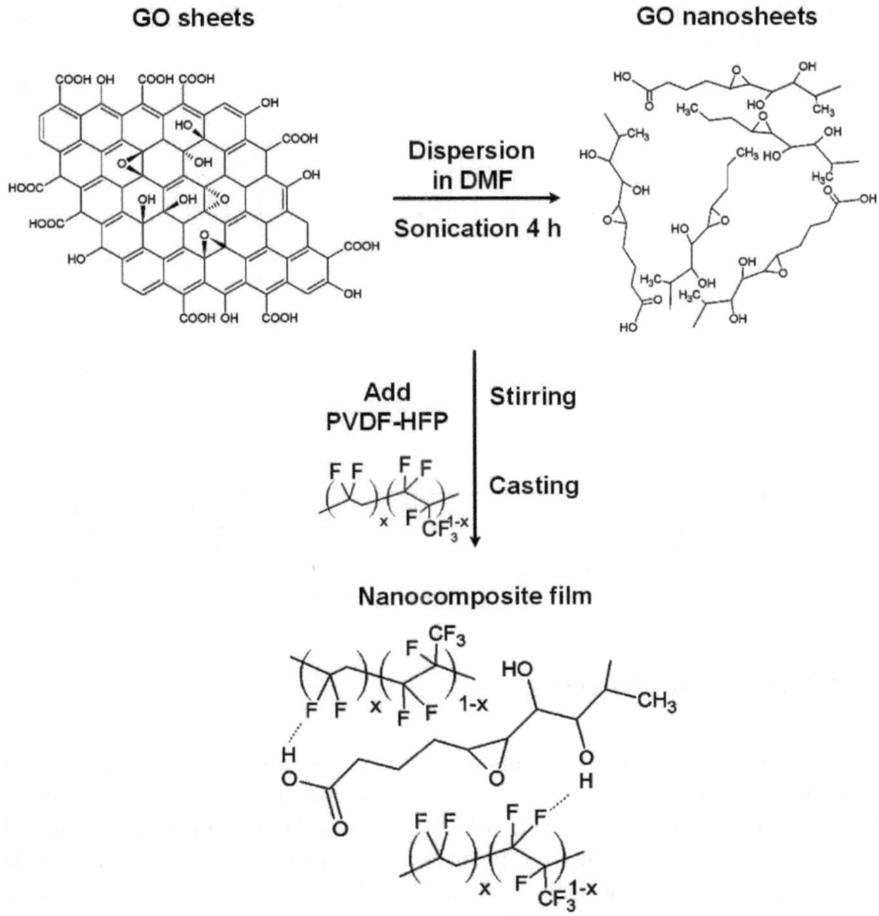

Fig. 55.1 Schematic illustration of nanocomposite formation

where A_α, A_β, and A_γ represent the absorbencies of α, β and γ-phase at 763, 839 and 510 cm^{-1}, respectively, K_α, K_β and K are the corresponding absorption coefficients, X_α, X_β and X_γ are the crystallinity of α, β and γ phase. $f(\beta)$ and $f(\gamma)$ represent the relative fraction to α-phase respectively. The value of K_α is 6.1×104 cm^2 mol^{-1}, K_β is 7.7×104 cm^2 mol^{-1}, and K_γ is 5.8×104 cm^2 mol^{-1} [6].

As listed in Table 55.1, with adding GO, the relative fraction of β-phase rises from 25.7 to 50.0 % and the polar phase content rises from 51.0 to 79.8 %. According to the adsorption energy for α and β-phase [9], the transformation from nonpolar to polar phase is due to adsorption of molecular chains of PVDF-HFP on the GO surface with the help of the strong interactions between the CF_2 segments of PVDF-HFP and the carbonyl groups of GO. In such films, 80 % polar phase was achieved at only 0.1 wt.% GO. It is much more effective and easier than what is reported in the literature by some groups using stretching method [10, 11].

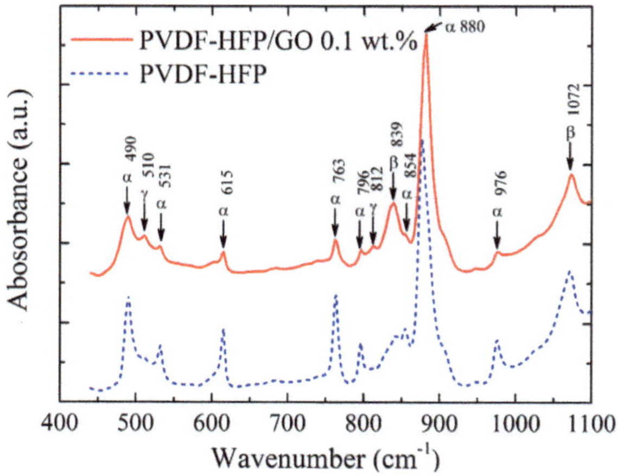

Fig. 55.2 FTIR spectra of PVDF-HFP/GO 0.1 wt.% nanocomposite film and pure PVDF-HFP film

Table 55.1 Relative fraction of α, β and γ-phase in pure and 0.1 wt.% GO/PVDF-HFP films

	$F(\alpha)$ (%)	$F(\beta)$ (%)	$F(\gamma)$ (%)
Pure	49.0	25.7	25.3
0.1 wt.%	20.2	50.0	29.9

55.3.2 Dielectric Properties

Figure 55.3a and b show the results of dielectric spectroscopy measurements on the nanocomposites with 0.1 wt.% over the frequency range of 0.1 Hz–10 MHz. The dielectric spectroscopy for the PVDF-HFP and PVDF-HFP/GO films reveal that the dielectric loss is due to dipolar relaxation of the polymer above 10 kHz and conduction loss of GO below 10 Hz. The dielectric loss of both are similar, from 0.04 at 1 kHz to 0.11 at 1 MHz. However, the 0.1 wt.% GO nanocomposite exhibits higher relative dielectric constants than the pure copolymer sample. The reason is that pure PVDF-HFP sample has less polar phases (β and γ-phase) than does PVDF-HFP/GO. The relative dielectric constants at 1 kHz are 11.5 for pure copolymer and 13.9 for 0.1 wt.% GO/PVDF-HFP.

55.3.3 Breakdown Properties

The breakdown strength of the dielectric films was evaluated by the two-parameter Weibull analysis [12, 13],

$$P(E) = 1 - \exp[-(E/E_b)^{\beta}] \quad (55.2)$$

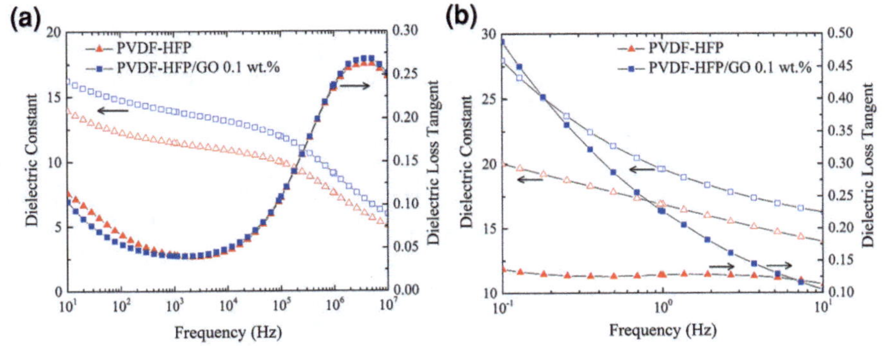

Fig. 55.3 Dielectric constant and loss tangent for pure PVDF-HFP and 0.1 wt.% GO nanocomposite films at 20 °C: **a** 0.1–10 Hz; **b** 10 Hz–10 MHz

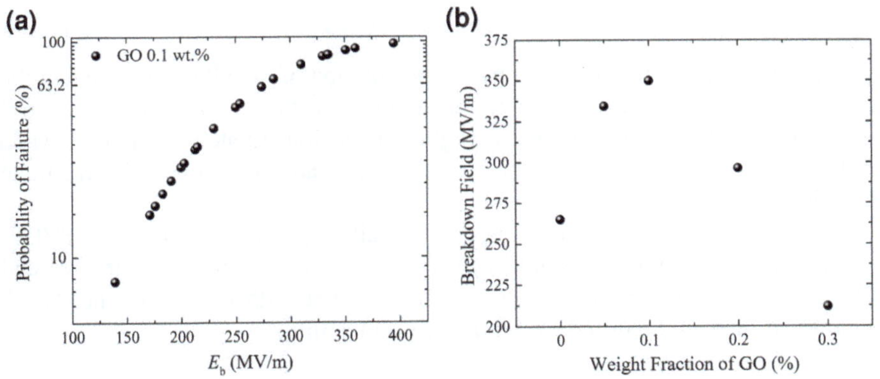

Fig. 55.4 **a** Percent cumulative distribution function of each nanocomposite as a function of applied electric field, and **b** the characteristic breakdown strengths (failure probabilities: 63.2 %) at each weight fraction as determined from the Weibull analysis

where E is the measured electrical breakdown strength, E_b is the characteristic breakdown field, and β is a shape parameter.The failure probability can then be expressed as

$$\lg\{-\ln[1 - P(E)]\} = \beta \lg E - \beta \lg E_b \tag{55.3}$$

The characteristic breakdown field for the data set is calculated when the failure probability is 63.2 % (i.e., $1 - 1/e$). The probability of failure as a function of applied field is showed in Fig. 55.4a. The results in Fig. 55.4b clearly show that breakdown strength decreases upon addition of GOn to the copolymer as expected. It should be noted that there is a sudden decrease in the breakdown strength when the weight fraction changes from 0.1 to 0.2 %.

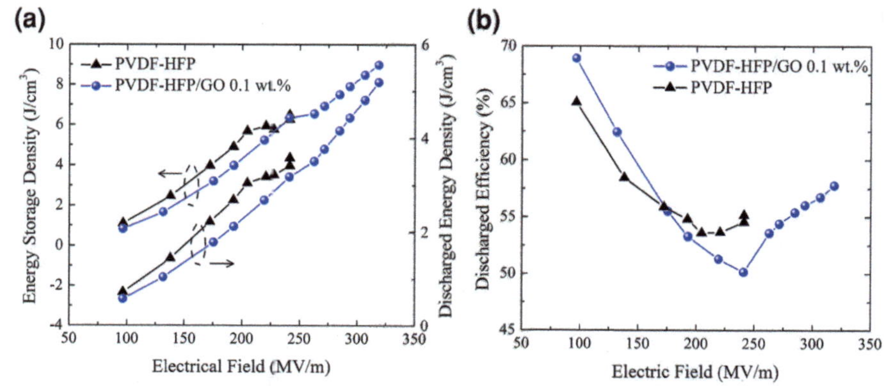

Fig. 55.5 **a** Energy storage density and discharged energy density of pure PVDF-HFP and 0.1 wt.% GO/PVDF-HFP at room temperature and, **b** energy discharged efficiency of 0.1 wt.% GO/PVDF-HFP

This behavior can be explained by the percolation of the GO nanosheets in the nanocomposite. There exists two different regimes: the "soft" and the "hard" percolation [14]. The breakdown strength can be immediately decreased when "soft" percolation reaches, because the charge pathways appear to facilitate breakdown.

Another interesting observation is that the failure probability of a pure PVDF-HFP film at moderate fields is higher than those of nanocomposites with 0.05 and 0.1 % GO. This implies that doping of the polymer with GO under the "soft" percolation threshold can reduce the failure probability.

55.3.4 Energy Storage Properties

As the electric displacement is not linearly dependent on electric field in ferroelectric materials, the energy density is calculated by integrating the electric field (E) with respect to the electric displacement (D):

$$U_e = \int EdD \tag{55.4}$$

Figure 55.5a presents the energy density of pure PVDF-HFP and 0.1 wt.% GO/PVDF-HFP at room temperature. Due to the higher breakdown strength of PVDF-HFP/GO, the maximum discharged energy can be 5.2 J/cm^3 under an applied field of 318 MV/m. It can be seen in Fig. 55.5b that the energy storage efficiency depends on the applied field and exists a critical point. This is because the energy density measurements are limited by the breakdown strength of samples.

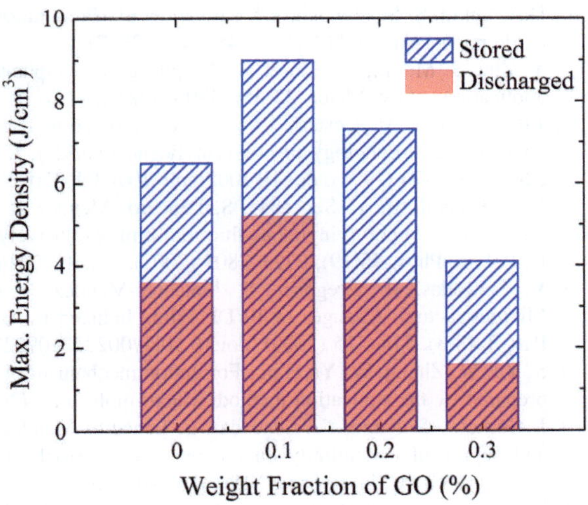

Fig. 55.6 Measured maximum energy storage density (*blue*) and discharged energy density (*red*) of PVDF-HFP/GO with different GO weight fractions

Actual energy densities of the nanocomposite films at each weight fraction of GO are compared in Fig. 55.6. Maximum energy densities measured showed a general trend of increase in energy density up to 0.1 wt.% followed by a decreasing trend due to the presence of air voids and lower breakdown strength. Therefore, the 0.1 wt.% GO/PVDF-HFP nanocomposites may be desirable for applications considering the actual energy storage capacity and breakdown strength. We need further investigation of the effect of the nanosheets weight fraction, thus provides a guideline to find an optimum weight fraction depending on the specific application.

55.4 Conclusion

In our study, GO can enhance the polymer to crystalize more polar phase ($\sim$80 %). PVDF-HFP with 0.1 wt.% GO shows higher breakdown strength than pure polymer. It can be explained by the soft percolation. Then such nanocomposite has a high energy storage capacity with 5.2 J/cm^3.

References

1. T. Furukawa, Ferroelectric properties of vinylidene fluoride copolymers. Phase Transitions: Multinational J. **18**(3–4), 143–211 (1989)
2. X. He, K. Yao, B.K. Gan, Phase transition and properties of a ferroelectric poly (vinylidene fluoride-hexafluoropropylene) copolymer. J. Appl. Phys. **97**(8), 84101 (2005)
3. M.E. Achaby, F.Z. Arrakhiz, S. Vaudreuil et al., Piezoelectric β-polymorph formation and properties enhancement in graphene oxide-PVDF nanocomposite films. Appl Surf. Sci. **258**(19), 7668–7677 (2012)

4. D.A. Dikin, S. Stankovich, E.J. Zimney et al., Preparation and characterization of graphene oxide paper. Nature **448**(7152), 457–460 (2007)
5. Y. Zhu, S. Murali, W. Cai et al., Graphene and graphene oxide: synthesis, properties, and applications. Adv. Mater. **22**(35), 3906–3924 (2010)
6. J.R. Gregorio, M. Cestari, Effect of crystallization temperature on the crystalline phase content and morphology of poly(vinylidene fluoride). J. Polym. Sci. Part B: Polym. Phys. **32**(5), 859–870 (1994). doi:10.1002/polb.1994.090320509
7. M.C. Branciforti, V. Sencadas, S. Lanceros-Mendez et al., New technique of processing highly oriented poly(vinylidene fluoride) films exclusively in the β phase. J. Polym. Sci. Part B: Polym. Phys. **45**(19), 2793–2801 (2007). doi:10.1002/polb.21239
8. V. Sencadas, R. Gregorio, S. Lanceros-Méndez, α to β Phase Transformation and Microestructural Changes of PVDF Films Induced by Uniaxial Stretch. J. Macromol. Sci. Part B **48**(3), 514–525 (2009). doi:10.1080/00222340902837527
9. S. Yu, W. Zheng, W. Yu et al., Formation mechanism of β-phase in PVDF/CNT composite prepared by the sonication method. Macro-molecules **42**(22), 8870–8874 (2009)
10. J. Gomes, J.S. Nunes, V. Sencadas, S. Lanceros-Mendez, Influence of the β-phase content and degree of crystallinity on the piezo-and ferroelectric proper-ties of poly (vinylidene fluoride). Smart Mater. Struct. **19**(6), 65010 (2010)
11. B.P. Neese, Investigations of structure-property relationships to enhance the multifunctional properties of PVDF-based polymers (2009)
12. R. Bartnikas, *Engineering Dielectrics: Electrical Properties of Solid Insulating Materials: Measurement Techniques*, vol. 2 (ASTM International, USA, 1987)
13. X. Zhou, X. Zhao, Z. Suo et al., Electrical breakdown and ultrahigh electrical energy density in poly (vinylidene fluoride-hexafluoropropylene) copolymer. Appl. Phys. Lett. **94**(16), 162901 (2009)
14. J.P. Calame, Finite difference simulations of permittivity and electric field statistics in ceramic-polymer composites for capacitor applications. J. Appl. Phys. **99**, 84101 (2006)

Chapter 56
Profitability Analysis of Residential Wind Turbines with Battery Energy Storage

Ying She, Ergin Erdem and Jing Shi

Abstract Residential wind turbines are often accompanied by an energy storage system for the off-the-grid users, instead of the on-the-grid users, to reduce the risk of black-out. In this paper, we argue that residential wind turbines with battery energy storage could actually be beneficial to the on-the-grid users as well in terms of monetary gain from differential pricing for buying electricity from the grid and the ability to sell electricity back to the grid. We develop a mixed-integer linear programming model to maximize the profit of a residential wind turbine system while meeting the daily household electricity consumption. A case study is designed to investigate the effects of differential pricing schemes and sell-back schemes on the economic output of a 2-kW wind turbine with lithium battery storage. Overall, based on the current settings in California, a residential wind turbine with battery storage carries more economical benefits than the wind turbine alone.

56.1 Introduction

To reduce fossil fuel consumption and pollution emissions, governments are adopting aggressive energy policies to encourage the installation of renewable energy systems. Incentive renewable energy policies have been developed to

Y. She (✉)
School of Economics and Management, Nanchang Hangkong University, Nanchang, 330063 Jiangxi, China
e-mail: sheying1980@hotmail.com

E. Erdem · J. Shi
Department of Industrial and Manufacturing Engineering, North Dakota State University, Fargo, ND 58108, USA
e-mail: ergin.erdem@my.ndsu.edu

J. Shi
e-mail: jing.shi@ndsu.edu

A. Y. Oral et al. (eds.), *International Congress on Energy Efficiency and Energy Related Materials (ENEFM2013)*, Springer Proceedings in Physics 155,
DOI: 10.1007/978-3-319-05521-3_56,

promote the utilization of renewable energy. Residential wind turbines (RWT) are one example. The favorable policies, along with the continuous cost reduction for manufacturing wind turbines, make it possible for wind energy from RWT to be competitive compared with conventional energy sources. In recent years, they have been widely adopted, especially in rural and remote areas. For instance, the number of RWT units sold in the U.S. market increased by 4 times in 2009 compared with that in 2001.

It is well known that the generation of wind power is intermittent. The intermittency may lead to black-out. To avoid this, one can connect the wind power generation system to the grid, and/or add an energy storage system. Adopting both methods for residential users ensure the highest reliability for uninterrupted electricity supply. The downside is the increase of the system cost, and the longer pay-back period. However, the downside can be mitigated when the excessive electricity generated from renewable sources is allowed to be sold to the utility. This is often called "buy-back" policy, which has been gaining momentum in many countries to encourage the development of renewable energy.

This paper investigates how an on-the-grid RWT system with battery energy storage can achieve the optimal operation schedule to maximize the profit. The optimal schedule is obtained by reacting to the changes of wind generation, household energy consumption, and electricity price. The problem is formulated into a mathematical programming model with the objective function of maximizing the monetary gains for the RWT owner. The key regulator in this situation, without any doubt, is the battery bank that stores the excessive generated energy or the purchased energy from grid when the electricity price is low, and release energy to meet the household demand when the electricity price is high. We intend to compare the scenarios of a grid connected RWT system with and without battery storage, and illustrate how the economic benefits will be affected by the electricity pricing schemes and renewable energy incentive policies.

56.2 Brief Literature Survey

The adoption of battery storage in renewable energy systems aims to mitigate the intermittence of renewable resources and improve the system stability. Motoshashi and Sasaki [7] developed a stand-alone wind power generation system with battery storage. A 4-month field test was conducted to illustrate the benefits of this system. Giraud and Salameh [3] reported the performance of a grid-connected residential wind-photovoltaic system with battery storage, with a focus on system reliability. Mohod and Hatwar [6] presented a variable speed type wind energy conversion scheme with battery energy storage to provide a choice to select the most economical power for the load amongst the wind, battery, and conventional resources, with the option of being stand-alone or grid-supported. Elhadidy and Shaahid [4] proposed a hybrid system, which is comprised of wind energy conversion systems, photovoltaic (PV) panels, battery storage unit and diesel back-up, to meet the power consumption of a commercial building in Saudi Arabia. The percentages of loads that need to be supplied by the battery and the diesel back-up were calculated under different scenarios.

More recently, Dufo-Lopez and Bernal-Agustin [1] presented an hourly management method for energy generated in a wind-battery system connected to the grid. The results were compared with a wind-hydrogen system. It was found that that the wind-battery system is more desirable both economically and technically. Sebastian and Alzola [11] modelled an off-the-grid wind diesel hybrid system. The performance of the hybrid system with/without battery storage were tested and compared by dynamic simulation. It was shown that the case with battery storage can reduce the fluctuations of wind power significantly. Yuen et al. [12] investigated two applications of battery connected to a wind system, namely, energy time shifting and capacity firming. Optimization approaches were employed to find the optimal schedule for charging and discharging of battery storage.

Economic scheduling is a problem existing in all power generation systems. Renewable energy systems are not an exception. The management of renewable power supply due to flexible load demand has been discussed in literature. Lu and Shahidehpour [5] developed a Lagrangian relaxation-based optimization algorithm to obtain the charging/discharging schedule of battery storage. Neely et al. [8] developed a Lyapunov optimization technique for fulfilling the load demand according to a time-varying and possibly unpredictable supply process. To reduce the unpredictable fluctuation of renewable energy, Papavasilious and Oren [9] developed a method to determine the optimal schedule for renewable energy supply for time flexible end-users. Two algorithms, namely, backward dynamic programming and approximate dynamic programming were compared. These studies mainly considered the impact of PV/battery on congestion mitigation, pricing, peak load reduction and the commitment of expensive thermal units.

56.3 RWT with Battery Storage

Figure 56.1 shows the schematic of an RWT system with battery storage. The RWT system generates AC power and is connected to an AC/DC converter which is fed to a common DC bus. This converter also rectifies the AC produced by the RWT and protects the battery storage from being overcharged. A bi-directional converter links the system to the grid. The power output of the RWT is mainly dependent upon wind speed by following the power curve of the RWT. In this study, the hourly wind power data are predicted 24 h in advance by converting the predicted hourly wind speeds according to the power curve of the RWT. The power generated by the RWT can be separated into three parts, namely (1) wind power sold to grid, $W_g(t)$, (2) wind power stored into the battery, $W_b(t)$, and (3) wind power consumed the household, $W_l(t)$.

The battery is charged from two sources: (1) the power generated from the RWT, $W_b(t)$, when the total output of the RWT, $W_r(t)$, is greater than the household load, $H(t)$; and (2) the power purchased from the grid, $G_b(t)$, when the

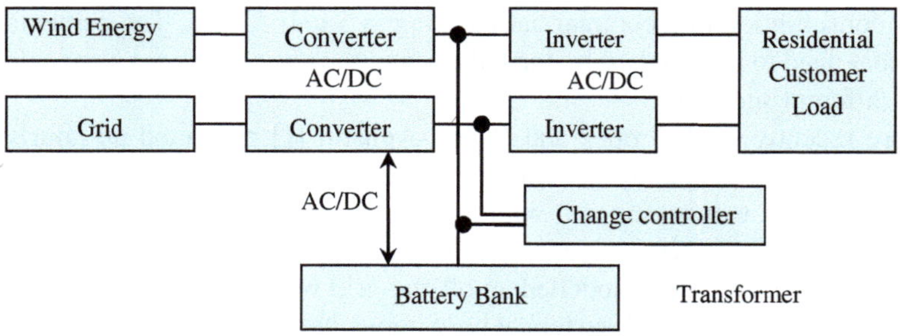

Fig. 56.1 Diagram of RWT system with battery storage

market electricity price is low. Also, during the peak hours of tiered electricity price structure, the residential customer can sell the amount of electricity, $B_g(t)$, to the grid from the battery storage by discharging. The battery also can discharge for satisfying residential customer load for the amount of $B_l(t)$.

The customer can both sell and buy electricity from or to the utility. Denote $C_s(t)$ the sales price when electricity is sold to the utility, and $C_b(t)$ the purchase price when the electricity is purchased from the utility. Both prices are a function of time, and they are determined based on the rates set by the utility.

The hourly load of one residential customer, represented by $H(t)$, includes the power consumption from a variety of sources. Based on the historical data, the daily load curve can be forecasted. $H(t)$ can be satisfied from three sources combined: (1) the power output of RWT and consumed by customer, $W_l(t)$ (2) the power supplied by grid and consumed directly, $G_l(t)$, and (3) the power discharged from battery storage, $B_l(t)$.

56.4 Model Formulation

The objective function is to maximize the overall profit, which is total revenue from wind power generation minus all the cost items of the RWT system and expenditures on electricity purchase from the utility. The total revenue has two parts: R_1 and R_2. R_1 represents the revenue obtained from the RWT system within a certain period of time t. It consists of two parts, namely, the implicit and explicit profits. The implicit profit, represented by $C_s(t)W_g(t) + C_s(t)B_g(t)$, is the profit gained from selling surplus wind power output to the grid. The explicit profit, represented by $C_s(t)W_l(t) + C_s(t)B_l(t)$, is the savings from consuming the power from the wind turbine and battery instead of purchasing electricity from the grid. R_1 can be described as follows:

$$R_1 = C_s(t)\left[W_l(t) + W_g(t) + B_l(t) + B_g(t)\right]. \tag{56.1}$$

The cost of power stored in battery at the initial hour is assumed to be insignificant, while the energy stored in the battery at the end of day is considered in the revenue. The average electricity selling price of 24 h is used to calculate the value of the energy stored in the battery at the 24th hour, represented by R_2,

$$R_2 = \frac{\sum_1^{24} C_s(t)}{24} \left[D(24) - B_g(24) - B_l(24) + G_b(24) + W_b(24)\right]. \tag{56.2}$$

The total cost consists of (1) the cost for purchasing electricity from the grid, S_1 (2) the cost of RWT, S_2, and (3) the cost of battery storage system, S_3. Here,

$$S_1 = C_b(t)[G_b(t) + G_l(t)]. \tag{56.3}$$

S_2, represents the cost for producing the wind power output at hour t.

$$S_2 = W(t)\frac{CO_w/Y_w}{E_{an}}, \tag{56.4}$$

where CO_w is the sum of capital cost and replacement/maintenance cost in the lifespan of the RWT system; Y_w is the lifetime year of the system; and E_{an} is the annual energy generated from the system. The battery life time is represented by the number of charge cycles. Therefore, the cost of battery used per kWh can be calculated as follows,

$$S_3 = \left[B_l(t) + B_g(t)\right] \frac{B_{\cos t}}{B_{\max}} \tag{56.5}$$

where B_{cost} is the cost of using battery bank per charge cycle, and $B_{\max}$ is the capacity of the battery bank.As such, the maximization of profit can be described as follows,

$$\text{Max profit} = \sum_t (R_1 - S_1 - S_2 - S_3) + R_2 \tag{56.6}$$

The above objective function is subject to the following constraints,

(a) The State of Charge of the battery at any time t can be calculated as,

$$D(t+1) = D(t) + [W_b(t) + G_b(t)] - [B_b(t) + B_l(t)] \tag{56.7}$$

(b) The maximum amount of battery energy sold to the grid and used for household consumption should be no greater than the stored energy, while the stored energy should be no greater than the battery capacity at each time period t

$$B_g(t) + B_l(t) \leq D(t) \leq B_{\max} \tag{56.8}$$

(c) The maximum battery energy sold to the grid and used for household consumption should also be no greater than the energy discharge at the maximum discharging rate, D_{max}, at each time period t.

$$0 \leq B_g(t) + B_l(t) \leq D_{\max} X(t); \quad X \in \{0, 1\} \tag{56.9}$$

where $X(t) = 1$ means that the battery is in discharging mode, while $X(t) = 0$ means the opposite.

(d) The maximum amount of energy charge to the battery from the grid and the wind turbine should also be no greater than the energy charge at the maximum charging rate, C_{max}, at each time period t.

$$0 \leq W_b(t) + G_b(t) \leq C_{\max} Y(t); \quad Y \in \{0, 1\} \tag{56.10}$$

where $Y(t) = 1$ means that the battery is in charging mode, while $Y(t) = 0$ means the opposite.

(e) The battery cannot be in both charge and discharge modes at the same time,

$$X(t) + Y(t) \leq 1 \tag{56.11}$$

(f) The electricity consumed by the household at time t,

$$H(t) = W_l(t) + B_l(t) + G_l(t) \tag{56.12}$$

(g) The total power generated by the RWT system at time t,

$$W_r(t) = W_l(t) + W_g(t) + W_b(t) \tag{56.13}$$

(h) The following variables are not less than zero,

$$W_g(t), W_b(t), W_l(t), G_b(t), G_l(t), B_g(t), B_l(t), D(t) \geq 0 \tag{56.14}$$

56.5 Case Study

In the case study, we model the operation of an RWT system with battery storage, owned by a household in California, USA. To find the optimal schedule of the RWT system and calculate the maximal profit, input information needs to be provided. A 2-kW residential wind turbine is chosen because its typical annual electricity output roughly aligns with the overall electricity consumption of a

Table 56.1 Costs of the RWT system components

	Initial cost ($)	Maintenance cost	Life time	Annual output	Unit cost	Adjusted cost (70 % unit cost)
Wind turbine	8,000	$120/year	30 years	6,000 kWh	0.064 $/kWh	0.0448 $/kWh
Battery bank	1,600	Null	3,000 cycle		0.53 $/cycle	0.371 $/cycle

Table 56.2 PG&E residential time-of-use electric rate schedule

Time-of-use period	Energy charge($/KWh)				
	Tier 1 Baseline	Tier 2 101–130 %	Tier3 131–200 %	Tier 4 201–300 %	Tier 5 >300 %
Peak ($)	0.28719	0.30529	0.46623	0.50623	0.50623
Partial-peak ($)	0.17528	0.19338	0.35432	0.39432	0.39432
Off-peak ($)	0.10074	0.11884	0.27978	0.31978	0.31978

typical California family. A lithium battery bank is chosen as the energy storage device. The costs of wind turbine and battery storage are listed in Table 56.1.

The electricity pricing information is obtained from Pacific Gas and Electric [10], which supplies electricity in California. The information consists of peak, partial-peak and off-peak spot market prices. Table 56.2 shows that the off-peak rate of Tier 1 is $0.09781 per kWh, while the peak rate of Tier 5 is $0.48653 per kWh. The baselines for different counties vary from 7.5 to 18.5 kWh. Under the renewable energy incentive policy of California, PG&E can purchase the excess electricity from the RWT owners.

We adopt the representative wind speed data from a site in California. The hourly wind speed is converted to the hourly wind power generation based on the RWT power curve. Meanwhile, we employ the average hourly use profile on a yearly base for a common residential customer in Southern California (NAHB, 2001). Figure 56.2 shows the hourly wind power output and the household load profile.

We adopt 3 scenarios to study the effects of electricity pricing for purchasing and selling. Table 56.3 summarizes the combination of renewable energy incentive policies and pricing schemes for the 3 scenarios.

- Scenario 1 assumes 30 % federal tax credit for the total initial installation cost, and Tier 1 retail pricing scheme for purchasing electricity from the grid and selling the electricity to the grid.
- Scenario 2 assumes the same 30 % federal tax credit, but it adopts Tier 4 retail pricing scheme for purchasing and selling the electricity.
- Scenario 3 also keeps the 30 % federal tax credit. It adopts Tier 1 pricing scheme to purchase electricity from grid, while it can only sell the electricity to the grid at the wholesale price. With a contract period of 20 years, the purchase price is $0.08956 per kWh.

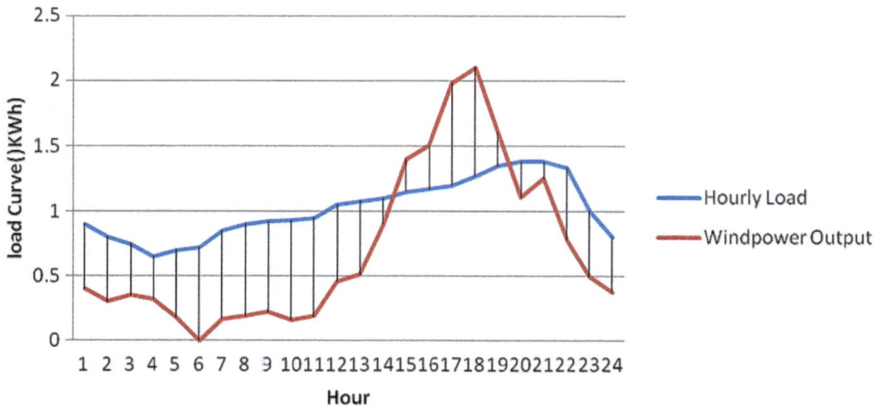

Fig. 56.2 Hourly household load and wind power output

Table 56.3 Parameter and policy selection for 3 scenarios

	Scenario 1	Scenario 2	Scenario 3
Tax incentive	30 % rebate	30 % rebate	30 % rebate
Purchase electricity	Tier 1 scheme	Tier 4 scheme	Tier 1 scheme
Sell electricity	Tier 1 scheme	Tier 4 scheme	Wholesale price

Table 56.4 Maximum profits per day under 3 scenarios

	Scenario 1	Scenario 2	Scenario 3
With battery($)	2.2920	4.7208	−0.4072
Without battery($)	1.6559	3.1372	−0.5238

The optimization model in (56.6)–(56.14) is a mixed integer and linear programming (MILP) model. The model is coded in GAMS platform [2], and solved by the embedded CPLEX solver for each scenario. The optimal trading benefits under all scenarios are compared in Table 56.4.

Based on the results, it is clear that the RWT system with battery storage can bring benefits to the residential customers. In all the scenarios, more monetary benefits can be achieved with the battery storage system installed. Overall, the strategy, which employs battery storage to harvest excess wind power and store energy from utility when the electricity price is low, and discharge or sell the energy when the price is high and wind production is low, works very well. Meanwhile, it can be observed that the profit outcome of the RWT system depends on the policy on pricing the excess energy generation to be sold to the utility. The profit of customers gained increases with the increase of selling price. In particular, under scenario 3, the customer sells the excessive electricity at a fixed wholesale price which is much lower than the retail price. Therefore, the profit customer gained from scenario 3 is $0.4072. This means the RWT owner will not be able to recover the investment on the system. In this case, the residential investment on RWT systems is discouraged. To improve this situation, the selling price needs to be increased, or the tax rebate rate needs to be increased.

56.6 Conclusions

This research presents a preliminary investigation how an on-the-grid RWT system with battery storage can be viable. Under the estimated load, electricity price, wind power output, cost of wind power and battery storage, we formulate a mixed integer linear programming model to obtain the optimal operation schedule for the RWT system. The objective function is to maximize the monetary benefits. In particular, we investigate whether the battery storage can bring more benefits to the residential users when the selling price for excess wind generation to the utility is adjusted. Three scenarios are designed in a case study and the results are discussed. It is discovered that the profitability of RWT system greatly depends on the pricing policy for selling the excess wind energy generation to the grid. Also, the adoption of battery storage to the RWT system can always bring more benefit to the customers underthe three scenarios. It is possible that with the decrease of battery cost, adding battery storage to the RWT systems will become even more attractive.

References

1. R. Dufo-Lopez, J.L. Bernal-Agustin, Generation management using batteries in wind farms: economical and technical analysis for Spain. Energy Policy **37**(1), 126–139 (2009)
2. GAMS Development Corporation, *GAMS-The Solver Manuals*. Washington, D.C., USA (2011)
3. F. Giraud, Z.M. Salameh, Steady-state performance of a grid-connected rooftop hybrid wind-photovoltaic power system with battery storage. IEEE Trans. Energy Convers. **16**(1), 1–7 (2001)
4. M.A. Elhadidy, S.M. Shaahid, Promoting applications of hybrid power systems in hot regions. Renewable Energy **29**, 517–528 (2003)
5. B. Lu, M. Shahidehpour, Short-term scheduling of battery in a grid-connected PV-battery system. Power Syst. **20**(2), 1053–1061 (2005)
6. S.W. Mohod, S.M. Hatwar, Grid support with variable speed wind energy system and battery storage for power quality. Energy Procedia **12**, 1032–1041 (2001)
7. H. Motoshashi, S. Sasaki, Field test of a small wind turbine generator system with storage battery. JSME Int. J. **35**, 325–329 (1992)
8. M.J. Neely, A.S. Tehrani, A.G. Dimakis, in *Efficient Algorithms for Renewable Energy Allocation to Delay Tolerant Consumers*. Proceedings of the First IEEE International Conference on Smart Grid Communications (IEEE, New York, 2010) pp. 549–554

9. A. Papavasilious, S. Oren, in *Supplying Renewable Energy to Deferrable Loads: Algorithms and Economic Analysis*. Proceedings of IEEE Power and Energy Society General Meeting, July 2010, (IEEE, New York, 2010) pp. 1–8
10. PG&E, Pacific Gas and Electric Company Rates and Tariff Library. http://www.pge.com/nots/rates/tariffs. Accessed 31 Oct 2012
11. R. Sebastian, R.P. Alzola, Simulation of an isolated wind diesel system with battery energy storage. Electr. Power Syst. Res. **81**, 677–686 (2011)
12. C. Yuen, J. Poland, A. Efinger, in *The Use of BESS for the Integration of Wind Energy*. Proceedings of the17th Power Systems Computation Conference, Stockholm, Sweden, August 2011

Chapter 57
Dynamic Model and Experimental Validation of a PEM Fuel Cell System

Younane Nassif, Emmanuel Godoy, Olivier Bethoux and Ivan Roche

Abstract Fuel cells are expected to become a challenging technology in terms of efficiency, and fitting the emission reduction schedules [Lemons, J. Power Sources, 29:251, 1] for the automotive application. Their fundamental component consists of two electrodes separated by a membrane. Fuel cells convert chemical energy into electrical energy while producing water and heat. To not disturb the transportation of the reactant gas, a proper membrane hydration needs to be maintained. Two different conditions can occur facing an inadequate water balance which decreases the performance of the stack. An insufficient removal of the accumulated water causes water flooding, decreasing reactant transport rate. Similarly, excessive water removal dries the membrane. To monitor the amount of water inside the cell, dynamic model based on the mass conservation principles and thermodynamic properties is developed in the form of nonlinear state space representation. Fick's law and Maxwell-Stefan model are used to describe multicomponent diffusion. Darcy's law is used to define the porous medium permeability. To demonstrate the accuracy of the proposed model, obtained results are compared with measured data at steady states operation mode. Investigation of the steady-state behavior is discussed in this paper.

Y. Nassif (✉) · E. Godoy
Automatic Control Department, Supélec Systems Sciences (E3S),
Gif-sur-Yvette Cedex, France
e-mail: Younane.nassif@supelec.fr

O. Bethoux
LGEP CNRS UMR 8507/SPEE Labs, Gif-sur-Yvette Cedex, France
e-mail: Olivier.Bethoux@lgep.supelec.fr

I. Roche
PSA Peugeot Citroën, 78943 Vélizy-Villacoublay Cedex, France
e-mail: Ivan.roche@mpsa.com

A. Y. Oral et al. (eds.), *International Congress on Energy Efficiency and Energy Related Materials (ENEFM2013)*, Springer Proceedings in Physics 155,
DOI: 10.1007/978-3-319-05521-3_57,

57.1 Introduction

Several models have already been proposed in the literature. Those can be divided in two main categories: the theoretical models and the empirical/semi-empirical ones. To diagnose, and to control systems, we use the empirical models. They are based on frequency representation in aim to characterize the electrical impedance or on time representation using ordinary differential equations [2–4]. On the other hand, the theoretical model is based on phenomenological equations such as mass transport, heat transport, and electrochemical phenomena. The focus of this paper is on presenting a relevant dynamic model. The zero-dimensional fluid flow model is developed using electrochemical, thermodynamic and fluid flow principles where we consider that the averaged stack temperature is constant. This contribution will lead us to describe and then analyze the steady-state behavior.

57.2 Dynamic Model Development

We present a mathematical approach for building the dynamic model of the PEMFC. All the variables associated with the lumped cathode/anode volume are denoted with a subscript (*ca*/*an*). Masses (kg) are denoted with (m), mass flows (kg/s) with *W*, pressure (Pa) with *P*, temperatures (K) with *T*, current (A) with *I*, current density (A/cm^2) with *i*, surface (cm^2) with *S*, molar masses (kg/mol) with M, volumes (m^3) with V, voltage (Volts) with v, Faradays constant with F, oxygen with O_2, nitrogen with N_2, hydrogen with H_2. The convective transport will be denoted with a superscript (*Conv*), diffusive transport with (*Diff*), through membrane (*mbr*), reacted mass flow (*rct*), and evacuated water with (*evac*), channel transport with (*ch*), and gas diffusion layer transport with (*gdl*).

We propose five control volumes (Fig. 57.1) where we are taking in consideration all the physical effects (including diffusive and convective transport).

- Cathode and Anode channel
- Cathode and Anode Gas diffusion layer
- Membrane.

57.2.1 Cathode/Anode Mass Flow Model

The mass conservation principle and the thermodynamic properties inside the cathode are used to balance the mass of three gases which are the oxygen, the nitrogen and the vapor water. The resulting cathode mass conservation equations are expressed by six state equations as follows

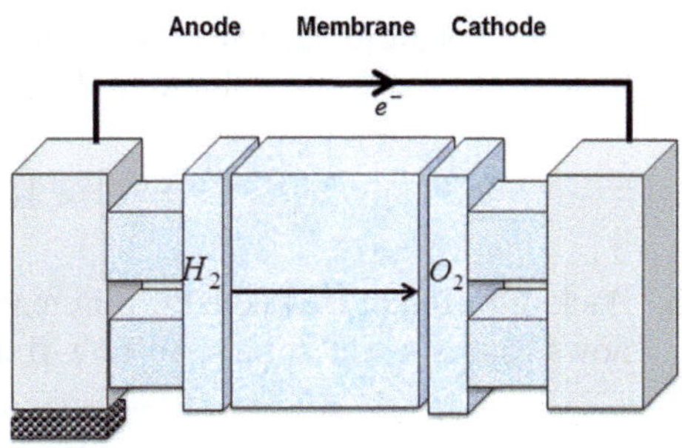

Fig. 57.1 Reactant supply subsystem model

$$
\begin{aligned}
\dot{m}^{ch}_{N_2} &= W^{in}_{N_2} - W^{out}_{N_2} - W^{Diff}_{N_2} - W^{Conv}_{N_2} \\
\dot{m}^{ch}_{O_2} &= W^{in}_{O_2} - W^{out}_{O_2} - W^{Diff}_{O_2} - W^{Conv}_{O_2} \\
\dot{m}^{ch}_{w,ca} &= W^{in}_{w,ca} - W^{out}_{w,ca} - W^{Diff}_{w,ca} - W^{Conv}_{w,ca}
\end{aligned}
\tag{57.1}
$$

$$
\begin{aligned}
\dot{m}^{gdl}_{N_2} &= W^{Diff}_{N_2} + W^{Conv}_{N_2} \\
\dot{m}^{gdl}_{O_2} &= W^{Diff}_{O_2} + W^{Conv}_{O_2} - W^{rct}_{O_2} \\
\dot{m}^{gdl}_{w,ca} &= W^{Diff}_{w,ca} + W^{Conv}_{w,ca} - W^{rct}_{w,ca} + W^{mbr}_{w,ca}
\end{aligned}
\tag{57.2}
$$

Similarly to the cathode, by applying the mass conservation balance, we obtain the four state equations:

$$
\begin{aligned}
\dot{m}^{ch}_{H_2} &= W^{in}_{H_2} - W^{Diff}_{H_2} - W^{Conv}_{H_2} \\
\dot{m}^{ch}_{w,an} &= W^{in}_{w,an} - W^{Diff}_{w,an} - W^{Conv}_{w,an}
\end{aligned}
\tag{57.3}
$$

$$
\begin{aligned}
\dot{m}^{gdl}_{H_2} &= W^{Diff}_{H_2} + W^{Conv}_{H_2} \\
\dot{m}^{gdl}_{w,an} &= W^{Diff}_{w,an} + W^{Conv}_{w,an} - W^{mbr}_{w,an}
\end{aligned}
\tag{57.4}
$$

where the rate of the water production, the rate of the oxygen and hydrogen reacted are:

$$
W^{rct}_{w,ca} = M_{H_2O} \cdot \frac{NI}{2F} \quad W^{rct}_{O_2} = M_{O_2} \cdot \frac{NI}{4F} \quad W^{rct}_{H_2} = M_{H_2} \cdot \frac{NI}{2F} \tag{57.5}
$$

57.2.2 Membrane

This block represents the water content and the rate of mass flow of water that pass through the membrane. Two distinct phenomena describe this flow.

- Electro-osmotic drag, where water molecules are dragged through the membrane from the anode volume to the cathode volume:

$$W_v^{osm} = M_v \cdot S \cdot n_d \cdot \frac{i}{F} \tag{57.6}$$

- Back diffusion drag, where the difference in humidity of the cathode and anode flows causes a water transport flow from the cathode to the anode:

$$W_v^{Diff} = D_w \cdot \frac{dc_v}{dz} \tag{57.7}$$

where n_d and D_w are the electro-osmotic and water diffusive coefficient. The positives values are taken in the direction from anode to cathode and combining these two transport phenomena, the total stack mass flow rate across the membrane (Nafion 117 [5]):

$$W_v^{osm} = M_v \cdot S \cdot n_{fc} \cdot \left\{ n_d \cdot \frac{i}{F} - D_w \cdot \frac{\left(c_v^{ca} - c_v^{an}\right)}{e_{mbr}} \right\} \tag{57.8}$$

57.2.3 PEM Fuel Cell Terminal Voltage Model

The fuel cell voltage is calculated by subtracting the fuel cell losses from the fuel cell open circuit voltage that is governed by the chemical potential [6]:

$$V_{fc} = E - V_{act} - V_{ohm} - V_{conc} \tag{57.9}$$

- Activation loss model: It is a need to cause an electron transfer and to break and form chemical bonds in the cathode and the anode [7–9].
- Ohmic loss model: This type of losses is due to the resistance of the polymer membrane, electrode and collector plate to the transfer of protons and electrons [10].
- Concentration loss model: Concentration overvoltage is caused by the change that occurs in the concentration of the reactants as they are consumed.

57.3 Diffusive and Convective Model

Through the gas channels, the gaseous reactants enter the fuel cell, and they are transported through the diffusion media into the catalyst layers where the reaction occurs. Two phenomena occur:

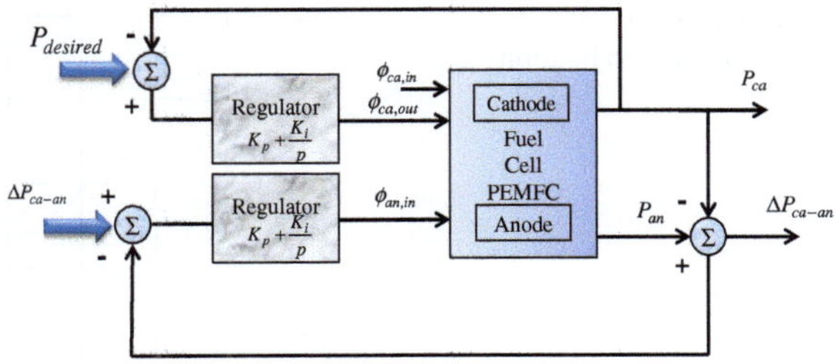

Fig. 57.2 Controlled fuel cell mass by pressure regulation

Diffusive transport: Fick's laws are used in aim to define and to give a molecular movement description for species in a mixture given. The Stefan-Maxwell is replaced when it is a multicomponent system because mutual interactions should be taken in consideration [11].

$$\nabla x_i = \frac{1}{c_t} \cdot \sum_{i=1, i \neq j}^{N} \frac{x_i W_j - x_j W_i}{D_{ij}} \tag{57.10}$$

Convective transport: Transport of fluids in porous media is described using Darcy's law [12]. For a multi-phase system, the mass transport is given by

$$W_j^{Conv} = \frac{S \cdot K_j^{Conv}}{e \cdot \vartheta} \cdot \Delta P_j(ch - gdl) \tag{57.11}$$

where e represents the thickness of the gdl, ϑ the dynamic viscosity and j represents the cathode or the anode.

57.4 Experimental Validation Test

In order to validate the developed model, a set of operational conditions has been investigated as the effect of the temperature, current, pressure and humidity on the performance and effectiveness of the fuel cell process, generated by a test bench presented by PSA Peugeot Citroën. The gas flow is regulated by a proportional–integral controller (Fig. 57.2) in order to assure the same pressure inside the anode and the cathode. The input to the model consists of stack current. The system output is the fuel cell stack voltage (Fig. 57.3).

Experimental data directly obtained from the industrial fuel cell by the finite element method were compared to the ODE model developed in this contribution and it validates the steady-state behavior and the accuracy of the model developed.

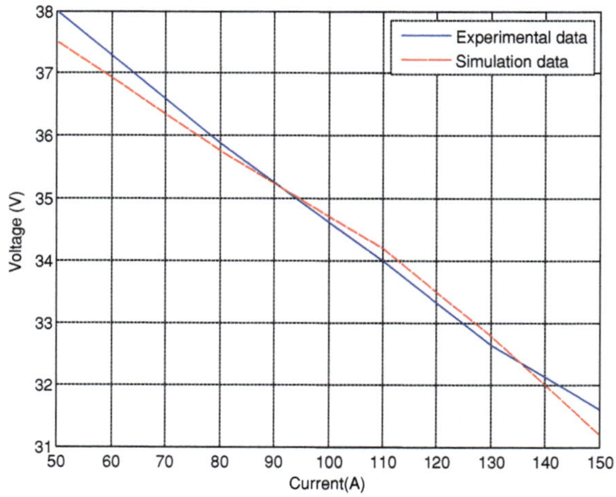

Fig. 57.3 Fuel cell voltage (Experimental validation)

57.5 Conclusions

This work has presented a complete dynamic isothermal model of a fuel cell. It is a control-oriented model that can be used for controller design and diagnosis fault system. The model has been validated on a 5 kW PEM fuel cell, showing that simulated data match experimental data quite closely. The next work will be a study of observer strategy in aim to estimate the internal states of the cell.

References

1. R.A. Lemons, J. Power Sources **29**, 251 (1990)
2. J. Amphlett, A. Dicks, *Fuel Cell Systems Explained*, 2nd edn. (Wiley, New York, 2003)
3. N. Fouquet, in Real time model-based monitoring of a PEM fuel cell flooding and drying out (2010)
4. J. Pukrushpan, Modeling and control of fuel cell systems and fuel processors. Ph.D. thesis, University of Michigan (2003)
5. T.E. Springer, T.A. Zawadzinski, S. Gottesfled, Polymer electrolyte fuel cell model. J. Electrochem. Soc. **138**(9), 1527–1538 (1991)
6. J.C. Amphmett, R.M. Bauùert, R.F. Mann, B.A. Peppley, P.R. Ronerge, Performance modeling of the Ballard Mark IV solid polymer electrolyte fuel cell. J. Electrochem. Soc. **142**(1), 1–8 (1995)
7. J.H. Lee, T.R. Lalk, A.J. Appleby, Modeling electrochemical performance in large scale proton exchange membrane fuel cell stacks. J. Power Sources **70**, 258–268 (1998)
8. J.C. Amphmett, R.M. Bauùert, R.F. Mann, B.A. Peppley, P.R. Ronerge, Performance modeling of the Ballard Mark IV solid polymer electrolyte fuel cell. J. Electrochem. Soc. **142**(1), 9–15 (1995)
9. K. Kordesch, G. Simader, *Fuel Cells and Their Applications*, vol. 117 (Wiley, New York, 1996)

10. J.C. Amphmett, R.M. Bauùert, R.F. Mann, B.A. Peppley, P.R. Ronerge, A. Rodrigues, Parametric modeling of the performance of a 5 kw proton exchange membrane fuel cell stack. J. Power Sources **49**, 349–356 (1994)
11. R. Taylor, R. Krishna, *Multicomponent Mass Transfer* (Wiley, New York, 1993)
12. H. Darcy, *The Public Fountains of the City of Dijon* (Kendall/Hunt Publication, Dubuque, 1856)

Chapter 58
First-Principles Structure Prediction of Dual Cation Ammine Borohydrides: $LiMg(BH_4)_3(NH_3)_x$

Yusuf Kışlak and Adem Tekin

Abstract On-board hydrogen storage for transportation applications continues to be one of the most technically challenging barriers to the widespread commercialization of hydrogen-fueled vehicles. In addition, hydrogen storage is also required for off-board purposes such as stationary power generation and hydrogen delivery and refueling infrastructure. After decades of extensive exploration, research into hydrogen storage materials based on metal borohydrides has become a highly active and exciting area owing to the high theoretical hydrogen capacities of these materials. However, they are thermodynamically too stable and therefore a very high temperature is required for their decomposition. This temperature can be lowered to the tolerable levels by the addition of ammonia and the resulting material is called as Ammine Metal Borohydrides (AMBs). In this study, we aim to search the ground state crystal structures of $LiMg(BH_4)_3(NH_3)_x$ [1] with $x = 2$, 3, 4 using CrystAl Structure Prediction via Simulated Annealing (CASPESA) method. This approach was successfully located the experimentally determined structure of $LiMg(BH_4)_3(NH_3)_2$ [1] and other interesting local minima. For $x = 3$ and 4 cases, our methodology also resulted new crystal phases.

58.1 Introduction

World's energy need is mostly satisfied by fossil fuels such as coal, natural gas and liquid fuels. Their reserves are very limited and their usage has certain harmful impacts on the environment, e.g., they are the primary reason of increasing amount

Y. Kışlak · A. Tekin (✉)
Informatics Institute, Istanbul Technical University, 34469 Maslak, Istanbul, Turkey
e-mail: adem.tekin@be.itu.edu.tr

Y. Kışlak
e-mail: yusufkislak@gmail.com

A. Y. Oral et al. (eds.), *International Congress on Energy Efficiency and Energy Related Materials (ENEFM2013)*, Springer Proceedings in Physics 155,
DOI: 10.1007/978-3-319-05521-3_58,

of CO_2 in the atmosphere. Therefore, we are required to find environmentally friendly alternative renewable energy solutions. Hydrogen is one the most promising and challenging possibility to be used as the energy source of the future. Storing hydrogen is one of the bottlenecks limiting its worldwide integration. Even though hydrogen can be stored in the conventional forms like a gas or liquid, arising handicaps e.g., high pressure and cryogenic temperature block its widespread usage. As an alternative, hydrogen may be stored in the solid state with a very high density. For this purpose, many distinct families of materials have been investigated such as carbon nanotubes, metal organic frameworks, metal hydrides, metal borohydrides and metal ammines (it is an indirect storage medium, since ammonia is stored instead of hydrogen). Among them, metal borohydrides and ammines took special interest due to their high gravimetric and volumetric hydrogen storage capabilities. However, metal borohydrides are very stable complexes and hence they require very high temperatures for their decomposition. This might be prevented by the inclusion of ammonia forming Ammine Metal Borohydrides (AMBs). In spite of their peculiar properties, they might lead to the undesirable release of ammonia during the dehydrogenation. This can be prohibited with the addition of a second metal atom into AMBs leading to dual cation AMBs with a general formula of $M_1M_2(BH_4)_y(NH_3)_x$. Very recently, Sun et al. [1] have successfully synthesized $LiMg(BH_4)_3(NH_3)_2$ and they found that this complex desorbs hydrogen below 200 °C and it does not emit NH_3 as a byproduct.

Since there is only a very limited experimental investigations carried out for dual-cation AMBs so far, little is known about their crystal structures. With this study, we aim to explore the crystal structures of these materials ($LiMg(BH_4)_y(NH_3)_x$ with y = 3 and x = 2, 3 and 4) using a hybrid methodology combining crystal structure prediction with density functional calculations. For the crystal structure predictions, we employed our in-house developed code CASPESA which was successfully applied to both metal ammines [2, 3] and borohydrides [4–7].

58.2 Computational Details

58.2.1 Crystal Structure Prediction

In some cases crystal structure of a material, which determines all the physical properties of the material, is not known. In such circumstances, computational approaches play a crucial role and therefore several crystal prediction algorithms based on genetic algorithm and swarm optimization have been developed such as USPEX [8], CALYPSO [9], XtalOpt [10] and GULP [11]. In this study, we used CASPESA approach due to its fast and robust nature.

CASPESA requires a unit cell and predefined bond constraints to start the crystal prediction. In the following details of the model used to run CASPESA has been exemplified for $LiMg(NH_3)_2(BH_4)_3$: (1) as a unit cell, two formula units of $LiMg(NH_3)_2(BH_4)_3$ has been employed, (2) a fixed coordinate system is used for $Mg(NH_3)_2(BH_4)_3$ (NH_3 and BH_4 groups are coordinated to Mg as a trigonal bipyramid), (3) a Li atom is placed to the origin and the spherical coordinates of the others are parameterized, (4) three Euler angle parameters (Θ, Φ, Ψ) are used to rotate the $Mg(NH_3)_2(BH_4)_3$ complex, (5) the lattice vectors are used as parameters, (6) the resulting 24 (48 for $LiMg(BH_4)_3(NH_3)_3$ and $LiMg(BH_4)_3(NH_3)_4$) parameters are globally optimized to maximize the number of interactions between hydrogens of BH_4 molecules and Li atoms. To avoid any unphysical structure, we employed the following bond constraints: Mg–Mg, Li–Mg and Li–Li distances must be longer than 6.00, 4.80 and 4.10 Å, respectively. As an objective function, the number of Li–B distance, which is between 2.85–3.31 Å, has been maximized. All these bond distance criteria were evaluated with the help of the experimental structure of Sun et al. [1]. The resulting most distinct structures have been chosen for the subsequent periodic DFT calculations.

58.2.2 DFT Calculations

We used Dacapo software package for the DFT calculations. As an exchange correlation functional RPBE [12] was employed. Planewave and density cutoff energies were set to 340 and 500 eV, respectively. The electronic Brioullin zones were sampled with $(2 \times 2 \times 2)$ k-points and the symmetry reduction was switched on. Structural optimizations were performed until all forces are smaller than 0.05 eV $Å^{-1}$ using a quasi-Newton method within the atomic simulation environment [13].

58.3 Results

Very recently, Sun et al. [1] synthesized $LiMg(BH_4)_3(NH_3)_2$ and elucidated its crystal structure with X-ray diffraction method. The structure as shown in Fig. 58.1a has a hexagonal symmetry (symmetry number 173, $P6_3$) with unit cell dimensions: a = 8.0002 Å and c = 8.4276 Å. In this structure, each Li atom is coordinated with H of six different BH_4 groups.

We found a structure, with space group $P6_3/m$, that is very similar to the experimentally determined one with cell dimensions of a = 8.5526 Å and c = 8.4012 Å. As shown in Fig. 58.1b, in this structure Li atoms coordinate with six BH_4 molecules similar to the experimental structure. Besides, we have also found other structures bearing different structural motifs as shown in Fig. 58.2. Among them, Fig. 58.2a was obtained as the lowest energy structure with a space group of $P6_3/m$ and it is lower in energy than Fig. 58.1b by 0.741 eV. In structures from Fig. 58.2a–d, each Li atom coordinated with three different BH_4 groups.

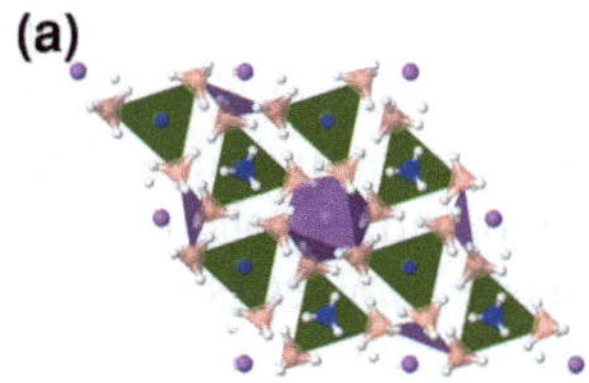

Fig. 58.1 **a** Experimental crystal structure of $LiMg(BH_4)_3(NH_3)_2$ [1]. **b** Similar structure, found by CASPESA, to the experimental one

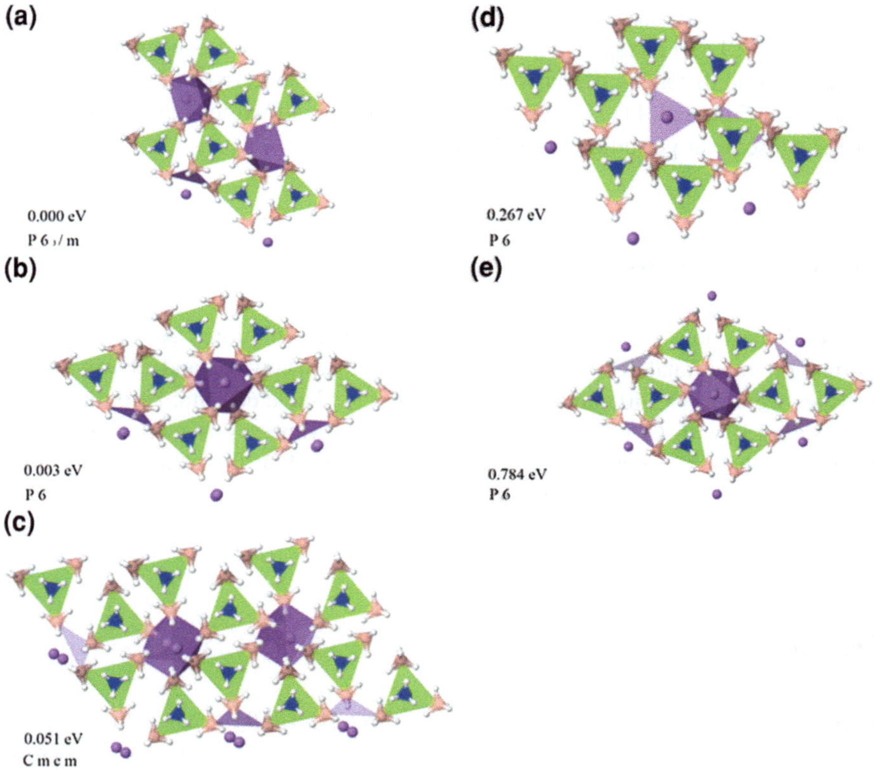

Fig. 58.2 Crystal structures of $LiMg(BH_4)_3(NH_3)_2$ predicted by CASPESA

However, the last structure (Fig. 58.2e) shows similar arrangements to the structures shown in Fig. 58.1.

We also employed CASPESA for $LiMg(BH_4)_3(NH_3)_3$ for which there is no reported experimental structure. Since three NH_3 and BH_4 groups can bind to Mg with a very different ways, we modelled several $Mg(BH_4)_3(NH_3)_3$ complexes with different coordination geometries (e.g., tetrahedral, trigonal bipyramidal, octahedral, trigonal prismatic) and numbers. We have found various structures bearing

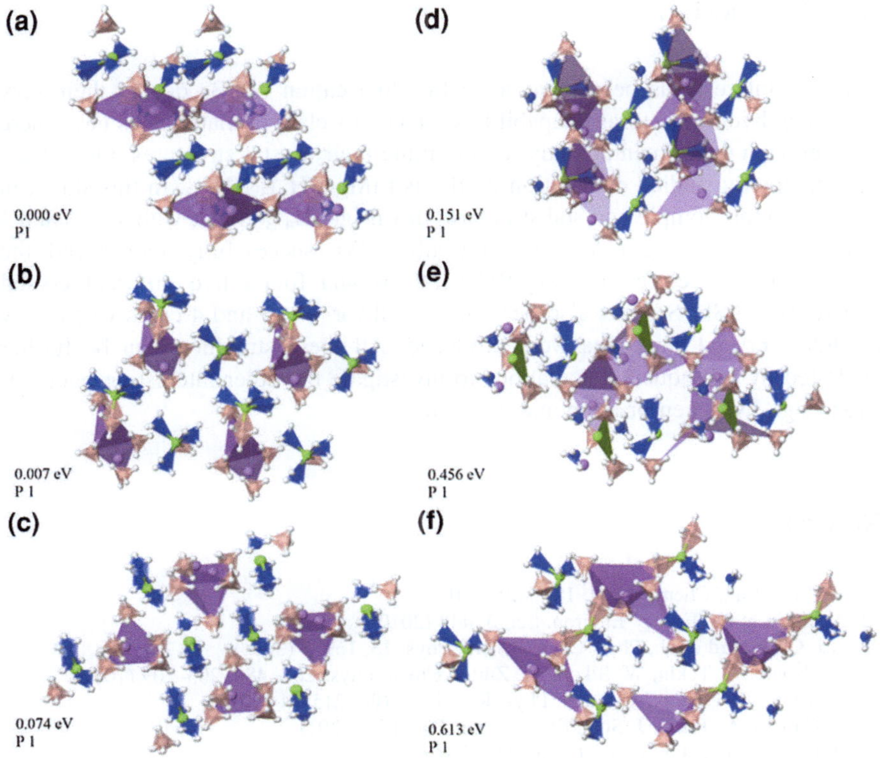

Fig. 58.3 Crystal structures of $LiMg(BH_4)_3(NH_3)_3$ predicted by CASPESA

different structural motifs as shown in Fig. 58.3. The first four isoenergetic structures (Fig. 58.3a–d) are lower in energy than Fig. 58.3e, f by 0.46 and 0.61 eV, respectively. In contrast to $LiMg(BH_4)_3(NH_3)_2$, here in the lowest energy structures, each Li atom is coordinated with four BH_4 groups. However, in the structures shown in Fig. 58.3e, f, each Li atom is coordinated with three different BH_4 groups.

Crystal structure of $LiMg(BH_4)_3(NH_3)_4$ is not yet experimentally determined. Therefore, employment of CASPESA can also be enlightening its structure. Here, only an octahedral arrangement of $Mg(BH_4)_2(NH_3)_4$ were considered and the remaining BH_4 was separately parameterized. It has been found that each Li atom coordinates with five different BH_4 groups in the lowest energy structure.

58.4 Conclusion

A recent interest has been devoted to the dual cation AMBs due to their very promising hydrogen storage capabilities. Since this class of materials is new, there is no enough experimental study to determine their crystal structures. One of the experimentally studied dual cation AMBs is $LiMg(BH_4)_3(NH_3)_2$. In this study, in addition to this complex crystal structures of $LiMg(BH_4)_3(NH_3)_x$ with $x = 3$ and 4 were predicted using CASPESA algorithm. We successfully reproduced the experimental structure of $LiMg(BH_4)_3(NH_3)_2$ and found five different crystal structures. CASPESA was also performed well for $x = 3$ and 4 cases where new candidate crystal structures were produced. All these structures can be further exploited by subsequent calculations to investigate the other interesting phenomenon e. g., hydrogen or ammonia dynamics.

References

1. W. Sun et al., Chem. Eur. J. **18**, 6825 (2012)
2. A. Tekin et al., Energy Environ. Sci. **3**, 448 (2010)
3. A.J. Churchard et al., Phys. Chem. Chem. Phys. **13**, 16955 (2011)
4. R. Caputo, A. Tekin, W. Sikora, A. Züttel, Chem. Phys. Lett. **480**, 203–209 (2009)
5. A. Tekin, R. Caputo, A. Züttel, Phys. Rev. Lett. **104**, 215501 (2010)
6. R. Caputo, A. Tekin, J. Solid State Chem. **184**, 1622 (2011)
7. R. Caputo, A. Tekin, Inorg. Chem. **51**, 9757 (2012)
8. A.O. Lyakhov, A.R. Oganov, H.T. Stokes, Q. Zhu, Comp. Phys. Comm. **184**, 1172 (2013)
9. Y. Wang, J. Lv, L. Zhu, Y. Ma, Comput. Phys. Commun. **183**, 2063 (2012)
10. D.C. Lonie, E. Zurek, Comput. Phys. Commun. **182**, 372 (2011)
11. J.D. Gale, JCS Faraday Trans. **93**, 629 (1997)
12. B. Hammer, L.B. Hansen, J.K. Nørskov, Phys. Rev. B: Condens. Matter Mater. Phys. **59**, 7413 (1999)
13. F. Shanno, Math. Comput. **24**, 647 (1970)

Chapter 59
Adsorption of Two Dyes by $Mg(OH)_2$: Procion Blue HB and Remazol Brilliant Blue R

Zohra Bouberka, Kahina Bentaleb, Khalil A. Benabbou and Ulrich Maschke

Abstract The use of low-cost and ecofriendly adsorbents has been investigated as an ideal alternative to the current expensive methods of removing dyes from wastewater. $Mg(OH)_2$ sludge was produced from precipitation of magnesium ions (Mg^{2+}) with NaOH in pH = 10 and investigated as a low-cost adsorbent. This paper deals with the removal of textile dyes from aqueous solutions by $Mg(OH)_2$. Reactive Procion blue HB (PR) and Acid Remazol brilliant blue R (RB) were used as model compounds. The adsorption capacity was found as 43.47 and 26.89 mg/g at initial pH 6.5.

59.1 Introduction

The adsorption technique has been proved to be an excellent method for the treatment of industrial wastewater containing colour, heavy metals and other inorganic and organic impurities. Adsorption is considered to be relatively superior to other techniques because of its low cost, simplicity of design, availability and ability to treat dyes in more concentrated form [1, 2]. Many adsorbents have been evaluated to reduce dye concentrations from aqueous solutions. Activated carbon [3–5] is regarded as an effective but expensive adsorbent due to its high costs of manufacturing and regeneration. Out of economic consideration, it's not practical to treat a large quantity of effluents by a great scale of activated carbon. In addition to activated carbon, some adsorbents including peat [6, 7], chitin [8, 9] and some

Z. Bouberka · K. Bentaleb · K. A. Benabbou
Laboratoire physico-chimie des matériaux-catalyse et environnement, Université des Sciences et de la Technologie d'Oran «USTO», BP 1505, El M'naouer, 31000 Oran, Algeria

U. Maschke (✉)
Unité Matériaux et Transformations—UMET (UMR CNRS N°8207), Bâtiment C6, Université Lille 1—Sciences et Technologies, 59655 Villeneuve d'Ascq Cedex, France
e-mail: ulrich.maschke@univ-lille1.fr

A. Y. Oral et al. (eds.), *International Congress on Energy Efficiency and Energy Related Materials (ENEFM2013)*, Springer Proceedings in Physics 155,
DOI: 10.1007/978-3-319-05521-3_59,

agriculture wastes [10–16] have also been reported. However, the adsorption capacities of the above adsorbents are not very high; some even less than 50 g/kg. In order to improve the efficiency of the adsorption processes, it is necessary to develop cheap and easily available adsorbents with high adsorption capacities.

Magnesium hydroxide has attracted much attention [17–22] because of its wide applications as flame-retardant, reinforcing agent, antacid and absorbent. Magnesium hydroxide is also being considered as a low cost candidate material for the possible sequestration of greenhouse gas, CO_2. The magnesium hydroxide sludge was produced from precipitation of magnesium ions (Mg^{2+}) with NaOH in pH = 10 [23]. Bivalent hydroxides, $M(OH)_2$ [M = Ca, Mn, Fe, Co and Ni] crystallize in the structure of mineral brucite, $Mg(OH)_2$ [24]. Brucite comprises a hexagonal packing of hydroxyl ions in which Mg^{2+} ions occupy alternate layers of octahedral sites. This results in a stacking of neutral charge layers of the composition [$Mg(OH)_2$]. The aim of the present work is to explore the possibility of utilizing $Mg(OH)_2$ sludge for the adsorption and removal of reactive and acid dyes from aqueous solutions. The effect of operating factors such as adsorbent dose, contact time, and initial concentration of pollutants was investigated. Influence of Cl^- and SO_4^{2-} on Reactive Procion blue HB (PB) and Acid Remazol brilliant blue R (RB) adsorption was studied using the same experimental conditions.

59.2 Experimental Part

A magnesium hydroxide slurry $Mg(OH)_2$ was precipitated at ambient temperature by simultaneous addition of 1M $Mg(NO_3)_2$ and 2M NaOH. A white precipitate was obtained at a pH value of about 10. The resulting was stirred for an hour then filtered and dried at 105 °C. This technique of $Mg(OH)_2$ synthesis is widely used in the literature and is fully described elsewhere [24].

$$Mg^{2+} + 2OH^- \rightarrow Mg(OH)_2 \quad (59.1)$$

Two dyes were used in the study. These selected dyes are regarded as dye contaminants in the discharged effluents. The dyes, Reactive Procion blue HB (PB), CI = 191430250, C.A.S. = 12236-82-7, M_W = 840.09 g/mole, λ_{max} = 610 nm, and Acid Remazol brilliant blue R (RP), CI = 207550250, C.A.S. = 2580-78-1, M_W = 624.52 g/mole, λ_{max} = 595 nm, were provided by Acros Organics. Dyes were used without further purification. The chemical structure of those dyes is given in Fig. 59.1. All other chemicals were of Analar grade and purchased from Merck (Germany). The pH measurements were made using a pH meter (Mettler Delta 350). A UV–V is spectrophotometer (SAFAS UV mc2) was employed for absorbance measurements. Only the linear range of the calibration curve was used in this research. The dye solutions were centrifuged for 10 min in a Jouan centrifuge at 5,000 rpm. In order to investigate the crystalline structure of the $Mg(OH)_2$

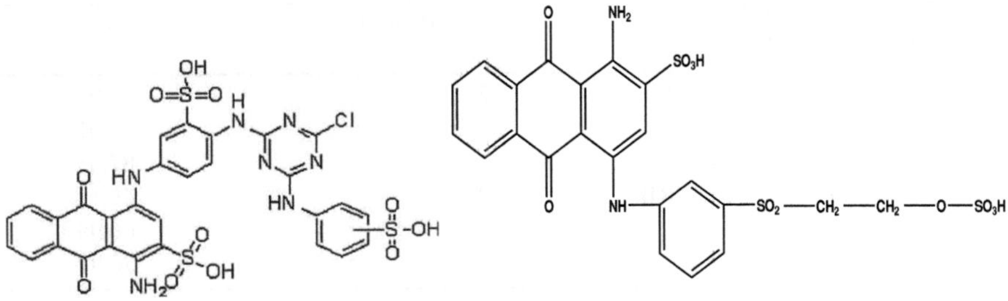

Fig. 59.1 Chemical structures of dyes employed in this study. On the *left* PB (Reactive Procion blue HB), and on the *right* RP (Acid Remazol brilliant blue R)

Fig. 59.2 X-ray powder diffraction (XRD) pattern of the $Mg(OH)_2$ sample showing a hexagonal phase

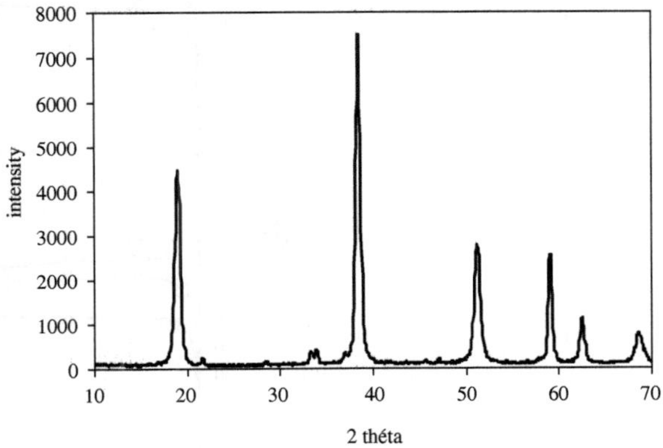

sample, RDX (Siemens D-5000 diffractometer with Cu–Kα radiation) was used. The X-ray tube was operated at 40 kV and a beam current of 30 mA, in the range of $2\theta = 10–70°$, at a scan speed of 2°/min. Figure 59.2 shows the RDX spectrum of $Mg(OH)_2$.

59.3 Results and Discussion

The effect of $Mg(OH)_2$ doses on the amount of dye adsorbed was investigated by contacting 25 ml of dye solution with initial dye concentration of 100 mg/l using closed Erlenmeyer flasks at room temperature (25 ± 1 °C) for 300 min at a constant stirring speed of 200 rpm. Different adsorbent dosage of $Mg(OH)_2$ (2–3 g/l) for PB and (2–6 g/l) for RB were examined. After reaching equilibrium, the samples were centrifuged and the concentration in the supernatant dye solution was analyzed. The dye removal (P %) versus adsorbent dosage (g/l) is shown in Table 59.1. From Table 59.1 one can conclude that the removal rate of PB from the solution is quite higher than for RB. This difference in the removal rate might

Table 59.1 Dye removal (P %) versus adsorbent dosage

Dyes	Adsorbent dosage m_s g/L	Dye removal %
PB	2	74.04
	2.5	82.01
	3	87.05
RB	2	49.1
	4	66.99
	6	70.97

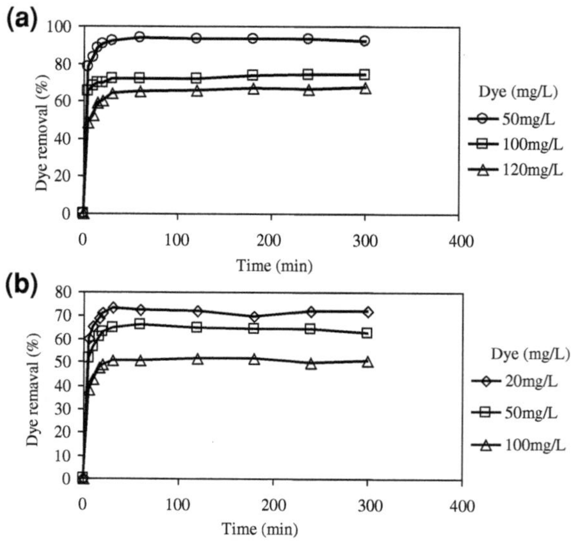

Fig. 59.3 The effect of time and concentration on **a** PB and **b** RB removal in $Mg(OH)_2$. Conditions: pH = 6.5, T = 25 ± 1 °C, S = 200 rpm, m_s = 2 g/l

be attributed to the difference of molecular weights and chemical structures of the dyes.

Figure 59.3 shows the effect of initial dye concentration on dye-% removal rate at different contact times. From Fig. 59.3 it was observed that the dye-% removal varies by changing the initial dye concentration and decreases with increasing initial dye concentration. For a contact time of 60 min, the dye-% removal decreases from 93.01 to 66.76 % for an increase of the initial concentration of PB from 50 to 120 mg/L and decreases from 69.83 to 51.60 % for an increase of the initial concentration of RB from 20 to 100 mg/L. Also from this figure, it was observed that at all initial dye concentrations, the dye-% removal was very rapid for the first 30 min with an average rate of 2.541 and 2.096 mg/min for PB and RB, respectively, and after 40 min the sorption rate decreased to 1.281 and 1.049 mg/min for PB and RB, respectively, and finally reached saturation after 60 min. A large fraction of dyes was thus removed within the first rapid uptake phase, i.e., the first 30 min. This is due to the decrease of the concentration gradient with time due to transfer of solute onto solid phase. The rapid uptake of dye particles for the first minutes is due to the occurrence of solute transfer related to adsorbate and adsorbent interactions with negligible interference of solute–solute interactions [25].

59.4 Conclusion

The present study attempted to show that $Mg(OH)_2$ is an effective adsorbent for the removal of PB and RB from their aqueous solutions. It was found that the dye removal capacity of $Mg(OH)_2$ towards PB was higher than for RB. The reactive dye PB, which possesses sulphonate groups, exhibited a high adsorption which can be attributed to both the higher molecular weight (840.09 g/mole) compared to RB (640.52 g/mole), and its large hydrophobic middle section.

References

1. N. Kannan, M.M. Sundaram, Kinetics and mechanism of removal of methylene blue by adsorption on various carbons—a comparative study. Dyes Pigm. **51**, 25–40 (2001)
2. V. Meshko, L. Markovska, M. Mincheva, A.E. Rodrigues, Adsorption of basic dyes on granular activated carbon and natural zeolite. Water Res. **35**, 3357–3366 (2001)
3. G. McKay, The adsorption of dyestuffs from aqueous solution using activated carbon: analytical solution for batch adsorption based on external mass transfer and pore diffusion. Chem. Eng. J. **27**, 187–196 (1983)
4. K.C.L.N. Rao, K.K. Ashutosh, Color removal from dyestuff industry effluent using activated carbon. Indian J. Chem. Tech. **1**, 13–19 (1994)
5. S.J. Allen, Types of adsorbent materials, in *Use of Adsorbents for the Removal of Pollutants from Wastewaters*, ed. by G. McKay (CRC Inc., Boca Raton, 1996), pp. 59–97
6. K.R. Ramakrishna, T. Viraraghavan, Dye removal using low cost adsorbents. Water Sci. Technol. **36**, 189–196 (1997)
7. Y.S. Ho, G. McKay, Sorption of dye from aqueous solution by peat. Chem. Eng. J. **70**, 115–124 (1998)
8. G. McKay, H.S. Blair, J.R. Gardner, Rate studies for the adsorption of dyestuffs on chitin. J. Colloid Interf. Sci. **95**, 108–119 (1983)
9. G. McKay, Analytical solution using a pore diffusion model for a pseudo-irreversible isotherm for the adsorption of basic dye on silica. AIChE J. **30**, 692–697 (1984)
10. K.S. Low, C.K. Lee, Quaternized rice husk as sorbent for reactive dyes. Bioresour. Technol. **61**, 121–125 (1997)
11. C. Namasivayam, D. Prabha, M. Kumutha, Removal of direct red and acid brilliant blue by adsorption on to banana pith. Bioresour. Technol. **64**, 77–79 (1998)
12. W.T. Tsai, C.Y. Chang, M.C. Lin, S.F. Chien, H.F. Sun, M.F. Hsieh, Adsorption of acid dye onto activated carbon prepared from agricultural waste bagasse by $ZnCl_2$ activation. Chemosphere **45**, 51–58 (2001)
13. R. Sivaraj, C. Namasivayam, K. Kadirvelu, Orange peel as an adsorbent in the removal of acid violet 17 (acid dye) from aqueous solutions. Waste Manag. **21**, 105–110 (2001)
14. G. Annadurai, R.S. Juang, D.J. Lee, Use of cellulose-based wastes for adsorption of dyes from aqueous solutions. J. Hazard. Mater. **92**, 263–274 (2002)

15. C. Namasivayam, D. Kavitha, Removal of Congo Red from water by adsorption onto activated carbon prepared from coir pith, an agricultural solid waste. Dyes Pigm. **54**, 47–58 (2002)
16. T. Robinson, P. Chandran, P. Nigam, Removal of dyes from a synthetic textile dye effluent by biosorption on apple pomace and wheat straw. Water Res. **36**, 2824–2830 (2002)
17. S. Utamapanya, K.J. Klabunde, J.R. Schlup, Nanoscale metal oxide particles/clusters as chemical reagents. Synthesis and properties of ultrahigh surface area magnesium hydroxide and magnesium oxide. Chem. Mater. **3**, 175–181 (1991)
18. O.B. Koper, I. Lagadic, A. Volodin, K.J. Klabunde, Alkaline-earth oxide nanoparticles obtained by aerogel methods. Characterization and rational for unexpectedly high surface chemical reactivities. Chem. Mater. **9**, 2468–2480 (1997)
19. J.A. Wang, O. Novaro, X. Bokhimi, T. Lopez, R. Gomez, J. Navarrete, M.E. Llanos, E. Lopez-Salinas, Structural defects and acidic and basic sites in sol-gel MgO. J. Phys. Chem. B **101**, 7448–7451 (1997)
20. J.A. Wang, O. Novaro, X. Bokhimi, T. Lopez, R. Gomez, J. Navarrete, M.E. Llanos, E. Lopez-Salinas, Characterizations of the thermal decomposition of brucite prepared by sol–gel technique for synthesis of nanocrystalline MgO. Mater. Lett. **35**, 317–323 (1998)
21. Y.D. Li, M. Sui, Y. Ding, G. Zhang, J. Zhuang, C. Wang, Preparation of $Mg(OH)_2$ nanorods. Adv. Mater. **12**, 818–821 (2000)
22. Y. Ding, G.T. Zhang, H. Wu, B. Hai, L.B. Wang, Y.T. Qian, Nanoscale magnesium hydroxide and magnesium oxide powders: control over size, shape, and structure via hydrothermal synthesis. Chem. Mater. **13**, 435–440 (2001)
23. A.V. Radhaa, Kamatha P. Vishnu, G.N. Subbanna, Disorder in layered hydroxides: synthesis and DIFFaX simulation studies of $Mg(OH)_2$. Mat. Res. Bull. **38**, 731–740 (2003)
24. H.R. Oswald, R. Asper, *Preparation and Crystal Growth of Materials with Layered Structures*, vol. 1 (D. Reidal Publishing Company, Dordrecht, 1977), p. 71
25. G. McKay, Y.S. Ho, The sorption of lead (II) on peat. Water Res. **33**, 578–584 (1999)

Chapter 60
Wireless Communications in Smart Grid

Zoran Bojkovic and Bojan Bakmaz

Abstract Communication networks play a crucial role in smart grid, as the intelligence of this complex system is built based on information exchange across the power grid. Wireless communications and networking are among the most economical ways to build the essential part of the scalable communication infrastructure for smart grid. In particular, wireless networks will be deployed widely in the smart grid for automatic meter reading, remote system and customer site monitoring, as well as equipment fault diagnosing. With an increasing interest from both the academic and industrial communities, this chapter systematically investigates recent advances in wireless communication technology for the smart grid.

60.1 Introduction

Generally, a fundamental evolution in electric power generation and supply system is needed to make the grids more reliable, efficient, and secure. This can be achieved by enabling the future generation electricity network smarter by embedding bidirectional communication architecture in power grids. The smart grid (SG) evolves the architecture of legacy grid which can be characterized as providing one-way flow of centrally generated power to end users into a more distributed, dynamic system characterized by two-way flow of power and information. The SG will involve networking vest numbers of sensors in transmission and distribution facilities, smart meters, back-office systems as well as home devices which will interact with the grid. Large amount of data traffic will be generated by meters, sensors and synchrophasors.

Z. Bojkovic (✉) · B. Bakmaz
University of Belgrade, Belgrade, Serbia
e-mail: z.bojkovic@yahoo.com

A. Y. Oral et al. (eds.), *International Congress on Energy Efficiency and Energy Related Materials (ENEFM2013)*, Springer Proceedings in Physics 155,
DOI: 10.1007/978-3-319-05521-3_60,

While communications technologies are seen as an essential enabling component of SG, there are a number of challenges that must be addressed in order to have fully robust, secure and functional SG networks [1]. In comparison with wireline networks, wireless networks have the advantage of low deployment cost and high flexibility, and are gaining more and more interest from both academia and industry for SG applications.

The remainder of this chapter is organized as follows. In the next section, the communication architecture for the SG is presented by decomposing it into three recognizable network segments. Cognitive radio technology is presented as perspective solution for SG communication infrastructure. Finally, the importance of heterogeneous network integration to coordinate the SG functions is invoked, too.

60.2 Smart Grid Communication Architecture

The SG is usually deployed in rather large geographical areas. Consequently, the communications infrastructure of the SG has to cover the entire region with the intention to connect a large set of nodes. The communications infrastructure is envisioned to be a multilayer structure that extends across the whole SG as shown in Fig. 60.1.

In particular, home area networks (HANs) gather sensor information from a variety of smart devices (e.g., meters, sensors, actuators, etc.) within the home and delivers control information to them for better energy consumption management. They can support functions such as cycling heaters or air conditioners off during peak load conditions, sharing consumption data with in-home displays, and enabling a card-activated prepayment scheme. Typically, HANs need to cover areas up to 200 m^2 and support data rates up to 100 kb/s.

Neighborhood area networks (NANs) connect multiple HANs to local access points. NAN endpoints are smart meters (SMs) that have been considered to be at the heart of the SG revolution. SMs involve energy consumption recording and real-time or near-real-time data acquisition and control for various SG applications, including distribution automation, power outage management, power quality monitoring, etc. [2]. The number of SMs in each NAN cluster varies from a few hundred to a few thousand depending on the grid topology and the employed communication protocol. NANs usually need to cover areas of several square kilometers, and each SM may need from 10 to 100 kb/s. It is important to note that a NAN is a critical segment of SG communications architecture since it is responsible for transporting a huge volume of different types of data and distributing control signals between utility companies and a large number of devices installed at customer premises.

Wide area networks (WANs) provide communication links between the NANs and the utility systems to transfer information. They also enable the long-haul communications among different data aggregation points of power generation plants, distributed energy resource stations, substations, transmission and

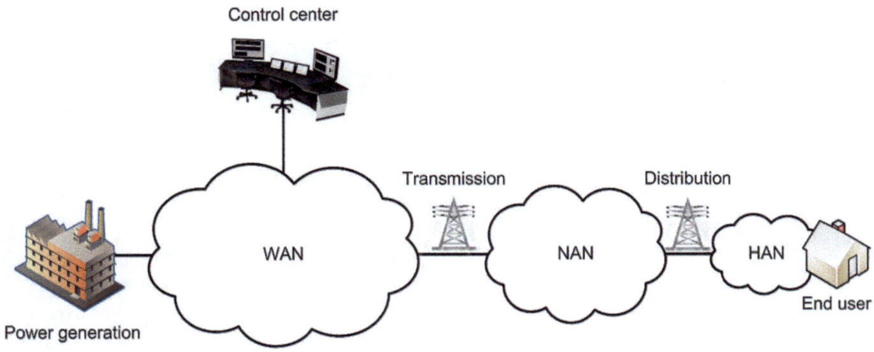

Fig. 60.1 Multilayer structure across the smart grid

distribution grids, control centers, etc. The utility company's WAN is also responsible for providing the two-way network needed for substation communications, distribution automation, power quality monitoring, and so forth, while also supporting data aggregation and backhaul for NANs. WANs may cover a very large area (thousands of square kilometers) and could aggregate a large number of supported devices that require 10–100 Mb/s of data transmission.

The three-layered structure of the communications networks provides a potential operation of the SG to operate economically, efficiently, reliably, and securely. On the other hand, there are many challenges imposed on the design of SG communications architecture like [3]: energy services, interoperability, tremendous data amount, highly varying traffic, quality of service (QoS), and security. A unique characteristic of the SG is the integration of distributed renewable energy sources (e.g., solar and wind power). For a NAN, there are two main power sources: the power from the utility and the distributed renewable energy. These power sources have two essential differences: price and availability. Balancing the usage of different energy sources will be very important for power grid stability, availability, and operation cost. Data will flow over generation, transmission, distribution, and user networks in the SG. One of the major problems of the multitiered topology of communications networks is interoperability among so many subnetworks. The amount of data generated by smart devices will experience explosive growth in the future. This tremendous data amount places considerable load on the communications infrastructure of the SG.

The amount of data varies tremendously during a day, so the traffic conditions change rapidly. During peak hours, the data communications requires higher data rate and more reliable services. Different categories of data have different QoS priorities in terms of bandwidth, latency, reliability, and security. A higher priority and guaranteed QoS should be provided to the meters data, while power price data used for accounting can have normal priority and QoS.

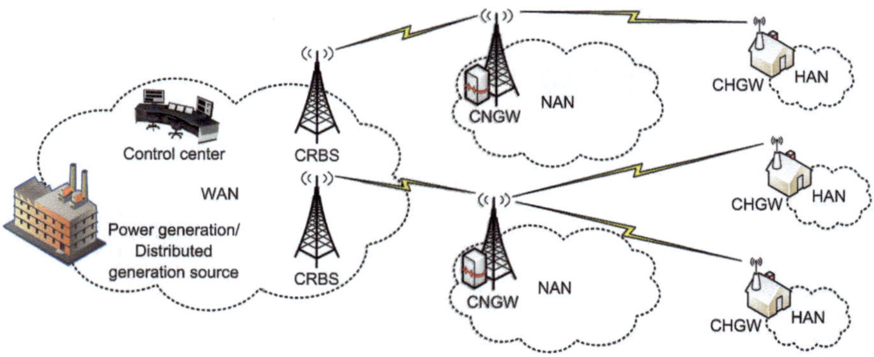

Fig. 60.2 Cognitive radio based communication architecture in the smart grid

60.3 Cognitive Radio Based Communications

Perspective cognitive radio based communication architecture for the SG [3, 4] is presented in Fig. 60.2. This architecture uses cognitive radio technology to enable the communications infrastructure more economically, flexibly, efficiently and reliably. Cognitive communications that operate in the unlicensed spectrum are applied in the HAN to coordinate the heterogeneous wireless technologies. On the other hand, cognitive communications that operate in the licensed spectrum are applied in the NAN and WAN to dynamically access the unoccupied spectrum.

The HAN consists of a cognitive home gateway (CHGW), smart meters, sensors, actuators, and other smart devices. It usually uses a star topology with either wired technologies (e.g., PLC) or different wireless technologies (e.g., ZigBee, Wi-Fi, etc.). In SG applications, the NANs collect energy consumption information from households in a neighborhood and distribute them to a utility company through WAN. Cognitive NAN gateway (CNGW) connects several CHGWs from multiple HANs. The CHGWs are the data access points of the HANs to the outside NAN, and they also act as the cognitive nodes located in the NAN. From the technology point of view, the CNGW can be considered as the cognitive point of attachment (PoA) to provide single-hop connection with CHGWs in a hybrid access manner. The CNGW manages the access of the CHGWs and distributes spectrum bands to them.

In the WAN, each CNGW is no longer a PoA, but a cognitive node with the capability to communicate with the control center through unused frequency space. This center is connected with cognitive radio base stations (CRBS) that are dispersed over a large area (e.g., a city). In conjunction with the control center, there is a spectrum broker that plays an important role in sharing the spectrum resources among different NANs to enable their coexistence. In a large geographical distribution of NANs, several NGWs may not be within the geographic area covered by base stations. These NGWs have to communicate in an ad hoc mode to share unoccupied spectrum by them.

60.4 Heterogeneous Networks Integration to Coordinate the Smart Grid Functions

Various communication technologies have been considered for implementing SG [5–7]. The main communications component of the SG is the sensor network which consists of a system of distributed nods that interact among themselves and with the infrastructure in order to acquire, process, transfer and provide information extracted from the physical world. The SM is the bridge between users' behavior and power consumption metering. Distribution management system (DMS) is required for analyzing, control and provides enough useful information to the utility. The SG is also composed of legacy remote terminal units (RTUs) that can perform sensor network gateway functions acting as intermediate points in the medium voltage network. The sensor network gateway is the bridge between the sensor network and the back-end system. Therefore, it provides necessary interfaces to other sensor nodes as well as interfaces to existing communications infrastructures.

Advanced metering infrastructure (AMI) consists of smart meters, data management, communication network and applications. Along with distributed energy resource (DER) and advanced distributed automation (ADA), AMI as one of the three main anchors of SG provides bidirectional communication between the utility and meters installed in residences. Usually, geographic information system (GIS) as well as consumer information system (CIS) contributes with tools and important processes. The information recollected and processed by DMS must be reported to a supervisory control and data acquisition system (SCADA).

Heterogeneous networks manage and collect information from established intelligent electronic devices (IEDs) for control and automation purposes in real-time. Thus, SG needs to communicate many different types of devices, with different needs for QoS over different physical media. The IEDs involved in these processes can be situated in different locations due to the decentralized architecture. As an example, electrical substation elements are connected to the substation's Ethernet. As for sensors, they can be installed along electrical cables communicated through wireless sensor standards (e.g., IEEE 802.15.4, IEEE 802.11s, etc.). Communications from the control center to energy meters and between substations can be carried out over a high variety of broadband technologies such as Long Term Evolution (LTE), Worldwide Interoperability for Microwave Access (WiMAX). Standardized, open information models and communication services for all data exchanges are needed in the SG. Thus, SG will be supported by a highly heterogeneous data network. Traditional wireless systems use point-to-point or point-to-multipoint technologies. Wireless mesh networks (WMNs) are an alternative to these topologies [8]. There are some reasons for it. For example, it is easy to associate new nodes in the network thanks to the fact of self-configuration and self-organization capabilities. Secondly, a WMN is a robust system as there will almost always be an alternative path to the destination. Taking into account the large scenario in which the SG is going to be deployed, different technologies will be needed in order to cover all the area.

60.5 Discussion and Conclusions

The success of the SG depends heavily on the communication infrastructure. In different segments of a power grid, various communication technologies are applied to meet the unique specific requirements. In a power transmission segment, wired communications over power lines or optical cables are adopted to ensure robustness of the backbone. However, in power distribution segment, both wired and wireless communications should be considered. In order to achieve cost-effective and flexible control and monitoring of end devices, efficient dispatching of power and dynamic integration of distributed energy resources with power grid, wireless communications and networking functionalities must be embedded into various electric equipment. Capability of wireless networking among various electric equipment is one of the key technologies that drive the evaluation of a conventional power distribution network into a SG. Our survey summarizes the current research status on wireless communications in the SG.

Different types of wireless networks are available, but which one is the suitable for a SG, depends on the system architecture and varieties of communication modules and links. The multihop networking is definitely necessary, as electric equipment out of range of each other need to exchange information. First of all, in order to simplify network organization and maintenance the entire environment needs to be self-organized. Moreover, communications modules may pertain heterogeneous nature in terms of coverage, computing power and power efficiency.

WMNs are considered as important networking for SG. They can easily integrate heterogeneous networks to fulfill different functions such as sensing, monitoring, data collection, control and pricing. However, when mesh topology is applied in SG wireless infrastructure, a few challenging issues still remains for future research (e.g. routing, QoS mechanisms, security level, etc.).

References

1. Z. Fan et al., Smart grid communications: overview of research challenges, solutions, and standardization activities. IEEE Commun. Surveys Tutorials **15**(1), 21–38 (2013). doi:10.1109/SURV.2011.122211.00021
2. Q.-D. Ho, Y. Gao, T. Le-Ngoc, Challenges and research opportunities in wireless communication networks for smart grid. IEEE Wirel. Commun. **20**(3), 89–95 (2013). doi:10.1109/MWC.2013.6549287
3. R. Yu et al., Cognitive radio based hierarchical communications infrastructure for smart grid. IEEE Network **25**(5), 6–14 (2011). doi:10.1109/MNET.2011.6033030
4. H. Wang, Y. Qian, H. Sharif, Multimedia communications over cognitive radio networks for smart grid applications. IEEE Wirel. Commun. **20**(4), 125–132 (2013). doi:10.1109/MWC.2013.6590059
5. A. Zaballos, A. Vallejo, J.M. Selga, Heterogeneous communication architecture for the smart grid. IEEE Network **25**(5), 30–37 (2011). doi:10.1109/MNET.2011.6033033

6. W. Wang, Y. Xu, M. Khanna, A survey on the communication architectures in smart grid. Comput. Netw. **55**(15), 3604–3629 (2011). doi:10.1016/j.comnet.2011.07.010
7. A. Usman, S.H. Shami, Evolution of communication technologies for smart grid applications. Renew. Sustain. Energy Rev. **19**, 191–199 (2013). doi:10.1016/j.rser.2012.11.002
8. Y. Xu, W. Wang, Wireless mesh network in smart grid: Modeling and analysis for time critical communications. IEEE Trans. Wireless Commun. **12**(7), 3360–3371 (2013). doi:10.1109/TWC.2013.061713.121545

Chapter 61
Innovative Solutions for Energetic Refurbishment of Historic Brick Buildings

Jurgis Zagorskas, Marija Burinskienė, Gražvydas Mykolas Paliulis and Jūratė Venckauskaitė

Abstract Building standards for energy effectiveness are increasing constantly and market follows these changes by construction of new buildings in accordance with standards and refurbishment of existing housing stock. Comprehensive trends of European construction market show tremendous increase in building retrofit works. It can be predicted that after this decade more than half of construction works in European cities will be taking place in existing buildings, pushing the construction of new buildings to the less important role. Such a growth in building refurbishment works is creating a demand for suitable materials, retrofitting techniques and research. The differences between refurbishment of new-build projects and historical or valuable buildings are insufficiently recognized—mostly the buildings without further cultural preservation requirements are studied. This article covers the theme of refurbishment measures in the historical buildings—the specific measures like inside insulation which is allowed due to the valuable façade or other heritage preservation requirements. An overview of other innovative methods for energy saving in existing buildings and their potential is given.

61.1 Introduction

Buildings are the localized tool of human activities; they support numerous functional needs, express historic preferences, contemporary choices and future visions of the population. As political, economic, and social contexts change, the built environment keeps track of a location's history: political collapses, wars, or forced migrations [1]. In recent decades, the role of culture and history has often become a driving factor in the process of urban regeneration. The focus on culture

---

J. Zagorskas (✉) · M. Burinskienė · G. M. Paliulis · J. Venckauskaitė
Department of Urban Engineering, Vilnius Gediminas Technical University, Saulėtekio al. 11, LT-10223 Vilnius, Lithuania
e-mail: jurgis.zagorskas@gmail.com

A. Y. Oral et al. (eds.), *International Congress on Energy Efficiency and Energy Related Materials (ENEFM2013)*, Springer Proceedings in Physics 155,
DOI: 10.1007/978-3-319-05521-3_61,

and history as factors in regional transformation has been particularly extensive in response not only to competitiveness among cities but also to sustainability requirements [2]. Creative cities are currently working on how to improve the interaction between building regeneration, economic development and social renewal in order to achieve more comprehensive development of the city [3–5].

Energy and sustainability are a hard challenge in building heritage, both the technical solutions in order to solve impact of energy conservation and aspect of conservation and maintenance of architectural heritage [6]. Energy efficiency and architectural heritage of brick buildings are two controversial topics. It is important to address these two issues so that obtained result can meet today's requirements of energy efficiency and, at the same time, promote the preservation of historic buildings for future generations [7]. Currently developed simulation and visualization methods and measurement technologies can assist energy managers at different stages of their activity and have the potential to achieve energy savings on a large scale [7–9].

61.2 Co2olBricks Project

"Co2olBricks—Climate Change, Cultural Heritage & Energy Efficient Monuments" is a project that is looking for solutions to increase the energy efficiency of historic buildings without destroying the historical value. The aims are to protect their historic value as well as reduce their energy consumption.

The project has been granted under the "Priority1: Fostering Innovations" of the Baltic Sea Region Programme 2007–2013. Co2olBricks project has started in December 2010 and runs until December 2013. During this time it has a total budget of 4.3 million euro, 18 partners from 9 countries with 9 languages are working together.

The conservation of heritage, in particular of historical buildings, is a common goal in the Baltic Sea Region. Due to the common identity in the Baltic Sea Region it is very important to protect the historical buildings with their individual characteristics which will preserve the attractiveness and competitiveness of the cities around the Baltic Sea. Project aims to find solutions to combine the necessary CO_2 reduction targets with technically, administratively and historically adequate approaches to heritage conservation. Especially the Baltic Sea Region's brick architecture in the former area of the Hanseatic League presents an excellent possibility to find transferable methods and solutions.

61.3 Specifics of the Baltic Sea Region

Around 50 % of the existing buildings in the Baltic Sea Region were built before 1970 and have lower energy efficiency and higher GHG emissions than modern buildings. Existing older buildings have a very low thermal standard compared with today's requirements—the energy consumption for heating is usually two and more times greater than in modern buildings built from the year 2000. E.g. in Denmark average energy consumption for heating of modern houses (constructed after 2000) is ~100 kWh/m^2 per year and less and in the houses built before 1980 it is ~200 kWh/m^2 per year and more [10]. In less economically developed Baltic Sea Region countries these numbers vary even greater—the difference is up to 3 times, because the old buildings are retrofitted very slowly and most of them are still waiting for energy saving measures to be applied.

In cities of Baltic Sea Region among other historical tracks the traces of Hanseatic League clearly denominate. The "German Hanse" greatly influenced development of Baltic Sea Region towns starting from XIV century leaving footprint of characteristic red brick buildings in the cities of most Baltic Sea Region countries [11].

Each historic building usually has its own specifics, but some general statements can be made about historical brick buildings in Baltic Sea Region:

- in the Baltic Sea Region most of historic buildings were built from bricks.
- the standard Baltic Sea Region historic building is 3–7 storey high;
- it usually has 45–90 cm thick solid brick wall, with outer walls and inside construction bearing wall also used for ventilation and smokestack;
- it usually has wooden ceilings and roof construction;
- the roof is usually covered with traditional tiling and has to have at least 35° incline.
- Baltic Sea Region is colder climate zone and the windows were traditionally small. Usually the biggest energy losses happen in the roof and walls.

61.4 The Most Effective Measures in Historic Buildings

In refurbishment process of historic buildings the compromise between payback period, comforts level, and preservation of historic authenticity of the building.

The energy consumption for heating and cooling the inside takes around 45 % of the energy consumed in buildings. Other great energy loses happen in heating the water, lights and electronics.

Energy efficiency can be greatly improved without touching the building construction—through optimization of the performance of the building envelope and intelligent operation and management of HVAC (Heating, Ventilation and Air Conditioning) system [12–14], changing the doors and windows, improving building air tightness, etc.

Co2olBricks project experts have made the list of most effective measures in energetic refurbishment of historic brick buildings (Table 61.1).

Table 61.1 The most effective measures for energetic refurbishment of historic brick buildings in Baltic Sea Region

Number	Possible energy savings, % from total energy use in building	Energy saving measure	Comments on applicability and other
1	17	Improving heat production/ boiler	Costly
2	15	Heat pumps	Costly
3	15	Post-insulation of roof and sloped ceilings	Costly, usually needs subsidizing
4	14	Post-insulation of exterior wall (from inside)	Costly, usually needs subsidizing
5	11	New windows or energy efficient secondary glazing on windows, shading the windows	Easily applied, short payback period
6	8	Increasing building airtightness	Very easily applied, short payback period
7	Increase of comfort, installed with 2a	Regulated ventilation system	Costly
8	7	Energy effective modern lighting system	Very easily applied, short payback period
9	3	Post-insulation of cellar ceiling, cellar walls	Costly, usually needs subsidizing
10	2	Home automation and smart applications	Moderate, small savings
11	Up to 30	Infrared heating	Can be applied and help to save energy only in special cases (i.e. churches)

The most effective energy effectiveness measures for historic buildings can be grouped in three categories:

1. Improving energy use for heating devices:

   - Improving heat production/boiler—modern heating system can increase heat production effectiveness and decrease energy consumption for heating by 40 %.
   - Installing heat pumps—can save up to 30 % of heating energy, or up to 10 % of overall energy use in historic building.
   - Infrared heating—radiant heating can be used in huge spaces, where great energy savings can be achieved because of local heat spot.

2. Increasing energy effectiveness of building envelope:

   - Post-insulation of roof and sloped ceilings—up to 42 % of heat is being transferred throw the roof of the building, therefore roof post-insulation is essential step in increasing energy effectiveness of existing building.

- Inside post-insulation of exterior wall—post-insulation of outer walls is the most challenging energy effective measure for historic building. The retrofit of interior insulation is commonly implemented to improve energy performance of these buildings, while maintaining their historic exterior appearance.
- Post-insulation of cellar ceiling, cellar walls—studies have shown that floor insulation provides a very small part of the total heat gains in comparison with roof and exterior wall insulation. However, if floor modification or reconstruction is planned the opportunity to improve the efficiency of it should be used.
- New windows or secondary glazing on windows, shading the windows.
- Increasing building airtightness and installing regulated ventilation system—air tightness is the fundamental building property that impacts infiltration. Due to poor air tightness in historic buildings heat loss increases up to 40 %. Reducing the fresh air inflow leads to the necessity of ensuring a comfortable indoor climate for the users of a building and for the preservation of the building. In order to ensure the comfortable indoor climate, the recommended air change per hour (ACH) should be 0.5 l/h.

3. Other technical measures:

- Heat recovery systems—the thermal efficiency of systems with heat recovery is between 50 and 90 % and annual efficiency between 60 and 95 %.
- Energy effective modern lighting system—lights use around 11 % of energy in historic building and it can be greatly reduced by using modern light bulbs and LED.
- Central control of electrical components—electronic devices consume around 7 % and lighting up to 11 % of energy used in houses and it can be greatly reduced by automation systems.

61.5 Conclusions

- There is a lack of understanding of historic building performance in industry and in policy, and a lack of connection between good research, standards, certification processes, guidance and practice.
- Most effective energy saving measures in historic buildings are improving heat production/boiler, installing heat pumps, post-insulation of roof and exterior wall, together these can increase energy effectiveness of old building by 60 %, but these measures are costly and complex to install.
- Among easily affordable measures, the most effective are new windows, increasing building airtightness and modern lighting system, but with these measures total energy saving can be only up to 18 %. However these are the most popular measures.

- Infrared heating can be used in huge spaces, where great energy savings can be achieved because of local heat spot.
- Some measures have to be implemented together (i.e. increased airtightness usually must be supplemented by mechanical ventilation system)
- Use of combination of measures can change the effectiveness of other measures
- Theoretical calculations usually tend to be slightly optimistic, the real energy saving usually happens to be less than theoretical by 10–15 %.

Acknowledgments This article was prepared within "Co2ol Bricks" project financed by Baltic Sea Region Programme 2007–2013.

References

1. C. Hein, Spaces of Identity in East European Cities (2012)
2. M. Smith, G. Richards, *Routledge Handbook of Cultural Tourism* (Routledge, London, 2013)
3. M. Sepe, Urban history and cultural resources in urban regeneration: a case of creative waterfront renewal. Plann. Perspect. (ahead-of-print), 1–19 (2013)
4. T. Lähdesmäki, Politics of cultural marking in Mini-Europe: anchoring European cultural identity in a Theme Park. J. Contemp. Eur. Stud. **20**(1), 29–40 (2012)
5. N.M. Martinez, City marketing and place branding: a critical review of practice and academic research. J. Town City Manage. **2**(4), 369–394 (2012)
6. K. Fabbri, Energy incidence of historic building: leaving no stone unturned. J. Cult. Heritage (2013)
7. J.E. Hensley, A. Aguilar, *Improving Energy Efficiency in Historic Buildings* (Government Printing Office, 2012)
8. A. Costa, M.M. Keane, J.I. Torrens, E. Corry, Building operation and energy performance: monitoring, analysis and optimisation toolkit. Appl. Energy **101**, 310–316 (2013)
9. Y. Heo, R. Choudhary, G. Augenbroe, Calibration of building energy models for retrofit analysis under uncertainty. Energy Build. **47**, 550–560 (2012)
10. J. Hensen, Building energy simulation: challenges and opportunities, in: SIMUREX 2012-Conception optimisée du bâtiment par la SIMUlation et le Retour d'EXpérience, EDP Sciences (2012), p. 00007
11. C. Jahnke, The city of Lübeck and the internationality of early hanseatic trade. Hanse Medieval Early Mod. Europe **60**, 37 (2012)
12. D.K. Jung, D.H. Lee, J.H. Shin, B.H. Song, S.H. Park, Optimization of energy consumption using BIM-based building energy performance analysis. Appl. Mech. Mater. **281**, 649–652 (2013)
13. Z. Xu, X. Guan, Q.-S. Jia, J. Wu, D. Wang, S. Chen, Performance analysis and comparison on energy storage devices for smart building energy management. IEEE Trans. Smart Grid **3**(4), 2136–2147 (2012)
14. N. Aste, R. Adhikari, M. Buzzetti, Energy retrofit of historical buildings: an Italian case study. J. Green Build. **7**(4), 144–165 (2012)

Errata to: International Congress on Energy Efficiency and Energy Related Materials (ENEFM2013)

Ahmet Yavuz Oral, Zehra Banu Bahsi and Mehmet Ozer

Errata to: *International Congress on Energy Efficiency and Energy Related Materials (ENEFM2013)*, DOI 10.1007/978-3-319-05521-3

1. Authors of Chapter 38 are Nabil Khelifati, Djoudi Bouhafs, Seddik-El-Hak Abaidia, Abd El-Ghani Boucheham and Baya Palahouane. Another name had been added erroneously.
2. The authors' acknowledgments are missing at the end of Chap. 58. It should read as:

Acknowledgments This work financially supported by the TUBITAK (112T988). Computing resources used in this work were provided by the National Center for High Performance Computing of Turkey (UYBHM) under grant number 20662009 and Informatics Institute of Istanbul Technical University.

The online version of the original book can be found under DOI 10.1007/978-3-319-05521-3

A. Y. Oral (✉)
Department of Materials Science and Engineering, Gebze Institute of Technology, Gebze, Kocaeli, Turkey
e-mail: aoral@gyte.edu.tr

Z. B. Bahsi
Department of Environmental Engineering, Gebze Institute of Technology, Gebze, Kocaeli, Turkey
e-mail: bbahsi@gyte.edu.tr

M. Ozer
Department of Physics, Istanbul Kultur University, Istanbul, Turkey
e-mail: m.ozer@iku.edu.tr

A. Y. Oral et al. (eds.), *International Congress on Energy Efficiency and Energy Related Materials (ENEFM2013)*, Springer Proceedings in Physics 155,
DOI: 10.1007/978-3-319-05521-3_62, © Springer International Publishing Switzerland 2014

MIX
Papier aus verantwortungsvollen Quellen
Paper from responsible sources
FSC® C105338

If you have any concerns about our products,
you can contact us on
ProductSafety@springernature.com

In case Publisher is established outside the EU,
the EU authorized representative is:
Springer Nature Customer Service Center GmbH
Europaplatz 3, 69115 Heidelberg, Germany

Printed by Libri Plureos GmbH
in Hamburg, Germany